DAMS AND DEVELOPMENT

DAMS AND DEVELOPMENT

A NEW FRAMEWORK FOR DECISION-MAKING

THE REPORT OF THE WORLD COMMISSION ON DAMS

November 2000

EARTHSCAN

Earthscan Publications Ltd, London and Sterling, VA

First published in the UK and USA in 2000
by Earthscan Publications Ltd

A catalogue record for this book is available from the British Library

ISBN: 1-85383-798-9 paperback
1-85383-797-0 hardback

Printed and bound in the UK by Thanet Press, Margate, Kent
Page design and layout by Page Arts, Cape Town
Cover design by Page Arts, Cape Town
Photo collages by Shane Smitsdorp, Cape Town

Cover photographs © Gallo Images. The Image Bank: Horowitz, Ross. I-Afrika: Miller, Eric.

Photo credits: Cape Argus: Schronen Johan, p226; Tromp, Dion, p66, 258; Yeld, John, p36. Cape Photo Library: Proust, Alain, p36; Stoffel-Wialliame, p72. Cosmi Corporation: p16, 19, 22, 49, 72, 80, 96, 108, 125, 134,157, 196, 250. I-Afrika: Bosch, Rodger, p14; Ingram Andrew, p96, 112; Miller, Eric, p2, 24, 96, 98, 104, 111, 114, 117, 129, 142, 153, 196, 201, 220; Pettersson, Anders, p243. IUCN -The World Conservation Union, p72, 85,87. Itaipú Binacional, p36, 68, 258. The Image Bank: Horowitz, Ross, p196, 206; Sund, Harald, p134, 149. United States Bureau of Reclamation, p36. World Bank, p96,100. World Wide Fund for Nature: Gunther, Michel, p72, 74,134,137; Harvey, Martin, p78. WWF-Canon; Burgler, Roel, p205, 212; Pratginestos, Juan, p72. Rautkari, Mauri; p145, Revesz, Tamás, p20, 210; Thorsell, James W, p212. Schafer, Kevin, p72, 92, 233; Torres, William H, p136.

Earthscan Publications Ltd
120 Pentonville Road, London, N1 9JN, UK
Tel: +44 (0)20 7278 0433
Fax: +44 (0)20 7278 1142
Email: earthinfo@earthscan.co.uk
http://www.earthscan.co.uk

22883 Quicksilver Drive, Sterling, VA 20166-2012, USA

Earthscan is an editorially independent subsidiary of Kogan Page Ltd and publishes in association with WWF-UK and the International Institute for Environment and Development

This book is printed on elemental chlorine-free paper

Chair's Preface

Globalisation From Below

Professor Kader Asmal
Chair, World Commission on Dams

If politics is the art of the possible, this document is a work of art. It redefines what is possible to all of us, for all of us, at a time when water pressure on governments has never been more intense. Consider: on this blue planet, less than 2.5% of our water is fresh, less than 33% of fresh water is fluid, less than 1.7% of fluid water runs in streams. And we have been stopping even these. We dammed half our world's rivers at unprecedented rates of one per hour, and at unprecedented scales of over 45 000 dams more than four storeys high.

As one who authorised the next stage of one of the largest dams in the Southern Hemisphere I can argue that nations build large dams for sound reasons. Dams store, use and divert water for consumption, irrigation, cooling, transportation, construction, mills, power and recreation. Dams remove water from the Ganges, Amazon, Danube, Nile or Columbia to sustain cities on their banks. For parting – or imparting – the waters, dams are our oldest tool. Yet are they our *only* tool, or our *best* option?

The World Commission on Dams has undertaken a rigorous, independent and inclusive global review, testing the waters to help you answer that question with authority. But just as water scarcity drove previous construction of dams, competition for water has underscored the Commission's work. As we seek water we face an escalating crisis, even of biblical proportions. In Ecclesiastes, recall the passage:

> *One generation passeth away,*
> *and another generation cometh:*
> *but the earth abideth always…*
> *All rivers runneth to the sea,*
> *yet the sea is not full…*

The words are beautiful, haunting and, suddenly, anachronistic. For they are not true due to demands and dams during our lives. Even degraded rivers seldom totally run, but loiter in a chain of reservoirs. In some years our mightiest rivers – Africa's Nile, Asia's Yellow, America's Colorado, Australia's Murray – do not reach the sea.

Compounding that shortage, one in five persons world-wide lacks access to safe drinking water. Half the world lacks sanitation; millions die from waterborne diseases. Farmers compete for water with booming but stressed cities. Towns drain aquifers that took centuries to fill. Saltwater pollutes groundwater miles from the sea. In China, Mexico and India water tables fall a metre a year. In a few decades, as we seek a fifth more water for 3 billion new people, one in three of us may struggle to drink or bathe.

Some see in our scarcity a harbinger of troubled waters to come. They believe water scarcity inevitably locks peoples, regions and nations in a fierce, competitive struggle in which restless millions race to the bottom in fear and self-interest. And thus, they maintain, when rivers cross borders within or between nations, water scarcity leads to water stress which leads to water wars.

Our Commission, and through it, this Final Report, contradicts that sentiment. We see water as an instrument, a catalyst for peace, that brings us together, neither to build dams nor tear them down but to carefully develop resources for the long term.

Easier said than done? Not necessarily. The hard part here may lie in what can be 'said' with intellectual honesty, vision and understanding. Surprisingly such a statement has rarely been attempted. Unlike every other aspect of our lives, large dams have long escaped deep and clear and impartial scrutiny into the process by which they emerge and are valued. This lapse is especially glaring when set against much smaller scale investments. We daily squeeze and weigh fruits and vegetables to ensure we get a fair return at the market. We rigorously test-drive and analyse the performance of motor cars before and after paying a few thousand dollars for one. We conduct thorough due-diligence before purchase of either house or business.

Yet this century we have collectively bought, on average, one large dam per day, and there have been precious few, if any, comprehensive, independent analyses as to why dams came about, how dams perform over time, and whether we are getting a fair return from our $2 trillion investment.

Until now.

Pioneer efforts are bound to be controversial. And while the World Commission on Dams is, by design, strictly advisory, make no mistake. Our genesis, work process and implications of this Final Report are by nature quite political. Our mandate involves the most precious element on earth, and that, of course, involves power: who wields it, how to share it, which ways the state may better balance it.

Some may feel this Report makes water use decisions even more difficult; by raising the bar higher, as we do, a government must exercise more energy and creativity to reach a sustainable result. But in truth we make those decisions easier; for we show clearly which, how, where and why decisions can either work well or fail to deliver.

For that reason I assert that we are much more than a 'Dams Commission'. We are a Commission to heal the deep and self-inflicted wounds torn open wherever and whenever far too few determine for far too many how best to develop or use water and energy resources. That is often the nature of power, and the motivation of those who question it. Most recently governments, industry and aid agencies have been challenged around the world for deciding the destiny of millions without including the poor, or even popular majorities of countries they believe to be helping.

To confer legitimacy on such epochal decisions, real development must be people-centred, while respecting the role of the state as mediating, and often representing, their interests. In the following pages we do not endorse globalisation as led from above by a few men. We do endorse globalisation as led from below by all, a new approach to global water policy and development.

In this approach, we must deal with the past before we can chart a course for the future. The integrity of our process determines the integrity of this product, which raises a key point. I am proud to sign this work, and to guide this project, but the document you hold is not, as are some Commission Reports, authored by its chair.

It comes instead from many authors who were originally separated by the cultural and philosophical divides of the debate. That is its true strength. Indeed, the assumption that the number of people who write something is inversely related to how much it says only goes so far. Hundreds of eloquent books and publications inform the dams and development debate from one side or the other. Written by single authors, they say much but resolve little. The WCD Final Report is meaningful, and will be remembered, not just for what it says, but for the fact that several hundred diverse men and women were directly involved in saying it. It is sculpted by the expertise of members of the Commission, many of whom have devoted their entire lives to engineering, environmental, social and institutional concerns of rivers and dams. All of us were sensitive to the needs of human development as we listened to the aspirations, the pain and the anguish of individual people.

Slow to speak, our Commission was quick to listen. Both sides of the debate gave their perspectives: from dam officials with an obligation to govern to dam affected people with stories to tell. By airing facts we reached a rapprochement that began in Gland, Switzerland and continued, non-stop, through hearings, consultations, case studies, submissions and reviews covering roughly one thousand large dams.

Through this process a shared understanding and truth began to emerge, and with it the thin thread with which to sew the stitches of reconciliation. On this Commission, the first stitches came, perhaps, as a woman who risks her life opposing a large dam threads the eye of the needle with an engineer who built his career designing them. Or when the leader of one the world's most powerful technology companies engaged with the leader of one of the world's proud but dispossessed peoples. As Seattle, Toronto, Washington, London and Berlin came unravelled by turbulent protests over globalisation and development, we quietly continued to apply stitch after stitch to sew a stronger, more resilient and colourful tapestry.

Our work is now over. With this document, yours has begun. I wish it could be as simple or catchy as popular manuals that offer simplistic solutions for complex situations. We recognise all those arguments, and absorb them. But we push beyond Declarations, which urge: 'Recognise human rights' or 'Sustain natural resources'.

Telling me, a harried public official who must answer to 48 million restless, hungry and thirsty people, to 'Ensure development is sustainable and humane' is like warning me, 'Operate, but don't inflict new wounds'. I know that. What I don't know is how to do it. To explain how to develop water in ways that do not exhaust either my constituents or the resources we all depend on, we must go beyond platitudes. Our healing must emerge not through anecdotes, but through a complex, coherent and cohesive argument that shows clearly where we have been, what happened, why we're in conflict, and how we can, with proper understanding, heal ourselves.

That involves first shedding misconceptions. Today's demands are too complex, our technology too advanced, our constituency too diverse, our options too numerous to allow just one solution. For example, imagine a typical dam.

Perhaps you see a smooth, parabolic, concrete structure. It seems to generate cheap electricity through turbines at its base. Engineers worship it, ecologists curse it, indigenous tribes lose their culture to it. Native fisheries plunge after construction, but floods decline as well. It pollutes neither air nor water, provides water for nearby towns, turns arid soils into rich farmland. People and animals were relocated, but the economic returns made doing so cost-effective. The dam embodies ambitions of statesmen, but when politicians approach with their ambitious plans, apprehensive peoples hold signs that say 'Save our beloved river'.

That image was my own. It was what I envisioned when I first took over the Ministry of Water Affairs and Forestry under Nelson Mandela. Five years of hands-on work tempered that vision. Chairing this Commission shattered it.

Instead of my archetype I saw: dams built of dirt and dams generating no electricity; dams praised by ecologists and dams despised by engineers; dams used for centuries by indigenous peoples, dams boosting fisheries, dams causing deadly floods; dams changing river chemistry or increasing net greenhouse gas emissions. I saw dam benefits by-pass thirsty adjacent communities en route to the city, dams exhaust and erode rich soils through water logging and salinity. I saw dams displace no one, dams create wetlands and work, dams cost thrice their budget, dams utterly abandoned and which had no symbolic value. Then I saw politicians approach rivers with ambitious, bureaucratic schemes, opposed by local activists shouting, 'Save our beloved *dam*'.

No matter how much you know, or think you know, about dams, you cannot read the following report and keep your assumptions intact. No matter how sceptical, you will come away changed, I think, for the better. For the truth is no typical dams exist.

Yet the decisions that led to those dams share a great deal. Clear patterns have emerged, and all parties have met. We have all reached agreement, established a healing process that we hope will work, and set this manual before you. Read it carefully – though not in one sitting – with an eye to where it may apply to your own specific needs and agenda. It is rigorous, without being rigid; it sees the State as an instrument of development yet recognises the necessity for popular participation; it is dispassionate and advisory in tone, but authoritative in its practical application.

It is said that if we do not master technology, technology will master us. In the past, our unrestrained reliance upon large dam technology weighed down upon us in all its unexamined mystery. It stood, like solid, divisive walls, between our left and right banks, between the upstream and downstream reaches of our rivers. The Commission's work is complete. And now, perhaps, technology can instead be kept under our united and democratic control, owned by all of us. In that way we can meet the coming water scarcity with confidence and assurance, knowing that water is not for fighting over. Water is for conserving. Water is for bathing. Water is for drinking. Water is for sharing. Water, through this report, can be our catalyst for peace.

Kader Asmal

Professor Kader Asmal,
Chair, World Commission on Dams

I would like to express on behalf of the Commission our particular appreciation to the following individuals. They have along with many other friends, partners and contributors to the WCD process, played a vital role in enabling the Commision to fulfil its mandate.

Bruce Babbitt, Sadi Baron, Ger Bergkamp, Richard Bissell, Robert Bos, Peter Bosshard, Rodney Bridle, John Briscoe, Ian Curtis, Shripad Dharmadhikary, Bert Diphoorn, Osmar Vieira de Filho, Luis Garcia, Raymundo José S. Garrido, Pham Hong Giang, Liane Greeff, George Green, Biksham Gujja, Geir Y. Hermansen, Kaare Hoeg, Ann Jennervik, Olav Kjorven, Jean-Etienne Klimpt, Manfred Konukiewitz, M.L Chanaphun Kridakorn, Maritta Koch-Weser, Nicholas Lapham, Donal O'Leary, Patrick McCully, David McDowell, Joseph Milewski, Reatile Mochebelele, Naoki Mori, Takehiro Nakamura, Peter van Niekerk, Raimundo Nonato do C. Silva, Tilak Ranaviraja, Frances Seymour, Aly Shady, Jaswant Singh, Jan Strömblad, Even Sund, Sardar Mohammed Tariq, Allan Taylor, Martin Ter Woort, Himanshu Thakkar, Klaus Töpfer, Dao Trong Tu, Mike J. Tumbare, Mumtaz Türfan, Michael Wiehen, James Wolfensohn, Mahmoud Abu Zeid, Tor Ziegler and Birgit Zimmerle.

Commissioners' Foreword

The World Commission on Dams (WCD) was born out of a small but significant IUCN-World Bank sponsored workshop in Gland, Switzerland in April 1997. Representatives of diverse interests came together to discuss the highly controversial issues associated with large dams. To the surprise of participants, deep-seated differences on the development benefits of large dams did not prevent a consensus emerging. That consensus included the proposal for a World Commission on Dams.

Professor Kader Asmal
Minister of Education,
WCD Chair, South Africa

Mr Lakshmi Chand Jain
Industrial Development Services,
WCD Vice-Chair, India

Dr Judy Henderson
Oxfam International,
Australia

Mr Göran Lindahl
ABB Ltd, Sweden

Prof Thayer Scudder
California Institute
of Technology, USA

As Commissioners, we have been honoured to serve on The WCD for the past two and a half years. Representing all sides of the debate, we have worked as an autonomous international team. Our task was to conduct a rigorous, independent review of the development effectiveness of large dams, to assess alternatives and to propose practical guidelines for future decision-making.

Since our work began in May 1998, we have met on nine occasions. We have listened to one anothers' different viewpoints in a genuine spirit of openness and desire to find a common understanding. All of us have found it a learning process, an enriching if sometimes uncomfortable one. This consensus document is the result of our work, but it would be unfair to the process to suggest that we have emerged completely of one mind. Individual differences still exist. However, we all agree on the fundamental principles and values that underpin this report and on the guidelines we offer for the way forward. In the final drafting of our report we have included "A Comment" by one Commissioner who was unable to attend the final meeting. It presents that Commissioner's additional views on the findings and recommendations we have developed as a group of Commissioners.

The WCD is delivering its product in a rapidly changing international environment. Debates proliferate about how to conserve the world's precious resource base while meeting the needs of growing populations hungry for economic progress. Terms of investment, terms of trade, democratisation, the role of the state, the role of civil society, the obligation to protect threatened ecosystems and preserve Planet Earth for future generations: all are part of the wider context. Any policy on large infrastructure projects – whether for dams, highways, power stations, or other mega-installations – has to be developed in this context.

At the same time, alternative perspectives on human rights and development are being more clearly expressed. The Right to Development, adopted by the UN General Assembly in 1986 argues that 'development is a comprehensive process aimed at the constant improvement of the well-being of the entire population…'. Recently, vocal condemnation of the globalisation process, pointing out that too many people are being left behind those forging ahead, has added support to this call for a better quality of life for all of humanity, not just for some. The United Nations Development Programme's Human Development Report 2000 has given us a timely reminder that the rights to security and basic freedoms, and to human development are two sides of the same coin and that when 'human rights and human development advance together, they reinforce one another.'

As an international Commission, our process has been unique in taking on board a range of interests and opinions previously held to be irreconcilable. We have examined evidence produced and opinions expressed by a wide range of stakeholders: government agencies, project affected people and non-governmental organisations, people's movements, the dam construction industry, the export credit agencies and private investors, and the international development community. Through this work programme the Commission has added a new body of knowledge to existing databases and information on large dams, looking at alternate ways of meeting irrigation, water supply, energy, flood control requirements and processes of development planning.

How did we achieve this? First by creating a 68 member Stakeholder Forum to act as a sounding board and advisory group for the WCD. The Forum has allowed us to reach other partners in the debate and to sound out those whose support for this report is essential to its effective use.

Second, by drawing on the wider stakeholder community for experts and analysts in developing the WCD Knowledge Base, and for funds to support the Commission's work. To date, 53 public, private and civil society organisations have pledged funds to the WCD. This independent funding model is unique for international commissions.

Third, by undertaking a programme of four Regional Consultations in different parts of the world that drew submissions from an even wider network of interested parties. This provided a platform for all voices to be heard. Altogether, 1 400 individuals from 59 countries and from every type of stakeholder group took part in these regional consultations. The WCD also participated in two hearings on large dams organised for its benefit by NGOs in Southern Africa and Europe. During its two-year lifetime the Commission has received 947 submissions from over 80 countries. We have listed all of them in a central database accessible via the Internet.

Fourth, by initiating eight independent in-depth Case Studies of specific large dam projects and two country studies (India and China). Using a common methodology, the case studies were conducted in a transparent and participatory manner drawing inputs from all stakeholders through an extensive consultation process. We have used the findings from all these studies and submissions to inform a central product of the Report: the Global Review of Large Dams.

Fifth, by undertaking 17 Thematic Reviews, employing experts from a wide range of disciplinary backgrounds, nationalities and institutional bases. These reviews fall into five thematic categories: Social Issues, Environmental Issues, Economic and Financial Issues, Options Assessment and Institutional Issues. Once again, we conducted these reviews in a participatory manner, commissioning over 100 papers with full peer review.

Lastly, by conducting a comprehensive global survey of 125 dams, which we used to 'cross-check', the findings of the individual studies. The Cross-Check Survey has provided a basic set of data on trends to complement the knowledge base.

We are fully aware that this body of data cannot and should not be seen as the 'final verdict' on the large dams story. The WCD has examined around 1 000 dams with varying degrees of intensity. This is a small fraction of the more than 45 000 large dams world-wide. There has been little systematic collection of data about dam projects in the past, and without baseline data we cannot arrive at definite conclusions about certain types of impacts. Albeit, the WCD report is the first comprehensive global and independent review of the essential aspects of the performance of dams. In many cases the impacts are still being played out, and will continue for many years to come. For this reason it is important for the future management of dams to have continued and systematic evaluation of their performance.

In carrying out our review we have not tried to judge individual dams. We have found that the unprecedented expansion in large dam building over the past century, harnessing water for irrigation, domestic and industrial consumption, electricity generation and flood control has clearly benefited many people globally. Nonetheless, this positive contribution of large dams to development has been marred in many cases by significant environmental and social impacts which, when viewed from today's values, are unacceptable.

We have sought to glean the lessons from the past in order to make recommendations for the future through a prism of equitable and sustainable development. Development based on five objectives: equity in resource allocation and in the spread of benefits; sustainability in the use of the world's diminishing resource-base; openness and participation in decision-making processes; efficiency in the management of existing infrastructural developments; and accountability towards present and future generations.

In today's rapidly globalising world the WCD process has pioneered a new path for global public policy making on issues of equitable and sustainable development. This

Ms Joji Cariño
Tebtebba Foundation, Philippines

Mr Donald Blackmore
Murray-Darling Basin Commission, Australia

Ms Medha Patkar
Struggle to Save the Narmada River, India

Prof José Goldemberg
University of São Paulo, Brazil

Ms Deborah Moore
Environmental Defense, USA

Mr Jan Veltrop
Honorary President,
ICOLD, USA

Mr Achim Steiner
WCD Secretary-General,
Germany

Report is the test against which that process will be measured. We hope that the policy framework and practical guidelines for its implementation presented here will add significant value to existing norms and form a basis for best practice in water and energy development. This is only a beginning, but we hope it is a dynamic beginning that others can take forward in the future. We also hope that the lessons learnt from our analysis of large dams will be seen as relevant for other large infrastructural projects, and that the framework of policy development and implementation we have identified will see wider application.

The life of the World Commission on Dams ends with the publication of this Report. For the Commissioners this has been an exciting, challenging and enriching process. It would not have been possible without the tireless commitment of an extraordinary team of professionals in the Secretariat. But more, we are indebted to hundreds of people around the world who, mostly at their own expense, have given us the bounty of their knowledge, expertise and life's experiences though discussion papers, submissions and presentations adding light and life to this report on large dams.

Professor Kader Asmal

Mr Donald Blackmore

Mr Lakshmi Chand Jain

Ms Medha Patkar

Dr Judy Henderson

Prof José Goldemberg

Mr Göran Lindahl

Ms Deborah Moore

Prof Thayer Scudder

Mr Jan Veltrop

Ms Joji Cariño

Mr Achim Steiner

Table of Contents

List of Tables

List of Figures

List of Boxes

Acronyms and Abbreviations

ADB	Asian Development Bank
AfDB	African Development Bank
CBA	cost-benefit analysis
CHIA	cultural heritage impact assessment
CHP	combined heat and power
DRD	Declaration on the Right to Development
DSM	demand-side management
DSS	decision support system
EA	environmental assessment
ECA	export credit agency
EFA	environmental flow assessment
EFR	environmental flow requirement
EIA	environmental impact assessment
EIRR	economic internal rate of return
EPA	Environmental Protection Agency
FIRR	financial internal rate of return
FSC	Forestry Stewardship Council
GATT	General Agreement on Tariffs and Trade
GDP	gross domestic product
GHG	greenhouse gas/es
HIA	health impact assessment
HLC	high level committee
IA	impact assessment
IADB	Inter-American Development Bank
ICID	International Commission on Irrigation and Drainage
ICJ	International Court of Justice
ICOLD	International Commission on Large Dams
IEA	International Energy Agency
IFC	International Finance Corporation
IFI	international finance institutions
IHA	International Hydropower Association
ILO	International Labour Organization
IRN	International Rivers Network
IRP	independent review panels
ISO	International Organization for Standardization
IUCN	World Conservation Union
LCA	life cycle assessment
MCA	multi-criteria analysis
MIGA	multilateral investment guarantee agency
MRDAP	mitigation, resettlement and development action plan
MW	megawatts
NGO	non-governmental organisation
NORAD	Norwegian Agency for International Co-operation
O&M	operation and maintenance
OECD	Organisation for Economic Co-operation and Development
OED	Operations Evaluation Division (of the World Bank)
PCB	polychlorinated biphenyls
PIC	prior and informed consent
PV	photovoltaic
R&D	research and development
SA	strategic impact assessment
SIA	social impact assessment
SIDA	Swedish International Development Agency
SRI	socially responsible investing
UNDP	United Nations Development Programme
USAID	US Agency for International Development
WCD	World Commission on Dams
WRI	World Resources Institute

Acknowledgements

The Forum

Coming from 68 institutions in 36 countries, members of the Forum reflect the diverse range of interests in the dams debate. The Forum acts as a 'sounding board' for the work of the Commission, and helps maintain two-way communication with the various far-flung constituencies. Forum members can help to build ownership of Commission work. The WCD is also conscious that reports in themselves have little impact if they are not firmly rooted in a process that enables all interest groups to develop an understanding of and confidence in the process itself. Membership of the Forum however does not imply endorsement of the Commission's report and findings.

Affected People's Groups

CODESEN, Co-ordination for the Senegal River Basin, Senegal
COICA, Federación de Indigenas del Estado Bolívar, Venezuela
Grand Council of the Cree, Canada
MAB, Movimento dos Atingidos por Barragens, Brazil
NBA, Narmada Bachao Andolan, India
Sungi Development Foundation, Pakistan
Cordillera People's Alliance, Philippines

Bilateral Agencies/Export Credit Guarantee Agencies

BMZ, Federal Ministry for Economic Co-operation and Development, Germany
NORAD, Norwegian Agency for International Co-operation, Norway
JBIC, Japan Bank for International Co-operation, Japan
SDC, Swiss Agency for Development and Co-operation, Switzerland
Sida, Swedish International Development Agency, Sweden
U.S. Export/Import Bank, USA

Government Agencies

United States Bureau of Reclamation, USA
LHWP, Lesotho Highlands Water Project, Lesotho
Ministry of Water Resources, China
National Water Commission, Mexico
Ministry of Mahaweli Development, Sri Lanka
Ministry of Water Resources, India

International Associations

ICID, International Commission for Irrigation and Drainage
ICOLD, International Commission on Large Dams
IHA, International Hydropower Association
IAIA, International Association for Impact Assessments

Multilateral Agencies

ADB, Asian Development Bank
AfDB, African Development Bank
FAO, United Nations Food and Agriculture Organisation
IADB, Inter-American Development Bank
UNDP, United Nations Development Programme
UNEP, United Nations Environment Programme
WB, World Bank

Non-Governmental Organisations

Berne Declaration, Switzerland
ENDA, Environmental Development Action, Senegal
Help the Volga River, Russia
IRN, International Rivers Network, USA
ITDG, Intermediate Technology Development Group, United Kingdom
IUCN, The World Conservation Union, Switzerland
Sobrevivencia-Friends of the Earth, Paraguay
WWF, World Wide Fund for Nature, Switzerland
DAWN, Development Alternatives with Women for a New Era, Fiji
TI, Transparency International, Germany
WEED, World Ecology, Environment and Development, Germany
Swedish Society for Nature Conservation, Sweden
Wetlands International, Japan

Private Sector Firms

Enron, USA
Harza Engineering, USA
Siemens, Germany
ABB, Switzerland
Saman Engineering Consultants, South Korea
Engevix, Brazil

Research Institutes/Resource Persons

Centro EULA, Ciudad Universitaria, Chile
Tropica Environmental Consultants Ltd., Senegal
WRI, World Resources Institute, USA
Water Research Institute, Israel
Winrock International, Nepal
Focus on the Global South, Thailand
ISPH, Institute of Hydroelectric Studies and Design, Romania
IWMI, International Water Management Institute, Sri Lanka
Worldwatch Institute, USA
Wuppertal Institute, Germany

River Basin Authorities

Confederación Hydrográfica del Ebro, Spain
Mekong River Commission, Cambodia
Volta River Authority, Ghana
Jordan Valley Authority, Jordan

Utilities

Eletrobras, Brazil
Hydro-Québec, Canada
Nepal Electricity Authority, Nepal
Mini Hydro Division, Philippines
Electricité de France, France

Partnerships and Co-operation

The Food and Agriculture Organisation

Transparency International

UNEP
The United Nations Environment Program

The Environmental Monitoring Group

CIGB ICOLD
International Commission on Large Dams

The World Archaeological Congress

The International Energy Agency

The World Health Organisation

International Institute for Sustainable Development

IAIA
International Association for Impact Assessment

World Resources Institute

International Rivers Network

IUCN
The World Conservation Union

The World Bank

Financial contributors

Financial support was received from 53 contributors including governments, international agencies, the private sector, NGOs and various foundations. According to the mandate of the Commission, all funding received by it was 'untied' – i.e. these funds were provided with no conditions attached to them.

AusAID

ABB

Germany (Bundesministerium für wirtschaftliche Zusammenarbeit und Entwicklung)

IUCN The World Conservation Union

WWF

DFID UK Department for International Development

Sida Swedish International Development Agency

National Wildlife Federation

Association for India's Development (AID)

C.S. Mott Foundation

People's Republic of China – Ministry of Water Resources

Finland – Ministry of Foreign Affairs

Ireland – Ministry of Foreign Affairs

South Africa – Ministry of Water Affairs and Forestry

Japan – Ministry of Foreign Affairs

Denmark – Ministry of Foreign Affairs

Netherlands – Ministry of Foreign Affairs

Norway – Ministry of Foreign Affairs

ENGEVIX

HARZA

VOITH SIEMENS HYDRO POWER GENERATION

Atlas Copco

UNEP United Nations Environment Programme

Packard Foundation

Tractebel

Swiss Agency for Development and Cooperation

Halcrow

Novib

Asian Development Bank

ENRON

U.S. Department of the Interior Bureau of Reclamation

Hydro Québec

Inter-American Develoment Bank

The World Bank

The British Dam Society

SNC·LAVALIN

LOTEK Wireless Fish & Wildlife Monitoring

Statkraft

NORPLAN A.S Consulting Engineers and Planners

EDF Electricité de France

Agence canadienne de développement international / Canadian International Development Agency (CIDA ACDI)

Deutsche Gesellschaft für Technische Zusammenarbeit (GTZ) GmbH

KfW Kreditanstalt für Wiederaufbau

United Nations Foundation

RBF Rockefeller Brothers Fund

Worley International

Paul C. Rizzo Associates, Inc. Consultants

BD Berne Declaration

Lahmeyer International Economics & Project Development

Manitoba Hydro

Coyne et Bellier Bureau d'Ingénieurs Conseils

Goldman Environmental Foundation

SKANSKA

WCD Case Studies – Team Leaders, Team of Authors, Additional Authors and General Acknowledgements

Grand Coulee Dam, Columbia River Basin, USA

TL: Leonard Ortolano, Katherine Kao Cushing **AA:** Nicole T. Carter, Harza Engineering, William Green, Carl Gotsch, Kris May, Tim Newton, Sophie Pierre, Josh Smienk, Michael Soules, Marilyn Watkins **GA:** BC Hydro, Bonneville Power Administration, Bureau of Reclamation, Canadian First Nations, Columbia Basin Trust, Colville Trust, Spokane Tribe, United States Army Corps of Engineers

Tarbela Dam, Indus River Basin, Pakistan

TL: Amir Muhammed Khan **AA:** Altaf Abro, Shahid Ahmed, Pervaiz Amir, Afzal Haq, Mehmooda S. Jilani, Riaz Ahmed Khan, Peter John Meynell, Javed Saleem Qamar, Riaz Hussain Qureshi, Riaz Nazir Tarar **GA:** Ministry of Water and Power, Water and Power Development Authority, Members of the Pakistan Network for Rivers, Dams and People, IUCN-Pakistan Office, World Bank-Pakistan Office

Aslantas Dam, Ceyhan River Basin, Turkey

TL: Refik Çölasan **AA:** O. Türker Altan, Okan Arihan, Çigdem Baykal, Ali Çaglar, Ahmet Eltekin, Nadir Izgin, Riza Kanber, Suhandan Karauz, Haluk Kasnakoglu, Birsen Gökçe, Zuhal Güler, Mete Kaan Kaynar, Şuha Şatana, Bora Sürmeli, Zeliha Ünaldi, Erdal Şekeroglu, Tuluhan Yilmaz, Recep Yurtal **GA:** Department of State Hydraulic Works (Ankara, Adana and Karamamaras), General Directorate of Rural Services, Water Users Associations of Ceyhan Aslantas Project, Dogal Hayati Koruma Dernegi, Ministry of Environment and Provincial Directorate of Health, Directorate of Agriculture, Forestry and National Parks, World Bank -Turkey Office

Kariba Dam, Zambezi River, Zambia/Zimbabwe

TL: Alois Hungwe **AA:** Julius Chileshe, Moses Chimbari, Dennis Chiwele, Paulman Chungu, Andrew Conybeare, Ezekiel Jonathan, Ronald Lwamba, Hillary Masundire, Dominic Mazvimavi, Ngonidzaishe Moyo, Herbert Mudenda, Fanuel Nangati, Daniel Ndlela, Elias Nyakunu, Norman Reynolds, John Santa Clara, Bennet Siamwiza, Steven Tembo **GA:** David Z. Mazvidza, Chris Magadza, Steve Rothert

Tucurui Dam, Tocantins River, Brazil

TL: Emilio Lèbre La Rovere, Francisco Eduardo Mendes **AA:** Bertha Becker, Gilberto Canali, Rosa Carmina Couto, Paulo Diniz, Iara Ferraz, Efrem Ferreira, José Alexandre Fortes, Maria das Graças da Silva, Marcia Ismerio, Ana Lacorte, Renato Leme Lopes, Adriana Neves Luna, Sandra Macedo, Rosa Acevedo Marin, Oscar de Moraes Cordeiro Netto, Sylvia Helena Padilha, Lúcio Flávio Pinto, Eneas Salati, Maria Nazareth da Silva, Wanderli Pedro Tadei **GA:** Electronorte, Agência Nacional de Energia Eléctrica, Centro Agroecologico de Assessoria y Educación Popular, Osmar Vieira Filho, Marcos V. Freitas, Raimundo Nonato do C. Silva, Sadi Baron, Henri Acselrad, Philip Fearnside, Birgit Zimmerle, Jean Remy D. Guimaraes

Pak Mun Dam, Mun-Mekong River Basin, Thailand

TA: Sakchai Amornsakchai, Philippe Annez, Sansanee Choowaew , Songkram Grachangnetara, Prasit Kunurat, Jaruwan Nippanon, Roel Schouten, Pradit Sripapatrprasite, Chayan Vaddhanaphuti, Chavalit Vidthayanon, Suphat Vongvisessomjai, Ek Watana, Wanpen Wirojanagud **GA:** Ammar Siamwalla, Chainarong Sretthachau, Darayes Mehta, Kitcha Polparsi, M. L. Chanaphun Kridakorn, Michai Veravaidhya, Prudhisan Jumbala, Sansern Wongcha-um, Vatana Meevasana, Vipada Apinan, Zakir Hussain

Glomma and Laagen Basin, Norway

TL: Jostein Skurdal **AA:** Øystein Aas, Tor Arnesen, Per Christian Bøe, Åge Brabrand,

Jon Arne Eie, Bjørn P. Kaltenborn, Svein Erik Hagen, Karine Hertzberg, Trygve Hesthagen, Arne Linløkken, Dan Lundquist, Pål Mellquist, Asbjørn Molle, Torbjørn Østdahl, Trond Taugbøl, Jens Kristian Tingvold **GA:** Arne Erlandsen Øyvind Fjeldseth, Geir Y. Hermansen, Thrond Berge Larsen, Kurt Ole Linn, Pål Mellquist

Orange River Pilot Study

TL: WCD Secretariat **AA:** Azghar Adelzadeh, Andrew Ainslie, Geoff Antrobus, Nicola Bergh, Bryan Davies, Chris de Wet, Tony Emmett, Muzi Muziya, Kyra Naudascher-Jankowski, Barry Nkomo, Maartin van Veelen **GA:** Thinus Basson, Fannie du Plessis, Mirriam Kibi, Mike Meuller, Geraldine Schoeman, Staff of Surplus Peoples' Project in Cape Town, Peter van Niekerk, Theo van Robbroek

Country Study: India

TA: Pranab Banerji, Ramaswamy R. Iyer, R. Rangachari, Nirmal Sengupta, Shekhar Singh **GA:** Som Pal, Z. Hasan, Raj Rajagopal, A. D. Mohile, A. S. Desai, B. G. Verghese, K. R. Datye, M. C. Gupta, M. Gopalakrishnan, N. C. Saxena, Pradeep K. Deb, Sunderlal Bahuguna, P. S. Raghavan

Country Study: China

TA: John Boyle, Richard Fuggle, Habib Khoury, Ismail Najjar, Sam Pillai, Bill Smith

Briefing Paper on Russia and NIS countries

TA: Elena A. Barabanova, Nikolai I. Koronkevich, Law and Environment Eurasia Partnership (Central Asia), Lilia K. Malik, Vladimir Smakhtin, Irina S. Zaitseva

Thematic Reviews - Lead Writers and Contributing Writers

- I.1 **Social impact of large dams: equity and distributional issues: LW:** William Adams **CW:** Adrian Adams; Hugh Brody; Dominique Egre; Carmen Ferradas; Pablo Gutman; Lyla Mehta; Joseph Milewski; Bina Srinivasan; Lubiao Zhang
- I.2 **Dams, indigenous people and vulnerable ethnic minorities: LW:** Marcus Colchester **CW:** Jaqueline Carino; Jaroslave Colajacomo; Andrew Corbett; Gabungan; Luke Hertlein; Manisha Marwaha; Lyla Mehta; Amrita Patwardhan; Ande Somby; Maria Stankovitch
- I.3 **Displacement, resettlement, rehabilitation, reparation and development: LW:** Leopoldo Bartolome; Chris de Wet; Harsh Mander; Vijay Nagaraj **CW:** Christine Danklmaier; Ravi Hemadri; Jun Jing; Scott Robinson
- II.1 **Dams, ecosystem functions and environmental restoration: LW:** Mike Acreman; Ger Bergkamp; Patrick Dugan; Jeff McNeely **CW:** Asheline Appleton; Edward Barbier; Garry Bernacsek; Martin Birley; John Bizer; Cate Brown; Kenneth Campbell; John Craig; Nick Davidson; Simon Delany; Charles DiLeva; Frank Farquharson; Nicholas Hodgson; Donald Jackson; Jackie King; Michel Larinier; Jeremy Lazenby; Gerd Marmulla; Don McAllister; Mathew McCartney; Steve Miranda; John Morton; Dianne Murray; Mary Seddon; Leonard Sklar; David Smith; Caroline Sullivan; Rebecca Tharme
- II.2 **Dams and global change: LW:** Nigel Arnell; Mike Hulme; Luiz Pinguelli Rosa; Marco Aurelio dos Santos **CW:** Albert Mumma
- III.1 **Economic, financial and distributional analysis: LW:** Alec Penman; Michelle Manion; Bruce McKenney; Robert Unsworth **CW:** Colin Green; Pablo Gutman; Anneli Lagman; Anil Markandya; David Mullins; Kyra Naudascher-Jankowski; Douglas Southgate
- III.2 **International trends in project financing: LW:** Per Ljung **CW:** Lily Donge; Chris Head; Michael Kelly; Hilary Sunman
- IV.1 **Electricity supply and demand management options: LW:** Maritess Cabrera; Anton Eberhard; Michael Lazarus; Thierry Lefevre; Donal O'Leary; Chella Rajan **CW:** Glynn Morris; Roger Peters; Bjorn Svenson; Rona Wilkinson
- IV.2 **Irrigation options: LW:** K. Sanmuganathan **CW:** Pablo Anguita Salas; S. Char; Keith Frausto; Alfred Heuper-

man; Khalid Hussain; ICRISAT; Hector Maletta; Dieter Prinz; Yehuda Shevah; Anupam Singh; Laurence Smith; Himanshu Thakkar

IV.3 **Water supply options: LW:** Colin Fenn; David Sutherland **CW:** Mary Dickinson; John Gould; Allan Lambert; Jon Lane; Guy Preston; Philip Turner

IV.4 **Flood control and management options: LW:** Colin Green **CW:** Luis Berga; Patrick Hawker; Minoru Kuriki; Dennis Parker; Sylvia Tunstall; Johannes van Duivendijk; Herb Wiebe

IV.5 **Operation, monitoring and decommissioning of dams: LW:** Peder Hjorth; Charles Howard; Kuniyoshi Takeuchi **CW:** K. Betts; Michael Falter; Enrique Garcia; Peter Goodwin; Brian Haisman; Joji Harada; V. Jauhari; Thomas Russo; Geoffrey Simms; James Westcoat; Rodney White

V.1 **Planning approaches: LW:** David Nichols; Theo Stewart; David von Hippel **CW:** Daud Beg; Catherine Fedorsky; Matthias Finger; J. Karmacharya; Miguel Nucete; Don Moore; Girish Sant

V.2 **Environmental and social assessment for large dams: LW:** Barry Sadler **CW:** Frank Vanclay; Iara Verocai

V.3 **River basins - institutional frameworks and management options: LW:** Peter Millington **CW:** Len Abrams; Enrique Castelan Crespo; Fiona Curtin; Luis Garcia; Raymundo Garrido; Ramaswamy Iyer; Erik Mostert; Cecilia Tortajada; Anthony Turton; Aaron Wolf

V.4 **Regulation, compliance and implementation: LW:** Angela Cropper; Mark Halle; John Scanlon **CW:** Daniel Bradlow; Gabriel Eckstein; Balakrishnan Rajagopal; Tom Rotherham; Lori Udall; Michael Wiehen

V.5 **Participation, negotiation, and conflict management: LW:** Bruce Stedman **CW:** Tisha Greyling; Anne Randmer; Vanchai Vatanasapt; Arch Isabel Viana

Working Papers – Writers

Dams and Human Health: Martin Birley, Robert Bos, M'barack Diop, William Jobin, P Unnikrishnan

Dams and Cultural Heritage Management: Steven Brandt, Fekri Hassan

Contributors to the WCD Cross-Check Survey

Keizrul Abdullah, M.A. Abrougui, Rocha Afonso, K. Akapelwa, Antonio Altadill, S.C Anand, Mike Anane, Valdemar Andrade, Carlo Angelucci, Cindy Armstrong, Alexander Asarin, Leo Atakpu, Paul Aylward, Riad Baouab, GR Basson, Mona Bechai, Edigson Perez Bedoya, Hans Helmut Bernhart, Carlos Bertagno, Isaac Bondet, Peter Bosshard, Rodney Bridle, Adelino Estevo Bucuane, Brian Davies, Cheickna Seydi A. Diawara, Shripad Dhamadhikary, Foto Dhima, Bob Douthwaite, Tim Dunne, Gary Ellis, Richmond Evans-Appiah, Daryl Fields, Renzo Franzin, Alejandro Garcia, Alfonso García, Luis Garcia, Refik Ghalleb, Ben Marcus Gillespie, Handan Giray, M. Gopalakrishnan, Liane Greef, Francis Grey, Jose Ramon Guifarro, Ronald Valverde Guillen, Brian Haisman, Phil Hirsch, Ku Hsu, Dan Hulea, Alois Hungwe, K.L. Karmacharya, Shaheen Rafi Khan, David Kleiner, Elena Kolpakova, Michael Kube, Minoru Kuriki, Hee Seung Lee, Knut Leitner, Melissa Loei, William Loker, Eleyterio Luz, Anil Markandya, Isidro Lázaro Martín, Aboubacry Mbodji, Patrick McCully, Brad McLane, Roland Mejias, Chantho Milattanapheng, Joseph Milewski, José Díaz Mora, Amir Muhammed, Jorge Carreola Nava, Eden Napitupulu, Humphrey Ole Ncharo, Luz Nereida, Nguyen Anh Minh, Ali Noorzad, Magdelena Nunez de Cordero, David Okali, Elizabeth Olsen, Torbjørn Østdahl, F.C Oweyegha-Afunaduula, Thomas Panella, Richard L. Pflueger, Hermien Pieterse, Julio Pineda, Jose Polanco, Ambriosio Ramos, Silvia Maria Ramos, Bernard Reverchon, Johan Rossouw, Paul Royet, Joseph Rückl, David Scivier, Jose Roberto Serrano, Thomas Siepelmeyer, Kua Kia Soong, Bjorn Svenson, Gustavo Tamayo, Himanshu Thakkar, Suresh Kumar Thapa, Dao Trong Tu, Jan Tosnar, Jaroslav Ungerman, Martin van Veelen, H.A. Wickramaratna, Brayton Willis, Ralph Wittebolle, Chusak Wittayapak, Patricia Wouters

Presenters at the WCD's Regional Consultations

South Asia

K B Chand, Giasuddin Ahmed Choudhury, K R Datye, Shripadh Dharmadhikary, Aly Ercelawn, Drona Ghimire, Ramaswamy Iyer, S Karunaratne, A R Karunawathie, Shaheen Rafi Khan, Tauhidul Anwar Khan, Ashish Kothari, Lakshman Mediwake, Alistair McKechnie, D K Mishra, Iswer Raj Onta, M G Padhye, Bikash Pandey, Tilak Ranaviraja, M S Reddy, Saleem Samad, Girish Sant, S Selvarajah, P C Senaratne, Sardar Muhammad Tariq, Himanshu Thakkar, B G Verghese, D C Wijeratna, Hemantha Withenage

Latin America

Aziz Ab' Saber, Carlos Avogadro, Ismael Aguilar Barajas, Sadi Baron, Celio Bermann, Ricardo Canese, Jorge Cappato, José Porfirio Fontanelle de Carvalho, Margarita Rosa de Castro Illera, Carlos Chen, Ivan Correa, Jorge Oscar Daneri, Luis Alberto Machado Fortunato, Cassio Viotti, Afonso Henriques Moreira Santos, Marcos Aurelio de Freitas, Philip Fearnside, Fabio Feldmann, Altino Ventura Filho, Hector Huertas, M. Kudlavicz, Jaime Millan, José Rodrigues, Humberto Marengo, Euclides Pereira Macuxi, Miguel Nucete, Cristian Opaso, Bonarge Pacheco, Elias Diaz Pena, Alan Poole, Grethel Aguilar Rojas, Teodoro Sanchez, Salomon Nahmad Sitton, Carlos Vainer

Africa and the Middle East

Adrian Adams, Tareq A. Ahmed, Cansen Akkaya, Leo Atakpu, Mohamed Lemine Ould Baba, H. El Badraoui, Mohammed N. Bayoumi, Kamau Bobotti, Geoffrey Chavula, Marwa Daoudy, Seydi Ahmed Diawara, Salif Diop, Stephanie Duvail, R. Evans-Appiah, Arif Gamal, Alioune Gassama, Munther J. Haddadin, Olivier Hamerlynk, Mohammed Jellali, E.A.K. Kalitsi, Elisabeth Khaka, Jacqueline Ki-Zerbo, O.M. Letsela, Claire Limbwambwa, Bazak Zakeyo Lungu, Bhekani Maphalala, Reatile Mochebelele, Anna Moepi, Frank Muramuzi, Martin Musumba, Mohammed Nabil, Takehiro Nakamura, Karim S Numayr, Hassan M. A. Osman, F.C. Oweyegha-Afunaduula, Levin Özgen, Guy Preston, Karen Ross, Mahammadou Sacko, Motseao Senyane, Noxolo Olive Sephuma, A. Shalaby , David Smith, Robyn Stein, David Syantami Syankusule

East and South East Asia

Keizrul Abdullah, Reiko Amano, Le Quy An, Vu Hong Anh, Vipada Apinan, Michael Bristol, Le Thac Can, Harvey Demaine, Hans Freiderich, Arnaldo Tapao Gapuz, Shalmali Guttal, Nguyen Dinh Hoa, Pham Thi Mong Hoa, Susannah Hopkins Leisher, Hiroshi Hori, Michael Horowitz, Tran Minh Huan, Tomoo Inoue, Sung Kim, Joern Kristensen, See-Jae Lee, Nguyen Duc Lien, Arthur H. Mitchell, Masaru Nishida, Alastair M. North, Do Hong Phan, Grainne Ryder, Shyama Shepard, Kua Kia Soong, Chainarong Sretthachau, Dao Trong Tu, A Rusfandi Usman, Mikhail Wakil, Wayne C. White, Lubiao Zhang

Executive Summary

The global debate about large dams is at once overwhelmingly complex and fundamentally simple. It is complex because the issues are not confined to the design, construction and operation of dams themselves but embrace the range of social, environmental and political choices on which the human aspiration to development and improved well-being depend. Dams fundamentally alter rivers and the use of a natural resource, frequently entailing a reallocation of benefits from local riparian users to new groups of beneficiaries at a regional or national level. At the heart of the dams debate are issues of equity, governance, justice and power – issues that underlie the many intractable problems faced by humanity.

The dams debate is simple because behind the array of facts and figures, of economic statistics and engineering calculations, lie a number of basic and easily understood principles. If adhered to and routinely applied, these principles would not only go a long way towards responding to the controversy surrounding dams, but would markedly improve decision-making on water and energy resources, achieving better outcomes. In identifying these principles, the World Commission on Dams (WCD) has not had to look far; they are the same principles that emerge from the global commitments to human rights, development and sustainability.

Our report tells a multifaceted story. But we draw from it some straightforward and practical advice to guide future decisions on water and energy resources development. The report sets out to distil more than two years of intense study, dialogue and reflection by the Commission, the WCD Secretariat, the WCD Stakeholders' Forum and literally hundreds of individual experts and affected people on every aspect of the dams debate. It contains all the significant findings that result from this work and expresses everything that the Commission believes is important to communicate to governments, the private sector, civil society actors and affected peoples – in short, to the entire spectrum of participants in the dams debate.

The evidence we present is compelling. We feel confident that the material collected and analysed by the Commission provides overwhelming support for the main messages in the report.

We believe there can no longer be any justifiable doubt about the following:

- Dams have made an important and significant contribution to human development, and the benefits derived from them have been considerable.
- In too many cases an unacceptable and often unnecessary price has been paid to secure those benefits, especially in social and environmental terms, by people displaced, by communities downstream, by taxpayers and by the natural environment.
- Lack of equity in the distribution of benefits has called into question the value of many dams in meeting water and energy development needs when compared with the alternatives.
- By bringing to the table all those whose rights are involved and who bear the risks associated with different options for water and energy resources development, the conditions for a positive resolution of competing interests and conflicts are created.
- Negotiating outcomes will greatly improve the development effectiveness of water and energy projects by eliminating unfavourable projects at an early stage, and by offering as a choice only those options that key stakeholders agree represent the best ones to meet the needs in question.

The direction we must take is clear. It is to break through the traditional boundaries of thinking and look at these issues from a different perspective. Our recommendations develop a rationale and framework that responds to this critical need and offers scope for progress that no single perspective can offer on its own. It will ensure that decision-making on water and energy development:

- reflects a comprehensive approach to integrating social, environmental and economic dimensions of development;
- creates greater levels of transparency and certainty for all involved; and
- increases levels of confidence in the ability of nations and communities to meet their future water and energy needs.

There are no shortcuts to equitable and sustainable development. The evidence of success and failure we present in this report provides the best rationale why the 'business as usual' scenario is neither a feasible nor a desirable option.

Water and Development – The Changing Context

The key decisions are not about dams as such, but about options for water and energy development. They relate directly to one of the greatest challenges facing the world in this new century – the need to rethink the management of freshwater resources. A number of global initiatives and reports have documented the dramatic impact of human-induced water withdrawals from the world's lakes, rivers and ground aquifers. Total annual freshwater withdrawals today are estimated at 3 800 cubic kilometres – twice as much as just 50 years ago.

The unfolding scenario for water use in many parts of the world is one of increasing concern about access, equity and the response to growing needs. This affects relations:

- within and between nations;
- between rural and urban populations;
- between upstream and downstream interests;
- between agricultural, industrial and domestic sectors; and
- between human needs and the requirements of a healthy environment.

The challenge is not to mobilise so as to compete successfully, but to co-operate in reconciling competing needs. It is to find ways of sharing water resources equitably and sustainably – ways that meet the needs of all people as well as those of the environment and economic development. These needs are all intertwined, and our challenge is to resolve competing interests collectively. Achieving equitable and sustainable solutions will be to the ultimate benefit of all.

The imperative to supply growing populations and economies with water in a context of depleting groundwater resources, declining water quality and increasingly severe limits to surface water extraction has brought sustainable water resources management to the top of the global development agenda. Although increasing competition for water suggests an expanding scope for conflict, it also provides an incentive for new forms of co-operation and innovation. Dire scenarios for water demand must not overshadow the fact that development paths that meet and manage the demand exist and are available for us to choose. History demonstrates that the path of co-operation has more often been followed than the path of conflict. The same must be true for our future.

During the 20th century, large dams emerged as one of the most significant and visible tools for the management of water resources. The more than 45 000 large dams around the world have played an important role in helping communities and economies harness water resources for food production, energy generation, flood control and domestic use. Current estimates suggest that some 30–40% of irrigated land worldwide now relies on dams and that dams generate 19% of world electricity.

From the 1930s to the 1970s, the construction of large dams became – in the eyes of many –synonymous with development and economic progress. Viewed as symbols of modernisation and humanity's ability to harness nature, dam construction accelerated dramatically. This trend peaked in the 1970s, when on average two or three large dams were commissioned each day somewhere in the world.

While the immediate benefits were widely believed sufficient to justify the enormous investments made – total investment in large dams worldwide is estimated at more than $2 trillion – secondary and tertiary benefits were also often cited. These included food security considerations, local employment and skills development, rural electrification and the expansion of physical and social infrastructure such as roads and schools. The benefits were regarded as self-evident, while the construction and operational costs tended to be limited to economic and financial considerations that justified dams as a highly competitive option.

As experience accumulated and better information on the performance and consequences of dams became available, the full cost of large dams began to emerge as a serious public concern. Driven by information on the impacts of dams on people, river basins and ecosystems, as well as their economic performance, opposition began to grow. Debate and controversy initially focused on specific dams and their local impacts. Gradually these locally driven conflicts evolved into a global debate about the costs and benefits of dams. Global estimates of the magnitude of impacts include some 40-80 million people displaced by dams while 60% of the world's rivers have been affected by dams and diversions. The nature and magnitude of the impacts of dams on affected communities and on the environment have now become established as key issues in the debate.

The World Commission on Dams was born from this debate. Established in February 1998 through an unprecedented process of dialogue and negotiation involving representatives of the public, private and civil society sectors, it began work in May of that year under the Chairmanship of Professor Kader Asmal, then South Africa's Minister of Water Affairs and Forestry and later the Minister of Education. The Commission's 12 members were chosen to reflect regional diversity, expertise and stakeholder perspectives. The WCD was created as an independent body, with each member serving in an individual capacity and none representing an institution or a country.

The Commission's two objectives were:

- to review the development effectiveness of large dams and assess alternatives for water resources and energy development; and
- to develop internationally acceptable criteria, guidelines and standards, where appropriate, for the planning, design, appraisal, construction, operation, monitoring and decommissioning of dams.

The decision to proceed with a large dam, the way the decision was made, the opinions and perspectives that were heard are at the heart of the current debate about dams. This same question of choice – of decision-making – also lay at the heart of the Commission's work. Our report is about improving the way such decisions are made.

The WCD Global Review of Large Dams

A large part of the Commission's work involved a broad and independent review of the experience with large dams. The resulting WCD Knowledge Base includes eight detailed case studies of large dams, country reviews for India and China, a briefing paper for Russia and the Newly Independent States, a Cross-Check Survey of 125 existing dams, 17 Thematic Review papers, as well as the results of public consultations and more than 900 submissions made available to the Commission. This provided the basis for the assessment of the technical, financial, economic, environmental and social performance of large dams, and the review of their alternatives. The review underlined the critical issues relating to governance and compliance that have come to be associated with large dams.

The evaluation was based on the targets set for large dams by their proponents – the criteria that provided the basis for government approval. In reviewing this experience the Commission has studied a broad spectrum of dams. Its analysis gave particular attention to understanding the reasons why, how and where dams did not achieve their intended outcome or indeed produced unanticipated outcomes that explain the issues underlying the dams debate. Presenting this analysis does not overlook the substantial benefits derived from dams but rather raises the question of why some dams achieve their goals while others fail.

Performance of large dams

The knowledge base indicates that shortfalls in technical, financial and economic performance have occurred and are compounded by significant social and environmental impacts, the costs of which are often disproportionately borne by poor people, indigenous peoples and other vulnerable groups. Given the large capital investment in large dams, the Commission was disturbed to find that substantive evaluations of completed projects are few in number, narrow in scope, poorly integrated across impact categories and scales, and inadequately linked to decisions on operations.

In assessing the large dams reviewed by the Commission we found that:

- Large dams display a high degree of variability in delivering predicted water and electricity services – and related social benefits – with a considerable portion falling short of physical and economic targets, while others continue generating benefits after 30 to 40 years.
- Large dams have demonstrated a marked tendency towards schedule delays and significant cost overruns.
- Large dams designed to deliver irrigation services have typically fallen short of physical targets, did not recover their costs and have been less profitable in economic terms than expected.
- Large hydropower dams tend to perform closer to, but still below, targets for power generation, generally meet their financial targets but demonstrate variable economic performance relative to targets, with a number of notable under- and over-performers.
- Large dams generally have a range of extensive impacts on rivers, watersheds and aquatic ecosystems – these impacts are more negative than positive and, in many cases, have led to irreversible loss of species and ecosystems.
- Efforts to date to counter the ecosystem impacts of large dams have met with limited success owing to the lack of attention to anticipating and avoiding impacts, the poor quality and uncertainty of predictions, the difficulty of coping with all impacts, and the only partial implementation and success of mitigation measures.
- Pervasive and systematic failure to assess the range of potential negative impacts and implement adequate mitigation, resettlement and development programmes for the displaced, and the failure to account for the consequences of large dams for downstream livelihoods have led to the impoverishment and suffering of millions, giving rise to growing opposition to dams by affected communities worldwide.
- Since the environmental and social costs of large dams have been poorly accounted for in economic terms, the true profitability of these schemes remains elusive.

Perhaps of most significance is the fact that social groups bearing the social and environmental costs and risks of large dams, especially the poor, vulnerable and future generations, are often not the same groups that receive the water and electricity services, nor the social and economic benefits from these. Applying a 'balance-sheet' approach to assess the costs and benefits of large dams, where large inequities exist in the distribution of these costs and benefits, is seen as unacceptable given existing commitments to human rights and sustainable development.

Options for water and electricity services

Today, a wide range of options for delivering water and electricity services exists, although in particular situations the cost and feasibility of these options will vary in the face of constraints such as natural resource endowments and site location. The Commission found that:

- Many of the non-dam options available today – including demand-side management, supply efficiency and new supply options – can improve or expand water and energy services and meet evolving development needs in all segments of society.

- There is considerable scope for improving performance of both dam projects and other options.
- Demand management, reducing consumption, recycling and supply and end-use efficiency measures all have significant potential to reduce pressure on water resources in all countries and regions of the world.
- A number of supply-side options at all scales (ranging from small, distributed generation sources or localised water collection and water-recovery systems to regional-interconnection of power grids) have emerged that – on their own or collectively – can improve or expand the delivery of water and energy services in a timely, cost-effective and publicly acceptable manner.
- Decentralised, small-scale options (micro hydro, home-scale solar electric systems, wind and biomass systems) based on local renewable sources offer an important near-term, and possibly long-term, potential particularly in rural areas far away from centralised supply networks.
- Obstacles to the adoption of these options range from market barriers to institutional, intellectual and financial barriers. A range of incentives – some hidden – that favour conventional options limit the adoption rate of alternatives.

Decision-making, planning and institutional arrangements

The decision to build a dam is influenced by many variables beyond immediate technical considerations. As a development choice, the selection of large dams often served as a focal point for the interests and aspirations of politicians, centralised government agencies, international aid donors and the dam-building industry, and did not provide for a comprehensive evaluation of available alternatives. Involvement from civil society varied with the degree of debate and openness to political discourse in a country. However, the WCD Global Review documents a frequent failure to recognise affected people and empower them to participate in the process. In some cases, the opportunity for corruption provided by dams as large-scale infrastructure projects further distorted decision-making.

Once a proposed dam project passed preliminary technical and economic feasibility tests and attracted interest from financing agencies and political interests, the momentum behind the project often prevailed over other considerations. Project planning and appraisal for large dams was confined primarily to technical parameters and the narrow application of economic cost-benefit analyses. Historically, social and environmental impacts were left outside the assessment framework and the role of impact assessments in project selection remained marginal, even into the 1990s.

Conflicts over dams have heightened in the last two decades due largely to the social and environmental impacts of dams that were either disregarded in the planning process or unanticipated. However, it also stems from the failure by dam proponents and financing agencies to fulfil commitments made, observe statutory regulations and abide by internal guidelines. Whereas far-reaching improvements in policies, legal requirements and assessment procedures have occurred in particular countries and institutions, in the 1990s it appears that business-as-usual too often prevailed. Further, past shortcomings and inequities remain unresolved, and experience with appeals, dispute resolution and recourse mechanisms has been poor.

Core Values for Decision-Making

As the Global Review of dams makes clear, improving development outcomes in the future requires a substantially expanded basis for deciding on proposed water and energy development projects – a basis that reflects a full knowledge and understanding of the benefits, impacts and risks of large dam projects to all parties. It also requires introducing new voices, perspectives and criteria into decision-making, as well as processes that will build consensus around the decisions reached. This will fundamentally alter the way in which decisions are made and, we are convinced, improve the development effectiveness of future decisions.

The Commission grouped the core values that informed its understanding of these issues under five principal headings:

- equity,
- efficiency,
- participatory decision-making,
- sustainability, and
- accountability.

These five are more than simply issues – they are the values that run through the entire report. They provide the essential tests that must be applied to decisions relating to water and energy development. If the report advances these values significantly we will emerge at our destination – improved decision-making processes that deliver improved outcomes for all stakeholders.

The debate about dams is a debate about the very meaning, purpose and pathways for achieving development. This suggests that decision-making on water and energy management will align itself with the emerging global commitment to sustainable human development and on the equitable distribution of costs and benefits. The emergence of a globally accepted framework of norms rests on the adoption of the Universal Declaration of Human Rights in 1948 and related covenants and conventions thereafter. These later resolutions include the Declaration on the Right to Development adopted by the UN General Assembly in 1986, and the Rio Principles agreed to at the UN Conference on Environment and Development in 1992. The core values that inform the Commission's shared understanding are aligned with this consensus and rest on the fundamental human rights accorded to all people by virtue of their humanity.

Rights, risks and negotiated outcomes

Reconciling competing needs and entitlements is the single most important factor in understanding the conflicts associated with development projects and programmes – particularly large-scale interventions such as dams. The approach developed by the Commission of recognising rights and assessing risks (particularly rights at risk) in the planning and project cycles offers a means to apply these core values to decision-making about water and energy resource management. Clarifying the rights context for a proposed project is an essential step in identifying those legitimate claims and entitlements that may be affected by the project or its alternatives. It is also a pre-condition for effective identification of legitimate stakeholder groups that are entitled to a formal role in the consultative process, and eventually in negotiating project-specific agreements relating, for example, to benefit sharing, resettlement and compensation.

The assessment of risk adds an important dimension to understanding how, and to what extent, a project may impact on people's rights. In the past, many groups have not had an opportunity to participate in decisions that imply major risks for their lives and livelihoods, thus denying them a stake in the development decision-making process commensurate with their exposure to risk. Indeed, many have had risks imposed on them involuntarily. Risks must be identified and addressed explicitly. This will require the notion of risk to be extended beyond governments or developers to include both those affected by a project and the environment as a public good. Involuntary risk bearers must be engaged by risk takers in a transparent process to negotiate equitable outcomes.

An approach based on the recognition of rights and assessment of risks can lay the basis for greatly improved and significantly more legitimate decision-making on water and energy resource development. It offers an effective way to determine who has a legitimate place at the negotiation table and what issues need to be included on the agenda. Only decision-making processes based on the pursuit of negotiated outcomes, conducted in an open and transparent manner and inclusive of all legitimate actors involved in the issue are likely to resolve the complex issues surrounding water, dams and development.

Recommendations for a New Policy Framework

Researching and analysing the history of water resources management, the emergence of large dams, their impacts and performance, and the resultant dams debate led the Commission to view the controversy surrounding dams within a broader normative framework. This framework, within which the dams debate clearly resides, builds upon international recognition of human rights, the right to development and the right to a healthy environment.

Within this framework the Commission has developed seven strategic priorities and related policy principles. It has translated these priorities and principles into a set of corresponding criteria and guidelines for key decision points in the planning and project cycles.

Together, they provide guidance on translating this framework into practice. They help us move from a traditional, top-down, technology-focused approach to advocate significant innovations in assessing options, managing existing dams – including processes for assessing reparations and environmental restoration, gaining public acceptance and negotiating and sharing benefits.

The seven strategic priorities each supported by a set of policy principles, provide a principled and practical way forward for decision-making. Presented here as expressions of an achieved outcome, they summarise key principles and actions that the Commission proposes all actors should adopt and implement.

1. Gaining Public Acceptance

Public acceptance of key decisions is essential for equitable and sustainable water and energy resources development. Acceptance emerges from recognising rights, addressing risks, and safeguarding the entitlements of all groups of affected people, particularly indigenous and tribal peoples, women and other vulnerable groups. Decision making processes and mechanisms are used that enable informed participation by all groups of people, and result in the demonstrable acceptance of key decisions. Where projects affect indigenous and tribal peoples, such processes are guided by their free, prior and informed consent.

2. Comprehensive Options Assessment

Alternatives to dams do often exist. To explore these alternatives, needs for water, food and energy are assessed and objectives clearly defined. The appropriate development response is identified from a range of possible options. The selection is based on a comprehensive and participatory assessment of the full range of policy, institutional and technical options. In the assessment process social and environmental aspects have the same significance as economic and financial factors. The options assessment process continues through all stages of planning, project development and operations.

3. Addressing Existing Dams

Opportunities exist to optimise benefits from many existing dams, address outstanding social issues and strengthen environmental mitigation and restoration measures. Dams and the context in which they operate are not seen as static over time. Benefits and impacts may be transformed by changes in water use priorities, physical and land use changes in the river basin, technological developments, and changes in public policy expressed in environment, safety, economic and technical regulations. Management and operation practices must adapt continuously to changing circumstances over the project's life and must address outstanding social issues.

4. Sustaining Rivers and Livelihoods

Rivers, watersheds and aquatic ecosystems are the biological engines of the planet. They are the basis for life and the livelihoods of local communities. Dams transform landscapes and create risks of irreversible impacts. Understanding, protecting and restoring ecosystems at river basin level is essential to foster equitable human development and the welfare of all species. Options assessment and decision-making around river development prioritises the avoidance of impacts, followed by the minimisation and mitigation of harm to the health and integrity of the river system. Avoiding impacts through good site selection and project design is a priority. Releasing tailor-made environmental flows can help maintain downstream ecosystems and the communities that depend on them.

5. Recognising Entitlements and Sharing Benefits

Joint negotiations with adversely affected people result in mutually agreed and legally enforceable mitigation and development provisions. These recognise entitlements that improve livelihoods and quality of life, and affected people are beneficiaries of the project. Successful mitigation, resettlement and development are fundamental commitments and responsibilities of the State and the developer. They bear the onus to satisfy all affected people that moving from their current context and resources will improve their livelihoods. Accountability of responsible parties to agreed mitigation, resettlement and development provisions is ensured through legal means, such as contracts, and through accessible legal recourse at the national and international level.

6. Ensuring Compliance

Ensuring public trust and confidence requires that the governments, developers, regulators and operators meet all commitments made for the planning, implementation and operation of dams. Compliance with applicable regulations, criteria and guidelines, and project-specific negotiated agreements is secured at all critical stages in project planning and implementation. A set of mutually reinforcing incentives and mechanisms is required for social, environmental and technical measures. These should involve an appropriate mix of regulatory and non-regulatory measures, incorporating incentives and sanctions. Regulatory and compliance frameworks use incentives and sanctions to ensure effectiveness where flexibility is needed to accommodate changing circumstances.

7. Sharing Rivers for Peace, Development and Security

Storage and diversion of water on transboundary rivers has been a source of considerable tension between countries and within countries. As specific interventions for diverting water, dams require constructive co-operation. Consequently, the use and management of resources increasingly becomes the subject of agreement between States to promote mutual self-interest for regional co-operation and peaceful collaboration. This leads to a shift in focus from the narrow approach of allocating a finite resource to the sharing of rivers and their associated benefits in which States are innovative in defining the scope of issues for discussion. External financing agencies support the principles of good faith negotiations between riparian States.

If we are to achieve equitable and sustainable outcomes, free of the divisive conflicts of the past, future decision-making about water and energy resource projects will need to reflect and integrate these strategic priorities and their associated policy principles in the planning and project cycles.

From Policy to Practice – The Planning and Project Cycle

The Commission's recommendations can best be implemented by focusing on the key stages in decision-making on projects that influence the final outcome and where compliance with regulatory requirements can be verified. Among the multitude of decisions to be taken, the Commission has identified five key decision points. The first two relate to water and energy planning, leading to decisions on a preferred development plan:

- Needs assessment: validating the needs for water and energy services; and
- Selecting alternatives: identifying the preferred development plan from among the full range of options.

Where a dam emerges from this process as a preferred development alternative, three further critical decision points occur:

- Project preparation: verifying that agreements are in place before tender of the construction contract;
- Project implementation: confirming compliance before commissioning; and
- Project operation: adapting to changing contexts.

Social, environmental, governance and compliance aspects have been undervalued in decision-making in the past. It is here that the Commission has developed criteria and guidelines to innovate and improve on the body of knowledge on good practices and add value to guidelines already in common use. Seen in conjunction with existing decision-support instruments, the Commission's criteria and guidelines provide a new direction for appropriate and sustainable development.

Bringing about this change will require:

- planners to identify stakeholders through a process that recognises rights and assesses risks;
- States to invest more at an earlier stage to screen out inappropriate projects and facilitate integration across sectors within the context of the river basin;
- consultants and agencies to ensure outcomes from feasibility studies are socially and environmentally acceptable;
- the promotion of open and meaningful participation at all stages of planning and implementation, leading to negotiated outcomes;
- developers to accept accountability through contractual commitments for effectively mitigating social and environmental impacts;
- improving compliance through independent review; and
- dam owners to apply lessons learned from past experiences through regular monitoring and adapting to changing needs and contexts.

The Commission offers its criteria and guidelines to help governments, developers and owners meet emerging societal expectations when faced with the complex issues associated with dam projects. Adopting this framework will allow states to take informed and appropriate decisions, thereby raising the level of public acceptance and improving development outcomes.

Beyond the Commission – An Agenda for Change

The Commission's report identifies the key elements of the debate on water and energy resources management and the role of dams in this debate. It summarises the lessons learned from our Global Review of experience with large dams. It elaborates the development framework within which the controversies and underlying issues can be understood and addressed and proposes a decision-making process anchored in a rights-and-risks approach and based on negotiated outcomes. It offers a set of strategic priorities, principles, criteria and guidelines to address the issues around existing dams and to use in exploring new water and energy development options.

The report is not intended as a blueprint. We recommend that it be used as the starting point for discussions, debates, internal reviews and reassessments of what may be established procedures and for an assessment of how these can evolve to address a changed reality. In looking at the future, the Commission proposes a number of entry points to help organisations identify immediate actions they might take in response to the Commission's report. Specific proposals are included for:

- national governments and line ministries;
- civil society organisations;
- the private sector;
- bilateral aid agencies and multilateral development banks;
- export credit agencies;
- inter-governmental organisations;
- professional associations; and
- academic and research bodies.

Engaging through these entry points will initiate permanent changes to advance the principles, criteria and guidelines we set out.

The trust required to enable the different actors to work together must still be consolidated. Early and resolute action to address issues arising from the past will go a long way towards building that trust in the future. So, too, will an assurance to countries still at an early stage of economic development that the dams option will not be foreclosed before they have had a chance to examine their water and energy development choices within the context of their development process.

The experience of the Commission demonstrates that common ground can be found without compromising individual values or losing a sense of purpose. But it also demonstrates that all concerned parties must stay together if the issues surrounding water and energy resources development are to be resolved. It is a process with multiple heirs and no clear arbiter. We must move forward together or we will fail. The Commission was given an exceptional opportunity and it has delivered a result reflecting our collective learning process and understanding. If our report does not win widespread support among participants in the dams debate, it is unlikely that there will be another such opportunity for a long time.

We believe that our report is a milestone in the evolution of dams as a development option. We have:

- conducted the first comprehensive global and independent review of the performance of essential aspects of dams and their contribution to development. We have done this through an inclusive process that has brought all significant players into the debate;
- shifted the centre of gravity in the dams debate to one focused on investing in options assessment, evaluating opportunities to improve performance and address legacies of existing dams, and achieving an equitable sharing of benefits in sustainable water resources development; and
- demonstrated that the future for water and energy resources development lies with participatory decision-making, using a rights-and-risks approach that will raise the importance of the social and environmental dimensions of dams to a level once reserved for the economic dimension.

We have told our story. What happens next is up to you.

WORLD COMMISSION ON DAMS
MINISTRY OF AGRICULTURE & RURAL DEVELOPMENT VIETNAM
FINAL REGIONAL CONSULTATION
Large Dams and their Alternatives in East & South East Asia:
Experiences and Lessons Learned
ỦY BAN QUỐC TẾ VỀ ĐẬP
HỘI THẢO TƯ VẤN KHU VỰC LẦN CUỐI
Hanoi, 26 - 27 February 2000
GABCIKOVO

Chapter 1:
Water, Development and Large Dams

Over 45 000 times in the last century, people took the decision to build a dam. Dams were built to provide water for irrigated agriculture, domestic or industrial use, to generate hydropower or help control floods. But dams also altered and diverted river flows, affecting existing rights and access to water, and resulting in significant impacts on livelihoods and the environment. Decisions to build dams are being contested increasingly as human knowledge and experience expand, as we develop new technologies, and as decision-making becomes more open, inclusive and transparent.

At the heart of the current debate on dams is the way choices are made, and the different opinions and perspectives that are expressed – or denied expression – in the process. The same question of decision-making is at the centre of the Commission's area of concern.

Dams are a means to an end, not an end in themselves. What is that end? How central are the challenges that large dams set out to meet? And how well can dams meet them?

The World Commission on Dams considers that the end of any dam project must be the sustainable improvement of human welfare. This means a significant advance of human development on a basis that is economically viable, socially equitable, and environmentally sustainable. If a large dam is the best way to achieve this goal it deserves our support. Where other options offer better solutions we should favour them over large dams. Thus the debate around dams challenges our view of how we develop and manage our water resources.

The World Commission on Dams considers that the end of any dam project must be the sustainable improvement of human welfare.

Water has attracted the attention of political leaders at the highest level and has triggered a series of strategic global initiatives such as the World Commission on Water.[1] The most powerful players in development have placed water at the top of their agendas. Why this sudden attention to a resource that was, after all, central to human needs long before the dawn of civilisation?

At one level, the water issue justifies the priority accorded it because demand for water is rapidly outstripping available supply in large parts of the world. As populations grow, and economic development leads to higher consumption, demand for water increases sharply putting intense pressure on available stocks. This can cause growing social tensions, or even lead to outright conflict. While commonly expressed predictions that future wars will largely be fought over water may be exaggerated, nobody doubts that access to both surface and ground water is an increasingly contentious issue.

Where co-operation gives way to unfair competition between different water uses or between communities and countries, a range of new issues come to the fore. These issues relate to the distribution of power and influence within societies and between countries. They concern the relative weight of the different factors that make up the decision-making mix. And they relate to how choices are made between available options.

The issues surrounding dams are the issues surrounding water, and how water-related decisions are made. There is little public controversy about the choice between an embankment dam or a gravity dam, or whether to use earth, concrete or rock-fill, possibly even over financing the development. The issues all relate to what the dam will do to river flow, to rights of access to water and river resources, to whether it will uproot existing human settlements, disrupt the culture and sources of livelihood of local communities and deplete and degrade environmental resources. Conflicts over dams are more than conflicts over water.

They are conflicts over human development and life itself.

If water is life, rivers are its arteries. Dams regulate or divert the flow through these arteries, affecting the life-blood of humanity. The fact that they aim to do so in the interests of humankind only makes the decision to build a large dam more sensitive, one that will trigger a range of apprehensions, hopes and fears, both rational and irrational.

This Chapter examines the context of the debate on large dams that led to the establishment of the Commission. It begins with a broad look at the past and present 'drivers' of the demand for water, and the role of large dams in meeting this demand. Then it presents general patterns and trends in the development of large dams during the 20th century, along with a brief description of the main purposes for building dams. The chapter also introduces the scale and significance of the benefits, costs and impacts of large dams, described in more detail in Chapters 2 to 6.

It goes on to look at the central issues in the large dams debate – the benefits and adverse impacts – and the positions taken by different constituencies on the past effectiveness of dams, and what may constitute good practice in future decision-making. Chapter 7 picks up many of these themes and also situates the dams debate in the emerging global development framework.

The final section of the chapter traces the establishment of the Commission and highlights the mandate given to it in response to the debate on large dams. It then lays out the methodology adopted and the process followed by the Commission.

Water and Development

Today, around 3 800 km^3 of fresh water is withdrawn annually from the world's lakes, rivers and aquifers.[2] This is twice the volume extracted 50 years ago.

Today, around 3 800 km^3 of fresh water is withdrawn annually from the world's lakes, rivers and aquifers. This is twice the volume extracted 50 years ago.

A growing population and a rising level of economic activity both increase human demand for water and water-related services. Development, technological change, income distribution and life-styles all affect the level of water demand.

How much water do we need?

World population has passed 6 billion. Although the annual increase probably peaked at about 87 million around 1990, the high proportion of young people in most developing countries means that global population will continue to increase significantly well into this century.

Recent projections suggest a peak of between 7.3 billion and 10.7 billion around 2050 before total population begins to stabilise or fall.[3] Predictions cannot be precise, because other dimensions of development such as access to health, education, income, birth control and other services influence the pace of population growth.

Despite the massive investment in water resource management and particularly in

Box 1.1 New paradigm for water use

Successfully meeting human demands for water in the next century will increasingly depend on non-structural solutions and a completely new approach to planning and management. The most important single goal of this new paradigm is to rethink water use with the objective of increasing the productive use of water. Two approaches are needed, increasing the efficiency with which current needs are met, and increasing the efficiency with which water is allocated among different uses.

Source: Gleick, 1998

dams, billions of children, women and men in rural areas lack access to the most basic water and sanitation services. Although problems of access are worst in rural areas, rapid urbanisation is also increasing the demand for water-related services. In 1995, 46% of the world's population lived in urban areas. If current trends persist (and they may accelerate), that figure could reach 60% by the year 2030 and over 70% by 2050.[4] Most of this growth will take place in developing countries where an estimated 25 to 50% of urban inhabitants live in impoverished slums and squatter settlements. Lack of access to water in both rural and urban areas is not just a question of supply. It is partly due to inequitable access to existing supplies.

Urbanisation implies an increasing concentration on water and energy demand in mega-cities, a switch to different lifestyles and consumption patterns, and a loss of productive agricultural land through urban expansion. It is a widely held view that lack of attention to development in rural areas is fuelling unsustainable forms of urban growth, shifting poverty from rural to urban areas, and contributing to rapidly growing demand for additional services. In heavily populated countries like China, India and Indonesia many question the sustainability of the high rates of urbanisation in mega-cities.

Economic growth and development

World economic activity has grown approximately five-fold since 1950 at a rate of about 4% per year.[5] The regional balance is changing, with significant growth in Asia over the past 25 years. At present, OECD countries continue to account for the largest share, amounting to 55% of world production at purchasing power parity, nearly 80% at market prices. [6]

Economic growth has two implications for water demand. The first is that increased economic activity will increase the demand for water-related services – regardless of whether the demand is satisfied by more efficient use of the existing supply, or by increasing the level of supply. The second is that both the development brought about by economic growth and the technological changes that accompany it will lead to structural changes in the pattern of goods and services that society produces, and in the way these services are provided. The water demand per unit of Gross Domestic Product (GDP) will depend on how these two components of economic growth are combined. Countries with the same product per capita but different production characteristics – for example, with large scale irrigated agriculture or water intensive industries – may consume three or four times more water per dollar of GDP. This is evident when comparing the United States and Canada with Germany or France, or India with China.

Development and technological change can also save water per dollar of output. Between 1950 and 1990 the world economy grew by a factor of five while world water withdrawals only grew by a factor of 2.[7] The last fifty years have seen a worldwide reduction in the amount of water per dollar of non-agricultural production as a result of improved technology, more recycling, enforcement of environmental standards, higher water prices, and industries moving away from natural resource intensive activities.[8] However, the water management practices and technologies that enable such advances are not widely available or promoted and are often absent where they are most needed. Despite the increasing number of options available, the total number of people without access to clean water is growing.

Income distribution and life-styles

Economic activity and development affect income, income distribution and lifestyles. These in turn affect the demand for water through changes in the level and composition of household consumption in areas such as diet, the use of household appliances and standards of sanitation.

How much water is needed for one more person? Although climate and culture influence what constitutes an appropriate level of domestic water consumption, several international agencies and experts have proposed 50 litres per person per day (or just over 18.25 m^3 a year) as an amount that covers basic human water requirements for drinking, sanitation, bathing and food preparation. In 1990 over a billion people were below that level.[9]

On the other hand, households in developed countries and better-off households in developing cities use from 4 to 14 times the threshold of 50 litres per person a day.

Drastically lower average figures for domestic consumption in developing countries reflect not only different life styles and lower incomes, but also a huge backlog of unsatisfied demand. The lower average also masks extremely high consumption among better-off urban households and acute deprivation among rural and urban poor.

Competing uses of water

Water analysts foresee increased competition among water users in meeting the growing demand. They predict that competition will increase among the three largest water users in global terms. Agriculture accounts for about 67% of withdrawals, industry uses 19% and municipal and domestic uses account for 9%.[10] Analysts foresee that these uses will continue to draw from the water needed to sustain natural systems. In dry climates, evaporation from large reservoirs, estimated at close to 5% of total water withdrawals, may also be a significant consumptive use of water.[11]

In 1990 over a billion people had access to less than 50 litres of water a day.

Regional trends vary widely as shown in Figures 1.1 and 1.2. Despite increasing urbanisation in Africa, Asia and Latin America, agriculture is the dominant water user in these regions, accounting for approximately 85% of all water used. In all regions of the world except Oceania, domestic or household water consumption accounts for less than 20% of water use.[12] In Africa, Central America and Asia, this is nearer to 5%. In the more developed regions of Europe and North America, industry is the major water consumer. The water use breakdown in a specific country influences where water demand management opportunities exist.

Figure 1.1: Annual fresh water withdrawals as a percentage of total resources withdrawn (1996)

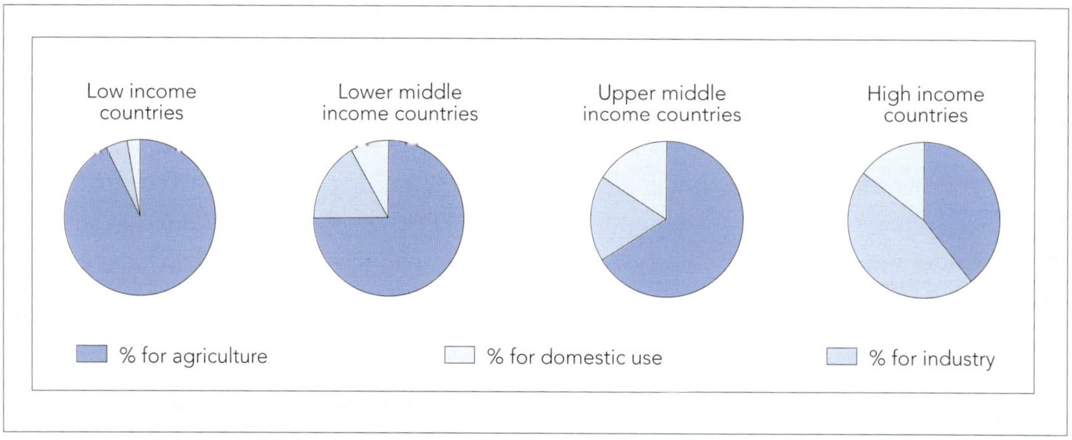

Source: World Bank 1999a.

For many water-stressed countries that are industrialising, and that have large urban populations, the present water crisis often revolves around industrial and domestic water supply and sanitation. This is evident in the significant trend over the past few decades towards diverting water from agricultural to municipal and industrial uses. As this is happening in the context of an overall increase in withdrawals, it will lead to increased competition for water, unless more efficient water use in both sectors accompanies the transition from agricultural to industrial based economies. [13]

Figure 1.2: Annual fresh water withdrawals per capita average (1987-95)

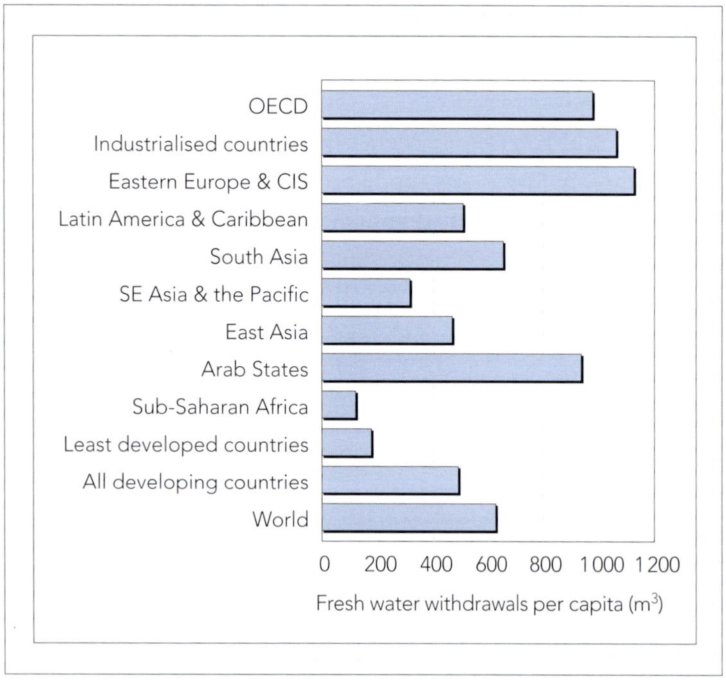

Source: UNDP, 1999.

There are additional challenges. To meet food requirements, water used in agriculture may have to increase 15 to 20% by 2025 even with improvements in irrigation efficiency and agronomic potential.[14] In addition to increasing food production in the face of water stress, distribution, equitable access, purchasing power and poverty are central issues in meeting food demand.

Beyond competing human demands, water for nature is an essential consideration. The fresh water ecosystems that provide the livelihoods of the world's riverine communities and many other goods and services to our societies depend on water. Arresting, and where possible reversing, the accelerating trend to increasing degradation of many of the world's watersheds caused by human activity have emerged as an urgent priority.

Availability and quality of water

Rainfall and other sources of freshwater (rivers, lakes, groundwater) are unevenly distributed around the world and are not always located where human water demand arises. Pollution threatens surface and ground water sources and may make them unfit for many uses, or require expensive treatment. Pollution is especially serious where ground water resources are over-exploited and suffer from reduced natural rates of recharge due to deforestation, land use changes and urbanisation.

Fully one third of the countries in water-stressed regions of the world are expected to face severe water shortages this century, and within these regions there are great disparities in access to fresh water. Not surprisingly, a significant number of less developed countries, including regions of India and China, are facing severe shortages.[15] With population growth, the number of countries in this category is increasing, and by 2025 there will be approximately 6.5 times as many people – a total of 3.5 billion – living in water-stressed countries.[16] Figures 1.3 and 1.4 show the distribution of world fresh water resources and selected water stressed countries. The uneven distribution of water supply means that countries may have water surplus and water deficit regions.

Not only surface water is under pressure. The growing rate of extraction of fresh water from rivers and lakes is matched by increasing extraction of ground water, with many aquifers now seriously depleted. The volume of ground water withdrawal, primarily for irrigation but also for municipal and industrial use, exceeds long-term recharge rates. In many parts of India, Pakistan and China, the water table is sinking at the rate of one to two meters a year.[17]

Climate change can also affect the seasonal distribution of rains and water availability. Studies and modelling exercises reviewed by the Commission, including those by the Intergovernmental Panel on Climate Change (IPCC), strongly suggest that global warming trends could significantly increase the variability of weather patterns.[18] Results could include a decline in rainfall in

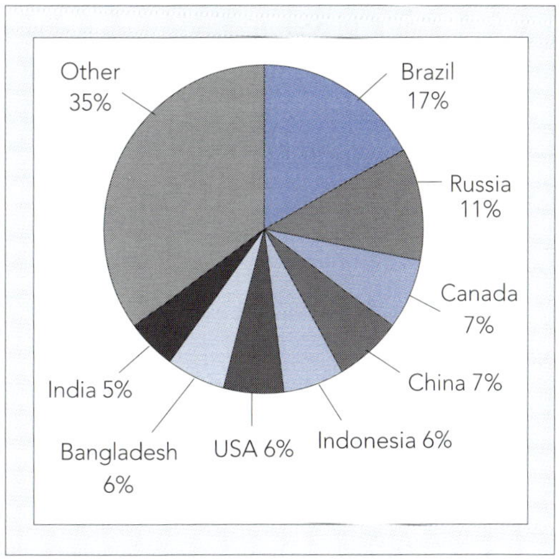

Figure 1.3: Distribution of the world's water

Source: Gleick, 1998.

Figure 1.4 : Selected water stressed countries

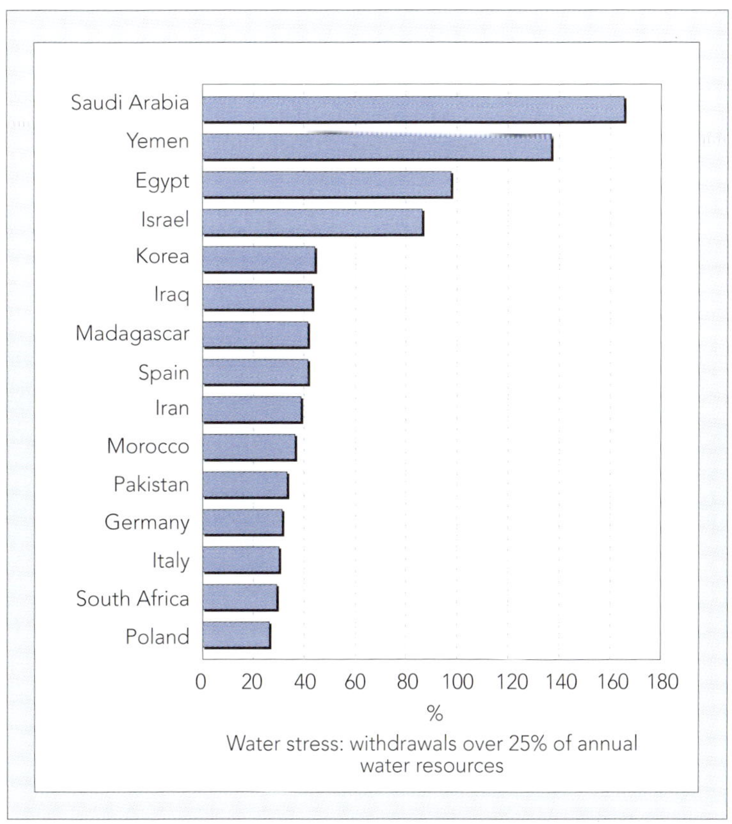

Source: Raskin et al, 1995.

arid and semi-arid regions, and an increase in seasonal variations in rainfall around the globe over the next 50 to 100 years.

Development and Large Dams

River basins are renowned as the cradles of civilisation and cultural heritage. Ancient and modern communities alike have depended on rivers for livelihood, commerce, habitat and the sustaining ecological functions they provide. Throughout history alterations to rivers – natural or human generated – have affected riverine communities in one way or another.

The earliest evidence of river engineering is the ruins of irrigation canals over eight thousand years old in Mesopotamia. Remains of water storage dams found in Jordan, Egypt and other parts of the Middle East date back to at least 3000 BC.

Historical records suggest that the use of dams for irrigation and water supply became more widespread about a thousand years later. At that time, dams were built in the Mediterranean region, China and Meso America. Remains of earth embankment dams built for diverting water to large community reservoirs can still be found in Sri Lanka and Israel.[19] The Dujiang irrigation project, which supplied 800 000 hectares in China, is 2 200 years old.[20] Dams and aqueducts built by the Romans to supply drinking water and sewer systems for towns still exist today.

The first use of dams for hydropower generation was around 1890. By 1900, several hundred large dams had been built in different parts of the world, mostly for water supply and irrigation.

Dams in the 20th century

The last century saw a rapid increase in large dam building. By 1949 about 5 000 large dams had been constructed worldwide, three-quarters of them in industrialised countries. By the end of the 20th century, there were over 45 000 large dams in over 140 countries.[21]

Figure 1.5: Regional distribution of large dams at the end of the 20th century

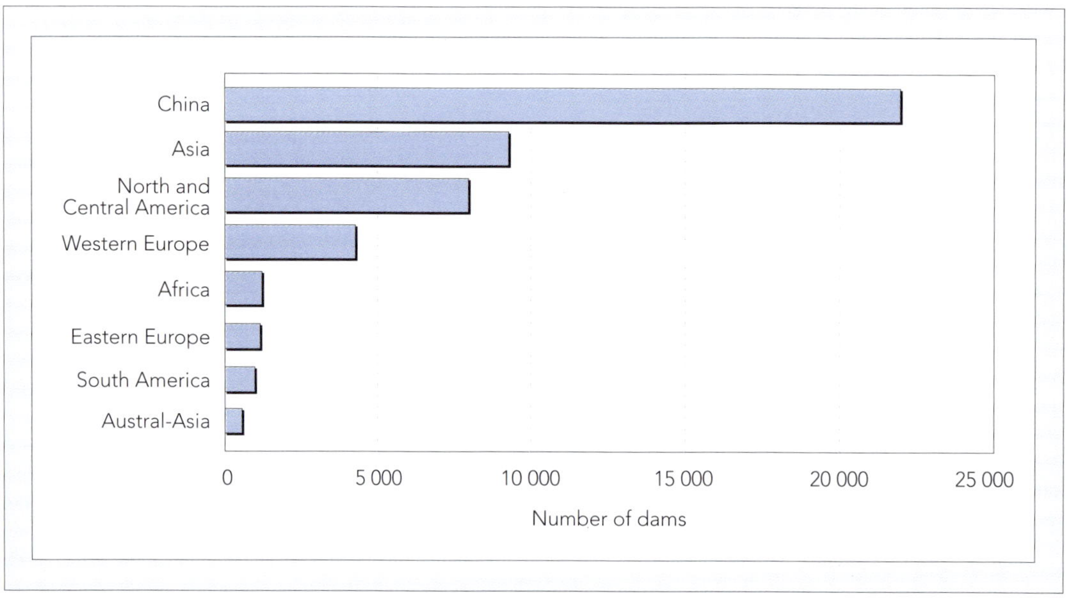

Source: WCD estimates based on ICOLD,1998 and other sources (See Annex V).

The period of economic growth following the Second World War saw a phenomenal rise in the global dam construction rate, lasting well into the 1970s and 1980s. At its peak, nearly 5 000 large dams were built worldwide in the period from 1970 to 1975. The decline in the pace of dam building over the past two decades has been equally dramatic, especially in North America and Europe where most technically attractive sites are already developed. The average large dam today is about 35 years old.

The top five dam-building countries account for nearly 80% of all large dams worldwide. China alone has built around 22 000 large dams, or close to half the world's total number. Before 1949 it had only 22 large dams. Other countries among the top five dam building nations include the United States with over 6 390 large dams; India with over 4 000; and Spain and Japan with between 1 000 and 1 200 large dams each.

Figure 1.6: Construction of dams by decade (1900-2000)

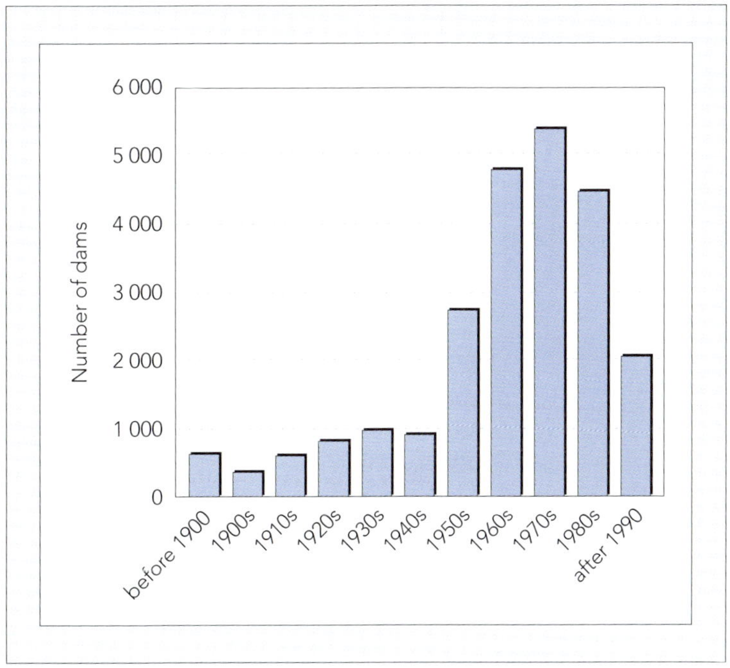

Source: ICOLD, 1998. Note: Information excludes dams in China.

Figure 1.7: Dams constructed over time by region (1900-2000)

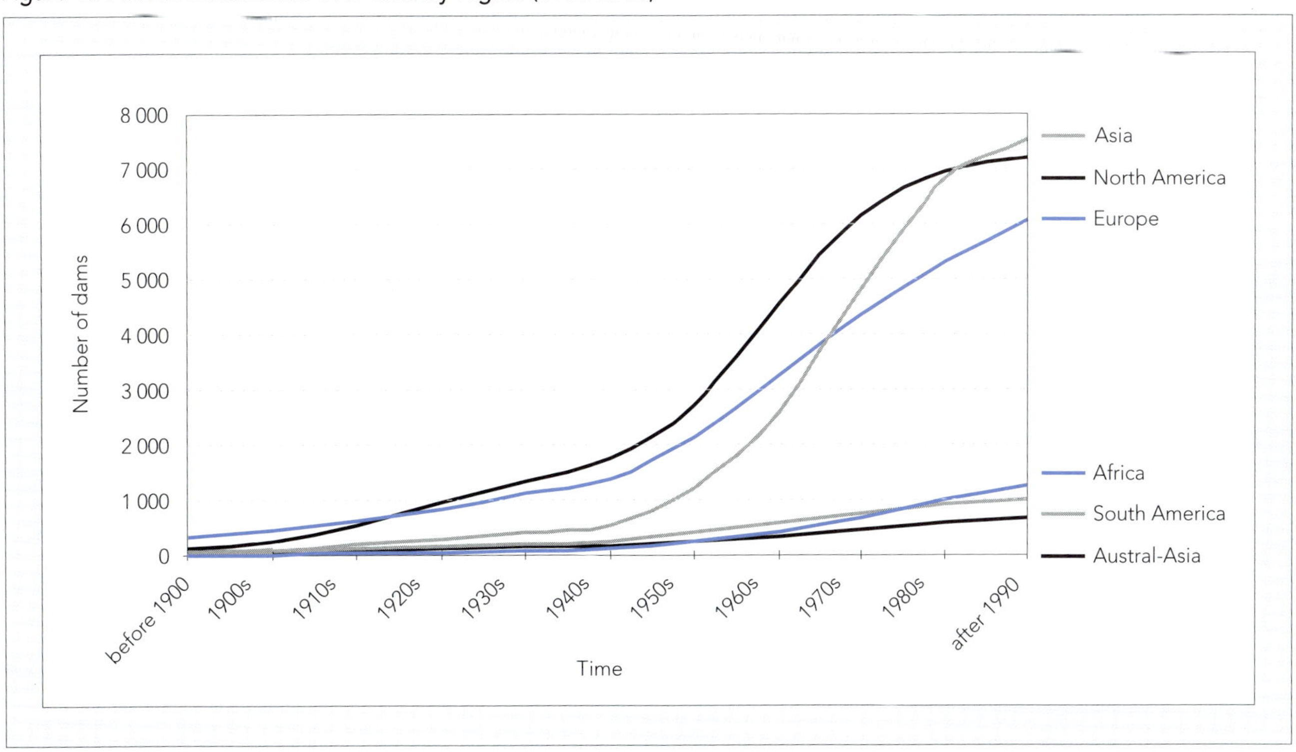

Source: ICOLD, 1998. Note: Information excludes the time-trend of dams in China.

Figure 1.5 shows the proportional distribution of large dams in different regions of the world. Approximately two thirds of the world's existing large dams are in developing countries. Figures 1.6 and 1.7 show the cumulative and time-line trends in the construction of large dams in the last century. The time-trend figures exclude data on most of the dams in China and therefore differ slightly from the trends described in the text.

Current regional focus for large dam construction

Construction of large dams peaked in the 1970s in Europe and North America. Today most activity in these regions is focused on the management of existing dams, including rehabilitation, renovation and optimising the operation of dams for multiple functions. An estimated 1 700 large dams have been under construction in other parts of the world in the last few years. Of this total, 40% are reportedly being built in India (see Table 1.1 and Annex V for details).

Since average construction periods generally range from 5 to 10 years, this indicates a worldwide annual average of some 160 to 320 new large dams per year.

Table 1.1: Dams currently under construction

Country	Number of dams	Purpose
India	Varies from 695 to 960 depending on the source of information	Irrigation, multipurpose
China	280	Flood control, irrigation, hydropower including pumped storage,
Turkey	209	Irrigation, hydropower, water supply
South Korea	132	Irrigation, hydropower, flood management, water supply
Japan	90	Mainly flood control
Iran	48 (above 60m)	Irrigation, multipurpose

Sources: ICOLD, 1997; International Journal of Hydropower and Dams, 2000; WCD India Case Study, 2000; Japan Dam Almanac, 1999, National Register of Large Dams for India cited in WCD India Case Study.

Decommissioning of large dams

The end of the 20th century saw the emergence of another trend relating to large dams – decommissioning dams that no longer serve a useful purpose, are too expensive to maintain safely, or have unacceptable levels of impacts in today's view. Momentum for river restoration is accelerating in many countries, especially in the United States where nearly 500 dams, mainly relatively old, small dams have been decommissioned. Since 1998, the decommissioning rate for large dams has overtaken the rate of construction in the United States.

Experience in North America and in Europe shows that decommissioning dams has enabled the restoration of fisheries and riverine ecological processes. However, dam removals without proper studies and mitigation actions cause public concerns and environmental problems. These include negative impacts on downstream aquatic life due to a sudden flush of the sediments accumulated in the reservoir. Where there has been industrial or mining activity upstream, these sediments may be contaminated with toxic substances. Another option is to open the floodgates of dams where this is possible, either as a form of decommissioning, or to allow natural river flows and unrestricted fish passage at critical times of the year. While decommissioning efforts in the United States and France have received public support thus far, there may be local opposition where changes in the flow and water levels affect services previously provided by the dam, or where development has taken place around the reservoir and downstream.

There is comparatively little experience with the removal of larger dams. The bigger the dam, the more problems decommissioning or removal are likely to face, and the more expensive they are likely to be. More studies are needed to address the costs, benefits and impacts of decommissioning as the stock of dams ages and choices must be made between refurbishing and decommissioning.

Large Dams as Instruments of Development

Dams have been promoted as an important means of meeting perceived needs for water and energy services and as long-term, strategic investments with the ability to deliver multiple benefits. Some of these additional benefits are typical of all large public infrastructure projects, while others are unique to dams and specific to particular projects.

Regional development, job creation and fostering an industry base with export capability are most often cited as additional considerations for building large dams. Other goals include creating income from export earnings, either through direct sales of electricity, or by selling cash crops or processed products from electricity intensive industry such as aluminium refining.

Water-rich countries such as Canada, Norway, Brazil and parts of Russia have developed large dams for hydropower generation where suitable sites were available. Governments in semi-arid countries such as South Africa, Australia and Spain have tended to build dams with large storage capacity to match water demand with stored supply, and for security against the risk of drought. For example, in Spain – one of the top five dam-building countries – rainfall is highly variable between seasons and from year to year.

In East and Southeast Asia, during the monsoon season, rivers swell to over 10

Table 1.2: Estimated annual investment in dams in the 1990s ($US billion per year)

<table>
<tr><th></th><th>Developing countries</th><th>Developed countries</th><th>Total</th></tr>
<tr><td>Dams for hydro power</td><td>12-18</td><td>7-10</td><td>19-28</td></tr>
<tr><td>Dams for irrigation</td><td>8-11</td><td rowspan="3">3-5</td><td rowspan="3">13-18</td></tr>
<tr><td>Dams for water supply</td><td>1.5</td></tr>
<tr><td>Dams for flood control</td><td>0.5-1.0</td></tr>
<tr><td>Total</td><td>22-31</td><td>10-15</td><td>32-46</td></tr>
</table>

Source: WCD Thematic Review III.2, Financing Trends. Note that these figures include generating installations for hydropower dams but not canals and piped distribution systems served by irrigation and water supply dams.

times the dry season flow. In these settings dams have been constructed to capture and store water during wet seasons for release during dry seasons.

Dams are promoted as an important way to meet water and energy needs and support economic development.

Large dams require significant financial investments. Estimates suggest a worldwide investment of at least two trillion US dollars in the construction of large dams over the last century. During the 1990s, an estimated $32-46 billion was spent annually on large dams, four-fifths of it in developing countries.[22] Of the $22-31 billion invested in dams each year in developing countries, about four-fifths was financed directly by the public sector.

Today the world's large dams regulate, store and divert water from rivers for agricultural

Box 1.2: Types of large dams

There are various definitions of large dams. The International Commission on Large Dams (ICOLD), established in 1928, defines a large dam as a dam with a height of 15m or more from the foundation. If dams are between 5-15m high and have a reservoir volume of more than 3 million m^3, they are also classified as large dams. Using this definition, there are over 45 000 large dams around the world.

The two main categories of large dams are reservoir type storage projects and run-of-river dams that often have no storage reservoir and may have limited daily pondage. Within these general classifications there is considerable diversity in scale, design, operation and potential for adverse impacts.

Reservoir projects impound water behind the dam for seasonal, annual and, in some cases, multi-annual storage and regulation of the river.

Run-of-river dams (weirs and barrages, and run-of-river diversion dams) create a hydraulic head in the river to divert some portion of the river flows to a canal or power station.

production, human and industrial use in towns and cities, electricity generation, and flood control. Dams have been constructed to a lesser extent to improve river transportation and, once created for other purposes, the reservoirs of many large dams have been used for recreation, tourism, and aquaculture.

Figure 1.8 shows that about one third of large dams serve two or more purposes. Recent trends have favoured multi-purpose dams. As the figure shows, there is considerable regional variation in the functions served by large dams and these functions have also changed over time.

The majority of large dams in Africa and Asia are for irrigation, though large dams are more often than not multi-purpose. There is growing interest in dams for flood protection and in pumped storage dams for power generation to meet peak demand in Asia. Single-purpose hydropower dams are most common in Europe and South America, whereas single-purpose water supply projects dominate in Austral-Asia. North America has a relatively even spread of large dam functions. All other potential purposes, including recreation and navigation, were found in less than 5% of projects. Overall, the proportional share of irrigation dams and multi purpose dams has been increasing over the last 20 years, while the share of hydropower dams has been decreasing.

Irrigation water supplied from large dams

Irrigation is the single largest consumptive use of fresh water in the world today. It is linked to food production and food security. About one fifth of the world's agricultural land is irrigated, and irrigated agriculture accounts for about 40% of the world's agricultural production.[23] The total area irrigated expanded dramatically during the first years of the green revolution in the 1960s, increasing yields and bringing down food prices. From 1970 to 1982, global growth in the irrigated area slowed to 2% a year. In the post green revolution period between 1982 and 1994 it declined to an annual average of 1.3%.[24]

Figure 1.8: Distribution of existing large dams by region and purpose

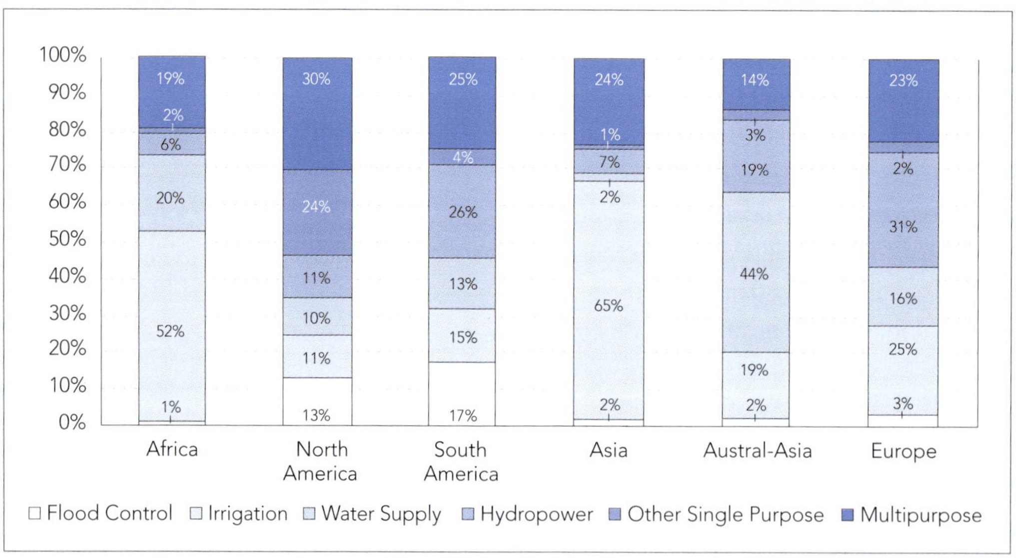

Source: Adapted from ICOLD, 1998 (See Annex V).

Half the world's large dams were built exclusively or primarily for irrigation, and an estimated 30 to 40% of the 268 million hectares of irrigated lands worldwide rely on dams. Discounting conjunctive use of ground water and surface water, dams are estimated to contribute to at most 12-16% of world food production. Ground water irrigation typically has higher yields, for a given amount of water, than surface water irrigation because of better control of the resource at farm level.[25]

Four countries – China, India, United States and Pakistan account for more than 50% of the world's total irrigated area. As Figure 1.9 shows, the scale and significance of large dams for irrigation varies significantly from country to country in terms of the percentage of agricultural land irrigated, and the proportion of the irrigation water supplied from large dams. Dams supply the water for almost 100% of irrigated production in Egypt – most coming from the Aswan High Dam – while in Nepal and Bangladesh dams provide only 1% of irrigation water. In the two countries with the largest irrigated areas – India and China – official statistics suggest large dams supply approximately 30 to 35% of irrigation water, with the balance coming mainly from ground water sources. There is some controversy in calculating the percentage of food production attributable to dams, and particularly on the methods used to account for conjunctive use of surface and ground water.[26]

Unsustainable irrigation practices have affected more than a fifth of the world's irrigated area in arid and semi-arid regions. As a result, soil salinity and waterlogging either make agriculture impossible, or limit yields and the types of crop that can be grown. In other regions, over-use of tubewells has depleted ground water aquifers, lowering water tables and making extraction increasingly expensive and especially difficult for small-holders. The absence of effective policies on conjunctive use of ground water and surface water resources is one of the most important concerns.

Water for industrial use and urban centres

Globally, urban water consumption accounts for 7% of total freshwater withdrawals from rivers, and 22% from lakes.[27] Many reservoirs

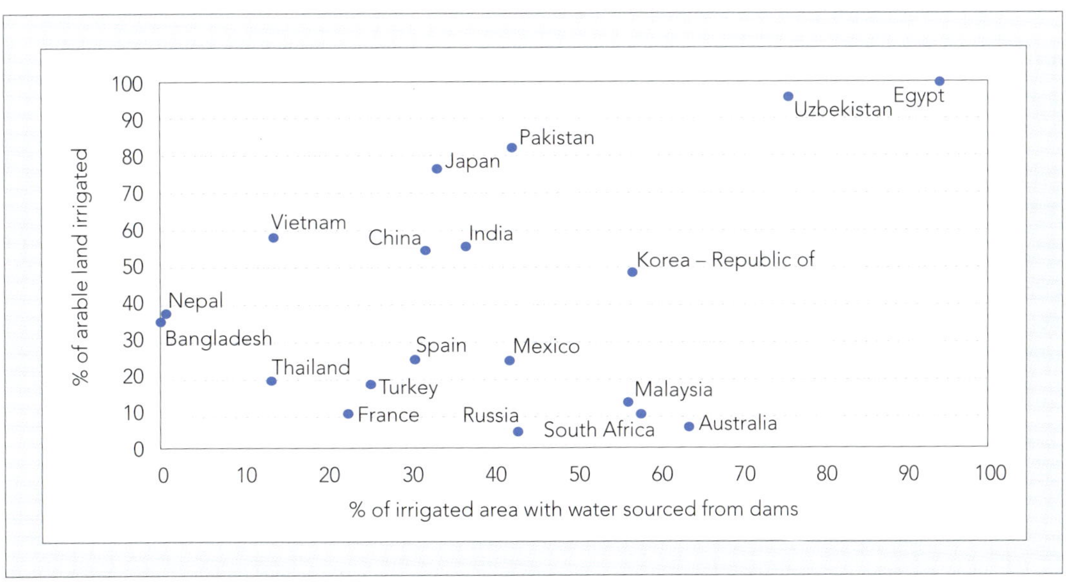

Figure 1.9: Agricultural land irrigated from dams

Source: WCD Thematic Review IV.2 Irrigation Options, Section 1.3.

were built to provide a reliable supply of water to meet rapidly growing urban and industrial needs, especially in drought-prone regions where natural ground water sources and existing lakes or rivers were considered inadequate to meet all needs. Globally, about 12% of large dams are designated as water supply dams. About 60% of these dams are in North America and Europe.

The extent to which cities rely on dams and reservoirs for urban and industrial water varies greatly even within countries. In the Saxony region of Germany, reservoirs provide 40% of the water supplied to two million people, while Los Angeles derives 55% of its water supply from local ground water resources and 37% from a system of reservoirs and pipelines that bring water from more distant locations. Ho Chi Minh City in Vietnam gets 89% of its water from surface sources, whereas Hanoi gets 100% from ground water.[28]

Electricity generation for the national grid

Electricity generation is an important reason for building large dams in many countries, either as the primary purpose, or as an additional function where a dam is built for other purposes. Over the last 22 years, global electricity production has more than doubled, though access is highly skewed between and within countries.[29]

Hydropower provides 19% of world's total electricity supply with 24 countries depending on it for more than 90% of their supply.

Hydropower currently provides 19% of the world's total electricity supply and is used in over 150 countries. It represents more than 90% of the total national electricity supply in 24 countries and over 50% in 63 countries. About a third of the countries in the world currently rely on hydropower for more than half of their electricity needs. Five countries – Canada, United States, Brazil, China and Russia – account for more than half the world's hydropower generation. Between 1973 and 1996 hydropower generation in non-OECD countries grew from 29 to 50% percent of world production, with Latin America increasing its share by the greatest amount in that period.[30]

Hydropower has been perceived and promoted as a comparatively clean, low-cost, renewable source of energy that relies on proven technology. Except for reservoir evaporation, it is a non-consumptive use of water. Once built, hydropower, like all renewable sources, is considered to have low operating costs and a long life, particularly for run-of-river projects and reservoir projects where sedimentation is no concern. In the past, hydropower was especially attractive to governments with limited fossil fuel resources, who would otherwise have had to import fossil fuels to sustain power generation. At the global scale, current levels of hydropower generation offset 4.4 million barrels of oil-equivalent (thermal electric generation) a day, roughly 6% of the world's oil production.

Protecting against floods

While natural floods have many beneficial functions they also pose a threat to life, health, livelihoods, and property. They remain among the world's most frequent and damaging disasters. Floods affected the lives, on average, of 65 million people between 1972 and 1996, more than any other type of disaster, including war, drought and famine[31]. During the same period, an estimated 3.3 million people were left homeless every year as a result of floods. In terms of its regional significance, flood-related economic losses in Asia exceeded those in North America and Europe between 1987 and 1996.[32]

About 13% of all large dams in the world – in more than 75 countries – have a flood management function.[33] Although dams have historically been extensively used as a defence against floods, recent approaches see flood protection as more than building defences against rising waters. For instance, reporting on the 1998 flood, Chinese authorities acknowledged that its severity was partly due to long-term environmental degradation and heavy logging throughout the affected watersheds.[34] There are also cases where dams have created or worsened floods due to dam breaks, poor reservoir operation and changed downstream sedimentation patterns that reduce river channel capacity.

The last two decades have seen a thorough re-evaluation of what constitutes the appropriate mix of prevention, defence and mitigation against flood disasters. As a result, the focus on controlling floodwaters, dominant in the 1950s–1960s, has lost ground to more environmentally based and integrated approaches. The reasons include frustration at floods occurring despite structures being in place, the high cost of engineering solutions, and a better understanding of how natural systems work.

Problems Associated with Large Dams

While dams have contributed to economic growth in the 20th century, the services they provide have come at a cost. This section gives an overview of the problems associated with large dams.

Physical transformation of rivers

Large dams have fragmented and transformed the world's rivers. The World Resources Institute (WRI) found that at least one large dam modifies 46% of the world's 106 primary watersheds.[35] The extent to which river flows have been changed varies around the world. The United States and the European Union regulate the flow of 60–65% of the rivers in their territories, though the amount varies from basin to basin. Spain's 53 km^3 of storage behind large dams regulates 40% of its river flow, varying from 71% in the Ebro river basin, to 11% in the basins on the Galicia coast.[36] In Asia, just under half the rivers that are regulated have more than one large dam.[37]

Large dams have fragmented and transformed the world's rivers, modifying 46% of primary watersheds.

The modification of river flows on transboundary rivers has particular implications. There are 261 watersheds that cross the political boundaries of two or more countries[38]. These basins cover about 45% of the earth's land surface, account for about 80% of global river flow and affect about 40% of the

Box 1.3: Changing physical attributes and impacts of large dams

The volume of the reservoir relative to the annual river flow is important in relation to the purpose of the dam and how it is operated. It is also a major factor in the scale of the ecological effects. The surface area of the reservoir, or the area flooded, points to the potential resettlement impact. The majority of large dams have reservoir surface areas of 0-1 km^2 (more than 60%). This includes run-of-river dams with no reservoir. A small percentage of dams (2%) have reservoir areas greater than 100 km^2.

During the first half of the 20th Century, the average height, reservoir volume and reservoir area increased in all regions. The average height of new dams was 30-34 meters from 1940 through 1990, but has increased to about 45 meters in the 1990's, largely due to trends in Asia. The average reservoir area dramatically increased from 1945 into the 1960's to 50 km^2; declined through the 1980's to and average of 17 km^2; and has again increased in the 1990's to around 23 km^2.

Source: WCD analysis of ICOLD, 1998

The construction of large dams has led to the displacement of some 40 to 80 million people worldwide.

world's population.[39] Issues and concerns facing riparian countries in these basins range from water quality to flow volumes.

The aggregate storage capacity of large dams, based on dam design, is about 6 000 km^3.[40] Assuming that half the design storage is achieved in reality, the aggregate real storage capacity of large dams globally is similar to total freshwater withdrawals estimated at around 3 800 km^3.[41]

An estimated 0.5–1% of the total fresh water storage capacity of existing dams is lost each year to sedimentation in both large and small reservoirs worldwide.[42] This means that 25% of the world's existing fresh water storage capacity may be lost in the next 25 to 50 years in the absence of measures to control sedimentation. This loss would mostly be in developing countries and regions, which have higher sedimentation rates.

Riverine ecosystems impacts

Growing threats to the ecological integrity of the world's watersheds come from rising populations, water pollution, deforestation, withdrawals of water for irrigation and municipal water supply and the regulation of water flows resulting from the construction of large dams. Among the many factors leading to the degradation of watershed ecosystems, dams are the main physical threat, fragmenting and transforming aquatic and terrestrial ecosystems with a range of effects that vary in duration, scale and degree of reversibility.

The watersheds of the world are the habitat of 40% of the world's fish species, and provide many ecosystem functions ranging from nutrient recycling and water purification to soil replenishment and flood control. At least 20% of the world's more than 9 000 fresh water fish species have become extinct, threatened or endangered in recent years.[43]

Fish are a critical source of animal protein for more that 1 billion people. In Africa, fish protein is 21% of the total animal protein in the diet, and in Asia it is 28%. While rivers supply about 6% of fish protein consumed by humans globally, it is often 100% of the supply for many inland riverine communities.

Ecosystem transformations do not only occur in the upper, lower and mid-reaches of watersheds, they also impact on river estuaries, which are frequently complex ecosystems. Closing the mouths of major rivers, salt intrusion, destruction of mangroves and loss of wetlands are among the many issues at stake.

Social consequences of large dams

While many have benefited from the services large dams provide, their construction and operation have led to many significant, negative social and human impacts. The adversely affected populations include directly displaced families, host communities where families are resettled, and riverine communities, especially those downstream of dams, whose livelihood and access to resources are affected in varying degrees by altered river flows and ecosystem fragmentation. More broadly, whole societies have lost access to natural resources and cultural heritage that were submerged by reservoirs or rivers transformed by dams.

The construction of large dams has led to the displacement of some 40 to 80 million people worldwide, as shown in Chapter 4.[44] Many of them have not been resettled or received adequate compensation, if any.[45] Between 1986 and 1993, an estimated 4 million people were displaced annually by an average of 300 large dams starting construction each year.[46] These figures only give an idea of the size of the problem, since there are enormous variations from case to case.

The scale and range of social issues encountered in the river basins altered by large dam construction vary from region to region. The number of people deriving their livelihood directly from the river and its ecosystem, and the overall population density in the river basin, gives an indication of the potential impacts. As table 1.3 illustrates, population density can vary significantly.

The world's two most populous countries – China and India, have built around 57% of the world's large dams – and account for the largest number of people displaced. In the late 1980s China officially recognised some 10.2 million people as 'reservoir resettlers', though other sources suggest the figure may be substantially higher.[47] Large dams and reservoirs already built in the Yangtze basin alone have led to the relocation of at least 10 million people.[48] In India, estimates of the total number of people displaced due to large dams vary from 16 to 38 million people.[49]

Resettlement caused by large dams has been a significant part of total resettlement for all public infrastructure development. In China, large dams are estimated to have displaced an estimated 27% of all people displaced by development projects (the total includes people displaced by urban expansion, roads and bridges).[50] In India the figure is 77% (this total excludes people displaced by urban development).[51] Among World Bank funded projects involving displacement, dams and reservoirs accounted for 63% of people displaced.[52] These figures are only estimates and do not include people displaced by other aspects of the projects such as canals, powerhouses, project infrastructure and associated compensatory measures such as bio-reserves.

The last three decades especially have seen dramatic and wide-reaching changes in perceptions of development and concepts of interdependence with other people and with nature.

These are not the only problem areas and Chapters 2 to 6 examine other reasons for dissatisfaction.

Understanding the Large Dams Debate

The huge growth of dam building in the 20th century took place against a backdrop of tremendous political, economic and technical transformations – while the world's population grew from 1.65 billion in 1900, to 6 billion by the end of the century. The last three decades especially have seen dramatic and wide-reaching changes in perceptions of development and concepts of interdependence with other people and with nature. This dynamic of change is also redefining the roles that governments, civil society and the private sector play in water

Table 1.3: Population density of selected river basins

Country	River basin	Population density (person/km2)
China	Yangtze	224
Brazil	Tocantins/Amazon	6
India	Ganges	375
Norway	Glomma and Laagen	26
South Africa	Orange River Basin	12
Thailand	Mun/Mekong	78
United States	Columbia/Snake	9
Zimbabwe and Zambia	Zambezi	35

Source: Revenga et al, 1998.

As dam building accelerated after the 1950 s, opposition to dams became more widespread, vocal and organised.

and energy resource development planning. And while the wider transformations have spurred the debate on large dams, that debate itself has become a major catalyst of change.

We begin by looking at the genesis of opposition to large dams from the perspective of social and environmental movements, before considering the main issues in the debate that the meeting in Gland Switzerland handed down to the Commission.

Genesis of the opposition to dams

In the previous section we saw that dam building has a long history. Conflicts too have a long history, though it is only in recent years that they have come to command wider attention. Conflicts over water and dams are probably as ancient as dam building itself. In medieval England, boat owners opposed millers blocking rivers to create millponds to turn their water wheels.

Records from the 17th century tell of Scottish fishermen trying to destroy a newly completed weir. In the 1910s, conservationist John Muir unsuccessfully lobbied public opinion and the US Congress against the building of O'Shaugnessy dam in Yosemite National Park in California.

Populations affected or threatened by dams have fiercely resisted dam building throughout the last century. Because they were often isolated, without help from outside sympathisers, affected people's resistance to dams often went unnoticed internationally and, in some cases, the states concerned used intimidation and violence to suppress it. Eight people died and over 30 were injured when colonial government forces fired on Tonga people resisting removal to make way for the Kariba dam reservoir. But international attention focused on the mission to rescue wildlife stranded in the impoundment area.[53] At around the same time in Mexico, the Papaloapan River Commission set fire to the houses of indigenous Mazatecs who refused to move for the Miguel Aleman dam. In 1978 police killed four people when they fired at an anti-resettlement rally at Chandil dam in the state of Bihar in India. In Nigeria in April 1980, police fired at people blocking roads in protest against the Bakolori dam.[54] And in 1985, 376 Maya Achi Indians, most of them women and children, were murdered in the course of clearing the area to be submerged by the Chixoy dam in Guatemala[55].

As dam building accelerated after the 1950's, opposition to dams became more widespread, vocal and organised. Conservationists in northern countries, especially in the United States, led the first notable successes for campaigns against large dams.[56] There, conservationists were able to stop the 175 metre-high Echo Park dam on a tributary of the Colorado River in the 1950s and two dams planned for the main stem of the Colorado River in the Grand Canyon in the following decade. A series of new laws (notably the 1969 National Environmental Protection Act, and the 1974 Endangered Species Act), together with growing public concern about environmental conservation, public subsidies and budget deficits served increasingly as grounds for halting expensive dams, canals and channelisation projects through the 1970s and early 1980s.

Over the past thirty years, the alliance of northern activist groups (environmental and human rights groups) with NGOs and affected groups' associations in the South has resulted in more vigorous and more co-ordinated opposition to dams worldwide. In many cases, the strength of these coalitions has had a major impact on dam-related planning and policy and at the level of individual dams. As a result of these concerted pressures the planning process, which until the 1970s was the restricted preserve of government agencies, engineers and economists, began to include environmental impact assessments and some public reviews. By the late 1980s environmentalists and sociologists began to play a more important role in the planning process, and by the mid-1990s the involvement of affected peoples and NGOs in the process became more significant.[57]

It is only fair to note that popular action has also supported dams. Farmers in Madrid recently marched to demand more water and more dams for irrigation.

The last few decades have seen a number of significant policy steps in response to civil society demands and changing values. The World Bank offers a good illustration. It is a priority target of dam critics, as it is often the first and single largest financier of large dams. In 1982, the Bank adopted an internal directive on indigenous peoples. Revisions of policies on resettlement and environmental assessment are other important milestones. More recently, in 1993, the World Bank established an appeals mechanism, the Inspection Panel. This allows citizens adversely affected by Bank funded projects to file claims regarding violations of its policies, procedures and loan agreements. At the same time the Bank promoted more flexible approaches to information disclosure.[58]

Often, these policy reforms have been closely related to the impacts of resistance from affected groups and international NGO campaigns around individual projects. In 1973-77, the resistance of indigenous peoples to four dams along the Chico River in the Philippines led the World Bank to withdraw from the project and resulted in the government postponing it indefinitely.[59] Other important milestones include the World Bank's withdrawal from the Sardar Sarovar project in India in 1993, and then from Arun III in Nepal in 1995 (although in the latter case public pressure may not have been the deciding factor)[60]. Sardar Sarovar, still on the agenda despite the World Bank withdrawal, continues to be the focus of local opposition and international support. More recent examples of campaigns include the Three Gorges dams in China, the Pak Mun in Thailand, Ilisu in Turkey, Ralco in Chile, Epupa in Namibia, the Lesotho Highlands Water

The last few decades have seen a number of significant policy steps in response to civil society demands and changing values.

Project involving Lesotho and South Africa, and Nam Theun II in Laos.

As earlier sections of this chapter showed, the pace of dam construction has slowed significantly in recent years. This is partly because industrialised countries have used most of their attractive sites, and because of other factors, such as shifts in dam financing from public to private sources, matched with the decline in donor funding and increasing costs of large dams. However, it also reflects the effectiveness of anti-dam strategies developed by environmental and human rights activist groups worldwide.

One frequently asked question is why the debate focuses so centrally on dams. The issues being debated, such as environmental sustainability, equitable development, transparent and participatory decision-making also apply to other large infrastructure projects and can only be addressed in the context of broad societal change.

The view that environmental and human rights groups have singled out large dams as their main target is misleading. One assessment found that, of the 36 World Bank supported projects that NGO activist groups have targeted with some success, only 12 are dam projects, compared to 14 forest and natural resource management projects, five mines or industrial management projects, and two urban infrastructure projects.[61] In fact, large dams, like many other industrial, commercial, and governmental facilities are increasingly contentious and difficult to site and build, as are hazardous waste decontamination facilities, solid waste landfills, hospitals, conservation areas, shopping complexes, highways, parking areas, and many more.

If dams are not the only infrastructure projects subject to rising criticism, why do they seem to be often at the centre of controversy, dispute, and even violent confrontation? There are many reasons largely related to the scale and scope of the dams and the impacts.

- Large dams represent major investments, and in some cases may be the largest single investment in a country. These investments are essentially irreversible and often highly politically charged.
- Large dams are generally justified by national or regional macro-economic benefits while their physical impacts are locally concentrated, mostly affecting those within the confines of the river valleys, and along the river reaches. The mismatch of benefits and costs translates easily into confrontational attitudes.
- Resettlement for large dams tends to be on a larger scale than resettlement for other types of physical infrastructure. Roads and thermal power stations can be sited on marginal land whereas dams generally flood rich and fertile agriculture land.
- Those resettled from dam or reservoir sites very often lose not only their homes but also their livelihoods. Relocation in rural settings where good land is already occupied can be problematic.
- Large dams affect critical, life sustaining resources, such as land, fisheries and the quality and allocation of freshwater, an increasingly scarce and coveted resource.
- Lack of adequate and accepted solutions to the social and environmental impacts of large facilities has resulted in increased social mobilisation around these issues.

- The lower than anticipated economic performance of many projects.

These factors aside, the perceived injustice in the distribution of the benefits and impacts, and the increased concern about the environmental implications indicate that the debates, controversies and conflicts surrounding large dams are not about dams alone. They are part of a wider debate about development, a debate where diverging views on the use of natural resources and public financial resources confront each other.

Disaggregating the debate

Clearly the dams debate has many layers and many interconnected issues. The debate is partly about what occurred in the past and continues to occur today, and partly about what may happen in the future if more dams are built – or are not built. The extent to which the debate is driven by social or environmental concerns, or by broader development considerations, varies from country-to-country. The dams debate in the United States, where the rate of decommissioning dams is greater than the rate of construction, is perhaps as intense as, but qualitatively different from, the debate in India which, along with China, is now building more dams than any other country.

The two principal poles in the dams debate, much in evidence at the Gland Workshop in 1997, help to define the range of views on past experience with large dams. The first focuses on the gap between the promised benefits of a dam and what happens on the ground. The review of global experience with dams set out in Chapters 2-6 confirms that these concerns are justified. Dams have often not met their targets.

The other pole looks at the challenges of water and energy development from a perspective of 'nation building' and resource allocation. To its proponents, the answer to the question of past performance is self-evident. Dams have generally performed well as an integral part of water and energy resource development strategies in over 140 nations and, with a few exceptions, dams have provided an indispensable range of water and energy services.

Beyond this rough characterisation, it is possible to disaggregate the debate along a few central themes. The way the debate is structured largely determines how it is conducted at national and international levels.

Performance: costs and benefits

Performance is often measured in terms of whether the project delivered the benefits that were the basis for approval and funding of the project. But there is more to judging performance than this. An assessment of overall performance, or performance over the full life-cycle of a dam, often yields many different results. So too with costs, much depends on how completely costs are internalised, and who bears particular costs compared to how the benefits are shared. There is no easy formula for calculating the costs and benefits to yield a quick and easy judgement on the overall balance.

Box 1.4: Central issues in the dams debate: past and present

Performance: costs and benefits
Environmental impacts and sustainability
Social impacts and equity
Economics and finance
Governance and participation
Wider development impacts of dams
Alternatives to dams
Cross-cutting issues

Environmental impacts and sustainability

At one level, the debate revolves around how to measure the scale of the impacts, whether these impacts can be avoided or successfully mitigated, and whether they are reversible or irreversible. A more fundamen-

tal controversy centres on how environmental considerations are valued against immediate human development needs. For communities who depend directly on river resources for their livelihoods, the environment is the basis for meeting their needs. An example is the value placed on biological diversity, or on the ecosystem functions that may be lost when the river flow is regulated. This debate becomes especially heated where other options are available. Other controversies concern wider regional or global environmental impacts, for example whether dams emit greenhouse gases or reduce acid rain, under what conditions, and to what extent relative to the alternatives. This aspect of the debate extends to whether large dams should be included in climate change protocols, and whether and how dams should be treated in future carbon emission trading schemes.

Social impacts and equity

This includes both the scale of the impacts and the distribution of costs, benefits and impacts, including those borne by relocated families, host communities where families are resettled, and riverine communities affected by the change in river flows and access to resources. Social issues go beyond equity in the distribution of benefits and impacts and relate to fundamental rights. They include:

- the burden placed on indigenous peoples and ethnic minorities and the degree of recognition of their distinctive status;
- the impact on gender and basic human rights; and
- the loss of livelihood and health impacts in rural areas.

Dealing with the legacy left by forced relocation under both authoritarian and democratic regimes and the need to allocate responsibility for redress are other issues in the debate. An issue that has caused a great deal of tension in the past concerns the basis on which trade-offs, such as the potential benefits to many at the cost of hardship for a few, are invoked and decided.

Economics and finance

Controversy also surrounds the limits and the ability of methods for economic assessment to fully capture and reflect the various social and environmental impacts and values. Governments and financial institutions continue to use traditional economic and financial analysis – rate of return, discount rates, sensitivity tests and the exclusion or inclusion of indirect costs – as primary decision criteria. The debate is how adequately these are applied in practice and how they are balanced against other development objectives or criteria. Related issues include the cost recovery levels for all types of dam projects, the implications for subsidised use of water and the equity dimensions of these subsidies.

Governance and participation

The principal considerations related to these issues centre on the transparency and openness of options assessment, and how planning and decision-making processes are conducted. Other issues relate to the methods used to reconcile local or community-led planning and consensus-building processes with more traditional and centralised planning approaches, access to information, and the dominance of single agencies in planning with multiple responsibilities for designing, construction and operating large infrastructure projects. At the heart of the debate on these issues is the degree of involvement of affected people and wider groups of stakeholders in needs assessment and project-level decision-making. Implementation creates its own set of governance challenges, including whether agreed standards are followed for social and environmental mitigation, compensation and enhancement.

Wider development impacts of dams

Many of the controversial issues go beyond the impact of the project itself and touch upon wider regional or national development choices. Examples include the proportion of the development budget allocated to large dams as opposed to other uses of public funds, the impact of an investment in a dam on the country's debt burden, and competitiveness considerations linked to subsidies. There are also more positive considerations, including the potential of dams to contribute to export earnings.

Alternatives to dams

The degree of even-handedness applied in considering alternatives to large dams is, perhaps, one of the most contested issues. It raises the question of whether dams are selected over other options that may meet the water development or energy objectives at lower cost, or that may offer more sustainable and equitable development benefits. This aspect of the debate extends to whether, and on what basis, dams should be considered complementary to, or mutually exclusive of other options of different scales and types. Whether to give primacy to options such as demand-side management or improving the efficiency of existing supply assets, and under what conditions, are also debated. The options debate connects with the political economy of decision-making, and therefore to the distribution of power and influence within societies. This includes how choices are made between available options, and the extent to which market or other institutional factors create barriers and incentives for different options that provide the same service.

The degree of even-handedness applied in considering alternatives to large dams is, perhaps, one of the most contested issues.

Cross-cutting issues

A range of cross-cutting issues turn on the role and influence of various public and private sector interests in the planning and decision making process. This includes the roles and influence of industry groups ranging from domestic and international consultants, to developers, contractors and suppliers, and extends to the financial service providers. The financing role is especially critical and includes the multilateral and bilateral development banks, insurers and export credit agencies, as well as the commercial banks. Issues raised in the debate range from harmonising standards for financing dam construction to steps to address corrupt practices that can distort decision-making. There are numerous other cross-cutting issues such as the transboundary implications of dams on shared rivers.

These examples illustrate the terrain, scope and complexity of the debate, and how it has become intertwined with wider development concerns.

Since the United Nations Charter (1945) and The Universal Declaration on Human Rights (1948), a globally accepted development framework setting out universal goals, norms, and standards has been gradually emerging.

Economic development during the first half of the 20th century was dominated by an approach that emphasised harnessing and appropriating water and other natural resources for economic activities. Since the United Nations Charter (1945) and The Universal Declaration on Human Rights (1948), a globally accepted development framework setting out universal goals, norms, and standards has been gradually emerging. These declarations have been augmented over time by the Convention

Concerning the Protection and Integration of Indigenous and Other Tribal and Semi-Tribal Populations in Independent Countries (1959), the International Covenant on Economic, Social, and Cultural Rights, (1966), the UN Declaration on the Right to Development (1986), and the Rio Declaration on Environment and Development (1992), among others. Together they form the current framework for sustainable human development.

This framework has been widely ratified by the world's nations, although differences emerge in setting priorities and emphases, and in fulfilling, implementing, and resolving conflicts between competing rights. The realisation of these rights presumes recognition of conflicts between competing rights and the setting up of mechanisms for negotiation and conflict resolution. We discuss the development framework as it relates to the WCD's analysis and methodologies later in this chapter, and Chapter 7 presents an approach for designing mechanisms to resolve and negotiate competing rights.

Where international laws, norms and aspirations are reflected in national legal and policy frameworks, a number of obstacles often prevent their translation into reality. These obstacles include lack of political will or incentive, especially where the democratisation process has not reached a level where citizens and civil society organisations are strong enough to influence decision-making. Another major obstacle, especially facing poor countries, relates to the lack of capacity (financial and human resources but also institutional capacity) to ensure that basic human rights and aspirations are fulfilled for their citizens. Compliance with existing laws and policies is another important issue.

Is there a way forward?

All parties involved in the debate have their own views on what needs to be done to address the problems that they have identified. Some of the main suggestions made by dam critics and proponents in various declarations and statements are as follows.

Dam critics point to:

- the need for more sustainable and appropriate alternatives to dams;
- the imperative for improved transparency, accountability and public participation in the planning of water and energy projects;
- the importance of prior project approval by potentially affected groups;
- the need for protecting and promoting the rights of potentially affected peoples, and for setting in place measures to reduce inequities; and
- the necessity of reparation measures to address the legacy of unfulfilled commitments and unresolved problems.

Dam proponents underline:

- the evolution and change in practices over time;
- the recognised need for social and environmental concerns to be elevated to the same level as safety concerns. (They believe these environmental and social costs must be avoided, or minimised and properly mitigated when they occur.);
- the importance of ensuring that affected people are better off as a result of dam development, and of considering them as

shareholders, partners, and therefore project beneficiaries;

- the imperative of participatory decision-making;
- the need to promote the principles of equity, efficiency and economic viability; and
- the importance of balancing the need for development with the requirement of ensuring environmental sustainability.

This broad taxonomy indicates that there are many areas for potential convergence, especially regarding what needs to be done in the future. All agree on the need:

- to take environmental and social costs of dams more seriously;
- for more systematic consultation with affected people;
- to ensure that affected people are better off as a result of the dam development; and
- to ensure that the costs and benefits of dams are shared more equitably.

However, unresolved issues still separate positions on a number of financial, economic, social, and environmental issues. The most intractable include:

- the extent to which alternatives to dams are viable for achieving various development goals, and whether alternatives are complementary or mutually exclusive;
- the extent to which adverse environmental and social impacts are acceptable;
- the degree to which adverse environmental and social impacts can be avoided or mitigated; and
- the degree to which local consent should govern development decisions in the future.

One of the central problems – and certainly one of the main causes of stalemate in the debate – is the lack of mutual trust between the key parties involved. For opponents, dams promoters too often agree to new policies and guidelines, but fail to comply with them.

Toward the establishment of the WCD and its mandate

The WCD can trace its roots back to the origins of the dams debate. In addition to the debate, two other related factors painted the backdrop to its conception. The first is the accelerating shift in accepted notions of the appropriate relationship between the State and its citizens. The second is the increased recognition of the negative environmental and social outcomes experienced with large dams. The result has been growing controversy and consequent delays in development decisions.

One of the central problems – and certainly one of the main causes of stalemate in the debate – is the lack of mutual trust between the key parties involved.

In Chapter 7, the section on development trends explores the shift in public support away from centralised decisions often taken on the basis of abstract notions of the greater common good. Instead the public is focusing more sharply on notions of equity in the distribution of the costs and benefits of development and participation in decision-making.

Growing disaffection by many groups in society with the way decisions are taken, with the distribution of costs and benefits, and with compliance and recourse mechanisms has provoked growing resistance to large dams and escalating tension and conflicts around individual dam projects. While much of this conflict has centred on the developing world, it nevertheless strikes at the dams industry as a whole. Because many of the most appropriate dams sites in the developed world have already been exploited, the future of the industry lies principally in the developing countries, whether the industry itself is based there or in the developed countries. Linked to this are considerations about the image of international companies, their standing in the community, their relationship with shareholders and their reputation for corporate social responsibility.

New approaches are not always well received by developing country governments. They often see them as a case of developed countries, having benefited from cutting corners themselves, turning to insist that developing countries meet higher standards. However, to obtain international financing – public or private – developing countries find themselves having to comply with new approaches, norms, and policies as a condition of financing or partnership.

By the early 1990s, it was becoming clear that the cost of controversy could seriously affect future prospects for dams and stall efforts to finance other non-dam water and energy development projects to serve rural or urban communities.

The net effect of opposition and the controversy over outcomes has been to increase the level of risk associated with projects – especially those that fail to recognise the need for a change in the way things are done. This has increased delays on dam projects in cases where controversial elements are contested in the courts. Both these factors directly or indirectly increase the costs of dams.

By the early 1990s, it was becoming clear that the cost of controversy could seriously affect future prospects for dams and stall efforts to finance other non-dam water and energy development projects to serve rural or urban communities. The stalemate did not benefit governments, dam builders, communities or the environment, as no actions or investments were considered attractive given the ongoing conflict. A new way had to be found.

The need of both dams proponents and opponents to negotiate a new, agreed basis for assessing options and for planning, deciding, implementing and operating them – created the conditions for setting up the WCD and giving it a mandate.

Beyond these general considerations, several specific milestones mark the road leading to the establishment of the WCD. These include:

- The 1992 Morse report.[62] This was an independent review of the Sardar Sarovar project, commissioned by the World Bank as a result of growing controversy over Bank funded projects and criticism of these projects at grassroots level and internationally. Instead of tempering the controversy, the Morse report fuelled deep criticism of the World Bank's internal decision-making.
- The Manibeli Declaration, signed in June 1994 by 326 activist groups and NGO coalitions from 44 countries, calling among other things for a moratorium on World Bank funded large dams until a comprehensive, independent review of all Bank funded dam projects had been conducted.[63]
- The 1996 report of the World Bank's Operations Evaluation Department (OED).[64] An internal review of the performance and impacts of a sample of 50 Bank funded large dams. This desk study observed that 90% of dams reviewed met the Bank's standards for resettlement at the time they were built, but 75% failed to meet the Bank's most recent standards. Another important

finding was that proper mitigation of the adverse environmental and social impacts of most of the dams reviewed would have been feasible without compromising the economic feasibility of the projects. The report and the process by which it was prepared were highly criticised by the NGO community.

- The soul-searching was not confined to the development finance community. Professional water and energy development associations also began assessing the causes of the growing controversy and reaching conclusions on what needed to be done. The International Commission On Large Dams (ICOLD) published its *Position Paper on Dams and the Environment* in 1997.[65] Similarly the International Commission on Irrigation and Drainage (ICID) initiated a process that resulted in a major statement, *The Role of Dams for Irrigation, Drainage and Flood Control*, in 2000.[66]
- The NGO community was also active in gathering case material on experience with large dams and drawing conclusions from it. Building on Goldsmith and Hildyard's 1984 report, *Silenced Rivers* by Patrick McCully of the International Rivers Network, published in 1996, depicts a particularly bleak record of the social and environmental impacts of dams and their underlying political dimensions.

By 1997, suspicion and mistrust between proponents and critics of large dams threatened to dominate and undermine wider discussion needed to reach agreements on ways to improve access to water and energy services. In response to this the World Bank and the World Conservation Union (IUCN), a global union of more than 800 governments, government agencies and NGOs – sponsored a meeting between the champions and the critics of large dams in Gland, Switzerland in April 1997. While the Gland workshop was focused on bringing a range of opinion around the table to discuss the implications of the World Bank/OED review of 50 Bank funded dams, it found sufficient common ground to set in motion the process that led to the formation of the WCD.

The Gland workshop brought together 39 participants representing governments, the private sector, international financial institutions, civil society organisations, and affected people in a balance later mirrored in both the WCD and the Stakeholders Forum.

The Gland workshop brought together 39 participants representing governments, the private sector, international financial institutions, civil society organisations, and affected people in a balance later mirrored in both the WCD and the Stakeholders Forum. In addition to assessing the OED report, they addressed three issues:

- Critical advances needed in knowledge and practice in relation to energy and water resources management.
- Methodologies and approaches required to achieve these advances.
- Proposals for a follow-up process involving all key players.

The workshop participants identified key issues relating to the social, environmental, technical, and financial aspects of dams that had to be addressed in reviewing the role of dams and their alternatives in sustainable development. They also formed an Interim Working Group composed of workshop participants and entrusted this group to establish the WCD. This task proved to be long and complex, in part because of the decision of the working group to consult all of the key stakeholder groups at each step, and also because of the time needed to build confidence in the good faith of all the parties.

The WCD was finally announced in February 1998, and began its work the following May, under the Chairmanship of Professor Kader Asmal, then South Africa's Minister of Water Affairs and Forestry and later Minister of Education. Its 12 members were chosen through a global search process to reflect regional diversity, expertise, and stakeholder perspectives. The Commission was independent, with each member serving in an individual capacity and none representing an institution or a country.

As defined by the Gland workshop, the Commission's two objectives were to:

- Review the development effectiveness of large dams and assess alternatives for water resources and energy development.
- Develop internationally acceptable criteria, guidelines and standards where appropriate, for the planning, design, appraisal, construction, operation, monitoring and decommissioning of dams.

The dual objectives are deliberate, reflecting to some extent the priorities of the different participants in the dams debate. The champions of large dams, while wishing to draw useful lessons from the review of past experience, tend to lay the emphasis on practical tools that will help overcome the controversy and set a foundation for more predictable scenarios. The opponents of large dams, on the other hand, tend to underscore the importance of the review, convinced that it will reveal the depth and persistence of the negative impacts that dams have caused. They want to see evidence that dams can be an acceptable option before giving too much attention to developing guidelines for building better dams in the future.

Fulfilling the WCD Mandate: Process and Methodology

To respond to both parts of the mandate it was given in Gland, the Commission began by developing an analytical framework and work programme to assemble a consolidated, shared knowledge base on the worldwide experience with large dams, that:

- is grounded in the accepted international norms of sustainable and equitable human development;
- aims to explore the key themes at the centre of the dams debate, especially those that are unresolved; and
- compares the planned performance and expectations of dams with the actual experience after project completion.

The Commission used both quantitative and qualitative methods to objectively evaluate and answer the key questions posed.

It did not set out to judge decisions on dam projects from 50 or 100 years ago, but rather to learn lessons about the outcomes of dams and how these lessons could work to change or affect outcomes in the future.

In order to ensure a solid foundation of material on which to base its analysis and conclusions, the WCD commissioned, organised or accepted:

- in-depth Case Studies of eight large dams on four continents, together with two country review studies;
- a Cross-Check Survey of large dams located in 52 countries across the globe;
- 17 Thematic Reviews grouped along five dimensions of the debate;
- four regional consultations; and

- over 900 submissions from interested individuals, groups and institutions.

Recognising the value of the perspectives provided by representatives attending the Gland meeting, in the first few months of its work the Commission decided that the group should be retained as a consultative body for the Commission process, to be known as the WCD Forum. It was composed of a mix of former Reference Group members from the Gland meeting, and new stakeholders and interest groups. In selecting the new members of the Forum the WCD was guided by criteria such as relevance, balance and representation of a diversity of perspectives and interest groups.

The Forum offers a consultation model that works at a somewhat different level than other forms of consultation. It has around 70 members and operates as a 'sounding board' for the work of the Commission. It is primarily a mechanism for maintaining a dialogue between the Commission and the constituencies of Forum members.

Since the Commission was facilitating debate on the complex issue of the development effectiveness of dams, input derived through consultation with these constituencies was essential for the understanding and acceptance of the Commission's final products. The Forum was one means of achieving these objectives.

Chapter 10 looks ahead to the dissemination and the adoption of the Commission's recommendations. It provides suggestions on the post-Commission role of stakeholder groups such as the Forum.

Developing the knowledge base

The Commission set out to develop a knowledge base that would give it access to the full range of issues and perspectives concerning large dams. The goal was to go beyond the realm of experts and intergovernmental processes to include constituencies with very different entry points into the dams debate. The process was designed to offer the opportunity for dialogue among the different interest groups, while providing a solid foundation for the Commission's findings.

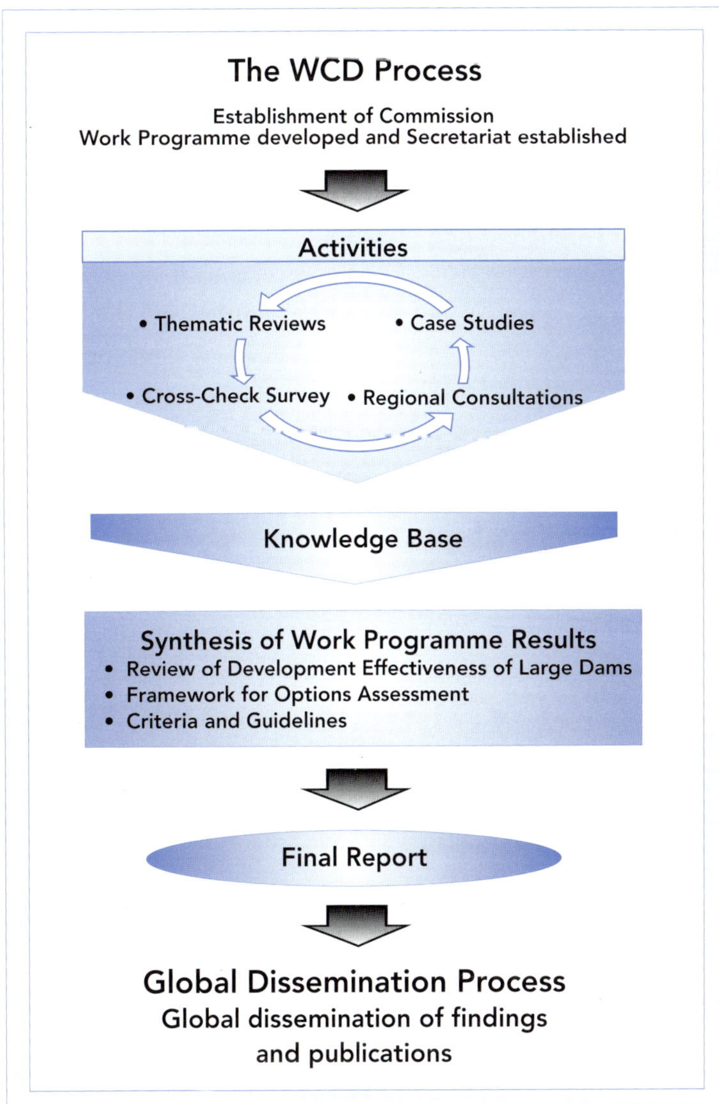

Main elements of the work programme

The Commission has based its report on a synthesis of information on past experience across all elements of the WCD work programme. A brief description of these main elements follows. Further details on the methodology are shown in Annex III.

Case Studies and Country Studies

The Commission undertook two case studies in OECD countries and six in developing countries. It prepared country review studies for India and China, and an issue paper for the Russian Federation and Newly Independent States. The case studies provide a thorough understanding of the challenges and dilemmas associated with specific dams set in the context of the development situation in specific countries, and the river basins. They were implemented in stages with stakeholder participation.

The Commission followed a standard framework for each case study, based on the model developed by the Secretariat. At the core were six questions:

- What were the projected versus actual benefits, costs and impacts?
- What were the unexpected benefits, costs and impacts?
- What was the distribution of costs and benefits – who gained and who lost?
- How were decisions made?
- Did the project comply with the criteria and guidelines of the day?
- How would this project be viewed in today's context in terms of lessons learned?

These questions served a much wider purpose than merely orienting the case studies. They helped to shape the global review, the element of the work programme used by the Commission to evaluate the key issues in the overall debate. The Commission also used the studies to focus more specifically on the development effectiveness issue from the viewpoints of the stakeholder group for each case study.

Cross-Check Survey of dams

The Commission developed the Cross-Check Survey to extend the analysis provided in the case studies to target a broader set of dams. Completed survey forms were received for 125 dams in 52 countries. The 125 dams included the case study dams, additional dams from the case study basins, dams from existing databases and a random selection of dams from the larger population to contribute to the overall diversity of the sample. The analysis aimed to detect broader patterns and trends in performance and decision-making relating to dams.

A variety of dams of different types (storage, run-of-river); ages (the 1930's through the 1990's), functions (water supply, irrigation, power, flood management, recreation and other); ownership structures (public, corporate and private); and regional locations were included in the survey. Data was verified by an internal review and by submitting contested and randomly selected data sets for review by constituencies other than the dams owners and operators. Annex III sets out the methodology and summary statistics. Figure 1.10 shows the location of the case study dams and cross-check survey dams.

Thematic Reviews

A total of 17 Thematic Reviews and some 130 papers were commissioned to address five major areas of concern identified in the WCD strategy and objectives paper:

Figure 1.10: World map showing the regional location of the Case Studies, Country Studies, Cross-Check Survey dams, Regional Consultations, submissions, Forum members, Commissioners and Secretariat

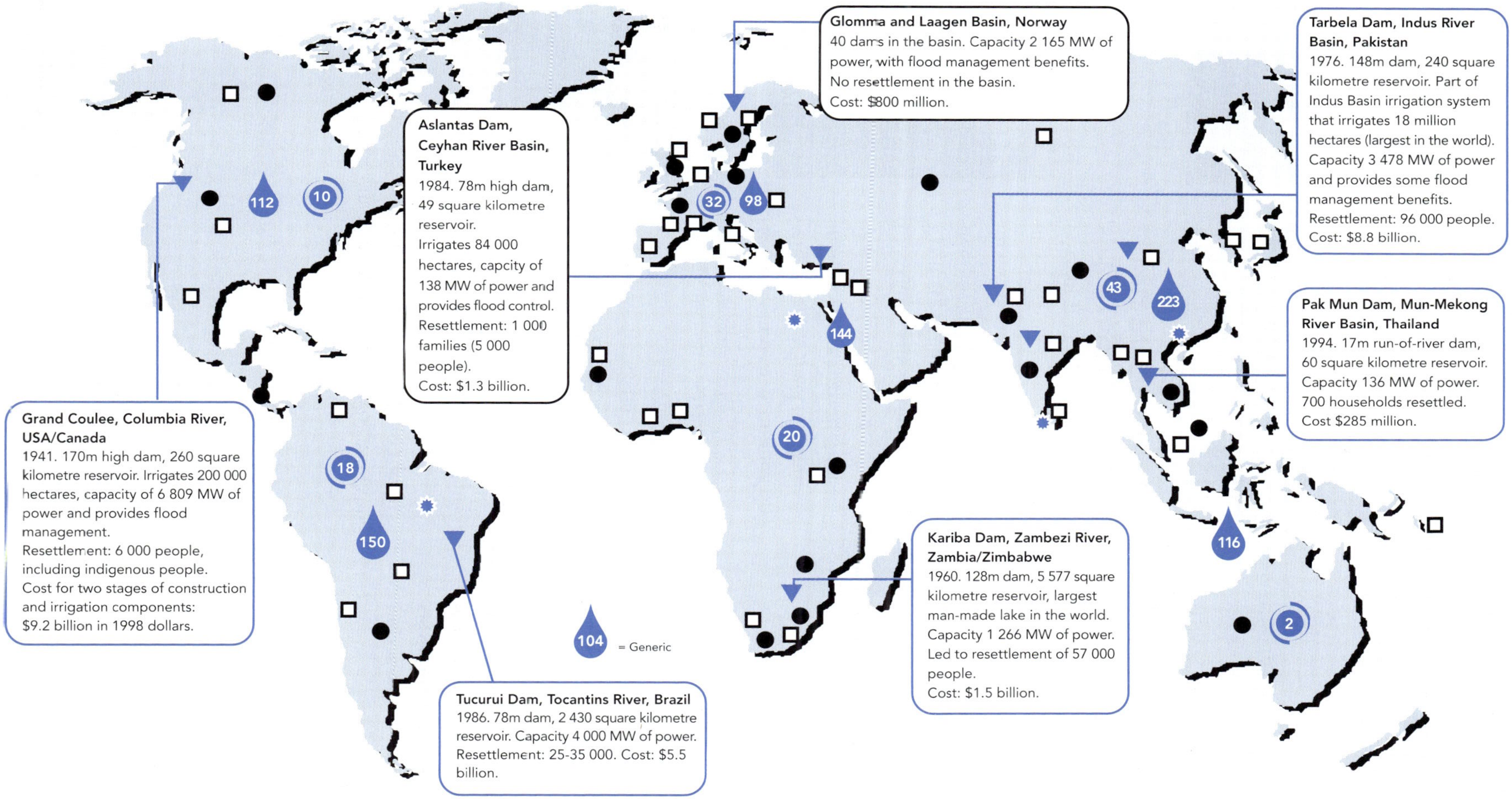

Commissioners and Secretariat
Argentina, Australia, Canada, China, Costa Rica, Germany, India, Japan, Kazakhstan, Kenya, Malawi, Netherlands, Pakistan, Philippines, Senegal, South Africa, Sri Lanka, Sweden, Switzerland, UK, USA, Vietnam, Zimbabwe

Country Case Studies
Brazil, China, India, Norway, Pakistan, South Africa, Thailand, Turkey, USA, Zambia/Zimbabwe

Regional Consultations
Africa-Middle East – Cairo; Latin America – Sao Paulo; South Asia – Colombo; South-East Asia – Hanoi

125 **Cross-Check Surveys by Region**
Africa, Asia, Australia, Europe, Latin America, North America (Canada and USA)

Forum Members
Brazil, Cambodia, Canada, Chile, China, Cote D'Ivoire, Fiji, France, Germany, Ghana, India, Israel, Japan, Jordan Kenya, Lesotho, Mexico, Nepal, Norway, Pakistan, Paraguay, Philippines, Romania, Russia, Senegal, South Africa, South Korea, Spain, Sri Lanka, Sweden, Switzerland, Thailand, UK, USA, Venezuela

947 **Submissions**

South Asia Consultation, Colombo, December 1998

Latin America Consultation, Sao Paulo, August 1999

East and South-East Asia Consultation, Hanoi, February 2000

Africa-Middle-East Consultation, Cairo, December 1999

- social and distributional issues;
- environmental issues;
- economic and financial issues;
- options assessment; and
- governance and institutional processes.

The Thematic Reviews provided the baseline information, analysis and recommendations on issues that cut across the central elements in the large dams debate. They consider past and current experience, as well as the forward-looking context, by synthesising the state-of-the-art knowledge, practices and key viewpoints on each topic. Within limits set by available resources and the Commission's schedule, the level of effort involved in preparing these review papers varies according to the complexity of the issue and the level of controversy surrounding it. Preparation of the review papers included setting up panels and procedures for broader peer review. This helped to bring together a wide spectrum of perspectives and approaches on the topic and to clarify the areas of potential agreement, and persistent disagreement, on highly controversial issues.

Regional Consultations

Over the course of two years, the Commission held four regional consultations, one each in South Asia, Africa and the Middle East, Latin America, and East and South East Asia. For each of the regional consultations, governments and non-government organisations, project-affected communities, financial institutions, industry and private sector representatives were invited to submit proposals on all aspects of their experience with dams and alternative options, and with water and energy resource development. Over 30 presentations were made at each consultation and participants engaged in debates on the issues raised. They gave the Commission a unique and unprecedented exposure to the many voices and perspectives in the dams debate in a number of countries and regions.

The Commission also had the benefit of listening to NGO and affected community presentations at two hearings organised by NGOs from Europe and from Southern Africa, as well as obtaining a wider range of inputs through attending and participating in meetings, workshops and conferences organised by a wide range of organisations and networks.

Endnotes

1 World Commission on Water in the 21st Century, 2000. Other examples of strategic initiatives in water include the United Nations Comprehensive Assessment of Freshwater Resources of the World, the United Nations Food and Agriculture Organisation Aquastat programme, and the World Water Vision conferences.

2 Gleick, 1998, p43.

3 UN, 1999, p1; also Raskin et al, 1998. Medium projection in 2050 was 8.9 billion.

4 UN, 1998, p2–29. In 30 years the largest population growth is projected to be in urban conglomerations; the 10 largest population centres in descending order of size would be Mumbai (Bombay), Shanghai, Chennai (Calcutta), Beijing, Delhi, Karachi, Tianjin, Metro Manila, Jakarta, and Dhaka.

5 IPCC, 1999. Purchasing power parity (PPP) uses accounting exchange rates to compare or add country economic data and avoid distortions introduced by market exchange rates. In 1995 world Gross Product reached a figure of US$33.4 trillion at purchasing power parity (PPP).

6 Raskin et al., 1998, p9. OECD Data; 78% at market prices and 55% at purchasing power parity.

7 Shiklomanov, 1998 in Gleick 2000, p51–54; Raskin et al, 1995; based on United Nations, World Bank, and World Health Organisation data.

8 Raskin et al, 1995.

9 Gleick, 1998, p44–45.

10 Seckler et al, 1998.

11 Shiklomanov, op cit.

12 WRI et al, 1998, p304–305.

13 Shiklomanov, op cit. Municipal and industrial use increased from 17 to 28% of fresh water withdrawals globally between 1950 and 1990.

14 Van Hofwegen and Svendsen, 2000.

15 Raskin et al, 1995, p9; Countries are considered to be water stressed when the annual renewable water supply drops below 1700 m^3 per capita, and water scarce when it drops below 1000 m^3.

16 Raskin et al, 1998.

17 Brown and Halweil, 1999.

18 WCD Thematic Review II.2 Global Change.

19 Schnitter, 1994; McCully, 1996.

20 Zhang, 2000, WCD Regional Consultation Paper.

21 ICOLD, 1998; see also Annex V; Compilation of estimates of the number of dams in the main dam building countries suggest there may be as many as 48 000 large dams.

22 WCD Thematic Review III.2 Financing Trends.

23 WCD Thematic Review IV.2 Irrigation Options. Yields from irrigated areas are on average double that of rainfed agriculture, and are generally higher on land irrigated by ground water than on land irrigated with surface water.

24 Cosgrove and Rijsberman, 1999, p40.

25 Ibid.

26 The WCD India Country Study noted that the official figures of the Central Water Commission indicate that 30% of irrigated land is supplied with water by dams; however, the study also suggested an alternative figure of 10%.

27 Shiklomanov, op cit.

28 Roo, 2000, pp1–31; McIntosh and Yñignez, 1997, p189.

29 IEA, 2000. The UN estimates that 2 billion people have no access to electricity. Electricity consumption per capita varies by more than a factor of 10 among different regions of the world: from approximately 10 000 kWh/year in North America to less than 1 000 kWh/year in Africa.

30 IEA, 1998.

31 IFRCRCS, 1998.

32 Berz, 2000.

33 ICOLD, 1998.

34 Lu, 2000.

35 Revenga et al, 1998. The WRI's 1998 study of the conditions of the major watersheds in the world focused on 150 watersheds, representing 55% of the world's land area.

36 Berger et al, 2000.

37 ICOLD, 1998; WCD Analysis.

38 Wolf, 2000, Contributing Paper for WCD Thematic Review V.3 River Basins.

39 Wolf et al, 1999.

40 Lecornu, 1998.

41 A considerable portion of the aggregate storage of large dams may be for non-consumptive hydropower releases. Analysis of all large dams in the ICOLD Register (1998), but excluding single purpose hydropower dams, shows 4 373 km^3 of designed storage capacity. This data set excludes many of the large dams in China, estimated to provide 451 km^3 of storage (IJHD, 1999)

42 Keller et al, 2000, p6–7.

43 Revenga et al, 1998.

44 The World Bank based on the review of resettlement experience estimated that between 1986-1993, an estimated 4 million people were displaced annually by the 300 large dams that entered on an average into construction every year. In the late 1980s some 10.2 million people were officially recognised as "reservoir resettlers" in China. This figure would be substantially higher if it is officially updated to include new figures of population resettlement. For example, dams and reservoirs already built on the tributaries and the main course of the Yangtze River alone have caused relocation of at least 10 million people. In India the estimates of people displaced due to large dams vary, from 21 million to 33 million people. Dams account for 34% of all people displaced by development projects in China (displacement due to city construction included in the total), 77% in India (urban displacement not included to total displaced) and 65% among the projects funded by the World Bank involving displacement. All these figures are at best only careful estimations and certainly do not include the millions who may have been displaced due to several others aspects of the projects such as canals, powerhouses, project infrastructure and associated compensatory measures such as bio-reserves etc. (See end notes 7 to 10 in chapter 4 for references to this endnote.)

45 Fox and Brown, 1998b.

46 World Bank, 1996a, p77.

47 Jing, 1999, Contributing Paper for WCD Thematic Review 1.3 Displacement, p2.

48 Wang, no date.

49 WCD Thematic Review I.3 Displacement.

50 ADB, 1999b, p1–2.

51 Fernandes and Paranjpye , 1997, p15–17.

52 World Bank, 1996a, p90–92.

53 WCD Kariba Case Study.

54 The protestors included both those to be resettled and farmers who were supposedly beneficiaries of the Bakolori irrigation scheme. According to the Nigerian government, 23 protesters were shot dead; unofficial estimates put the death toll at more than 126.

55 Stewart et al, 1996; World Bank, 1996b; Chen, 1999, WCD Regional Consultation Paper. At the WCD Regional Consultation at Sao Paulo one of the survivors narrated how his wife and children were shot dead before his eyes when he merely enquired of the authorities 'where do you want us to move to?' The answer came in the form of four rapid bullets.

56 McCully, 1996, p281-282.

57 Goodland, 2000.

58 Udall, 1998, p392.

59 Gray, 1998, p269-270.

60 Moore and Sklar, 1998, p286, WCD Submission eco048.

61 Fox and Brown, 1998a, p489.

62 Morse and Berge, 1992.

63 Manibeli Declaration, 1994.

64 OED, 1996a and 1996b.

65 ICOLD, 1997.

66 ICID, 2000.

Part One:
The WCD Global Review of Large Dams

The World Commission on Dams was charged with reviewing the development effectiveness of large dams and assessing alternatives for water and energy resources management. Part One of the report contains the results of our Global Review of large dams. It consists of five chapters.

- Chapter 2 provides the findings of the Commission's independent review of the technical, financial and economic performance of large dams.
- Chapter 3 examines the environmental performance of large dams, including ecosystem and climate impacts.
- Chapter 4 evaluates the social performance of large dams, looking especially at the displacement of people, and the distribution of gains and losses from dams projects.
- Chapter 5 assesses the scope of various alternatives to large dams for meeting the needs of irrigation, drinking water, electricity, and flood management in terms of both the opportunities they provide and the obstacles they face.
- Chapter 6 considers the planning, decision-making and institutional arrangements that guided the development of water and energy resources and the selection, design, construction and operation of dams.

Chapter 2:
Technical, Financial and Economic Performance

Any development project – particularly a large infrastructure one such as a large dam – is conceived, planned, and designed to achieve a set of objectives that will enhance the welfare of society. In assessing the performance of large dams the Commission first assessed the dams in the WCD Knowledge Base against the targets set by those planning and designing the facilities. These projections of costs and benefits provided the rationale and basis for project approval and funding. In most cases, project proponents set firm physical, financial, economic and, increasingly, social and environmental performance targets. The Commission has also examined past experience with the benefit of hindsight, that is, within the context of presently available information. Here the Knowledge Base is used to assess the impacts of large dams that were not explicitly targeted and planned for – such as cost recovery and adverse impacts on indigenous peoples – but that are important factors in assessing the contribution of large dams to development. This chapter begins the analysis with a focus on technical, financial and economic performance.

The evidence and findings presented below indicate that there is considerable scope for improving the selection of projects and the operation of existing dams, even prior to consideration of the social and environmental impacts. The performance of large dams in terms of achieving technical, financial, and economic targets is marked by a high degree of variability, with a considerable portion of dams failing to deliver on their overall objectives and many falling short of specific targets. Still, a substantial number have met or exceeded their targets and continue generating benefits after 50 years or more. Nonetheless, the emerging message from the WCD Knowledge Base is that project selection, design, and operations could be improved substantially relative to past performance.

The emerging message from the WCD Knowledge Base is that project selection, design, and operations could be improved substantially relative to past performance.

Structure and Methodology

The analysis presented here draws on several independent samples of the experience with large dams, oriented along different axes and supplemented by Commissioners' personal experiences. The information on dams in the WCD Knowledge base is reported first. Results of the Cross-Check Survey are used to indicate the direction, variability and extent of broad trends and patterns. These broad findings are then supplemented by experience captured in Case Studies, Thematic Review papers, contributions to the Regional Consultations, and submissions. The Case Studies, in particular, are used to provide indicative illustrations of these broad findings and their immediate causes. Existing literature and perspectives are then cited to verify the original findings, and also to clarify new directions emerging from the Commission's work.

The performance indicators for the implementation phase of dam projects are the same for dams with different purposes. Thus this section begins by reporting on the degree to which dams have met targets for capital costs of projects and proceeded according to schedule. It then proceeds sector-by-sector to provide insight into the relative performance of dams built and operated for different purposes. Large dams built for irrigation, hydropower, water supply and flood control have separate objectives, involve different components, respond to different markets and are operated in different ways. The inquiry is driven by consideration of the delivery of benefits measured in physical terms (such as water, power and crops) and then in terms of financial and economic profitability. Where relevant, sectoral performance on costs and time schedules is brought into the analysis.

The extent of cost recovery is also discussed, as it provides a window not just on financing issues but also on the effect of subsidies on the efficient allocation and use of resources. In addition, it informs the issues of distribution and equity treated in Chapter 4.

The secondary economic impacts generated by large dams (such as multiplier effects) are also treated in Chapter 4. Of course, many dams fulfil not just one but a number of purposes; these multi-purpose dams are examined in an additional section and their performance is contrasted with that of single-purpose dams. Finally, the sustainability of dam operations is considered by reviewing evidence on dam safety, sedimentation, waterlogging and salinity.

Construction Costs and Schedules

During the implementation period of a dam project – that is during construction – two key performance indicators are the extent to which projects come in on time and on budget. Large dams in the WCD Knowledge Base have demonstrated a marked tendency towards schedule delays and cost overruns.

Capital costs of large dam projects

Assessment of actual versus predicted performance on capital costs is important for a number of reasons.[1] First, dam projects are typically approved on the basis of a financial budget for the investment. If they end up costing substantially more than expected, additional funds have to be found. As large dams and their associated infrastructure may cost billions of dollars – three of the Case Study dam complexes cost over $6 billion each – financial overruns have important consequences for public and private budgeting. Second, projects also often derive tariff charges based on cost estimates, so underestimates will undermine financial viability or efforts to recover costs.

Cost performance data in the WCD Knowledge Base suggest that large dam projects often incur substantial capital cost overruns. The average cost overrun of the 81 large dam projects included in the WCD Cross-Check Survey was 56%. Variability was high. Of the total sample, one-quarter of the dams achieved less than planned capital cost targets whilst almost three-quarters of the dams exhibit capital cost overruns (see Figure 2.1).

Within the Cross-Check sample, multi-purpose dams demonstrated high variability in performance compared with single purpose dams, ranging from as low as 22% underruns to 180% overruns of planned cost targets. Furthermore, the average cost overrun was 63% for the 45 multi-purpose projects – three times that of the single-purpose hydropower dams in the sample. Comparatively, the cost overrun for single-purpose projects was greatest for water supply dams, with all but one project showing a 25 to 100% overrun and the average for this category being twice that of

Cost performance data in the WCD Knowledge Base suggest that large dam projects often incur substantial capital cost overruns.

Figure 2.1 Cost overruns on large dam projects

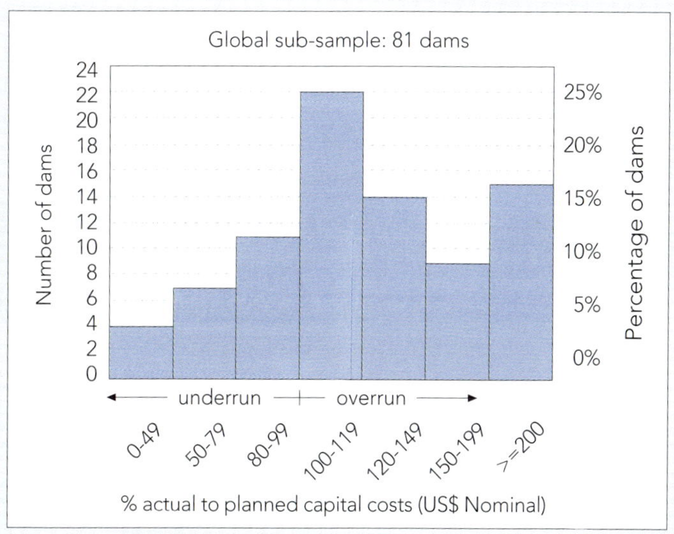

Source: WCD Cross-Check Survey.

single-purpose irrigation or hydropower dams. Interestingly, performance was worst in the sub regions of Latin America, Europe, Central Asia, and South Asia, with cost overruns averaging 53%, 69%, 108%, and 138% respectively.

The WCD Case Studies – viewed as a sub-sample separate from the Cross-Check Survey – suggest a more marked tendency towards cost overruns. Of these, only Stage 1 of the Kariba dam (built in the late 1950s) came in on target (a 3% underrun). Note also that the experience is not limited to very large billion-dollar projects, as the four smaller, million-dollar projects in the Glomma and Laagen basin in the WCD Case Studies cost some 60 to 185% more than projected. The Pak Mun dam in Thailand, a medium-size hydropower dam financed by the World Bank and commissioned in 1994, had a 68% overrun.[2]

The causes of cost variations fall into four categories:

- poor development of technical and cost estimates and supervision by sponsors;
- technical problems that arose during construction;
- poor implementation by suppliers and contractors; and
- changes in external conditions (economic and regulatory).[3]

Part of the difficulty in developing accurate projections for construction costs of large dams is that the geotechnical conditions at a site (the quality of the rock for the foundations of the major structure and for tunnels), and the quality of the construction materials cannot be determined precisely until construction is under way. Discovery during construction of less favourable site conditions than those assumed in the engineering designs and construction plans can be a significant contributor to cost overruns and delays in time schedules. Despite being a common factor in causing overruns, little to no provisions have been made to improve the estimates in this regard.

Figure 2.2 Average cost overruns for large dams

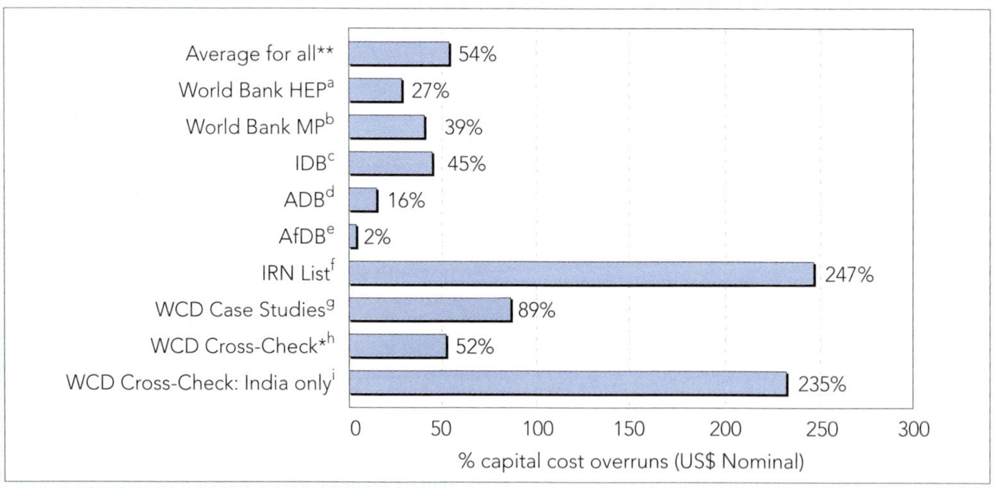

Sources: [a]*Bacon and Besant-Jones, 1998;* [b] *OED, 1996a;* [c]*IDB, 1999;* [d]*Lagman, 2000;* [e]*AfDB, 1998;* [f]*McCully, 1999 eco061;* [g]*WCD Thematic Review III.1 Economics;* [h, i] *WCD Cross-Check Survey*
*Notes: *Excludes Case Study and India dams. **Weighted averages based on number of dams in each sample. HEP refers to hydropower projects and MP to multi-purpose projects.*

Additional estimates of cost performance gathered or submitted as part of the WCD process include an Inter-American Development Bank review of its portfolio of large dam projects from 1960 to 1999. The results suggest an average cost overrun of 45% (see Figure 2.2).[4] In a similar exercise, the WCD reviewed the portfolio of large dam construction projects financed by the Asian Development Bank (ADB) between 1968 and 1999.[5] Of the 23 completed ADB large dam projects with available data, there was an average cost overrun of 16%. This masks considerable variability, as the majority of large dams in the ADB sample actually had cost underruns. The International Rivers Network (IRN) submitted a list of 14 large dams with cost performance data that showed a 242% overrun, with eight projects in India dominating the results with an average overrun of 262%.[6] The results from these latter dams, as well as the results for the Indian dams in the Cross-Check (235% overrun) confirms the serious overruns reported in the India Case Study.[7]

The data on cost performance reported by WCD confirm the results found in other studies. Perhaps the most cited study is that of 70 World Bank financed hydropower projects commissioned between 1965 and 1986, where costs at completion were on average 27% higher than estimated at appraisal. This, compared to average cost overruns of just 6% for a sample of 64 thermal power projects, and an 11% overrun for a sample of over 2 000 development projects of all types.[8] An analysis of the data on cost indicators for 10 multi-purpose dams included in the World Bank Operations and Evaluation Department's (OED) 1996 report on large dams yields an average cost overrun of 39%.[9] Finally, a 1998 review of 10 projects by the African Development Bank (AfDB) found an average cost overrun of only 2%.[10]

The evidence gathered by WCD strongly confirms the view that there is a systematic bias towards underestimation of the capital costs of large dams.[11] Aggregating all the aforementioned samples yields an average cost overrun for all 248 dam projects of just over 50% (or 40% without the dams from the IRN list and the Indian Cross-Check). The implication is that large dams have performed poorly relative to budgetary targets.

Poor prediction of inflation is often an important component of these overruns. As a consequence, when the figures are adjusted for inflation, the overrun in terms of the real economic costs of the materials and resources used is likely to be lower than that reported here.[12] For the 81 large dams in the Cross-Check subsample the cost overrun as measured in constant 1998 dollar terms came to 21%, a significant drop from the 56% cost overrun obtained in current dollar financial terms, but still large enough to significantly affect the economics of these projects.

Delays in commissioning projects affect the delivery of services, increase interest payments and delay revenue generation.

Project implementation schedules

A second important indicator of performance during the implementation phase of a large dam project is the extent to which project time schedules are met. Delays in the date at which a project is commissioned lead to increases in interest accumulated on funds borrowed for construction activities and to delays in revenues accruing to the owner from the completed project.[13] For consumers, delays mean additional periods of not being served with electricity or water. Delays thus affect the delivery of benefits, as well as the financial and economic performance of a project.

The WCD Knowledge Base suggests a marked tendency towards schedule delays for large dam projects compared with the planned time to implementation. Of the 99 projects included in the analysis of project schedule performance in the Cross-Check Survey, only half the projects came in on schedule (see Figure 2.3). Approximately 30% of the other half were delayed for one or two years, and about 15% were delayed between three and six years. Four projects were delayed more than 10 years.

The WCD Case Study dams also display a range of results in achieving project schedules. Stage 1 of Kariba dam came in on schedule, whereas Tarbela took two extra years to finish and Aslantas four. Following the initiation of construction in the late 1970s, financing difficulties led to a nine-year delay in the case of Tucurui. This led to much higher than expected payments of interest during construction. Not counting interest the cost overrun was 51% but this rises to 77% once the comparison includes actual and predicted interest costs. Other factors that can lead to schedule delays are late delivery to the site of essential equipment, unrealistic construction schedules, contractor and construction management inefficiencies, labour unrest and protests and legal challenges by affected groups.

Figure 2.3 Project schedule performance

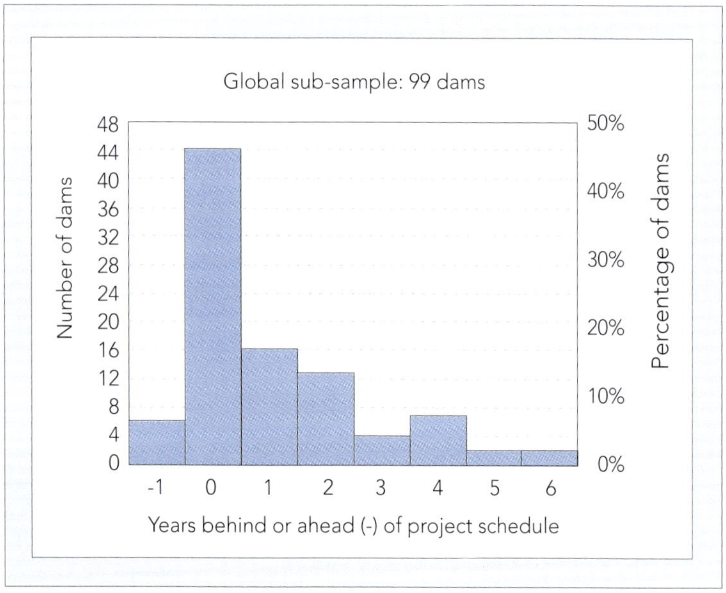

Source: WCD Cross-Check Survey.

The existing literature on large dams and related projects confirms this finding: large dams tend to be subject to significant schedule slippages. A recent study of World Bank financed hydropower projects reports a 28% delay on average. While this is a considerable slip, it is no different than that recorded in the same study for thermal power projects (30%).[14]

Irrigation Dams

Large dams and irrigation projects are a nested set of sub-systems involving the dams as source of supply, the irrigation system (including canals and on-farm irrigation application technology), the agricultural system (including crop production processes), and the wider rural socio-economic system and agricultural markets.

Potential performance indicators for large dam irrigation projects include:

- physical performance on water delivery, area irrigated and cropping intensity;
- cropping patterns and yields, as well as the value of production; and
- net financial and economic benefits.[15]

Large irrigation dams in the WCD Knowledge Base have typically fallen short of physical targets, failed to recover their costs, and been less profitable in economic terms than expected. The secondary benefits of irrigation projects were rarely specified as targets (these are discussed later, in Chapter 4).

Irrigated area and cropping intensity

Irrigation components of large dam projects in the WCD Knowledge Base fell well short of targets in terms of development of command area (and infrastructure), area actually irrigated, and to a lesser extent the intensity with which areas are actually irrigated. With respect to the achievement of command area targets, the Cross-Check Survey demonstrates a general pattern of underachievement, with almost half of the 52 dams in the sub-sample falling short of the planned target (see Figure 2.4).[16] Poor performance is most noticeable during the earlier periods of project life, as the average achievement of irrigated area targets compared with what was planned for each period increased over time from around 70% in year five to approximately 100% by year 30 (see Figure 2.5). Nevertheless, a characteristic pattern observed in the sample is variability of performance between projects. In particular, one-quarter of the projects achieved less than 35% of their target irrigation areas during the first five years.

The 52 projects in the survey sub-sample likewise underachieve in terms of cropping

Figure 2.4 Achievement of command area

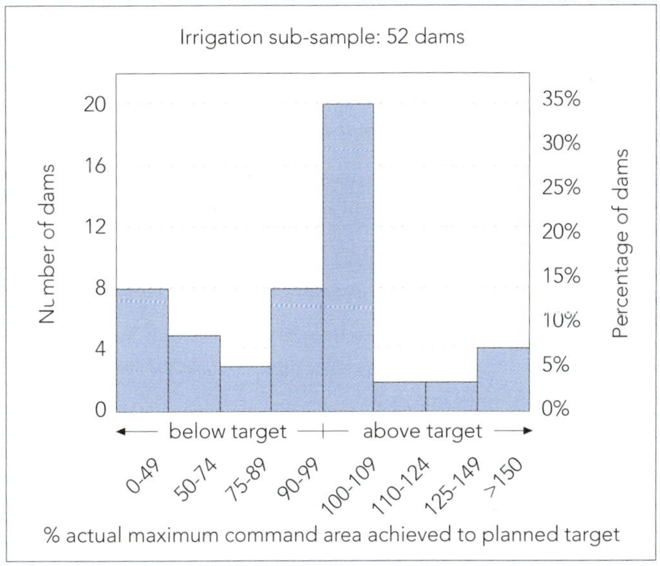

Source: WCD Cross-Check Survey.

Figure 2.5 Actual irrigated area compared to planned targets over time

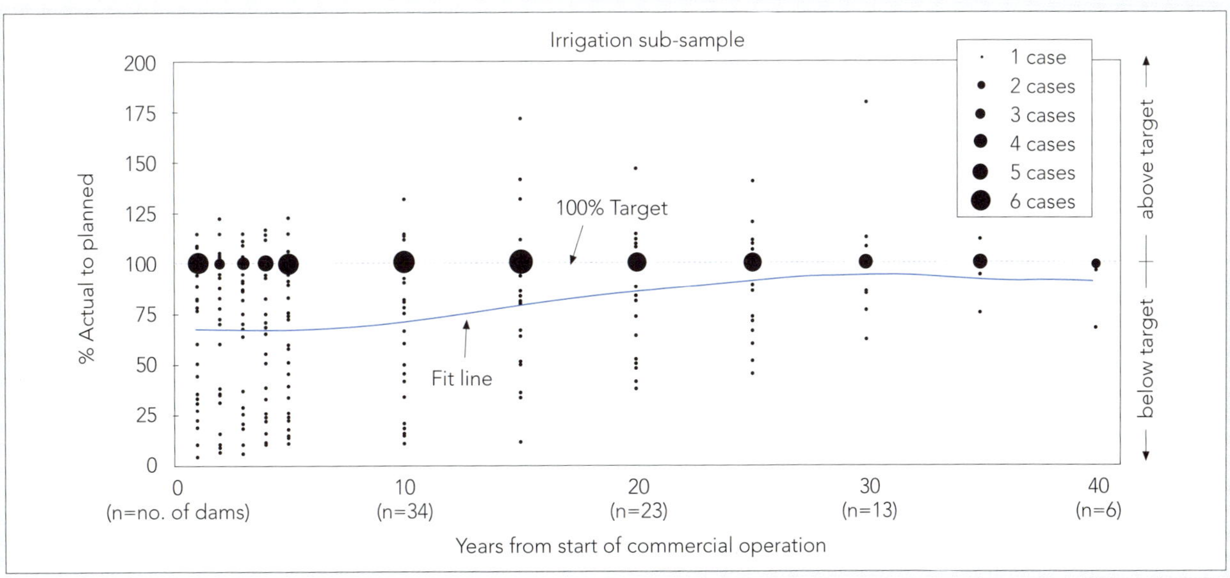

Source: WCD Cross-Check Survey.

intensity targets. However, cropping intensity performance comes closer to targets than irrigation area development.[17] Less than half of the projects in the sub-sample consistently achieve or exceed planned cropping intensities from year one of irrigation. Of the remainder, 20% achieve less than three-quarters of targets and the other 40% fall between 75 and 100%.

With regard to cropping intensity there is little difference in the average values for single and multi-purpose dams in the survey. Actual values of maximum irrigated area achieved by single-purpose irrigation projects are consistently better on average by a margin of around 10%. Single-purpose irrigation schemes also show more of a spread in the values around the average, with one-quarter of the 21 single-purpose projects achieving more than 115% of planned targets.

When compared with larger projects in the survey, dams with heights less than 30 meters and reservoir areas of less than 10 km² tended to be closer to the predicted targets and demonstrated less variability for command area development, actual irrigated area and actual crop intensity. All projects below 90% of area and intensity targets were larger than 10 km² and higher than 30 meters.

When compared with larger projects in the survey, dams with heights less than 30 meters and reservoir areas of less than 10 km² tended to be closer to the predicted targets

The WCD Case Studies corroborate the results of the Cross-Check Survey and display a similar breadth of results with respect to irrigation area and cropping intensity targets. In the case of Grand Coulee, commissioned in 1941, only about half of the predicted area in the Columbia Basin Project was eventually developed; the same can be said of the Gariep dam on the Orange River in South Africa. As Tarbela dam is an integral part of the Indus Basin Irrigation System (IBIS), it is not possible to specify the achievement of area targets. However, the amount of water released for irrigation from Tarbela has exceeded predictions by 20% over its 25-year life. This is mainly due to a lower-than-expected rate of reservoir sedimentation. Shortfalls of 9 to 60% in achieving cropping intensity targets are observed in the provinces of Punjab and Sindh. Meanwhile, the India Case Study reports that the level of under-utilisation of irrigated areas is between 13 and 25%.

In the case of the Aslantas dam in Turkey, 96% of the irrigated area was developed by the end of the implementation period, but this has now fallen to 87% due to the conversion of agricultural land to urban and industrial purposes. Additional irrigation area foreseen under the original basin plan has subsequently been implemented through other projects, although delayed from original projections. Cropping intensity has increased from 89% to 134% of the target, but such figures mask a return to growing wheat, a crop that requires little irrigation in Turkey.

The under-achievement of targets for irrigated area development from large dams has a number of causes. Institutional failures have often been the primary causes, including inadequate distribution channels, over-centralised systems of canal administration, divided institutional responsibility for main system and tertiary level systems, and inadequate allocation of financing for tertiary canal development. Technical causes include delays in construction, inadequate surveys and hydrological assumptions, inadequate attention to drainage, and over-optimistic projections of cropping patterns, yields and irrigation efficiencies, including the late realisation that some areas were not economically viable. Also, a mismatch between the static assumptions of the planning agency and the dynamic nature of the incentives that govern actual farmer behaviour has meant that projections quickly became outdated.

Performance data on irrigation projects as reported elsewhere support the original findings of the Commission, although these are not solely linked to large dams. An ADB evaluation of 35 irrigation projects found actual cropped areas generally at 60–85% of appraisal estimates, with only four exceeding targets.[18] A World Bank study of seven irrigation projects found all but one with crop intensities less than expected (in the range of 65–91%, with one at 107%).[19] A 1990 evaluation by the World Bank of 21 irrigation projects 5–12 years after completion showed that irrigated area had fallen in 11 of the projects and that cropping intensity was lower than at completion in 18 of the projects (at 85%).[20]

Cropping patterns, yields, agricultural production, and gross value of production

Crop yields and the gross value of production from large irrigation dams in the WCD Knowledge Base have often varied significantly from those predicted at the outset of the projects. Lower yields are often observed for crops specified in planning documents – which emphasise food grain production for growing populations – than for the crops actually selected by farmers. This occurs as farmers respond to the market incentives offered by higher-value crops – either seasonal or longer-term orchard-based crops – and allocate available resources to these crops. This implies higher-than-expected gross value of production per unit of area, with the caution that such increases have varied with the long-term real price trend of the relevant agricultural commodities. But when changes in cropping patterns are combined with shortfalls in area developed and cropping intensity, the end-result is often a shortfall in agricultural production from the scheme as a whole. Gross value of production is higher where the shift to higher-value crops offsets the shortfall in area or intensity targets.

The WCD Case Studies provide examples of both underestimation and overestimation of the quantity and value of agricultural production. In the case of the Columbia Basin Project, yields have been 30–50% higher than predicted in planning reports in 1932. Partly because of a shift to higher-value crops, the gross monetary value of actual agricultural production per unit of area almost doubled from that anticipated at the time. For example, the area planted to fruit and vegetables in 1992 was 60% compared to 20% thirty years earlier. Yields of fodder and cereal crops also increased significantly due to improved varieties and mechanisation.

Similarly, in the case of the Aslantas dam in Turkey, the cropping pattern changed dramatically from that planned – partly retaining the staple wheat crop and growing second high-value crops such as soybean, rather than cotton as predicted. Yields of wheat and cotton reached 75% of predictions, while watermelon and maize exceeded targets by 50–100%. Overall, the gross monetary value of agricultural production reached 71% of that predicted. In the case of Tarbela, yields of wheat, rice and cotton are between 9% and 50% lower than predicted in the feasibility report.

Lower than expected crop yields have been caused by agronomic factors, including cultivation practices, poor seed quality, pest attack and adverse weather conditions, and by lack of labour or financial resources. Physical factors such as poor drainage, uneven or unsuitable land, inefficient and unreliable irrigation application, and salinity also hinder agricultural production (see below). The efficiency of water use affects not only production but also demand and supply of irrigated water (see Box 2.1).

Box 2.1 Efficiency in the use of irrigation water

Surface irrigation systems of the type supported by large dam projects have tended to use water inefficiently. Surface water irrigation efficiency is in the range of 25-40% in India, Mexico, Pakistan, the Philippines and Thailand; 40-45% in Malaysia and Morocco; and 50-60% in Israel, Japan and Taiwan. The Aslantas Case Study suggests that the overall irrigation efficiency of the Aslantas project is 40%. These examples suggest that the ratio of water consumed by crops to the water delivered from the source varies from 25-60%.

Sources: WCD Thematic Review IV.2 Irrigation Options, Chapter 3; WCD Aslantas Case Study

An important cause of the difficulties in achieving targets for gross value of production is the decline in commodity prices. For example, in real terms, world prices for grains in the 1990s were of the order of half those in the 1950s. Although there were significant annual variations, the average price of rice from 1950 to 1981 was $850/ton (in 1997 prices), compared with $350/ton from 1985 to date.[21] Wheat prices have shown a similar decline, but less severe ($330/ton from 1950 to 1981 in 1998 prices, compared with $140 from 1985 to 1999).[22] The observed fall in prices stems in part from the increase in food production that was spurred by irrigated agriculture and the Green Revolution, but also from production subsidies and other incentives used by many countries to support agriculture. This fall in prices has contributed to lowering the value of production achieved as against predictions.

A general pattern of shortfalls and variability in agricultural production from irrigation projects in developing countries is also revealed by other sources. In the 1990 World Bank OED study on irrigation cited earlier, 15 of 21 projects had lower than planned agricultural production at completion. Evaluations of 192 irrigation projects approved between 1961 and 1984 by the World Bank indicated that only 67% performed satisfactorily against their targets.[23]

Financial and economic profitability

Since the 1930s in the industrial countries and from the 1970s in developing countries, financial and economic profitability have become an important, if not the dominant, decision criteria in water projects.[24] Consequently, approval of many large dam projects was contingent upon estimates of their predicted profitability. The measures typically used to assess profitability are the financial internal rate of return (FIRR) and economic internal rate of return (EIRR) as determined through cost-benefit analyses. The FIRR tells the project owner if the project is profitable, while the EIRR is intended to tell society if the project improves the overall economic welfare (or well-being) of the nation. Under-performance relative to targets does not necessarily imply that a project is unprofitable in economic terms, as the rate of return of a project may fall short of its target but still exceed the opportunity cost of capital to the economy.

Typically, an EIRR of over 10% is judged acceptable in the context of a developing economy.[25] On this basis, irrigation dam projects in the WCD Knowledge Base have all too often failed to deliver on promised financial and economic profitability – even when defined narrowly in terms of direct project costs and benefits.

Given the lack of evaluation studies on large dams for irrigation purposes, the WCD compiled financial and economic performance data from a series of project appraisal, completion, and audit reports on large dam projects funded by the World Bank and ADB (see Figure 2.6).[26] The average EIRR at appraisal for the 14 irrigation dams was slightly above 15%, and at evaluation it was 10.5%, a significant shortfall in economic performance for the group. Whereas 12 projects had expected returns of over 12% at appraisal, this number had fallen to five by evaluation. In four cases, the EIRR at evaluation fell below the cut-off rate of 10%.

The results extracted from the ADB and World Bank reports are of course based only on evaluation studies undertaken at completion of the implementation phase or just a few years after commissioning. They incorporate the effects of cost overruns and initial operating results, but are not long-term or comprehensive in nature. They typically only consider the direct project costs and benefits of the project and do not account for the social and environmental impacts associated with the dam or agricultural production. In the case of the Columbia Basin Project, even a cursory analysis of the long-term performance data available from the WCD Case Study which shows that the large cost overruns and lower-than-expected delivery of benefits raise questions about the economics of the project (see Box 2.2).

In post-evaluation studies of irrigation and rural development projects by the World

Figure 2.6 Economic performance of multilateral-financed irrigation dams

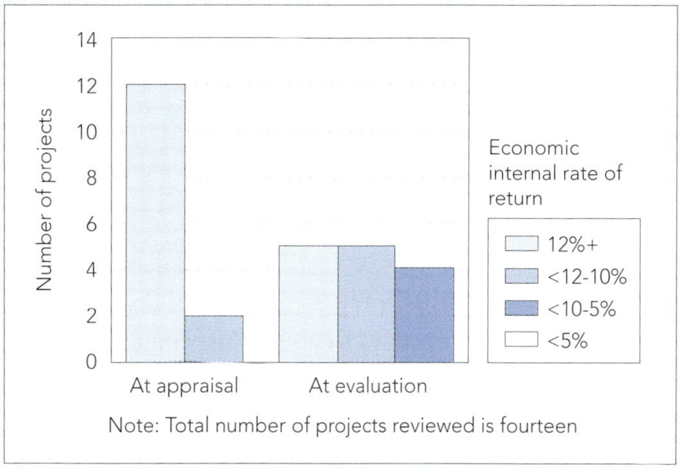

Source: WCD Thematic Review III.1 Economic Analysis, Chapter 10.

Box 2.2 Economic and financial performance of the Columbia Basin Project

The Columbia Basin Project (CBP) was never expected to cover its costs, and was criticised by early opponents because it would not be economically profitable. Yet the WCD Case Study reports that a $2 150/ha increase in assessed land values has occurred for irrigated land within the project area. When applied to the 268 000 hectares under cultivation, the capitalised increase in the value of the land would be $575 million in 1998 dollars. Even if the full value of this increase was due to the CBP alone, it is clear that it would not come close to the real cost of the CBP of $3.6 billion, as reported in the Case Study.

A better approach to assessing the project is to compare the present value of costs of the project with the benefits. The Case Study calculated the present value of CBP costs at $1.47 billion at a 10% discount rate.[27] Two simple assumptions may be used to generate a best-case estimate of benefits. First, the full value of the average net production value of $500/ha per year as reported by a recent study of the CBP is attributed to irrigation (that is as a return to irrigated water and not other farming or capital inputs). Second, these benefits are considered to occur for all years since 1945 (project start-up) and to all of the 268 000 hectares. The resulting net present value of benefits from 1945 to 2010 at 10% discount rate is $1.32 billion. Note that the assumptions made are generous, given that actual acreage and gross value of production increased only gradually over time, and thus the actual magnitude of early economic benefits would have been considerably less. In addition, the opportunity cost of water diverted to the CBP of $39 million per year is left out of the calculation.

As the benefits are less than the costs, the results suggest that the CBP did not achieve a 10% rate of return (when measured on an incremental basis, that is apart from the Grand Coulee dam (GCD)). While this simplistic analysis provides just an indication of possible returns from the project, it illustrates the type of information provided by ex-post evaluation of economic profitability. Further, it demonstrates the concerns raised about multi-purpose projects, where the irrigation component fails a cost-benefit test. Would the United States economy have been better off investing only in GCD – and using the revenues from GCD to invest in other profitable opportunities in the economy – rather than reinvesting the profits in the CBP?[28]

Source: calculations based on WCD Grand Coulee Case Study

Bank and the ADB, half the projects were judged as unprofitable in economic terms, that is having EIRRs of less than 10%. More than three-quarters of the projects returned less than expected at appraisal. In fact, the average of the EIRRs for the World Bank irrigation projects was 17.7% at appraisal. At completion, this dropped to 14.8%, and at the time of impact evaluation – typically six to eight years after completion – the EIRRs had fallen to 9.3%. This rather strong downward trend indicates that many irrigation projects – not just large dams projects – suffer from a tendency to overstate project returns at the outset, also referred to as 'appraisal optimism.'[29]

Cost recovery

Public agencies have not always attempted to recover costs associated with public infrastructure projects. Where the services provided by large dams are valued as consumption goods or productive inputs, the absence of cost recovery by the sponsoring agency is often equivalent to a subsidy in the sense that the large dam project provides a benefit for which no fees are paid. Which participants in large dam projects receive these subsidies is examined further in Chapter 4. A lack of cost recovery is not just a matter of subsidy, however. Provision of free services and subsidised inputs often leads to misallocation of resources and inefficient production. Further, it may lead to perverse behaviour as people direct their efforts to obtaining such subsidies (rent-seeking behaviour) rather than productive activities.

The analysis here assesses the extent to which cost recovery is an explicit objective in large dam irrigation projects, and the extent to which it has met expectations or, if cost recovery was not anticipated, the extent to which it has occurred in any event. Not surprisingly, recovery of capital costs for irrigation has rarely been a target and is even more rarely achieved. Performance in recovering operational and maintenance (O&M) costs is typically poor, although increasing recognition of the importance of recovery to performance led to institutional innovations that increased collection of O&M charges in the 1990s.

Recovery of operational and maintenance costs

There was mixed performance on the recovery of O&M costs in the three WCD Case Study dams that involved irrigation. In the case of the Aslantas dam, recovery of costs was expected but only made progress after 1995 with the adoption of a water user association (see Box 2.3). In the Indus Basin Irrigation System, where the Tarbela Dam

Box 2.3 Cost recovery for the Aslantas Dam

In negotiations leading up to the Aslantas dam project, the World Bank was assured that arrangements would be made for recovering O&M costs for the dam and irrigation systems, capital costs for the dam and irrigation system (over 50 years), and on-farm development costs (over 15–20 years). An annual recovery charge of TL 1 250 per hectare to cover the first two items was proposed in the 1972 appraisal by the World Bank, and assurance was received that this would be reviewed every five years, when the rates would be adjusted for inflation. This comes to roughly $350 per hectare in 1998 dollars and would have represented half of the expected farming profits per hectare as estimated at appraisal. Even under such a repayment program, the Bank estimated that farmers would pay back only about one-quarter of the total capital and O&M costs.

Actual recovery of capital costs appears to have fared quite poorly. The capital cost recovery charge rate was increased in 1981, 1982 and 1986. Due to rapid local inflation and subsequent devaluation of the Turkish lira, however, the TL 7 500 charge set in 1986 came to just $0.03 per hectare in 1998. Collection rates have been a continual problem. In 1984, the collection rates on assessed charges for capital costs were only 63% at which time penalties were introduced. More recently, the WCD Case Study suggests that the transfer of O&M responsibilities to the local water user association in 1995 resulted in greatly increased recovery of annual O&M costs. Indirect cost recovery for the sector as a whole in the form of a withholding tax of 5% on agricultural sales was instituted country-wide in 1981, yielding on average $93 of revenue per hectare of agricultural holding. Still, the project has a very low rate of capital cost recovery and has not come close to the 25% recovery agreed at appraisal.

Source: WCD Aslantas Case Study; World Bank, 1973; World Bank, 1985.

plays a pivotal role, revenues from irrigation fees basically covered operation and maintenance costs up until 1970. Subsequently, revenues have declined, and the gap between O&M expenditure and recoveries reached 44% by 1992 in Punjab and 30% in Sindh. At the same time, an increasing proportion of the revenues was being allocated to agency costs as opposed to maintenance activities.

A similar situation occurs in India, where the gross receipts from irrigation charges are considerably less than the recurrent costs of O&M. In the 1960s receipts covered expenditures, however, by the end of the 1980s receipts were only of the order of 10% of expenditures.[30] Annual operational losses became a huge fiscal liability, with annual operational losses exceeding $1 billion by the mid-1990s. The water charges collected represent about 2% of the incremental benefits of irrigation.[31] In the Columbia Basin Project, irrigators pay only a very small portion of the costs of pumping water into the CBP system and nothing for the water itself – which has a large opportunity cost in terms of foregone hydropower production at Grand Coulee.

The information gathered by the WCD on O&M cost recovery is confirmed by the literature in this field. A survey done since 1992 of 18 irrigation systems worldwide shows considerable variation of recovery rates, with public agency systems in the range of 30–50% and some locally managed systems reaching full cost recovery.[32] The 1990 evaluation by the World Bank reports that in 11 of 21 cases, recovery rates were too low to cover irrigation O&M costs. In Mexico, water user associations have proved effective in improving cost recovery and management.[33]

Recovery of Capital Costs

The tendency to poor financial and economic performance and the failure to recover O&M costs suggest that even where it is an explicit objective, recovery of capital costs will be limited. The Aslantas dam provides a fairly stark example of the failure not only to recover these costs but also to stick to agreements made in this regard (see Box 2.3). In the World Bank's 1990 evaluation report on irrigation projects even the cases of 'excellent' cost recovery resulted in only partial recovery of capital costs.

In sum, the assessment of large dam irrigation schemes reveals that while there is considerable variability in performance, such schemes have all too often fallen short of physical targets and failed to recover their costs in cases where that was the intention indicated in the project document. Further, in many cases the economic justification for the approval of the project has not been borne out by actual experience in implementation and operation due to cost overruns and shortfalls in net benefits of agricultural production.

The bulk of projects have delivered power within a close range of pre-project targets but with an overall tendency to fall short of targets.

Hydropower Dams

Large hydropower dams in the WCD Knowledge Base can be divided into three groups. The first two groups represent the extremes: a number of projects that have greatly exceeded their targets and a few notable under-performers. The bulk of projects have delivered power within a close range of pre-project targets but with an overall tendency to fall short of targets. Hydropower projects – as with other large dams – have incurred cost overruns and schedule delays. Limited available evidence suggests that hydropower projects often diverge substantially from their economic targets, in either a positive or negative direction. Financial performance is more consistent and with less downside variability. Finally, a number of older projects continue generating benefits even after a half-century or more of operation.

Delivery of services and benefits

Large dams in the WCD Knowledge Base that were designed to deliver electric power have on average met expectations for the delivery of power but with considerable variability, much of it on the downside. A number of projects have far exceeded their technical, financial and economic targets, whereas others have fallen well short. Delivery of services and benefits are examined by assessing performance to targets for installed capacity and delivery of power. Hydropower also offers ancillary services to the power grid.

Figure 2.7 Project averages for actual versus planned hydropower generation

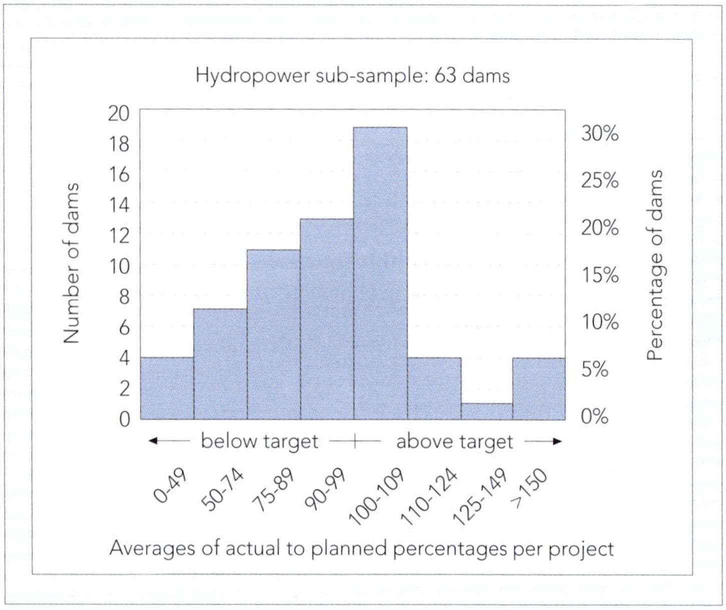

Source: WCD Cross-Check Survey.

In contrast to irrigation, the hydropower performance of 63 large dams in the WCD Cross-Check Survey was on average closer to target (see Figure 2.7). But as with irrigation dams, the variance in performance across the projects was extremely large (see Figure 2.8). On average, almost half of the sample exceeded the set targets for power generation – with about 15% exceeding targets by a significant amount. It also shows that around one-fifth of the projects in the sample achieve less than 75% of the planned power targets. Aside from these marked extremes, Figure 2.7 shows that over half of the projects in the sample fall short of their power production targets. Thus average performance in the sample is sustained by a few over-performers and should not mask the variance in performance that is weighted towards shortfalls in power delivery.

The higher-than-expected output in hydropower generation from almost half of the large dam projects in the Cross-Check Survey is due only in small part to the addition of extra installed capacity prior to commissioning, but more especially since commissioning. One-quarter of the large dams with higher-than-expected output had installed more than 100% of the capacity planned in the feasibility study.

The Tucurui dam diverged from feasibility design when initial installed capacity was raised from 2 700 to 4 000 MW before commissioning. Tarbela, Grand Coulee, and the Glomma and Laagen dams have all seen subsequent installation of significant amounts of additional capacity that were not foreseen at feasibility. Furthermore, both Kariba and Tucurui are multi-stage projects that involve doubling of capacity. The profiles of hydropower performance of the Case Study dams in Figure 2.9 illustrate how installation of more than initially expected capacity leads to higher-than-expected power output (Grand Coulee and Tarbela).

However, energy output is often also lower than initially estimated. The Victoria dam in Sri Lanka had a predicted energy generation of 970 GWh/annum, but in reality only produces an average of 670 GWh, a shortfall of over 30%.[34] Higher-than-expected upstream irrigation abstractions and lower-than-predicted natural stream flows were the causes in this case. Case Study results from Pak Mun, which is a run-of-river project with peaking capacity, showed that in the first four years of operation after 1994 the installed capacity and total annual energy generation were as expected. However, the ability of the project to deliver energy for its primary purpose of a planned four-hour peak period was considerably less than expected in the dry season, leading to questions regarding its economic viability (as discussed below).

Further analysis of the Cross-Check data reveals that the average generation in the first year of commercial operation is 80% of the targeted value for large hydroelectric dams (see Figure 2.8). In years two to five, the average percentage realisation of targets rose to near 100%, however this improvement in the average for any time period masks considerable variation in the sub-sample with half or more of projects still falling short of predicted power generation. Delays in the construction phase of projects (as documented earlier), in reservoir filling (if low rainfall prevails), and in installing and bringing turbines on-line explain shortfalls in performance in the early years

Figure 2.8 Actual versus planned hydropower generation over time

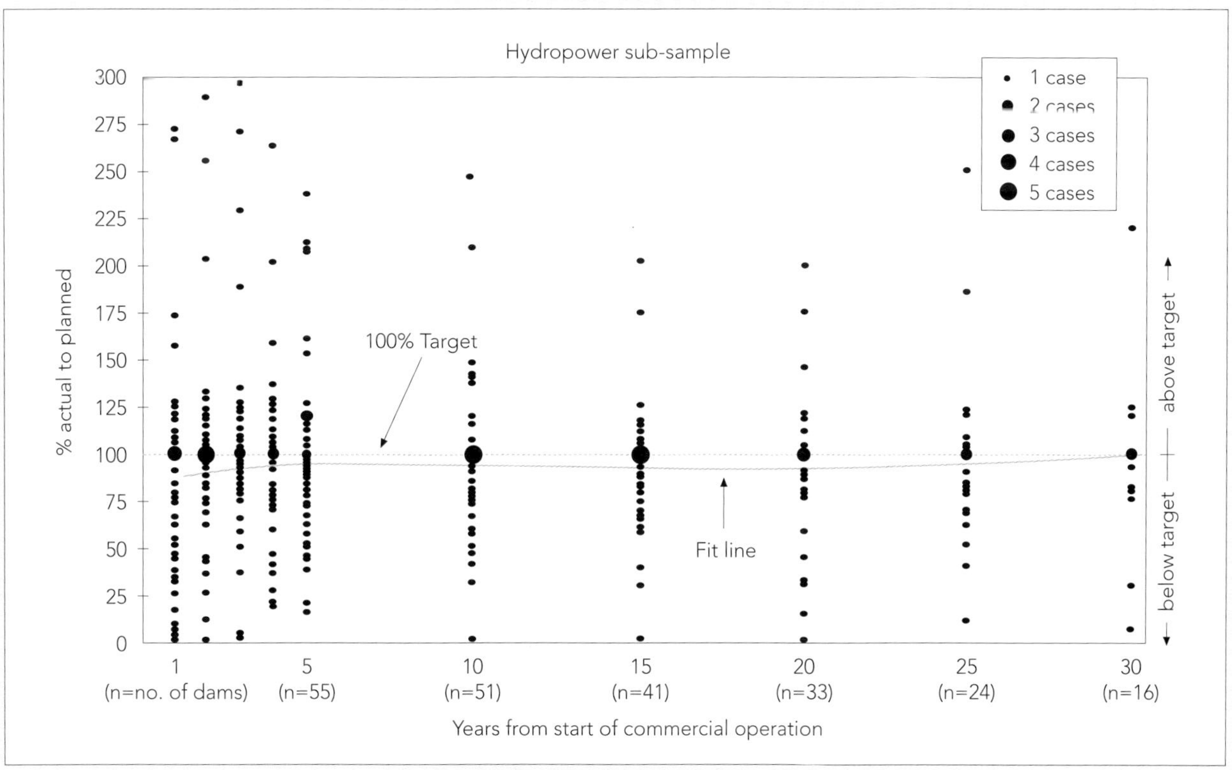

Source: WCD Cross-Check Survey.

of commercial operation. Delays in installing capacity and subsequent delays in meeting power targets are noted in two case studies (Tarbela and Kariba). Tarbela met and exceeded predicted levels in 1992, which coincided with the commissioning of the full complement of planned capacity (see Figure 2.9).

Figure 2.9 WCD case study hydropower performance: capacity and power generation

Source: WCD Grand Coulee, Kariba and Tarbela Case Studies.

The WCD Case Studies show that unexpected events and design changes during the project development stage may lead to delays in achieving power generation targets. For example, Tarbela experienced major structural damage in commissioning trials that led to a two-year loss of power generation, and in South Africa the decision to increase power generation by increasing the height of the Van der Kloof dam delayed its commissioning. Where delay is not the result of slower than expected growth in demand for power, such delays in reaching targets may have important consequences for providing power to consumers and securing early economic benefits from a project.

A further observation on timing of power delivery from the Cross-Check Survey is the consistency in the distribution of the sample over the period surveyed (see Figure 2.8). The WCD Case Studies provide a series of examples (Grand Coulee, Tarbela, and Tucurui) in which power output from very large dams increases over long periods (in absolute terms and relative to predictions). Power production at Grand Coulee, for example, has trended upward for the last 60 years – though marked by significant interannual variability (see Figure 2.9).

This variation in power production over time within a single project is noted in the WCD Cross-Check Survey and additional Case Studies. Normal variations in weather and river flows dictate that virtually all hydroelectric projects will have year-to-year fluctuations in output. The effect of drought years can be easily seen in the large swings in annual power generation from Grand Coulee and Kariba, particularly over the last two decades. Whether changes in regional and global climate are exacerbating normal weather-related interannual variation remains to be seen. Such variation may also reflect changes in other demand and supply factors.

Delays in reaching targets may have important consequences for providing power to consumers and securing early economic benefits from a project.

Technical efficiency and ancillary services

The Glomma and Laagen and Grand Coulee Case Studies illustrate how unplanned improvements in generation were achieved by a combination of factors such as adding new powerhouses to the same reservoir, adding additional turbines, upgrading existing turbine and generation equipment, or optimising reservoir operations to improve performance.

These experiences of improving the performance of hydropower generation over a project's life are not confined to industrial countries. In Nepal, modifications to the intake, provision of an extra desander, dredging the forebay and refurbishing the generators/turbines and power house control systems at the Trushuli-Devighat hydropower station in 1995 improved average annual power generation by 46% – from 194 to 284 GWh a year.[35] In other cases the optimised operation of reservoirs has led to increased generation during a project's life.[36] The trend in the industry in Europe and North America is to optimise reservoir operations and power dispatch schedules to improve performance by using more sophisticated 'decision-support' systems (see Box 2.4).

Box 2.4 Optimising operations with the aid of a computerised decision-support system

Since 1987, two hydroelectric plants in the coastal mountains of British Columbia have used a computerised decision support system (DSS) to guide weekly reservoir release decisions. Studies of 1970–74 operations (before the support system became operational) showed that the rule-curve-based operation had produced 83% of the maximum attainable energy. With the DSS, the actual energy produced each year between 1989 and 1993 rose to 100, 93, 98, 94, and 96% respectively of the maximum possible.

Source: WCD Thematic Review IV.5 Operations

The ancillary services associated with hydropower generation – for example, reactive power generation and fast starting reserve generation – reduce or even defer other investments in the electrical system.

Financial and economic profitability

Schedule delays, cost overruns and variability in delivery of power suggest a broad variation in economic performance for hydropower projects. Further, evidence from North America suggests that the O&M costs of hydropower rise over time.[37]

Large hydropower dams in the WCD Knowledge Base confirm that there is considerable variability in meeting economic targets and achieving economic profitability (when defined narrowly in terms of project costs and direct benefits). Unlike large dam irrigation projects the variability is not only to the down-side, with some projects performing better than expected. Much less variation is observed in financial performance relative to targets, although there is a wide spread in terms of actual financial rates of return.

Evidence compiled by WCD from multilateral bank appraisals and evaluation studies shows that although a number of hydropower projects fall short of their financial and economic targets and can be considered economically unprofitable, others meet their targets or even exceed expected profitability (see Figure 2.10). Of 20 World Bank, AfDB, and ADB projects reviewed, 11 fell below initial targets and seven rose above these targets; overall, nine projects had returns below 10% but only six of the projects actually fell to this level (the other three already had low rates of return at appraisal). These evaluation studies do not reflect long-term performance data, but rather the effect of cost overruns and initial lags in performance.

Figure 2.10 Multilateral bank evaluation results on the economic performance for hydropower dams

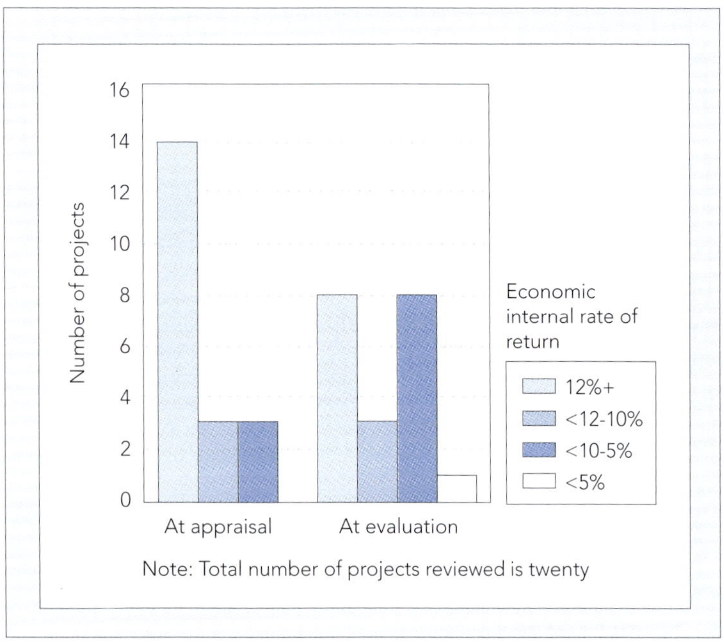

Source: WCD Thematic Review III.1 Economic Analysis, Chapter 10.

The nine evaluation studies that included data on financial performance (all from the World Bank and ADB) showed much less downward variability, with only one project falling short of target by a significant amount (namely from 11% down to 6%). Three projects improved their performance by 5% over expected returns. The results may reflect the administrative nature of tariff setting, which enables tariffs to be adjusted to suit the financial requirements of a project.

Long-term data gathered by the Commission through the Case Studies illustrates the range of profitability actually achieved by hydropower projects. Indeed, three of the Case Studies provide a succinct illustration of projects that are superlative, respectable, or borderline in terms of profitability. A fourth demonstrates how a project may be approved, but in the end fall short in terms of profitability.

The financial and economic performance of Grand Coulee as a stand-alone hydropower project can only be described as superlative (see Box 2.5). Further, after 60 years of operation it shows no sign of slowing down. In the case of the Kariba dam, the Case Study reports on a cost-benefit calculation that took the original economic assumptions and developed both predicted and actual EIRRs. Based on 40 years of operating data, and including the cost overruns under the second stage of the project, the actual EIRR is a respectable 14.5%, down from the 16.5% calculated for the project as planned.

In the case of Tucurui dam, the project was planned under a military dictatorship and there was little concern during planning for economic profitability or cost recovery. The WCD Case Study suggests that the project is marginally profitable relative to the consumer price of electricity, but this omits the costs of transmission and distribution (see Box 2.6). Perhaps the simplest observation is that had the project met its original cost target it would have produced, in financial terms, low-cost power. However, the large cost overrun – $3.3 billion more than expected – effectively erased the competitive advantage that the site may have had. Further, the relatively high unit cost of electricity from the facility and the subsidies to industry established at commissioning imply a failure to recover costs directly, as the operating utility continues to require state subsidies.

In the case of Pak Mun, installed hydropower capacity and generation have met actual targets for the few years of operations to this point. However, the WCD Case Study suggests that the failure to produce a firm supply of peaking power during low flow months implies that the alternative thermal power plant capacity used in the appraisal of the project was too generous. The Case Study goes on to value the benefits of Pak Mun based on the avoided costs of the alternative. The large reduction in the alternative power plant from 150 MW gas turbine to a 21MW gas turbine when combined with the 68% cost overrun reduces the EIRR from 12.1% to 7.9%, which is below the opportunity cost of capital in Thailand. It is worth noting that this calculation simply demonstrates that the energy supplied by Pak Mun could have been supplied in a more inexpensive fashion, for example by using the 21 MW gas turbine and secondary power from other plants.

Few formal and comprehensive post-evaluations of the financial and economic profitability of large hydropower dams exist for comparison with the WCD analysis.

Few formal and comprehensive post-evaluations of the financial and economic profitability of large hydropower dams exist for comparison with the WCD analysis. Despite

Box 2.5 Financial and economic performance of hydropower at Grand Coulee dam

Financial profitability. Profitability of the Grand Coulee dam (GCD) was not an explicit objective. Still, even at a real tariff of $0.02/kWh the GCD will have generated real revenues of around $15 billion over the last 50 years. These can be compared with real project costs of $5.7 billion. While the inclusion of operating costs and discounting would reduce this gap, it seems that GCD is quite profitable in financial terms.

Economic profitability. No cost-benefit analysis was undertaken in the 1932 Butler Report. Prior to installation of the third power plant, an analysis by the US Bureau of Reclamation, yielded a benefit-cost ratio of over 3:1. When the cost overruns on the third plant are accounted for in the calculation, the benefit-cost ratio drops – but at 2:1 it still indicates that the plant would remain a very profitable undertaking.

Economic efficiency. The current operating costs of GCD compare favourably (more than 17 times less expensive) with next best current alternatives. This is best stated in terms of today's decision to continue producing power from GCD. Compared with natural gas at $25 MWh, current production levels of 20 000–25 000 GWhs are responsible for saving $475–600 million a year in real costs to the economy.

Of course, these figures reflect only the effects of direct project costs and benefits, not the external social and environmental impacts of the project.

Source: WCD Grand Coulee Case Study

Box 2.6 Economic performance and cost recovery of hydropower at Tucurui dam

The Tucurui dam is a single-purpose hydropower facility developed and operated by Eletronorte, the public utility for electricity in the north of Brazil. The WCD Case Study calculates generating costs at $40–58/MWh (for 8% and 12% discount rates). Of the total power produced in 1998 by Tucurui dam, about half (12 000 GWh) went to industry at a price of $24/MWh – from $16 to $34/MWh below cost. Using these below-cost sales figures as an indication of the subsidy provided to industry yields a range from $190 million per year (for the cost at 8%) to over $400 million (at 12%). Eletronorte calculates its annual subsidy from the public purse in 1998 at $194 million, and Tucurui is the largest of Eletronorte's projects. Failing more precise data it appears that Tucurui could recover its costs but that to date it may not have achieved this, in part due to continued subsidies to industrial producers. Indirect cost recovery through taxation of these industries was not documented in the study.

Source: WCD Tucurui Case Study

being probably the single largest financier of dam projects in the post-war period, the World Bank did not undertake a dam-specific review of its portfolio until the mid-1990s. Even then its 1996 OED study did not draw on actual performance of the dams in its sample and provides little evidence on economic performance.[38] The African Development Bank recently reviewed its experience with six hydroelectric dams and found that only four passed the economic viability test using a 10% discount rate.[39]

Thus the WCD Knowledge Base shows that a considerable number of hydropower projects fall short of their initial economic targets, although only a smaller number can be classified as economically unprofitable (falling short of the rate of return target for the economy as a whole). Meanwhile there are almost as many projects in the Knowledge Base that actually outperform their economic targets. Finally, it is worth emphasising that cost recovery has not been a substantial problem for hydropower projects; indeed, the focus is more on profitability in the current context of trends towards private-sector participation in electric power production.[40]

Water Supply Dams

Water supply dams in the WCD Knowledge Base have generally fallen short of intended timing and targets for bulk water delivery and have exhibited poor financial cost recovery and economic performance. These results reflect the longer development horizon of such dams, as well as over-estimates of demand, and are similar to the general direction of results in the water supply and sanitation sector.

Delivery of bulk water supply

The Cross-Check Survey found that one-quarter of the 29 dams with a water supply function have delivered less than 50% of target. Furthermore, on average 70% of the sample did not reach their targets over time for the delivery of bulk water supply (see Figures 2.11 and 2.12). Much of the overall sample variability is due to the multi-purpose, single-purpose distinction in conjunction with reservoir size. The data suggests that all delivery of bulk water in excess of planned targets can be ascribed to multi-purpose dams. This is contrary to performance trends for other purposes where single purpose dams generally come out better. Regarding reservoir size, it is clear from the survey that the smaller the reservoir area, the closer to target results have been – with the exception of the 11 reservoirs larger than 100 square kilometres. These very large reservoirs exhibit extreme variability, ranging from under-performance to considerable over-performance and delivering maximum achieved bulk water supplies up to 2.5 times planned targets.

The principal message with regard to water supply emerging from the Case Studies is that even when it is not planned, demand for water supply from dams built for other purposes emerges over time. In the case of

Tarbela, water from the river system downstream of the dam is diverted through irrigation canals to Karachi to supplement other sources of municipal water supply. In the case of the Aslantas, the growth of local districts has prompted an application to the authorities for the supply of over 400 million litres per day from the reservoir.

The Cross-Check sample showed a tendency towards under utilisation of capacity that is reflected more generally in a 1994 post evaluation synthesis study of 31 water supply and sanitation projects by ADB. The utilisation of capacity varied from 33–80%. Apart from the fact that evaluations were carried out at early stages in the life of the projects, a number of causes were suggested for low rates of utilisation, including lower

Figure 2.11 Project averages for actual versus planned bulk water supply delivery

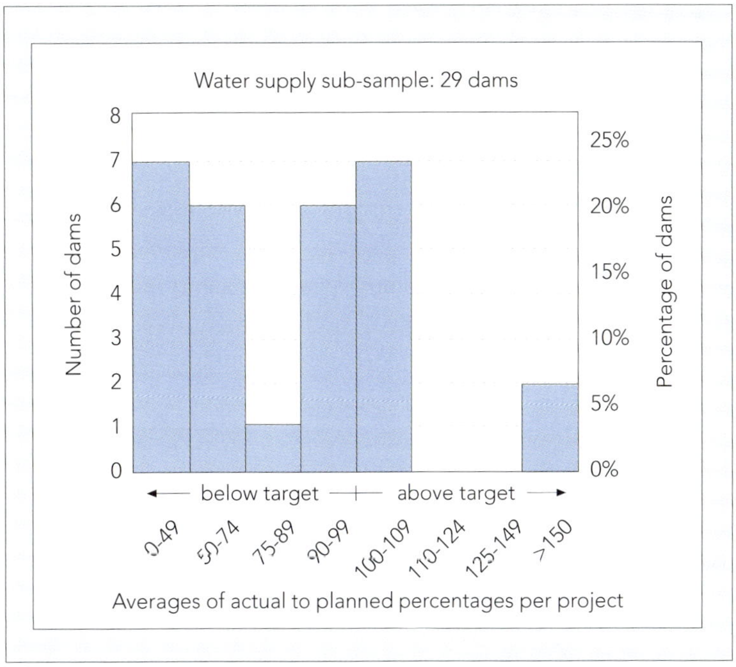

Source: WCD Cross-Check Survey.

Figure 2.12 Actual versus planned bulk water supply delivery over time

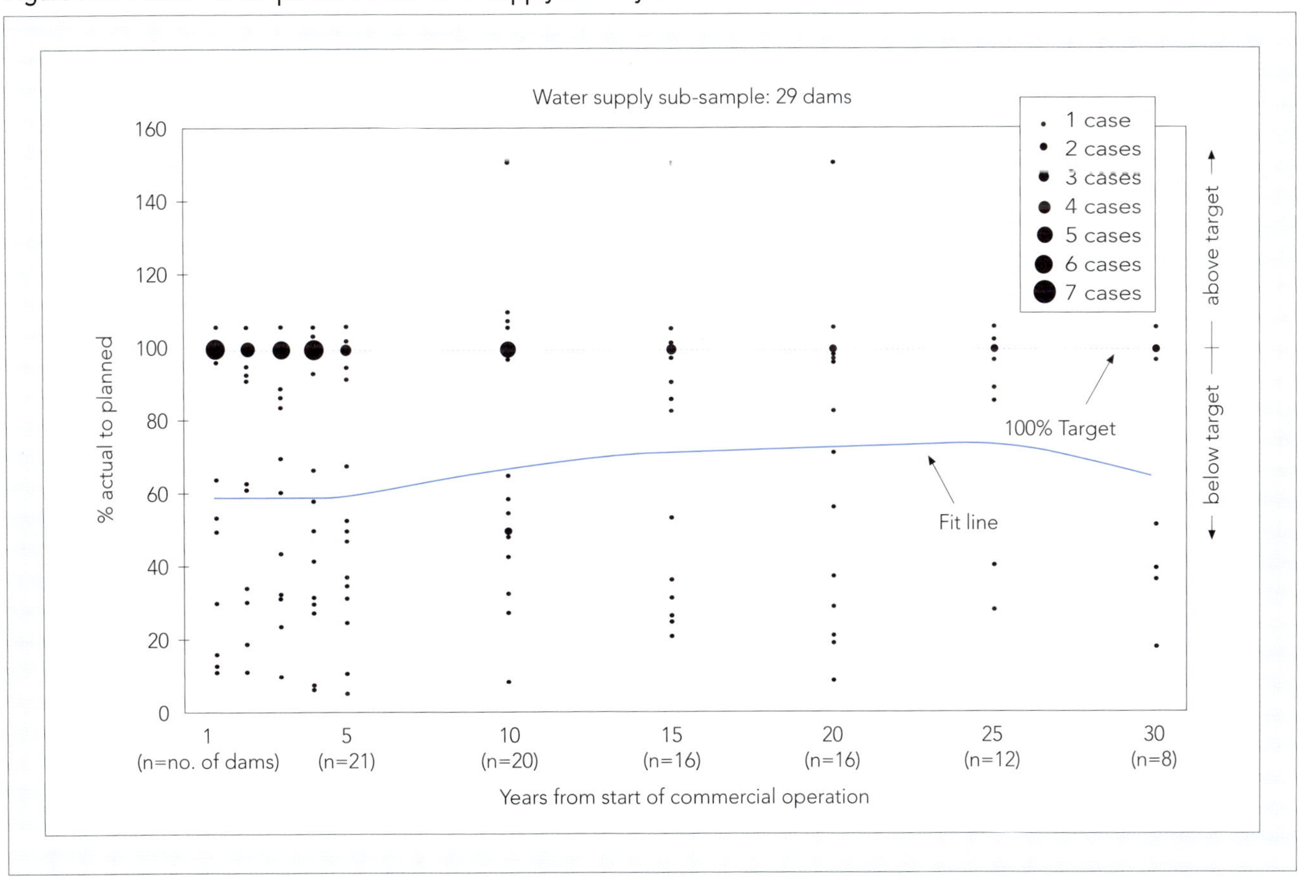

Source: WCD Cross-Check Survey.

At current rates, water fees are rarely sufficient to recover both capital and recurrent costs for water supply systems in many developing countries.

than expected growth in both population and in per capita consumption.[41]

Financial and economic profitability

Single-purpose water supply dams in the WCD Knowledge Base had poor performance in financial and economic terms. Examination of appraisal and evaluation figures for four World Bank and ADB water supply dams show that three of them dropped from EIRRs above 10% to well below that. As a whole, the EIRRs of the group fell by over 6%.

Poor financial and economic performance for large dams roughly approximates that for the sector as a whole. A sector synthesis of post-evaluation findings from 20 ADB water supply and sanitation projects found that 18 projects had actual FIRRs that failed to meet projected FIRRs and 17 of these were below 10%.[42] These types of sectoral results are confirmed by other sources. A similar World Bank analysis found that almost all of 129 water supply and sewerage projects reviewed had EIRRs below 10%.[43]

Growing concern over the cost and effectiveness of large dams and related structural measures as long-term responses to floods has led to support for integrated flood management as opposed to flood control.

At current rates, water fees are rarely sufficient to recover both capital and recurrent costs for water supply systems in many developing countries. Average tariffs in an ADB survey ranged from very low in Calcutta ($0.01 per cubic metre) to $0.66 per cubic metre in Cebu, Philippines. Thirty five of the 50 utilities in the survey covered O&M costs through their tariff collections. The ratio of billings to O&M costs for 37 utilities increased from 1.03 to 1.12 during 1991 to 1995, indicating an increase in financial viability for recurrent costs.[44]

Still, a number of studies by the World Bank have demonstrated that people, even those who are less well off in developing countries, are often willing to pay for improved water supply services.[45] For example, a study in Nigeria demonstrated that on an annual basis households pay water vendors over twice the O&M costs of a piped distribution system.[46]

Flood Control Dams

For centuries societies have built levees and embankments along riverbanks to contain and control the effects of floods. The purpose has been to occupy floodplains for agricultural, urban, and industrial uses and to reduce any resulting threat to lives and property. The WCD Knowledge Base highlights two very different perspectives on the past performance of dams in this regard. The first is a narrow focus on the role of dams in flood control and the second is a broader, more integrated approach to flood management as an objective. Evidence in the Knowledge Base confirms that while dams have provided important flood control benefits, some dams have increased the vulnerability of riverine communities to floods. Growing concern over the cost and effectiveness of large dams and related structural measures as long-term responses to floods has led to support for integrated flood management as opposed to flood control.

Flood control benefits

Large dams are used to control floods by storing all or a portion of the flood waters in the reservoir and then releasing the water slowly over time. Typically, the principal use of such dams is to store a portion of the flood in order to delay or manage when the peak occurs. This minimises the chance of coincident peaks from floods in different tributaries arriving at the same time in the main-stem of the river, reducing the probability of breaching dykes and overwhelming other flood defences. The main performance parameter in assessing flood control benefits is therefore the extent of reduction of the flood peak. Indicators of the benefits derived from flood control include reductions in the area flooded and prevention of any consequent loss of life, social disruption, health impacts and property and economic losses.

The Aswan High dam is an example of a dam that stores the flood. It can store 1.5 times the average annual flow of the Nile River and has provided a high degree of protection to the lower Nile simply by retaining the whole flood. At the same time the beneficial aspects of natural flooding – for example restoring the fertility of the floodplain – have been lost, a point returned to below and again in Chapter 3.[47]

Four of the WCD Case Studies provide other additional examples of flood control, although none of these dams were built primarily for this purpose. The Tarbela dam regulates about 16% of the annual flow of the Indus River. Analysis in the Case Study suggests that the early season flood peak in the Indus was reduced by 20%, however there was little reduction in the downstream flow during the late season flood event of 1992. Similarly, the reservoirs in the Glomma and Laagen basins in Norway regulate about 16% of the flow in that basin, where 20% mitigation in the peak flood level has also been achieved. A major flood event in the Glomma and Laagen basin in 1995 prompted a Royal Commission of enquiry. Extensive studies undertaken at that time confirmed the operation of the reservoirs reduced the flood peak by 2 meters, leading to a corresponding reduction in flood emergency relief and compensation payments required by the state.

Large dams in Japan have dramatically reduced the sudden arrival of floods in populated areas where the rivers are exceptionally steep and short, and susceptible to flash floods.

For Grand Coulee flood control was not an initial objective and the inability of the dam to control the damaging 1948 flood led to the construction of more reservoirs in Canada to store snowmelt. Although a precise breakdown of the individual contribution of the Grand Coulee is hard to isolate because a whole cascade of dams contribute to flood control in the Columbia basin, estimates indicate its contribution may be in the order of $20 million a year.

Other examples in the Knowledge Base provide similar indications of the performance of dams in reducing peak flood levels. For example, during the monsoon period in 1995, the Nam Ngum River experienced a 50-year flood three times in the same season. The reservoir absorbed the first two but was then full. When the third flood occurred the spillway gates were opened and

Box 2.7 Flood protection in Japan

Japan is one of the top five dam-building countries in the world, and flood control is the major purpose for many dams in Japan. An estimated 50% of Japan's population lives in flood-prone areas, and floods have affected 80% of municipalities in the last 10 years. The Chikugo River in the Kyushu District in southern Japan has a main river channel of 143 km and a catchment area of 2 860 km^2. A flood in 1953 caused dyke breach in many places, causing great damage in the area (147 people dead, approximately 74 000 houses flooded, and one-fifth of the catchment inundated). Matsubara and Shimouke Dams were subsequently built for flood control and power generation. The dams have performed well; for example, during a flood in 1982 the Matsubara dam reduced the peak river discharge 64% from a high of 2 900 to 1 040 m^{3}s.

Source: Berga, 2000, Contributing paper to Thematic IV.4 Flood Management Options; Takeuchi and Harada, 1999: p4.

The India Case Study shows the potential conflict between keeping space in the reservoir for flood control and storing as much water as possible for hydropower and irrigation.

the water level rose above the full supply level. Due to the retention effect of the reservoir, the peak of the third flood was reduced by 20%.[48] Large dams in Japan have dramatically reduced the sudden arrival of floods in populated areas where the rivers are exceptionally steep and short, and susceptible to flash floods (see Box 2.8). Delay in flooding provides time for public warning and evacuation where necessary.

Limitations of flood control operations

There are also clearly problems that emerge from the operation of large dams for flood control. Some dams have increased the vulnerability of riverine communities to floods. For example:

- While rare, dam breaks have and do occur and usually during exceptional storms; when they do, communities downstream are subject to extreme floods amplified by the dam break.
- Significant downstream damage to communities has resulted where reservoirs have not been operated properly in times of emergency or rapidly developing situations or when floodgates have mechanical failures at critical times. Often communities have adapted to the level of protection normally provided and the contingency plans – or their implementation – have been inadequate.

There have also been cases where peaking operation of hydropower stations has caused an unexpected surge of water in the river; lives have been lost when measures to warn populations downstream have not been effective or heeded. Local flooding can be caused in a similar fashion when gates are opened to release water at peak rainfall periods. One such serious incident was reported in Nigeria where a delay in warning inhabitants led to a flood that overran approximately 200 communities, submerging 1 500 houses and killing over 1 000 people.[49]

As shown in the WCD Knowledge Base, flood management may only be needed a few days or weeks in any particular year. Thus large dams used for flood control virtually always have another function, such as power generation or irrigation. The India Case Study portrays the potential for conflict between flood control objectives for operation of the reservoir (where storage space in the reservoir is required) and hydropower and irrigation (where it is desirable to store as much water as possible). According to the Case Study most of the complaints about dams aggravating floods downstream stem from this situation. The Case Study goes on to document the lack of co-ordination or real-time information exchange between the upstream Tenughat reservoir and the operation agencies of the downstream Damodar Valley Corporation, which put the downstream river reach and reservoirs at risk.

Another aspect is that with climate variations, the frequency, duration and intensity of storm events that lead to floods appear to be changing. If so, there is a risk that a changing climate will modify the hydrological basis on which many flood control dams were designed. This raises concerns about the physical adequacy of many dams to perform their flood management functions, as well as the adequacy of spillways to handle higher flood volumes likely in a changed climate. United States dams are being reviewed for their flood design and in some cases the spillway capacity is being increased.[50] The technical literature on dams is also increasingly focusing on the role of dams in flood management, looking at both the safety considerations and the means to improve the flood performance.[51]

From flood control to integrated flood management

The Knowledge Base identifies a series of concerns over the effectiveness of dams and related structural measures as an element in a broader flood management strategy including the realisation:

- that dams have encouraged settlement in areas that are still subject to floods that exceed the maximum design flood;
- that the costs of ensuring complete protection against all floods are exceptionally high;
- that the effectiveness of structural measures are reduced over time due to the accumulation of sediment in river beds and reservoirs; and
- that floods have many beneficial uses in different river basins and that the elimination or reduction of natural flooding has led to the loss of important downstream ecosystem functions, as well as loss of livelihood for flood-dependent communities (see next two chapters on environmental and social performance of large dams).

These concerns indicate the difficulty of fully controlling floods and of managing the relationship between floods and people. This underpins the shift to an emphasis on integrated flood management – that is the need to set objectives in terms of predicting, managing and responding to floods – rather than simply in terms of flood control.

Large dams with a flood control component may provide an increased feeling of security, leading to settlement of flood-prone areas. When the exceptional flood finally arrives, there are more people and higher-value property at risk than there otherwise would have been. Damages may therefore be larger than if floods had continued to be normal events within the range of regular experience and awareness.

In the Nam Ngum case referred to earlier, the dam was blamed for a major inundation of downstream agricultural areas despite its alleviation of two earlier peaks and reduction of the third peak. Since the reservoir had not spilled for many years people had developed a false sense of security and drains were not maintained. In the end the floodwaters only dispersed slowly and standing crops did not survive. Another example comes from Poland where the 1997 floods covered an area only half as large as that in 1934 yet three times as many buildings were

Box 2.8 From flood control to flood management in the United States

Between 1960 and 1985, the United States Federal government spent $38 billion on flood control, mostly on structural responses such as large dams. Yet average annual flood damage, adjusted for inflation, continued to increase – more than doubling. Average flood damage in the United States, adjusted for inflation, was as follows:

1903–33:	$1.7 billion
1934–63:	$2.8 billion
1964–93:	$4.6 billion
1994–97:	$5.1 billion

The United States Army Corps of Engineers points out that its dams and 8 500 miles of levees have saved $387 billion in damages since 1928, but it has no figures for damages in areas where the Corps' projects encouraged development that was later inundated. After the 1993 Mississippi flood, congressional testimony pointed to dams and levees as exacerbating the problem. Subsequently, the Corps has called for more consideration of non-structural flood management methods, including restoration of wetlands and riparian habitat, limits on development in floodplains, and farm policy that discourages conversion of wetlands to cropland. Likewise, communities from Rapid City, South Dakota, to Valmeyer, Illinois, to St. Charles, Missouri, to Napa, California, have opted to pursue non-structural approaches to alleviate recurring floods.

Source: Schildgen, 1999

flooded, 38 times as many bridges and 134 times more kilometres of road.[52]

The high cost of flood control and flood damage in the United States, as well as the emerging shift from a reliance on flood control to an approach based on flood management is described in Box 2.8. In many countries reliance on structural measures, including dykes, leads to a need to continually invest in additional measures as sedimentation decreases their effectiveness over time. For centuries, dykes in Vietnam have been progressively increased in height as the river bed gradually rises due to accumulating sediments deposited by floods.[53] The same situation prevails in China, where after centuries of raising dykes, the height of the dykes at places on the Yangtze River is over 16 metres above the floodplain.[54]

Multi-Purpose Dams

Many large dams fulfil a number of purposes with a single facility. Multi-purpose projects in the WCD Knowledge Base display many of the same performance shortfalls experienced by single-purpose projects and, in a number of cases, achieved less relative to targets than their single-purpose counterparts.

The WCD Cross-Check Survey shows that multi-purpose dams have had a high degree of variability in achieving physical targets across most benefit streams. As indicated earlier, single-purpose projects in the Cross-Check sub-sample tend to cluster closer to planned targets for project schedule, hydropower, and water supply performance than multi-purpose dams. The exceptions were irrigation projects which showed little difference between single and multi-purpose projects in variability of performance.

The Cross-Check Survey also suggests that multi-purpose projects have higher cost overruns and higher variability in these overruns than single-purpose projects. A small sample of 12 multi-purpose projects funded by the World Bank, AfDB, and ADB examined by the Commission indicates that estimates of the EIRR at evaluation were about 4% below those projected at appraisal. This figure masks large variability, with four of the projects moving up and down by almost 10%.[55] In the AfDB's review of four projects, only one was financially and economically viable.[56]

As noted in previous sections, hydropower projects tend to perform relatively well in financial terms while irrigation projects typically fail to recover O&M and capital costs. In practice this has often led to the use of hydropower facilities in conjunction with a project designed for irrigation as a way to cover the costs of the irrigation facility. The WCD Case Study of Grand Coulee dam and the Columbia Basin Project provides an illustrative example of the cross-subsidies that often result from such arrangements (see Box 2.9).

Box 2.9 Cost recovery in a multi-purpose scheme: Grand Coulee and the Columbia Basin Project

In the case of the Grand Coulee dam and the accompanying Columbia Basin Project, the intention was for hydropower revenue to subsidise irrigation capital costs. The Bonneville Power Administration (BPA), which is responsible for GCD, is charged with paying the United States Treasury for the costs associated with hydropower development (the dam and three powerhouses) plus the share of allocated irrigation costs not paid by irrigators. These amounts are calculated in nominal dollars but include interest. As of 1998, the total capital costs for GCD and CBP came to $1.93 billion. Of this, the BPA has already repaid the hydropower costs of $1.1 billion.

The share of the remaining costs allocated to irrigation is $674 million. The last portion is for non-reimbursable capital costs such as flood control, as covered directly by the United States Treasury. Of the irrigation share, BPA is responsible for 87% ($585 million) and the irrigators are to cover the remaining 13% ($89 million). As of 1998, irrigators had paid in $51 million (in dollars uncorrected for inflation). BPA is scheduled to make payments on its share of the irrigation costs during 2009–45. As hydropower revenues from the sale of GCD power come close to $500 million, it is clear that the project could easily cover costs under such terms.

Source: WCD Grand Coulee Case Study.

In other cases, adding hydropower to an irrigation facility is simply a way of increasing the overall economic profitability of a scheme. The design for the Aslantas Dam proceeded on this basis. Analysis of the irrigation component as a stand-alone project indicated a 13% rate of return. A stand-alone hydropower project was judged not a least-cost source of power, but when analysed in terms of just the incremental costs and benefits of being built into an irrigation facility, the hydropower component yielded a 15.7% rate of return. Combined, the rate of return for the multi-purpose project came to 13.4%. Following a cost overrun in nominal terms of 37%, the project completion team reported a recalculated EIRR for the project in 1985, including flood control, of 13.6%, although the validity of the irrigation benefit calculations was questioned by a subsequent post-project audit report.[57]

In summary, these emerging trends and patterns of higher variability and lower average performance of multi-purpose versus single-purpose projects are not surprising. While single-purpose dams are designed for optimal delivery of a particular targeted benefit, multi-purpose projects are designed for sub-optimal outputs of all intended benefits. They aim to maximise economic efficiency achieved through shared costs and infrastructure of the proposed scheme. In doing so, multi-purpose schemes are inherently more complex, and many experience operational conflicts that contribute to under-performance on financial and economic targets.

What emerges from the WCD Knowledge Base is that these sub-optimal targets set for multi-purpose projects were still not achieved to the level desired. This suggests that the extent to which conflict arising from multi-benefit operation will affect performance is probably under-estimated.

Physical Sustainability Issues

Many factors affect the physical sustainability of the benefits and services provided by dams. The following section provides brief findings from the Knowledge Base on three of these issues: dam safety, sedimentation and waterlogging and salinity.

Dam safety

Dam failure is defined by the International Commission on Large Dams (ICOLD) as the 'collapse or movement of part of a dam or its foundation, so that the dam cannot retain water.'[58] In general, a failure results in the release of large quantities of water, posing serious risks for the people or property downstream. The findings of a recent global compilation of information about the failure of dams by ICOLD are as follows:

Multi-purpose schemes are inherently more complex, and many experience operational conflicts that contribute to under-performance on financial and economic targets.

- The failure rate of large dams has been falling over the last four decades. Of dams built before 1950, 2.2% failed, while the failure rate of dams built since 1951 is less than 0.5%.
- The proportion of dams failing varies little with the height of the dam and so most failures involve small dams.
- Most failures involve newly built dams. Some 70% of failures occur in the first ten years of life of the dam and proportionately more during the first year after commissioning.
- The highest failure rate is found in dams built in the ten years 1910-1920.
- Foundation problems are the most common cause of failure in concrete dams, with internal erosion and insuffi-

cient shear strength of the foundation each accounting for 21% of failures.

- The most common cause of failure of earth and rockfill dams is overtopping (31% as primary cause and 18% as secondary cause). This is followed by internal erosion in the body of the dam (15% as primary cause and 13% as secondary cause) and in the foundation (12% as primary cause and 5% as secondary cause).
- With masonry dams, the most common cause is overtopping (43%) followed by internal erosion in the foundation (29%).
- Where other works were the seat of the failure, the most common cause was inadequate spillway capacity (22% as primary cause and 30% as secondary cause).
- The post-failure action most frequently reported was scheme abandoned (36%), construction of a newly designed dam (19%) and overall reconstruction with the same design (16%).[59]

Box 2.10 Dam safety in the United States

The American Society of Civil Engineers (ASCE) gave dams in the United States a poor grade (a 'D') in their '1998 Report Card for America's Infrastructure' – citing age, downstream development, dam abandonment, and lack of funding for dam safety programmes as major unaddressed problems.

The Association of State Dam Safety Officials (ASDO) concluded that:

- Current levels of dam safety expenditures are insufficient. There are instances when dam safety items are given a lower priority than water or power delivery.
- Some $40 billion is needed to maintain and improve current dams.
- Little if any information is provided to downstream populations at risk; less than 10% of jurisdictions surveyed had warning or evacuation plans.
- In many cases irrigation districts – which are required by law to share dam safety expenses – withheld funds to pay for necessary safety items. Continued deferment could result in unacceptable public risks.
- Maintenance costs continue to escalate because of the ageing structures. With constant or declining maintenance funds from federal and irrigation district sources, new sources of revenue are needed.

According to the United States Federal Emergency Management Agency:

- Dam breaks are rare and there have been few deaths in recent years. There have been 1 449 dam failures over the past 150 years in the United States. The annual dam failure rate was 29 in 1996 and 1997.
- In massive floods like Hurricane Floyd's inundation of North Carolina in 1999, 36 dams failed but most were small and claimed no lives.
- Failure of the Buffalo Creek dam in West Virginia in 1972 killed 125 people, while failure of the Teton dam in 1976 killed 11 and of the Kelly Barnes dam in Georgia killed 39.
- Of the 80 000 small and large dams in the United States, 9 326 are rated as 'high hazard', meaning that if they should fail, loss of life and serious property damage would result. Some 1 600 significant hazard dams are within one mile of a downstream city.
- Less than 40% of high hazard dams have an emergency action plan for people to follow.

Sources: ASCE 1998; Knudsen and Vogel, 1997; Schmid, 2000; ASDSO 2000.

When most large dam projects are built, the assumption is that river flows in the future (total runoff and severe floods) will be much like those in the past. In some cases the historical time series of hydrological data is too short and may not reflect cyclical phenomena. Climate change has introduced another level of uncertainty about changing flows within the life span of most dams. The safety of large dams is affected by changes in the magnitude or frequency of extreme precipitation events. These changes are highly uncertain, but climate change is expected to lead (and perhaps already has led) to larger and more frequent extreme precipitation events. One of the first studies in this area concluded that the discharge of the 50-year flood on the River Severn, in the United Kingdom, may increase by around 20% by 2050.[60] There is concern whether existing spillways can evacuate such floods in future.

The WCD Cross-Check Survey shows a trend towards increased attention to the assessment of dam safety, although around 20% of dams in the sub-sample that were built in the last three decades do not report undertaking a safety assessment (see Figure 2.13). A report card on the United States experience with an ageing stock of dams provides a sobering indication of the importance of dam safety concerns (see Box 2.10). A key element in keeping dams safe is providing finance for proper and regular maintenance work. A recent Ontario Hydro Study of several hundred North American dams indicates that, on average, hydropower operating costs rise significantly after 25 to 35 years of operation due to the increasing need for repairs.[61]

Figure 2.13 Trends in dam safety assessments

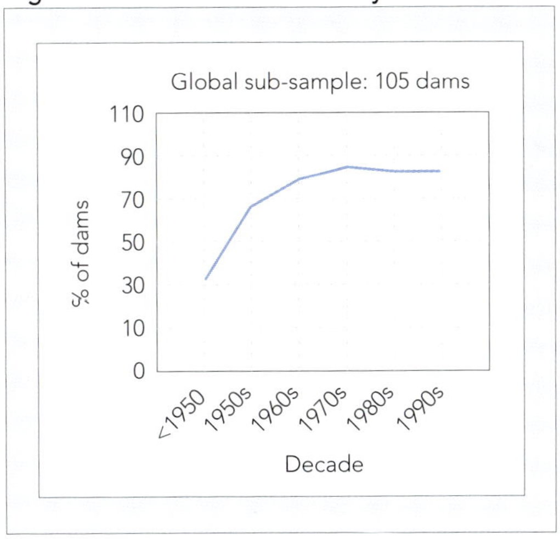

Source: WCD Cross-Check Survey.

Sedimentation

Many reservoirs are subject to some degree of sediment inflow and deposition. It is estimated that some 0.5–1% of the world reservoir volume is lost from sedimentation annually.[62] Sedimentation of the active storage affects physical and economic performance, but only where design storage is more or less fully used.[63] Sediment may also cause erosion of turbines if it reaches power intakes. Eventually, sedimentation will affect project life by silting up the dead storage, leading to intake blockage.

The WCD Knowledge Base indicates that while sedimentation potentially undermines the performance of a large dam project, the conditions – and therefore the frequency of occurrence of this phenomenon – are project and site-specific. For example,

Figure 2.14 Loss of active storage due to sedimentation

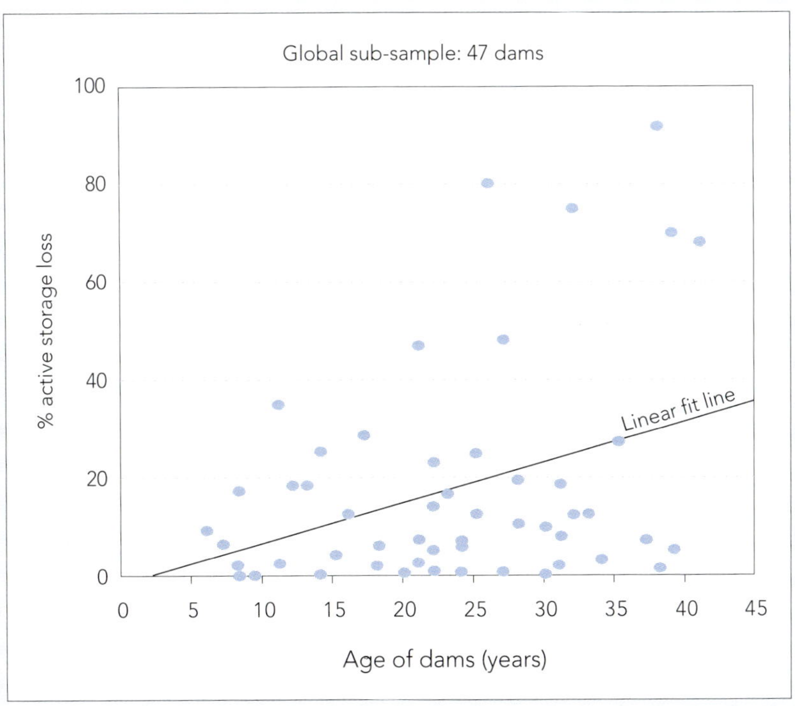

Source: WCD Cross-Check Survey.

Figure 2.15 Loss of active storage due to sedimentation by reach of river

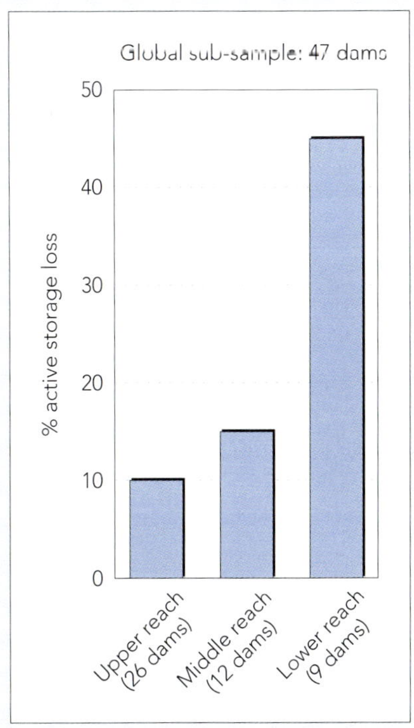

Source: WCD Cross-Check Survey.

higher sedimentation rates are observed for smaller dams and for dams located in the lower reaches of rivers.

Analysis of the Cross-Check Survey shows that more than 50% of active storage was lost due to sediment deposition for 10% of projects in the sample, all of which were in operation for at least 25 years (see Figure 2.14). But there is great variability within this average global figure (see Figure 2.15). In the sample, the average loss of active storage was greatest for reservoirs in the lower reach of rivers.

Of the WCD Case Study dams, only Tarbela faces a sedimentation problem. In this case, the reservoir has lost 18% of live storage after 25 years. Although this is less than predicted, the loss of storage capacity reduces the dam's capacity to store water for irrigation; in addition, the build-up of sediment close to the dam is threatening operation long before the end of its design life. Conversely, sedimentation is not an issue in the temperate dams in the Glomma and Laagen basin and Grand Coulee, where sediment concentrations are negligible.

A survey of 547 U.S. dams found that:

- one-quarter of U.S. hydropower reservoirs have some form of sedimentation problem;
- 15% of owners and operators consider these problems serious; and
- the frequency and perception of sedimentation as a problem is higher for smaller-volume reservoirs.[64]

Other studies confirm a number of these findings. Data from 42 dams in Morocco indicate storage is depleting at the rate of 1.1% for reservoirs with storage volumes less than 500 m^3 and 0.6% for larger dams.[65]

Waterlogging and salinity

Salinisation of agricultural land is caused by the rise of groundwater brought about by surface irrigation and is related to the problem of waterlogging. When water tables are close to the surface, capillary action draws salts naturally occurring in the soil profile to the surface. Even where groundwater use is controlled, secondary salinisation can occur due to poor-quality groundwater. Salinisation reduces yields of crops that do not tolerate high salinity levels to the point of eventually rendering the land unproductive. In India, the yields of rice and wheat on salt-affected land were approximately half those on unaffected land.[66]

The WCD Knowledge Base indicates that problems of waterlogging and salinity for irrigation systems have reached serious levels globally and have severe, long-term and often permanent impacts on land, agriculture and livelihoods. In the Cross-Check Survey approximately one-fifth of the large dam projects with an irrigation component reported impacts from waterlogging. Data from 11 major irrigation countries indicate that approximately 20% of irrigated land is affected by salinity.[67] But the variation across countries in the share of irrigated land affected by salt is also large, ranging from 15% in China and 33% in Egypt to 80% in Turkmenistan (see Figure 2.16).

The phenomena of waterlogging and salinity are not new. In the case of Grand Coulee dam, the unexpected rise in groundwater levels was recognised in the early 1950s, requiring considerable additional expenditure on drainage to control increased water levels. To date, 7 300 ha have been taken out of production through a government-sponsored set-aside programme. In the Indus Basin, 38% of the irrigation system is classified as waterlogged and production is estimated to be 25% lower than potential as a result of salinity. Major engineering works to remove saline effluent have recently been implemented, but it is too early to evaluate the effectiveness of this approach. In justifying a major drainage project in Pakistan, the World Bank ascribed a production decline of approximately 25% to salinity and waterlogging, with specific cases reaching 40–60%.[68]

Although the need for drainage has been evident for some time, proponents have often omitted the necessary infrastructure from project plans.[69] Drainage facilities have been difficult to justify at the outset of a project under prevailing economic analysis as the main benefit of drainage is realised only after some time, namely when the groundwater levels have risen close to the surface. Regardless of the timing of actual investment in drainage facilities, the exclusion of drainage facilities from the original project design – only to require remedial action later – may lead to the over-estimation of project net benefits. Resolving waterlogging and salinity problems entails significant rehabilitation costs (underestimation of project costs) and loss of productivity over time (over-estimation of benefits).

Data from 11 major irrigation countries indicate that approximately 20% of irrigated land is affected by salinity which makes land increasingly unproductive.

A further difficulty is that predictions may underestimate the time required for such problems to appear. One cause of this problem is over-irrigation. For example, in the Chashma project in Pakistan there was a shift to more water-demanding crops such as rice and sugarcane, and excessive irrigation during the early stages of project development when water was abundant. As a result, water tables rose more quickly than expected leading to the need to invest in drainage works at an earlier date than anticipated.[70]

Figure 2.16 Waterlogging and salinity

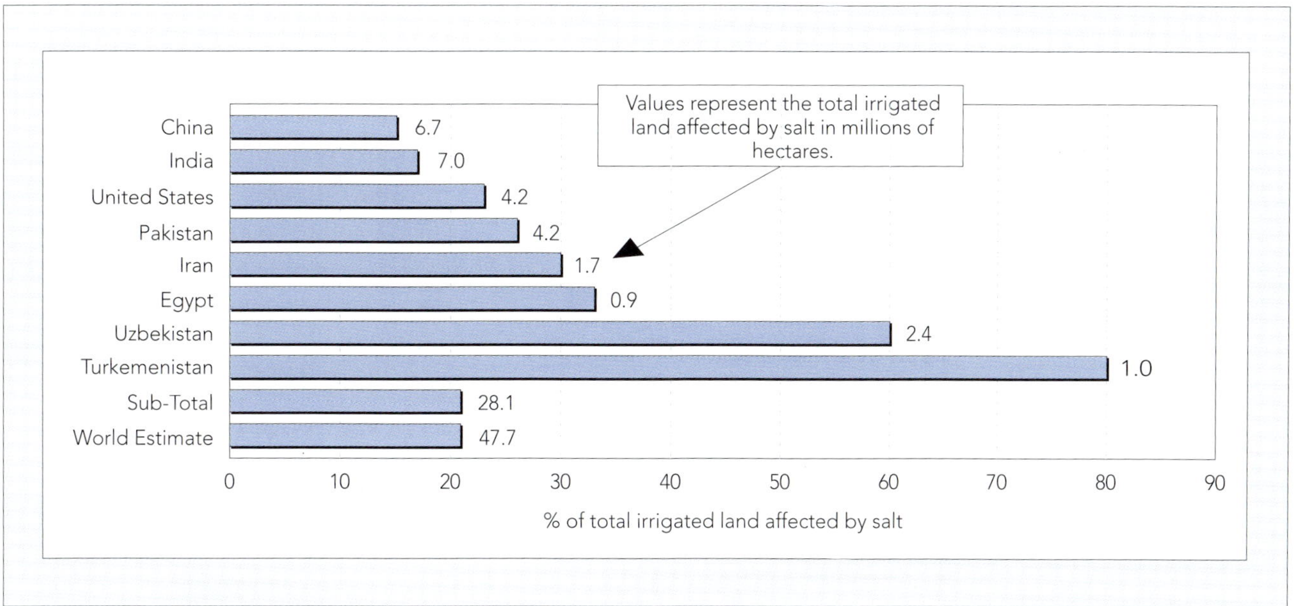

Source: Postel, 1999.

Once drainage is in place, recurrent costs of operation and maintenance are often not recovered, leading to early degradation and a reduction in the effectiveness of drainage systems. A report by the International Water Management Institute notes that despite improved understanding of the process involved, the area adversely affected by waterlogging and salinity is increasing at a rate faster than reclamation, pointing to conflicting sector interests and priorities.[71]

Findings and Lessons

The degree to which large dams in the WCD Knowledge Base have delivered services and net benefits as planned varies substantially from one project to the next with a considerable portion falling short of physical and economic targets. In spite of this, the services produced by dams are considerable – in the order of 12-16% of world food production, 19% of world electricity supply amongst others, as presented in Chapter 1. In addition, the WCD Knowledge Base confirms the longevity of large dams, as many continue generating benefits – even if less than planned – after 30 to 40 years of operation.

A sectoral review of technical, financial and economic performance suggests that of the dams in the Knowledge Base, those:

- designed to deliver irrigation services have typically fallen short of physical targets, did not recover their costs and have been less profitable in economic terms than expected;
- built to deliver hydropower tend to perform close to but still below targets for power generation, generally meet their financial targets but demonstrate variable economic performance relative to targets, and include a number of notable under- and over-performers;
- built for municipal and industrial water supply have generally fallen short of intended targets for timing and delivery of bulk water supply and have exhibited poor financial cost recovery and economic performance;
- with a flood control component have provided important benefits in this regard, but at the same time have led to an increased vulnerability to flood hazards due to increased settlement in areas still at risk from floods, and in some cases have worsened flood damages for a number of reasons, including poor operation of dams.
- which serve a number of purposes also under-achieve relative to targets, in some cases exceeding the shortfalls registered by single purpose projects, demonstrating that the targets were often over-optimistic.

The review of performance suggests two further findings:

- large dams in the Knowledge Base have a marked tendency towards schedule delays and significant cost overruns; and
- growing concern over the cost and effectiveness of large dams and associated structural measures have led to the adoption of integrated flood management that emphasises a mix of policy and non-structural measures to reduce the vulnerability of communities to flooding.

The review also examines threats to the physical sustainability of large dams and their benefit flows. Based on the extent and nature of these threats the findings suggest:

- Ensuring the safety of dams will require increasing attention and investment as the stock of dams ages, maintenance costs rise and climate change alters the hydrological regime used as a basis for the design of dam spillways.

- Sedimentation and the consequent long-term loss of storage is a serious concern globally and the effects will be particularly felt by basins with high geological or human-induced erosion rates, dams in the lower reaches of rivers and dams with smaller storage volumes.
- Waterlogging and salinity affect one-fifth of irrigation land globally – including land irrigated by large dams – and have severe, long-term and often permanent impacts on land, agriculture and livelihoods where rehabilitation is not undertaken.

Using the information on the performance of large dams collected in the WCD Knowledge Base this chapter shows that there is considerable scope for improving the selection of projects and the operation of existing large dams and their associated infrastructure. Considering the vast amounts of capital invested in large dams, substantive evaluations of project performance are few in number, narrow in scope and poorly integrated across impact categories and scales. The resounding message is that we need better and continued monitoring of technical, financial and economic performance. The next chapters turn to the environmental and social dimensions of large dams.

Endnotes

1 The emphasis here is on the direct capital costs of a project. Direct costs are those costs that are incurred by public or private project owners as part of investment and operations, and not those that remain outside the owner's perspective (the latter being external costs).

2 WCD Thematic III.1 Economic Analysis, Chapter 2.

3 Bacon et al, 1996, p7-8.

4 WCD Thematic III.1 Economic Analysis, Chapter 2; IDB, 1999.

5 Lagman, 2000, in WCD Thematic III.1 Economic Analysis, Chapter 2.

6 In nominal US dollar terms. Of the 35 dams in the IRN list, 11 were already accounted for in the World Bank and IDB samples and 10 had incomplete information to calculate the nominal cost overruns in dollars. WCD Thematic III.1 Economic Analysis; McCully, 1999, eco061, WCD Submission.

7 The India Case Study reports on a series of historical reports of overruns in India, including a 1983 study of 159 projects showing 232% overruns. These overruns are calculated in local currency units not US dollars and, thus, follow a different method then the other studies reported in the text.

8 Bacon et al, 1996, p30; Bacon and Besant-Jones, 1998, p321.

9 WCD Thematic III.1 Economic Analysis, based on OED, 1996a, p57-67.

10 AfDB, 1998, in WCD Thematic III.1 Economic Analysis, Chapter 2.

11 Young, 2000, eco066, WCD Submission, p3, suggests that US Bureau of Reclamation projects cost on average three times as much as is planned.

12 Financial costs of projects are determined by projecting quantities of goods and services at current year prices and then applying a price escalation factor to account for expected inflation. When actual inflation exceeds expected inflation, this contributes to overruns in terms of cash flows (as against original budgets) but it does not necessarily alter the underlying real cost of the goods and services as employed in economic cost-benefit analysis. The real price of goods and services used in construction - and not the general price level - must increase at a higher than expected rate for the real costs to 'overrun'.

13 Delays due to difficulty in reaching financial closure do not add to interest during construction changes.

14 Bacon et al, 1996, p30.

15 Note that although regional development is often an objective of irrigation projects, the Commission has not found cases where it is made an explicit objective with firm performance targets. Such benefits are also inherently distributional in nature and are treated in the later section on this topic.

16 Actual command area figures are directly comparable with ultimate command area targets in the Survey sub-sample as all projects had passed the planed time frame for achieving full command area development.

17 Cropping intensity describes the extent of land utilisation in a year and reflects the degree of multiple cropping. It is the ratio of total area cropped per year to the irrigation command area.

18 ADB, 1995, p5.

19 WCD Thematic IV.2 Irrigation Options, Table 3.13, Section 3.4.2.

20 OED, 1990, p4-2.

21 Barker and Dave, in print.

22 World Bank Commodities Prices Data.

23 OED, 1990, p v.

24 Eckstein, 1958, p2.

25 World Bank appraisals typically cite 10% as the 'hurdle rate' which a project must exceed to be deemed worth undertaking. The ADB only approves projects that have an EIRR over 12%, although in exceptional cases those with an EIRR of less than 12% but still more than 10% are approved; ADB, 1997, p37, in WCD Thematic Review III.1 Economic Analysis, Technical Annex I, p10-14.

26 The sample included all 13 ADB-funded large dams where both completion and evaluation reports exist and a sampling of 27 large dams funded by the World Bank, and eight projects funded by the African Devel-

opment Bank. For the full analysis see WCD Thematic Review III.1 Economic Analysis, Chapter 10.

27 The present value is less than the full cost due to the long development period and the effect of discounting.

28 This assumes that there was no general or sector-specific excess capacity in the economy. As the project began in 1945 it can only be said that the likelihood of consistent excess capacity in the US economy during the project period is low given the high growth rates observed during this period.

29 ADB, 1995, p7; OED, 1990, p4–2.

30 Thakkar, 1999, p17, Contributing paper to WCD Thematic IV.2 Irrigation Options.

31 WCD Thematic Review IV.2 Irrigation Options, Section 4.2.2.

32 Molden et al, 1998, p15.

33 Johnson, 1997, p28.

34 ADB, 2000.

35 NEA, 1997.

36 WCD Norway Case Study; WCD Grand Coulee Case Study.

37 Wong, 1994, in WCD Thematic Review III.1 Economic Analysis, Chapter 2.

38 McCully, 1997a, p5, in WCD Thematic Review III.1 Economic Analysis, Chapter 10.

39 AfDB, 1998, p6, in WCD Thematic Review III.1 Economic Analysis, Chapter 2.

40 WCD Thematic Review III.2 Financing Trends, Chapter 2.

41 ADB, 1994, p3.

42 ADB, 1994, p17.

43 OED, 1992, p1 Executive Summary.

44 McIntosh and Yñiguez, 1997, p16.

45 Whittington et al, 1991, p193; World Bank Water Demand Research Team, 1993, p47-52.

46 Whittington et al., 1991, p193.

47 Shalaby, 1999.

48 Oud, pers. comm. 2000.

49 Atakpu, 1999, WCD Regional Consultation Paper, p1.

50 Townshend, 2000, p81.

51 MacDonald and McInally, 1998, p183.

52 WCD Thematic Review IV.4 Flood Management Options, section 1.2.1.

53 World Bank et al, 1996, Annex 1, p51.

54 WCD China Country Study, Chapter 2.

55 WCD Thematic Review III.1 Economic Analysis, Chapter 10.

56 AfDB, 1998, p6.

57 WCD Aslantas Case Study; World Bank, 1987.

58 ICOLD, 1995.

59 ICOLD, 1995.

60 Tedd, 2000.

61 McCully 1997b, p2.

62 Mahmood, 1987.

63 Active storage is the reservoir volume above the outtake, whereas dead storage is the reservoir volume below the outtake.

64 Dixon, 2000.

65 Smith, 1999, p13.

66 Kijne et al, 1998, p26.

67 Heuperman, 1999, Contributing Paper to WCD Thematic Review IV.2 Irrigation Options, Section 4.2.

68 World Bank, 1997, p3.

69 Young, 2000 eco066, WCD Submission, p3.

70 ADB, 1984.

71 Kijne et al, 1998, p26.

Chapter 3:
Ecosystems and Large Dams:
Environmental Performance

The nature of the impacts of large dams on ecosystems is generally well known and scientists, NGOs and professional groups such as the International Commission on Large Dams (ICOLD), International Hydropower Association (IHA) and the International Energy Agency (IEA) have written extensively on them.[1] *A useful indicator of the scale of human intervention in this regard is a recent estimate that dams, inter-basin transfers, and water withdrawals for irrigation have fragmented 60% of the world's rivers.*[2]

Ecosystem impacts can be classified according to whether they are:

- first-order impacts that involve the physical, chemical, and geomorphological consequences of blocking a river and altering the natural distribution and timing of streamflow;
- second-order impacts that involve changes in primary biological productivity of ecosystems including effects on riverine and riparian plant-life and on downstream habitat such as wetlands; or
- third-order impacts that involve alterations to fauna (such as fish) caused by a first-order effect (such as blocking migration) or a second-order effect (such as decrease in the availability of plankton).

In addition, modifying the ecosystem changes the biochemical cycle in the natural riverine system. Reservoirs interrupt the downstream flow of organic carbon, leading to emissions of greenhouse gases such as methane and carbon dioxide that contribute to climate change.

The current state of knowledge indicates that large dams have many mostly negative impacts on ecosystems.[3] These impacts are complex, varied and often profound in nature. In many cases dams have led to the irreversible loss of species populations and ecosystems. Because the ecosystem impacts are many and complex it is hard to give a precise and detailed prediction of the changes likely to result from the construction of a dam or series of dams. Based on the geographical location of a dam and the natural river regime it is possible to give a broad indication of the type and direction of impacts with decreasing certainty, from first- to third-order impacts. To date efforts to counter the ecosystem impacts of large dams have had only limited success. This is due to limited efforts to understand the ecosystem and the scope and nature of impacts, the inadequate approach to assessing even anticipated impacts and the only partial success of minimisation, mitigation and compensation measures.

This chapter describes the nature of the impacts in general, supported by the Case Studies, the Cross-Check Survey and the Thematic Reviews on Ecosystems (II.1) and on Global Change (II.2). Given that the WCD Knowledge Base describes a large number of ecosystem impacts, the focus is on summarising these impacts by grouping them as follows:

- the impacts of reservoirs on terrestrial ecosystems and biodiversity;
- the emission of greenhouse gases associated with large dam projects and their reservoirs;
- the impacts of altered downstream flows on aquatic ecosystems and biodiversity;
- the impacts of altering the natural flood cycle on downstream floodplains;
- the impacts of dams on fisheries in the upstream, reservoir and downstream areas;
- the enhancement of ecosystems through reservoir creation and other means; and
- the cumulative impacts of a series of dams on a river system.

The extent to which efforts to reduce or eliminate these impacts were undertaken in the past is described at the end of each discussion. The final subsection contains a further assessment of past experience with efforts to avoid, mitigate, minimise or compensate these impacts. Current efforts to restore environmental function through decommissioning are also reviewed.

Terrestrial Ecosystems and Biodiversity

The construction of a storage dam and subsequent inundation of the reservoir area effectively kills terrestrial plants and forests and displaces animals. As many species prefer valley bottoms, large-scale impoundment may eliminate unique wildlife habitats and affect populations of endangered species.[4] Efforts to mitigate the impacts on fauna have met with little success (see Box 3.1). Construction of irrigation infrastructure may have similar impacts.

Flooding a reservoir may lead to the occupation and clearing of upstream catchment areas as replacement for land lost to the reservoir. Land use change provoked in this manner not only has direct effects in terms of habitat loss, elimination of flora and fauna and, in many cases, land degradation, but also feedback effects on the reservoir through alterations in hydrologic function. The resulting loss of vegetative cover leads to increases in sedimentation, stormflow, and annual water yield; decreases in water quality; and variable changes in the seasonal timing of water yield.[5]

Greenhouse Gas Emissions

The emission of greenhouse gases (GHG) from reservoirs due to rotting vegetation and carbon inflows from the catchment is a recently identified ecosystem impact (on climate) of storage dams.[6] A first estimate suggests that the gross emissions from reservoirs may account for between 1% and 28% of the global warming potential of GHG emissions.[7] This challenges the conventional wisdom that hydropower produces only positive atmospheric effects, such as a reduction in emissions of carbon dioxide, nitrous oxides, sulphuric oxides and particulates when compared with power generation sources that burn fossil fuels.[8] It also implies that all reservoirs – not only hydropower reservoirs – emit GHGs. Consequently, reservoir and catchment characteristics must be investigated to find out the likely level of GHG emissions.

All large dams and natural lakes in the boreal and tropical regions that have been measured emit greenhouse gases (carbon dioxide, methane, or sometimes both) (see Figure 3.1 and 3.2).[9] Figure 3.1 shows the range of values recorded by field measurements and models of GHG emissions for 15 reservoirs.[10] Some values for gross GHG emissions are extremely low and may be 10 times less than the thermal option. Yet in some circumstances the gross emissions can be considerable, and possibly greater than the thermal alternatives.[11] These emissions may change significantly over time as the

Box 3.1 Mitigating and compensating for terrestrial impacts

Some large dam projects have tried to mitigate terrestrial impacts on biodiversity by physically rescuing animals from the area to be flooded or by anticipating that mobile species will simply move to neighbouring areas. Operation Noah and Operação Curupira are two examples undertaken at Kariba and Tucurui dams. The respective WCD Case Studies show that neither programme yielded tangible benefits for the wildlife involved. This may be a result of the implicit and probably incorrect assumption that the recipient habitat was not already at carrying capacity for the species concerned.

An alternative to mitigation is a compensatory project approach, or environmental 'offsets'. For example, in India there is a legal requirement that forests flooded by reservoirs must be replanted elsewhere. However, the India Case Study found that only half of the required forest has typically been planted – and even this is poorly managed, yielding little in the way of comparable benefits or services. Additional compensatory measures may include either trust funds established through grants from developers (for example Harvey Basin Restoration Trust, Australia) or trust funds that manage parts of the revenue stream and use it for environmental purposes. This latter model is proposed for the planned Nam Theun II dam in Laos, with the intention of creating and managing a National Park in the catchment. The plan has the potential to benefit both forest ecosystems and the lifespan of the dam through reduced sedimentation.

Sources: WCD Kariba, Tucurui and India Case Studies; Bizer, 2000

biomass decays within the reservoir during the first few years of impoundment. In other cases the emissions may depend more on carbon inflows from the catchment in the longer term and have greater stability over time, subject to catchment conditions.

Figure 3.1 Gross greenhouse gas emissions from reservoirs

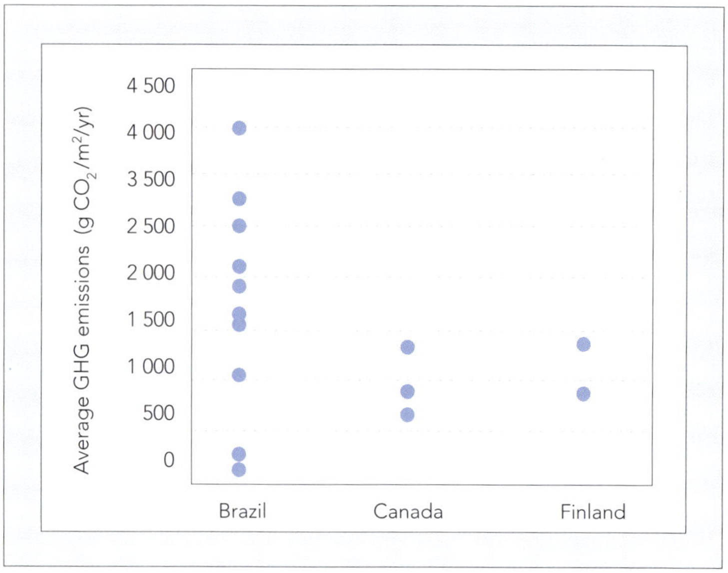

Source: WCD Thematic Review II.2 Global Change
Note: Average measured emissions of greenhouse gases from 15 reservoirs in boreal and tropical regions show large variations within countries and between regions. These averages mask strong seasonal and annual variations.[12]

Establishing that a reservoir emits GHGs is not enough to assess the impact of a dam on climate change. Natural habitats (undisturbed by dams) may also emit GHGs (see Figure 3.2). Alternatively they may store carbon or act as a net carbon sink. For example, a floodplain tropical forest in Amazonia may emit methane from soils and, at the same time, absorb carbon dioxide in leaves. The balance of all these potentially counteracting effects would determine the profile of GHG emissions from the natural habitat without the dam. A further complication is that the land use change induced by displacement of people, resource extraction and other economic activities may also form part of the net contribution to GHG emissions associated with the construction of the dam. Calculations of the contribution of new reservoirs to climate change must therefore include an assessment of the natural pre-dam emission or sink in order to determine the net impact of the dam.[13]

The WCD Case Studies only provide data on carbon dioxide and methane emissions from the Tucurui reservoir (see Box 3.2).

Figure 3.2 Greenhouse gas emissions from natural habitats

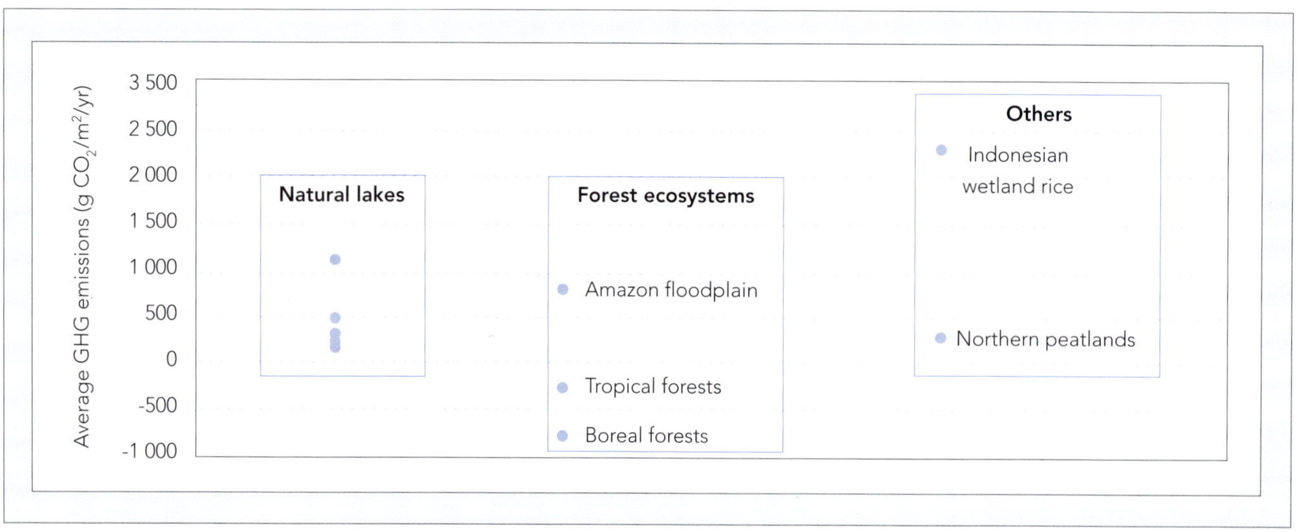

Source: WCD Thematic Review II.2 Global Change.
Note: Natural habitats may be sources (positive values) or sinks (negative values) for carbon. Emissions from dam reservoirs in Canada and Finland (Fig 3.1) are at similar levels to natural lakes.

Even in this case there is no data on emissions without the dam, making a conclusion on the net effect impossible. This applies more generally to other dams that have even less information available. Current understanding of emissions suggests that shallow, warm tropical dams are more likely to be major GHG emitters than deep cold boreal dams.

In the case of hydropower dams, tropical dams that have low installed capacity and large shallow reservoirs are more likely to have gross emissions that approach those of comparable thermal alternatives than those with small, deep reservoirs and high installed capacity.[14]

To date, no experience exists with minimising, mitigating, or compensating these impacts. Pre-inundation removal of vegetation is one alternative, but the net effects of such an activity are not well understood. The outcome of global negotiations on climate change may bear on future penalties and incentives for net GHG emissions from dams.

Downstream Aquatic Ecosystems and Biodiversity

Storage dams are intended to alter the natural distribution and timing of streamflow. They compromise the dynamic aspects of rivers that are fundamental to maintain-

Box 3.2 Greenhouse gas emissions at Tucurui, Brazil

Recent monitoring in the 2 600 km² reservoir of Tucurui show that greenhouse gas emissions are substantial and highly variable from year to year. Values in 1998 exceeded those measured in 1999 by more than a factor of 10 for methane and by 65% for carbon dioxide (see table below).[15]

Total Gross Emissions (tons/km²/ year)

Year	Methane	Carbon dioxide
1998	76.36	3 808
1999	5.33	2 378

Modelling taking into account emissions from water passing through the turbines or over the spillway leads to higher estimates of total emissions.[16] The figure below compares these gross emissions to those of alternative technologies for large-scale power generation.[17] Background emissions from natural pre-impoundment habitats have not yet been measured for Tucurui, so true comparisons of net emissions with alternatives remain elusive.

The alternative technology for large-scale electricity generation required for aluminium smelting (the main consumer of electricity) was thermal power employing diesel fuel when the project was built in the 1970s. Today the alternative would be gas combined cycle plants.

Source: WCD Tucurui Case Study.

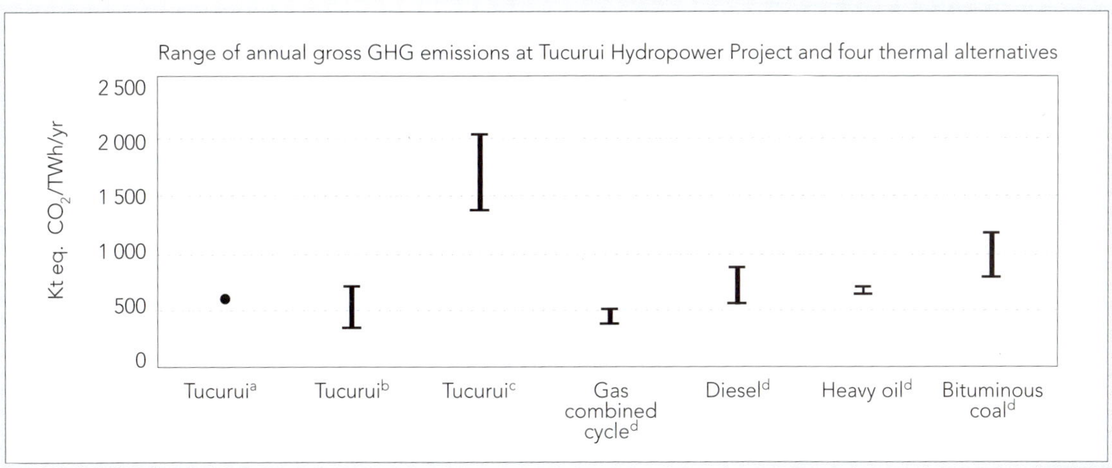

Sources: [a]Fearnside, 1995; [b]Rosa et al, 1999; [c]Fearnside, 2000; [d]IEA, 2000.

ing the character of aquatic ecosystems. Natural rivers and their habitats and species are a function of the flow, the quantity and character of the sediment in motion through the channel, and the character or composition of the materials that make up the bed and banks of the channel. The defining river discharge includes both high- and low-flow elements. These dynamics, not the average conditions of controlled dam operations, determine a stream's physical foundation, which in turn ensures ecosystem integrity.[18]

The extent of impacts will also depend on whether water is extracted or diverted for consumption, or left instream. Introduction of non-native species, modified water quality (temperature, oxygen, nutrients), loss of system dynamics, and loss of the ability to maintain continuity of an ecosystem result in ecologically modified river systems. The establishment of a new dynamic has positive effects on some species and negative effects on others (see Box 3.3).

Box 3.3 How one dam has affected two different species in opposite ways

Before it was dammed, the Waitaki River in New Zealand was highly unstable, flooded frequently, and had a constantly changing channel. After damming the river flood runoff is now stored in the reservoir to produce electricity. This increased the stability of the sandbars downstream, allowing colonisation by vegetation, which further stabilised the channel. The increased flow stability has benefited chinook salmon (*Oncorhynchus tshawytscha*) populations, an exotic species introduced in the early 1900s, because the more stable channels provide more shelter for fry at high flows as well as a larger area of spawning gravel. Currently the Waitaki has the largest population of salmon in New Zealand. However, the beneficial change for salmon has been detrimental to the black stilt (*Himantopus novaezealandiae*) – a native species. This bird is so endangered that fewer than 100 individuals remain. They nest exclusively on the large exposed sandbars isolated from the shore, a habitat that was maintained by the unstable nature of the river. The vegetation that has proliferated and stabilised the gravel bars has increased the cover for predators, which in turn have exacted a significant toll on adult stilts, eggs, and nestlings.

Source: Ligon et al, 1995, cited in WCD Thematic Review II.1 Ecosystems

Impacts of changes in flow regimes

Flow regimes are the key driving variable for downstream aquatic ecosystems. Flood timing, duration and frequency are all critical for the survival of communities of plants and animals living downstream. Small flood events may act as biological triggers for fish and invertebrate migration: major events create and maintain habitats by scouring or transporting sediments. The natural variability of most river systems sustains complex biological communities that may be very different from those adapted to the stable flows and conditions of a regulated river. Finally, water temperature and chemistry are altered as a consequence of water storage and the altered timing of downstream flows. Algal growth may occur in the reservoir and in the channel immediately downstream from dams because of the nutrient loading of the reservoir releases. Self-purification processes diminish this effect downstream.

Storage dams, particularly hydropower peaking plants, can significantly disrupt the whole flow regime, resulting in both high seasonal and day-to-day fluctuations that differ greatly from natural flow levels. As shown in Figure 3.3, the construction of Glen Canyon dam on the Colorado River in the United States reduced daily average flows during the annual September peak from about 2 000 m^3/sec to about 700 m^3/sec. In addition, as shown in Figure 3.4, streamflow can fluctuate daily more than 425 m^3/sec due to dam releases for electricity generation during the peak daytime demand periods. These changes in flow have dramatically altered the riverine environment, creating consistently colder temperatures due to release of water from the bottom of the reservoir. A general decline in native fish abundance in the Colorado River is attributed specifically to the cold-water release from large dams there.[19] The population of the fish *Tandanus tandanus* in Australia's Murray River disappeared due to short-term fluctuations in water level caused by reservoir releases in response to downstream water-user requirements.[20]

Figure 3.3 Modification of annual flow regimes due to a hydropower dam, Colorado River at Lee's Ferry, United States

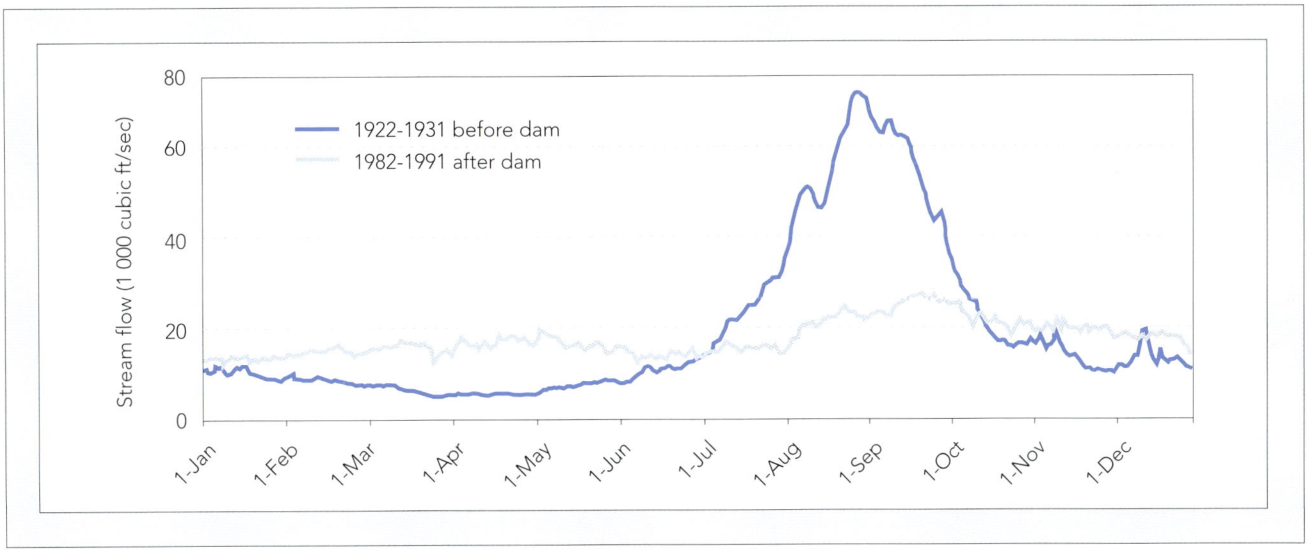

Source: Data from United States Geological Survey, 2000.

Figure 3.4 Fluctuation of daily streamflow regime due to hydropower peaking operations, Colorado River at Lee's Ferry, September

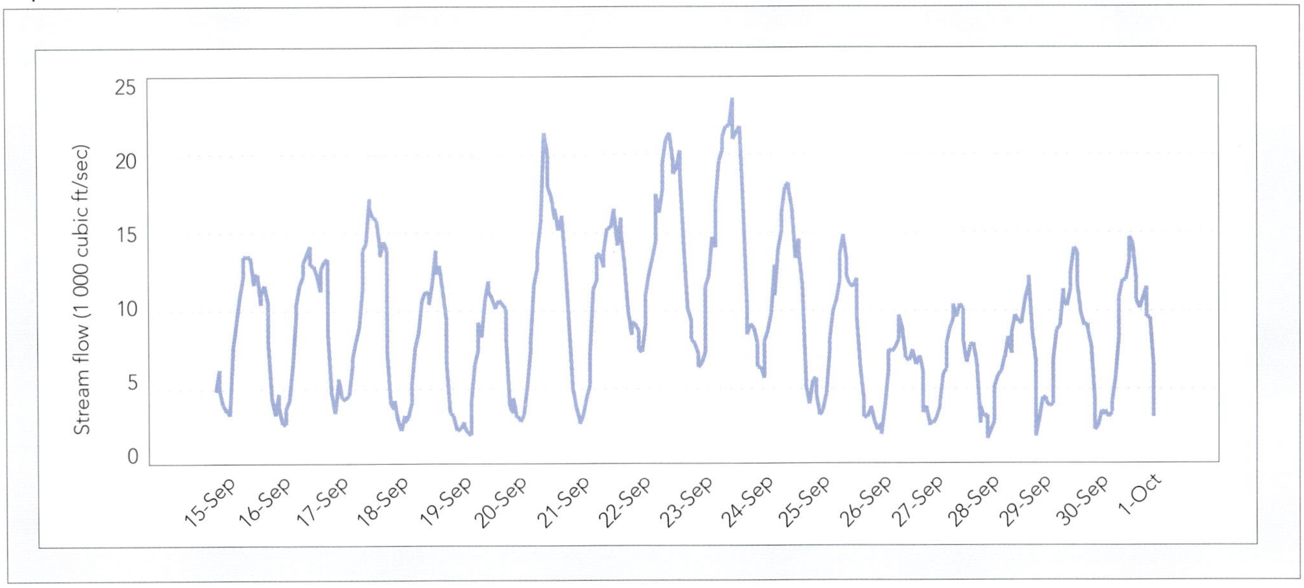

Source: Data from United States Bureau of Reclamation, 2000b.
Notes: Peak flows are associated with the power generation between 14.00 and 19.00 daily, with minima at 04.00 am, and the fluctuation in demand also varies from day to day.

The modified habitats resulting from large dams often create environments that are more conducive to non-native and exotic plant, fish, snail, insect, and animal species.

Particularly high hydropower dams cause gas to become supersaturated when water flows over the spillway. This causes fish deaths due to a condition similar to the bends that can affect divers who dive too deep for too long. The Grand Coulee Case Study reports that this is a particular problem on the Columbia River in the United States, where regulators have fixed a maximum total dissolved gas concentration to reduce impacts on migratory fish.

The modified habitats resulting from large dams often create environments that are more conducive to non-native and exotic plant, fish, snail, insect, and animal species.[21] These resulting non-native species often out-compete the natives and end up modifying ecosystems that may become unstable, nurture disease vectors, or are no longer able to support the historical environmental and social components.

Compared with a natural river, the root systems of plants in rivers below dams experience reduced effects of scour, the plants themselves suffer less stress from high discharges and the rate of channel migration is reduced, so that an area of the channel-bed available for the development of aquatic plants can be stabilised. In both Africa and Australia, the elimination of high discharges to flush systems has allowed the extensive development of the aquatic weeds Water Hyacinth (*Eichhornia crassipes*) and Water Fern (*Salvinia molesta*).[22] The Orange River Pilot Study documents the colonisation by reeds (*Phragmites australis*) of 41 000 ha of riverbed has occurred as a result of stabilised flows on the Orange River.

Biological linkages also extend laterally away from and parallel to the river, pushing the effect of river changes onto a band of varying width. As long as the river flow is sufficient, other wildlife such as deer, antelope, and elephants will come to the water, especially in the dry/hot season, to drink. Hippos in Africa will use water of sufficient depth as a daytime refuge, emerging to forage at night. Many species of birds and bats fly in to drink. These lateral movements can extend to several kilometres from the river. Thus many wildlife species in a fairly wide strip of land on either side of the river depend upon it, and they may all be affected when the flow of the river is disrupted by the construction of a large dam. Equally, long reservoirs that extend many kilometres up river valleys present a barrier for terrestrial species inhabiting each riverbank that were previously able to cross the river.

When waters of one basin are diverted into another one, changes in volume and seasonality of flow result. New biota from the source basin may invade the recipient basin and compete with the native species. If all the water is diverted from the source basin, this will clearly have serious impacts on any unique species or genetically different stocks. In South Africa, for example, the diversion of the Orange River into the Great Fish River resulted in a six-fold increase in flow, making the Great Fish River permanent rather than intermittent. The Orange River Case Study reports that one beneficiary is the larvae of the blackfly *Simulium chutteri*, which does not tolerate desiccation. In the absence of control measures, this biting fly causes livestock losses, reduces recreational use, and irritates local farmers.

Efforts to minimise the impacts of changes in flow regime have relied on measures to restore the streamflow regime through the setting of environmental flow releases (EFR) (see Box 3.4)

Impacts of trapping sediments and nutrients behind a dam

The reduction in sediment and nutrient transport in rivers downstream of dams has impacts on channel, floodplain and coastal delta morphology and causes the loss of aquatic habitat for fish and other species. Changes in river water turbidity may affect biota directly. For example, plankton production is influenced by many variables, including turbidity. If this is reduced due to impoundment, plankton development may be enhanced and may occur in new sections of a river.

Reduction in sediment moving downstream from the dam leads to degradation of the river channel below the facility.[23] This can lead to the elimination of beaches and backwaters that provided native fish habitat, and the reduction or elimination of riparian vegetation that provides nutrients and habitat for aquatic and waterfowl species, among others. Impounding rivers invariably results in increased degradation of coastal deltas due to reduction in sediment input. For example, the slow accretion of the Nile Delta was reversed with the construction of the Delta barrage in 1868. Today, other dams on the Nile, including the Aswan High Dam, have further reduced the amount of sediment reaching the delta. As a result, much of the delta coastline is eroding by up to 5–8 metres per year, but in places this exceeds 240 metres per year.[24]

The consequence of reduced sediment also extends to long stretches of coastline where the erosive effect of waves is no longer sustained by sediment inputs from rivers. It is estimated that the coastlines of Togo and Benin are being eroded at a rate of 10-15 metres a year because the Akosombo dam on the Volta River in Ghana has halted the sediment supply to the sea.[25] Another example is the Rhone River in France, where a series of dams has reduced the quantity of sediment transported by the river to the Mediterranean from 12 million tons in the 19th century to only 4 to 5 million tons today.[26] This has led to erosion rates of up to 5 metres per year for the beaches in the regions of the Camargue and the Languedoc, requiring a coastal defence budget running into millions of dollars.

Measures for mitigating the impacts of trapping sediments and nutrients are limited. Where feasible, flushing sediments can be part of a programme of managed flood releases.

Box 3.4 Minimising impacts of changes in streamflow regimes: environmental flow requirements

At least twenty-nine countries seek to minimise ecosystem impacts from large dams by using an EFR to meet predetermined ecosystem maintenance objectives. The practice of EFRs began as a commitment to ensuring a 'minimum flow' in the river (often arbitrarily fixed at 10% of mean annual runoff). It has since grown to include a definition of ecosystem requirements and a planned flow release programme, which may vary annually or seasonally, to meet downstream needs for both the environment and people. The level of EFR required is determined by the need to maintain particular ecosystem components downstream, often with reference to national legislation. The countries that use this method have recognised that a short-term reduction in financial returns from a project often leads to improved long-term sustainability and attainment of broader societal objectives for a healthier environment. Still, this represents a re-distribution of the benefits of a dam project and thus existing beneficiaries such as irrigators and operators of hydropower facilities may resist EFRs.

Amongst the WCD Case Study large dams, only Grand Coulee has an environmental flow requirement, in this case consisting of a specially designed release for flow augmentation for salmon, while avoiding high total dissolved gas concentrations. Implementation of a planned release to maintain downstream ecosystems is being considered for the Orange River in South Africa.

Sources: Brown et al, 1999, Contributing Paper for WCD Thematic Review II.1 Ecosystems; Tharme, 2000

Blocking migration of aquatic organisms

As a physical barrier the dam disrupts the movement of species leading to changes in upstream and downstream species composition and even species loss. River-dwelling species have several migratory patterns. These include anadromous fish such as salmon and catadromous fish such as eels. Adults of the former migrate up rivers to spawn and the young descend, while the reverse occurs with the latter. But many other freshwater fish move up rivers or their tributaries to spawn, while the glochidia larvae of freshwater mussels are carried by host fish. To help counteract the drift downstream of their larvae, aquatic insect adults such as mayflies and stoneflies move upstream to lay their eggs.[27] Dams block these migrations to varying degrees.

As a physical barrier the dam disrupts the movement of species leading to changes in upstream and downstream species composition and even species loss.

Box 3.5 Mitigation measures: fish passes

Fish passes are often used as an engineered mitigation measure for reducing impacts on fish. However, very few of the over 400 large dams in Australia have fish passes of any description, only 16 had been constructed on the 450 large dams in South Africa by 1994, and only 9.5% of 1 825 hydropower dams in the United States have an upstream fish pass facility. An example is Idaho Power Company which built fish passes into each of its dams in the Hells Canyon Complex. However, all were unsuccessful and salmon no longer migrate above Hells Canyon Dam.

The Glomma and Laagen Case Study reports that there are 34 fish passes on the 40 dams in this Norwegian basin. Of these only 26% work with 'good efficiency,' 41% work less well, and as many as 32% are not working at all. In general, the efficiency is considered low, and fish migrations are severely affected. At Pak Mun Dam in Thailand, the case study documents the ineffectiveness of the fish pass, especially for the large migratory species in the Mekong that may be up to two metres long and cannot fit through the 15x20 cm slots. Grand Coulee, Tucurui, Tarbela and Aslantas have no fish passes despite having migratory fish species in the river.

Even when fish passes have been installed successfully, migrations can be delayed by the absence of navigational cues, such as strong currents. This causes stress on the energy reserves of the fish, as anadromous fish such as salmon do not feed during migration.

Recent research in Australia, the United States, and Japan has shown that fish passes need to be modified to meet the needs of each species and the particular situation at each dam. They cannot simply be considered an easily transferable technology, as shown by the Pak Mun fish pass, which used a design appropriate for leaping trout and salmon in mountain streams, but which was ineffective for species living in the slower-flowing Mekong.

Sources: Australia in Blackmore, pers. comm. 2000; South Africa in Benade, pers. comm. 1999; USA in Francfort et al, 1994, Executive Summary pviii;Collier et al, 1996, p22; Larinier, 2000, Contributing Paper for WCD Thematic Review II.1 Ecosystems

The WCD Cross-Check Survey found that impeding the passage of migratory fish species was the most significant ecosystem impact, recorded at over 60% of the projects for which responses on environmental issues were given. In 36% of these cases, the impact of the large dam on migratory fish was not anticipated during project planning.

Migratory fish require different environments for the main phases of their life cycle: reproduction, production of juveniles, growth, and sexual maturation. Many anadromous fish populations such as salmon and shad have died out as a result of dams blocking their migratory routes. The sturgeon populations in the Caspian Sea now rely on stocking from hatcheries (mainly in Iran), as dams built by the former Soviet Union on rivers entering the sea halted natural spawning migrations.[28]

Detailed studies in North America indicate that dam construction is one of the major causes for freshwater species extinction. For example, a study of the threatened fish of Oklahoma suggested that the loss of free-flowing river habitat due to reservoirs had led to 55% of the human-induced species loss, while a further 19% was caused by dams acting as barriers to fish migration.[29] The best-documented examples of disrupted fish migrations are from the Columbia River in the United States, where many stocks of salmon have been lost. The impact of these disruptions on the productivity of the fishery are described below.

Fish passes are typically used where efforts are made to mitigate the effect of dams in blocking migrations of fish (see Box 3.5).

Floodplain Ecosystems

Reduction in downstream annual flooding affects the natural productivity of riparian areas, floodplains and deltas. The characteristics of riparian plant communities are controlled by the dynamic interaction of flooding and sedimentation. Many riparian species depend on shallow floodplain aquifers that are recharged during regular flood events. Dams can have significant and complex impacts on downstream riparian plant communities. High discharges can retard the encroachment of true terrestrial species, but many riparian plants have evolved with and become adapted to the natural flood regime.

Typically, riparian forest tree species are dependent on river flows and a shallow aquifer, and the community and population structure of riparian forests is related to the spatial and temporal patterns of flooding at a site. For example, the Eucalyptus forests of the Murray floodplain, Australia, depend on periodic flooding for seed germination, and headwater impoundment has curtailed regeneration.[30] Conversely, artificial pulses generated by dam releases at the wrong time – in ecological terms – are recognised as a cause of forest destruction. For example, *Acacia xanthophloea* is disappearing from the Pongolo system below Pongolapoort dam, South Africa, as a result of the modified flood regime.[31]

The control of floodwaters by large dams, which usually reduces flow during natural flood periods and increases flow during dry periods, leads to a discontinuity in the river system. This, together with the associated loss of floodplain habitats, normally has a marked negative impact on fish diversity and productivity. The connection between the river and floodplain or backwater habitats is essential in the life history of many riverine fish that have evolved to take advantage of the seasonal floods and use the inundated areas for spawning and feeding. Loss of this connection can lead to a rapid decline in productivity of the local fishery and to extinction of some species. Additionally, dewatering of stream channels immediately downstream from dams can be a serious problem.

In Africa, the changed hydrological regime of rivers has adversely affected floodplain agriculture, fisheries, pasture and forests that constituted the organising element of community livelihood and culture.

The direct loss of annual silt and nutrient replenishment as a consequence of upstream impoundment is thought to have contributed to the gradual loss of fertility of formerly productive floodplain soils as used in agriculture and flood-recession agriculture. Dramatic reductions in bird species are also known, especially in downstream floodplain and delta areas, where wetlands may not be replenished with water and nutrients once a dam is installed. Finally, recharge of groundwater in floodplain areas is severely diminished once floods are eliminated.[32]

In Africa, the changed hydrological regime of rivers has adversely affected floodplain agriculture, fisheries, pasture and forests that constituted the organising element of community livelihood and culture. Economically important wetlands in Africa include river floodplains, freshwater lakes and coastal and estuarine environments. In the Sahel, there are major wetlands in the Delta Intérieur of the River Niger in Mali and

Lake Chad (on the border between Niger, Chad, Cameroon and Nigeria). These have counterparts elsewhere in semi-arid Africa, notably the Sudd in Sudan and the Okavango Delta in Botswana, and in humid areas, such as the swamps of eastern Zaire.

Some of these wetlands cover extensive tracts of land. The fringing floodplain of the Senegal River covers some 5 000 km^2 in flood, and shrinks to 500 km^2 in the dry season. The fringing floodplain of the Niger covers about 6 000 km^2 in the flood season, shrinking to about half that at low water, while the Niger Inland Delta extends to 20 000 to 30 000 km^2 in the flood season, shrinking to 4 000 km^2 at low water. In the Logone-Chari system, flooding covers some 90 000 km^2.[33]

Efforts to restore floodplain ecosystem functions rely on reversing the effects of the dam through a program of managed floods designed to simulate the floods that occurred prior to the dam (Box 3.6).

Box 3.6 Restoring ecosystem function through managed floods

The WCD Knowledge Base includes a number of cases where artificial floods have been released from large dams to regenerate the natural resource base of downstream floodplains for local livelihoods (for example Manantali dam in Mali and Senegal and the Pongolapoort dam in South Africa). Managed floods generate economic benefits when downstream communities depend on natural, flood-maintained resources such as grazing, flood-recession agriculture and fishing (see Chapter 4). For example, on the Tana River, Kenya, a released flood from the planned Grand Falls scheme would have a net present value of at least $50 million for the downstream floodplain economy. Managed floods also entail an opportunity cost which may be greater or lesser depending on the value of the released floodwaters to the dam for irrigation, hydropower or other uses. A set of preliminary studies show that in some cases there are clear net economic benefits to these releases and in other cases the opportunity costs exceed the value of downstream benefits that were identified, quantified and valued in economic terms. The potential for managed floods is often constrained by the design of the sluice-gates, sedimentation in the reservoir and in downstream channels and the development of infrastructure on areas previously prone to flooding. Another constraint may be the political will to support traditional means of livelihood at the expense of benefits from the dam.

Source: Acreman et al, 2000, Contributing Paper to WCD Thematic Review II.1 Ecosystem Impacts; Grand Falls in Emerton, 2000

Fisheries

As indicated earlier, the blockage of sediment and nutrients, the re-regulation of streamflow, and elimination of the natural flood regime can all have significant, negative effects on downstream fisheries. Marine or estuarine fisheries are also negatively affected when dams alter or divert freshwater flows. Still, productive reservoir fisheries can follow from dam construction, although they are not always anticipated or part of project design proposals.

Substantial losses in downstream fishery production as a result of dam construction are reported from around the world. Along with subsistence agriculture, fisheries constitute an important livelihood activity among large rural populations in the developing world. Many of these households depend on fisheries either as a primary or supplementary source of livelihood. For example, the partial closing of the river channel by Porto Primavera dam in Brazil blocked fish migration and diminished upstream fish catch by 80%, affecting livelihoods.[34] In areas of rich fish species diversity, such as the lower Mekong region in East Asia, community livelihoods and culture are woven around fisheries. The Pak Mun Case Study reports a drastic decline in upstream fish catch once the dam had effectively blocked fish migration from the Mekong River upstream into tributaries of the large Mun River watershed.

Data on the losses to downstream fishery production as a result of dam construction are reported from river basins in Africa as well. For example, 11 250 tonnes of fish per year from the Senegal River system were lost following dam construction.[35] Studies on the Niger have shown that fish productivity increases linearly with the volume of river flow.[36] Other basins in the WCD Knowledge Base reporting loss in fish production include the *yaeres* in Cameroon, the Pongolo flood plain in South Africa, and the Niger in West Africa below Kainji dam.[37] Adverse impacts have been felt in the delta and estuary areas in the lower Volta region, in the Nile delta, and on the Zambezi in Mozambique.[38]

Freshwater flows also help support marine fish production as many marine fish spawn in estuaries or deltas. A decrease in freshwater flow and in nutrients due to dam construction affects the nursery areas in a number of ways, including increasing salinity, allowing predatory marine fish to invade, and reducing the available food supply. These impacts are well illustrated by the effect of the Aswan High Dam on the coastal waters of the Mediterranean, where reduction in nutrients transported to the sea has lowered production at all trophic levels, resulting in a decline in catches of sardines and other fish.[39] In the Zambezi delta, the impact of modified seasonal flows on local shrimp fisheries has been estimated at $10 million per year.[40]

Dams can enhance some riverine fisheries, particularly tailwater fisheries immediately below dams that benefit from discharge of nutrients from the upstream reservoir. If discharge is from the lower layer of water in the reservoir, lowered temperatures in the receiving tailwater can curtail or eliminate warmwater river fisheries and require stocking with exotic coldwater species such as salmonids (assuming that the water is sufficiently oxygenated). Productive tailwater fisheries targeting these coldwater fish can result but generally require supplemental hatchery programmes and the introduction of coldwater invertebrates to serve as food for these fish.[41]

Productive reservoir fisheries often follow from dam construction, although they are not always anticipated or part of project design proposals. While practically all the WCD Case Study dams have reservoir fisheries, predictions of fish production were made in only three cases. In the case of Aslantas, consultants estimated a production of 580 tonnes, and actual figures are 86–125 tonnes (for 1987–95). At Pak Mun, actual production came to only one-tenth of prediction. In both cases targets were not met as the predictions depended partly on a functional stocking programme that was not fully implemented. At Kariba estimates were made for the artisanal fishery but not for the more productive open-water commercial fishery.

On the other hand, an unanticipated but productive fishery has developed in Tucurui. In addition to commercial production at Tucurui, a local sport-fishing industry has developed around the sought-after Peacock Bass, *Cichla ocellaris* (known in Tucurui as *tucunaré*), Kariba has seen a similar development of a vibrant sport fishery. At Grand Coulee, the lack of a fish pass deprived salmon of over 1 000 kilometres of upstream spawning grounds, and affected First Nations tribes in the United States and Canada, while a fish hatchery largely maintained salmon numbers (but not genetic diversity) in downstream runs in the United States.

Data from before and after the construction of Tucurui illustrate the changing nature of species composition and fish production in the downstream, reservoir and upstream areas. The experimental catch data document that the number of species found in each of the three areas has declined significantly (by 30–50 species) following impoundment. The data suggest that in total 11 species are no longer found in these areas (see Figure 3.5a). Within the reservoir and to a lesser degree upstream, species now include more piscivorous species at the expense of the detritivores that were more common prior to construction. In production terms, the harvest upstream of the reservoir remained stable for the first 10 years or more, but now appears to be increasing (see Figure 3.5b). Meanwhile, the downstream fishery has shown a continued downward trend. However, the reservoir fishery has expanded tenfold in the last 20 years, with the result that the total fishery (upstream, downstream and in the reservoir) has tripled in size to 4 700 tonnes per year since the dam was created.

Figure 3.5 Decline in species numbers but increase in fisheries productivity, Tucurui (a&b)

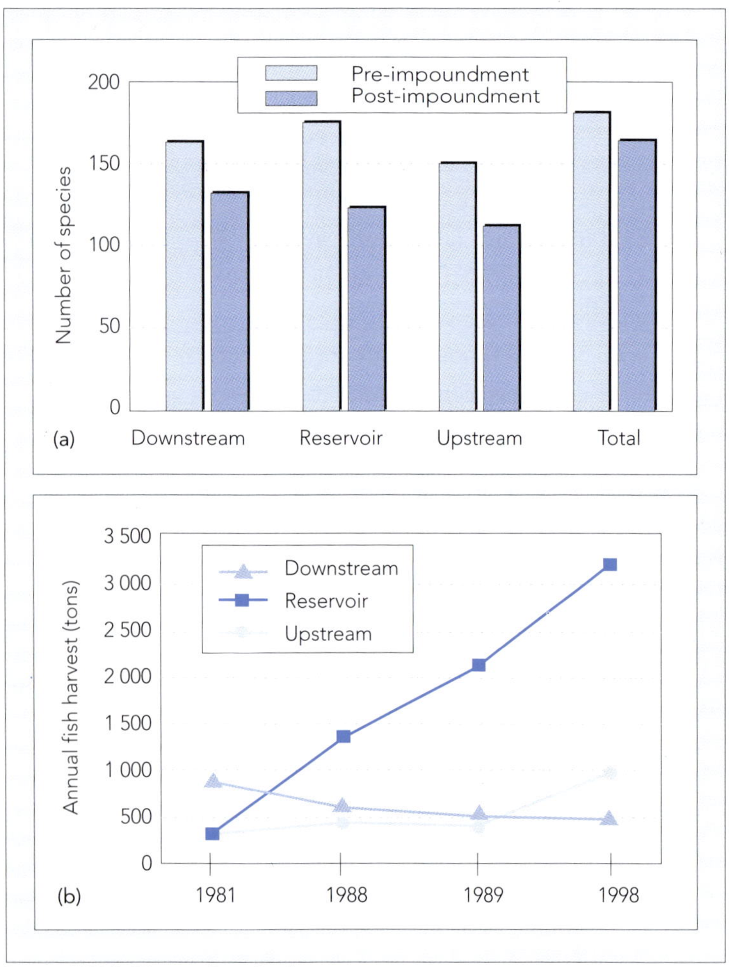

Source: WCD Tucurui Case Study.

Mitigation or compensation measures have been used to reduce the impacts of changes in fisheries. Fish passages are the most prevalent measure and have been of limited applicability and usefulness (see Box 3.5). Compensation measures consist of fish hatcheries and stocking programs designed to reproduce the productivity of the fishery.

Ecosystem Enhancement

The WCD Knowledge Base provides a number of examples of the ecosystem enhancement effects of large dams. The Case Studies show, for example, that productive wetlands have been created by pumping Grand Coulee water through a previously dry area in the Columbia River Basin, and along the shores of Lake Kariba, with considerable wildlife and tourism values resulting.

Some reservoirs support globally threatened reptiles (for example Hillsborough dam, Trinidad), and others have been declared Ramsar sites of international importance for birds. Indeed, one measure of the environmental value of water bodies is to consider the list of sites designated as internationally important for waterfowl under the Ramsar Convention on Wetlands. Of 957 sites designated by December 1998, only 10% included artificial wetland types, while 25% included natural lake types.[42] Many of the designated artificial wetlands are dammed sites: of the almost 100 artificial wetlands designated as internationally important, 78 include water storage areas.[43]

A study by Wetlands International for WCD showed that the wintering waterfowl assembled on natural and dammed lakes in Switzerland are broadly similar, and the same species occur most abundantly on both types of lake. The study also showed that the situation in South Africa is very different, as it lacks natural permanent water bodies and almost all permanent water bodies are reservoirs. At least 12 reservoirs support major and important concentrations of waterbirds. Large dams in South Africa have provided generally beneficial conditions for pelicans, darters and cormorants. They provide suitable moulting sites for waterfowl: for example, at least 70% of the global population of the South African Shelduck, *Tadorna cana,* moults at only 23 localities in South Africa, 21 of which are large reservoirs. Dam reservoirs provide dry-season or drought refuges for many waterfowl species in the semi-arid parts of the country.[44]

But productive wetlands are most likely to be created around reservoirs where these are shallow or have shallow margins and limited reservoir drawdowns. Where sediment inflows from the catchment are heavy, small deltaic wetlands may evolve at the inflow.

In general, deeper reservoirs that have steep sides or high seasonal water-level fluctuations are unlikely to support major wetland habitats.

Dammed lakes support a more restricted range of species, and several common and uncommon species from natural lakes were not recorded on dammed lakes by Wetlands International. The damming of rivers has increased the number of open water sites available to wintering waterfowl in Switzerland and provided a more suitable habitat for these birds than the generally fast-flowing stretches of river in-between. However, these sites support only relatively small numbers of birds, of mostly common and widespread species and do not appear to provide as diverse a waterfowl habitat as the natural lakes in the area. Eleven of the 2 596 British reservoirs support substantial and important wintering waterfowl populations, while 60 natural inland waters and 52 estuaries support populations of international importance.[45]

Figure 3.6 Fragmentation in 225 large river basins

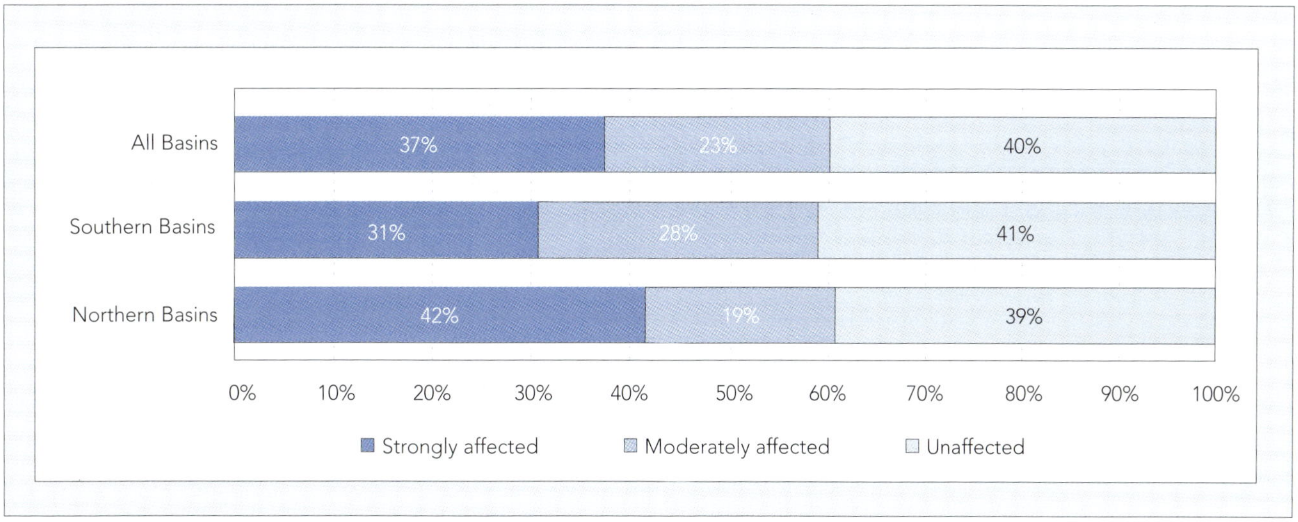

Source: Revenga et al, 2000.

Cumulative Impacts

Many of the major catchments in the world now contain multiple dams. Within a basin, the greater the number of dams, the greater the fragmentation of river ecosystems. An estimated 60% of the world's large river basins are highly or moderately fragmented by dams (see Figure 3.6). The magnitude of river fragmentation can be very high. In Sweden, for example, only three major rivers longer than 150 km and six minor rivers have not been affected by dams.[46]

Problems may be magnified as more large dams are added to a river system, resulting in an increased and cumulative loss of natural resources, habitat quality, environmental sustainability and ecosystem integrity.

Although seldom analysed, cumulative impacts occur when several dams are built on a single river. They affect both the physical (first-order) variables, such as flow regime and water quality, and the productivity and species composition of different rivers. The problems may be magnified as more large dams are added to a river system, resulting in an increased and cumulative loss of natural resources, habitat quality, environmental sustainability and ecosystem integrity. The cumulative impacts of interbasin water transfers can be of special concern, as this often involves the transfer of species into new watersheds.

Box 3.7 Cumulative impact of dams: the Aral Sea

The Aral Sea, fed by the Amu Darya and Syr Darya, was once the fourth largest inland body of water in the world, ranking just behind Lake Superior. It supported 24 species of fish and a fishing population of 10 000 people. A series of dams was built on the rivers to feed an immense irrigation system and grow cotton on 2.5 to 3 million hectares of new farmland. The withdrawal of water has reduced the Aral Sea to about 25% of its 1960 volume, quadrupled the salinity of the lake and wiped out the fishery. Pollutants that had formerly fed into the lake became airborne as dust, causing significant local health problems. The environmental damage caused has been estimated at $1.25 to $2.5 billion annually.

Source: Anderson, 1997, Section 1 pii, Section 6 pii.

The WCD Knowledge Base documents a number of cumulative impacts that include water quantity, water quality and species impacts. Flood regimes are clearly affected as increasing the total storage volume by adding additional dams reduces the flood flows downstream.

In Pakistan, the Tarbela Case Study reveals that only 21% of the historical dry season flow of the Indus reaches the delta; the rest is diverted for irrigation and water supply by 22 dams and barrages,. Since the Kotri barrage was commissioned in the early 1960s, the average number of days with no river flow downstream in the dry season increased from zero to 85 (the average from 1962 to 1997). Similar impacts have occurred around the Aral Sea (see Box 3.7) and in Australia where 80 years of river regulation, construction of additional storages, and diversion of water from the Murray Darling River have reduced the median flow reaching the sea to 21% of the pre-regulated flow.[47]

Water quality parameters recover only slowly when water is released from a dam. Oxygen levels may recover within a kilometre or two, while temperature changes may still exist 100 km downstream. Where the distance between dams does not allow recovery to natural levels, the biology of many hundreds of kilometres of river may be affected by a handful of dams. Examples from the WCD Case Studies include the Orange-Vaal river in South Africa, where the impacts of 24 dams may have led to 2 300 km (63%) of the river having a modified temperature regime. On the Columbia River, Grand Coulee dam receives water that is already high in total dissolved gasses as a result of upstream Canadian dams. Before the levels can recover to natural values, spill at Grand

Coulee increases them again, passing the problem further downstream.[48] Construction of a series of dams may therefore have increasing impacts on downstream ecosystems and biodiversity.

Also on the Columbia River, the cumulative impact of an additional dam on salmon migrations is significant. It is estimated that 5–14% of the adult salmon are killed at each of the eight large dams they pass while swimming up the river.[49]

What is not well researched is the change in the magnitude of the incremental response of ecosystem function and biodiversity as a river is increasingly fragmented. Thus it is not known if there is some threshold level at which the marginal impacts of the addition of one or more dams to a particular cascade of dams will begin to decline. It is therefore a case by case call whether the ecosystem impacts of further modifying a particular river may at some point be of less consequence than, for example, putting the first dam on a free-flowing river.

Anticipating and Responding to Ecosystem Impacts

Examination of efforts to counter the ecosystem impacts of large dams in the WCD Knowledge Base indicates that they have met with limited success owing to:

- the lack of attention paid to anticipating and avoiding impacts;
- the poor quality and uncertainty of predictions;
- the difficulty of coping with all impacts; and
- partial implementation and success of mitigation measures.

Anticipating and predicting ecosystem impacts

In order for ecosystem impacts to be addressed properly, they have to be understood and predicted. The Cross-Check Survey found that for the 87 projects that provided data on ecosystem impacts, almost 60% of the impacts identified were unanticipated prior to project construction, largely due to inadequate studies. While the sample size is small for some time periods, the Cross-Check also suggests that over time the trend is increasingly to anticipate impacts (see Figure 3.7). This confirms the expectation that the trend towards the use of environmental impact assessments (EIA) would result in improved identification of potential impacts (see Chapter 6 for discussion of EIAs).

Anticipating an impact is, however, not synonymous with accurately predicting the direction and magnitude of its effect on ecosystems and biodiversity. Nor does it guarantee understanding of the further impact of such changes on the livelihoods and economic welfare of affected people. While the generalised impacts of reservoir creation on terrestrial ecosystems and biodiversity are well-known, specifics,

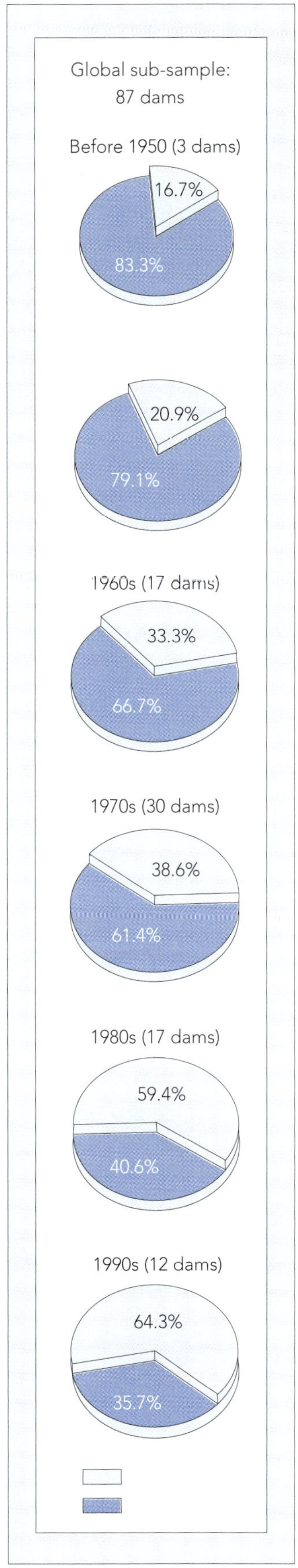

Figure 3.7 Anticipated and unanticipated ecosystem impacts

Source: WCD Cross-Check Survey.

such as the net emissions of greenhouse gases from a particular dam site, cannot be predicted with any certainty at present. Further research into the factors determining net emissions may reduce this uncertainty over time.

Downstream impacts on aquatic ecosystems and biodiversity and on floodplain ecosystems represent the sum of many complex interactions and thus are inherently difficult to predict where baseline data are absent or unreliable. However, the overall direction of the impacts is generally negative. As shown for the case of floodplain effects, the impact of large dams on these ecosystems will vary. With regard to fisheries, while it appears that the effect on species composition is generally negative at all levels (upstream, reservoir and downstream), downstream losses in productivity may be accompanied by increases in reservoir fishery production. Finally, the nature of cumulative impacts as additional dams are added to a river system may be significant, but a lack of research on the topic makes predictive assessment difficult.

Downstream impacts on aquatic ecosystems and biodiversity are inherently difficult to predict where baseline data are absent or unreliable.

In sum, past anticipation and prediction of ecosystem impacts was limited, in part due to a lack of reliable baseline data, scientific uncertainty regarding the nature of the interactions, inadequate attention paid to these issues and a correspondingly limited ability to model these complex systems. While improvements in measurement, scientific understanding and modelling capability have occurred over time, most ecosystem impacts remain site-specific. Their exact nature cannot be predicted in the absence of appropriate field studies of individual river systems.

Avoidance, minimisation, mitigation, compensation and ecosystem restoration

The WCD Knowledge Base reveals that efforts to avoid or minimise impacts through choice of alternative projects or alternative designs were more successful than efforts to manage the impacts once they were built into the design of the dam. Avoidance and minimisation of impacts, by their very nature, reduce ecosystem impacts on the site concerned. But where alternative sites or designs have been chosen, the net consequences for ecosystems have rarely been recorded.

Project planners and proponents have employed five principal measures to respond to ecosystem impacts:

- measures that avoid the anticipated adverse effects of a large dam through the selection of alternative projects;
- measures to minimise impacts by altering project design features once a dam is decided upon;
- mitigation measures that are incorporated into a new or existing dam design or operating regime in order to reduce ecosystem impacts to acceptable levels;
- measures that compensate for unavoidable residual effects by enhancing ecosystem attributes in watersheds above dams or at other sites; and
- measures to restore aspects of riverine ecosystems.

The primary option for avoiding ecosystem impacts from large dams has been not to build the dams in the first place. This is given a legal basis in Austria, Finland, France, Norway, Sweden, Switzerland, the United States and Zimbabwe, where legal 'set-aside' provisions to protect particular river segments or basins from regulation or development have been established.[50]

Good site selection, such as not building large dams on the main-stem of a river system, and better dam design also played significant roles in avoiding or minimising impacts in a number of cases in the WCD Knowledge Base.[51] The International Energy Agency also supported such policies in its recent policy paper *Hydropower and Environment*.[52] As reported earlier, an increasing number of countries are using environmental flow requirements to minimise downstream impacts (see Box 3.4) sometimes in the form of managed floods (see Box 3.6).

Mitigation was the most widely practised response to ecosystem impacts for the large dams in the WCD Knowledge Base. As noted earlier, mitigation has failed or worked only sporadically in the case of wildlife rescue operations and fish passes (Box 3.1 and Box 3.5). In the Cross-Check sub-sample of 87 projects for which ecosystem impacts were recorded, mitigation was undertaken for less than one-quarter of the anticipated ecosystem impacts (10% of all ecosystem impacts that occurred). Of these projects, 47 also recorded the effectiveness of mitigation measures implemented. Respondents stated that about 20% worked effectively, 40% did not mitigate the impact, and 40% were moderately effective. The conclusion can be drawn that only a small percentage of ecosystem impacts that occurred were actually mitigated effectively, while the relative significance of these impacts remains unknown.

While there are cases of good mitigation, the success is nevertheless contingent upon stringent conditions including:

- a good information base and competent professional staff available to formulate complex choices for decision-makers;
- an adequate legal framework and compliance mechanisms;
- a co-operative process with the design team and stakeholders;
- monitoring of feedback and evaluation of mitigation effectiveness; and
- adequate financial and institutional resources.[53]

Good site selection, such as not building large dams on the main-stem of a river system, and better dam design also played significant roles in avoiding or minimising impacts.

If any one of these conditions is absent, mitigation is unlikely to succeed. Mitigation, though often possible in principle, presents many uncertainties in field situations and is therefore at present not a credible option in all cases and all circumstances. In addition, the weaknesses of the EIA process for many projects reduces the possibility of positive outcomes.[54] This supports the use of alternative strategies rather than simply one of mitigation.

Compensation for lost resources may be 'in-kind' (for example construction of a fish hatchery for lost fish spawning areas) or 'out-of kind' (for example watershed protection in the upper catchment for loss of riverine or wetland habitat). Compensation may also be paid 'in-basin' (for example restoration of forest area within the river basin for forest lost to inundation) or 'out-of-basin' (for example assistance in expanding management capability at similar locations in another river basin). These are applied to offset ecosystem and biodiversity loss, as well as to replace lost productive use

of natural resources (as with fish hatcheries). Concerns with the effectiveness of compensation include questions about the possibility of 'replacing' ecosystem functions and species (for example, are fish raised in a hatchery equivalent to native fish stocks?) and the consequences of such efforts, for example whether fish hatcheries actually damage native fish stocks through disease and competition.

Restoring ecosystem through decommissioning

Ecosystem restoration has been undertaken in a range of countries where evolving national legislation has required higher standards of environmental performance (see Box 3.8). In the United States and France, dams have been decommissioned to restore key environmental values, often related to migratory fish (salmon), and often as a condition of project relicensing.[55]

Box 3.8 Ecosystem restoration through decommissioning in the United States

A total of 467 dams have been removed to date in the United States, 28 of these are large dams higher than 15 metres. Reasons for removal have included safety concerns, the restoration of riverine fisheries, financial considerations, or removal of unauthorised structures.

One example of a removal is the Grangeville dam on Clearwater Creek, Idaho. Built in 1903, it housed a 10-MW power plant. The removal was motivated by excessive sedimentation in the reservoir and blockage of migratory fish following the collapse of the fish pass in 1949. The dam was removed in 1963, and the river washed out the accumulated sediment within six months, with no recorded downstream effects. Removal restored access for salmon and steelhead runs to 67 km of main stem river and over 160 km of tributary habitat in the upper reaches of the Clearwater River. It also allowed members of the Nez Perce tribe to regain a traditional fishery long denied them despite the provisions of the 1855 treaty with the United States.

Source: Bowman et al, 1999, pxix, 27–31.

Ecosystem impacts of decommissioning are also complex and site-specific. One major issue in dam decommissioning is what to do with possibly polluted sediment accumulated behind the dam. The fate of this sediment when the dam is removed is frequently a major obstacle to restoration.

Current large dam designs are often not sufficiently flexible to allow for changed operating regimes to meet environmental (or other) goals. Global experience shows that these long-lived structures may be called on to operate differently in the future than in the past as society's needs and values evolve and as other dams are added in the catchment area.

In some cases, the dam design is completed before the environmental flow needs are determined, and cannot accommodate water releases of the required quantity and quality. Five dams on the Colorado River have now been retroactively fitted with variable level offtakes to draw off surface water, increase the temperature of the downstream river, and satisfy the needs of native fish.[56]

Findings and Lessons

Large dams generally have extensive impacts on rivers, watersheds and aquatic ecosystems. From the WCD Knowledge Base it is clear that large dams have led to:

- the loss of forests and wildlife habitat, the loss of species populations and the degradation of upstream catchment areas due to inundation of the reservoir area;
- emissions of greenhouse gases from reservoirs due to the rotting of vegetation and carbon inflows from the basin;
- the loss of aquatic biodiversity, upstream and downstream fisheries and the services of downstream floodplains, wetlands

and riverine estuarine and adjacent marine ecosystems;

- the creation of productive fringing wetland ecosystems with fish and waterfowl habitat opportunities in some reservoirs; and
- cumulative impacts on water quality, natural flooding, and species composition where a number of dams are sited on the same river.

The ecosystem impacts are more negative than positive and they have led, in many cases, to irreversible loss of species and ecosystems. In the Cross-Check Survey 67% of the ecosystem impacts recorded were negative. The social consequences of environmental impacts are examined in Chapter 4.

Efforts to date to mitigate the ecosystem impacts of large dams in the WCD Knowledge base have met with limited success owing to the lack of attention given to anticipating and avoiding impacts, the poor quality and uncertainty of predictions, the difficulty of coping with all impacts and the only partial implementation and success of mitigation measures. More specifically:

- it is not possible to mitigate many of the impacts of reservoir creation on terrestrial ecosystems and biodiversity, and efforts to 'rescue' wildlife have met with little sustainable success;
- the use of fish passes to mitigate the blockage of migratory fish has had little success, as the technology has often not been tailored to specific sites and species;
- good mitigation results from a good information base, early co-operation between ecologists, the dam design team and affected people, and regular monitoring and feedback on the effectiveness of mitigation measures;
- environmental flow requirements (which include managed flood releases) are increasingly used to reduce the impacts of changed streamflow regimes on aquatic, floodplain and coastal ecosystems downstream; and
- avoidance or minimisation of ecosystem impacts can be achieved through legislative or policy measures that set-aside particular river segments or basins, or through good site selection (such as avoiding main stem dams).

Finally, a number of countries, particularly the United States, are making efforts to restore ecosystem function and native fish populations by decommissioning large and small dams.

Endnotes

1 ICOLD, 1981 and 1988 cited in WCD Thematic Review II.1 Ecosystems; IEA, 2000, cited in WCD Thematic Review II.1 Ecosystems; see also other papers in WCD cited Thematic Review II.1 Ecosystems.

2 Revenga et al, 2000.

3 Respondents to the Cross-Check Survey found that 67% of the recorded ecosystem impacts were negative.

4 WCD Thematic Review II.1 Ecosystems, section 3.

5 Where there is a loss of vegetation cover there will be an increase in annual yield, but the direction of change in dry season flows will depend on the balance between evapotranspiration and infiltration effects (Bruijnzeel, 1990). In most cases it is expected that the evapotranspiration effect will dominate leading to lower dry season baseflow (Lamb and Gilmour, 2000).

6 The WCD Thematic Review II.2 Global Change and a WCD Workshop on the topic provide reviews of the literature and the perspectives of scientists working in this field.

7 The authors stress that the large range underscores the need for further quantification in order to better understand the contribution of reservoirs to global GHGs. St Louis et al, in press.

8 Bosi, 2000, p12.

9 WCD Thematic Review II.2 Global Change, Executive Summary, pv.

10 WCD Thematic Review II.2 Global Change; IEA, op cit.

11 To make the comparison with the thermal alternatives requires the measured reservoir emissions to be converted to emissions per kWh generated (see Box 3.2).

12 Values for methane are converted to a carbon dioxide equivalent using a Global Warming Potential of 21 and expressed as grams of CO_2 equivalent (IPCC, 1996, cited in WCD Thematic Review II.2 Global Change).

13 WCD Thematic Review II.2 Global Change.

14 Ibid.

15 Field measurements from Rosa et al, 1999.

16 Fearnside, 2000, develops a mathematical model.

17 Emissions per km^2 are converted to emissions/TWh using the mean annual generation 1995-1999.

18 Dietrich, 1999 env082, WCD Submission.

19 Holden and Stalnaker, 1975, p217, 229.

20 Walker, 1979, p156-57.

21 Thomas, 1998, p2.

22 WCD Thematic Review II.1 Ecosystems, section 3.6.2.2.

23 Collier et al, 1996, p56-58.

24 Abdel Megeed and Aly Makky, 1993, p298; Stanley and Warne, 1993, p628, 630.

25 Bourke, 1988, p117.

26 Balland, 1991.

27 Hynes, 1970, p422-423.

28 Jackson and Marmulla, 2000, Contributing Paper for WCD Thematic Review II.1 Ecosystems, p12–13; Larinier, 2000, Contributing Paper for Thematic Review II.1 Ecosystems, pii.

29 Hubbs and Pigg, 1976, p115.

30 Walker, op cit, p152.

31 Furness, 1978.

32 For example: Hadejia Nguru in Hollis et al, 1993.

33 WCD Thematic Review I.1 Social Impacts, Annex I.

34 Kudlavicz, 1999 env129, p3 and 2000 env063, p1, WCD Submissions.

35 Jackson and Marmulla, op cit, p8.

36 Welcomme, 1976, p361.

37 Benech, 1992, p161; Jubb, 1972; Lowe-McConnell, 1985, p120.

38 Lower Volta in Adams, 1992, p145-146; Kassas, 1973; Gammelsrød, 1996, p120.

39 Aleem, 1972, p205; Drinkwater and Frank, 1994, p141.

40 Gammelsrød, op cit, p123.

41 Jackson and Marmulla, op cit, Executive Summary piv.

42 Frazier, 1999, p17-18.

43 Ramsar Convention Database, 1999.

44 Davidson and Delany, 2000, Contributing paper for WCD Thematic Review II.1 Ecosystems, p4, 13.

45 Ramsar Convention Database, op cit; Bridle and Sims, 1999, p3.

46 Lovgren, 1999 env136, WCD Submission, p2, 8.

47 Crabb, 1997, p42.

48 This problem is lessened when the water all passes through the turbines, and there is no spill.

49 Eley and Watkins, 1991, p21.

50 WCD Thematic Review II.1 Ecosystems, section 4.2.1.

51 WCD Thematic Review II.1 Ecosystems, section 6.2.

52 IEA, op cit, p27-29.

53 WCD Thematic Review II.1 Ecosystems, Executive Summary, p xii.

54 WCD Thematic Review V.2 Environmental and Social Assessment, section 2.5.

55 Bowman et al, 1999, Executive Summary p xii; Epple, 2000 opt136, WCD Submission, p3.

56 US Bureau of Reclamation, 2000a.

CEAP
"Vejo o Futuro Repetir o Passa

Chapter 4
People and Large Dams – Social Performance

The social impacts of large dams – their impact on people's livelihoods, health, social systems, and cultures – are an integral part of their performance record. Dams are one of a series of public infrastructure projects aimed at the economic development of a region, nation, or river basin. The direct benefits they provide to people are typically reduced to monetary figures for economic analysis and are not often recorded in human terms. In addition, simply accounting for these direct benefits often fails to capture the full social benefits associated with

providing water, electricity, and flood control, and including any indirect economic benefits or multiplier effects.

At the same time, however, dams have negatively affected many people and societies. This is clear throughout the WCD Knowledge Base, most poignantly through many of the presentations made by dam-affected peoples in the WCD Regional Consultations and the non-governmental organisation (NGO) hearings in Europe and South Africa. Globally, the overall magnitude, extent and complexity of these adverse social impacts for the displaced and for those dependent on the riverine ecosystem – both upstream and downstream from a dam – are of such significance as to merit detailed consideration in any assessment of the rationale for dam construction. Further, it is apparent that these impacts are – even today – often not acknowledged or considered in the planning process and may remain unrecognised during project operations. Where measures are put in place to mitigate impacts on affected people they typically fail to address adequately the problems caused by the decision to build a large dam.

Just as with the economic impacts of large dams, the social and environmental impacts of dams can be classified as gains or losses accruing to different social groups – now and in the future. Analysis of the WCD Knowledge Base, and in particular the WCD Case Studies, indicates that the poor, vulnerable groups and future generations are likely to bear a disproportionate share of the social and environmental costs of large dam projects without gaining a commensurate share of the economic benefits. Where the broader costs and benefits – economic, environmental and social – fall unequally within society, decision-making on project appraisal and selection based simply on summing up the positives and negatives is inadequate on equity grounds. At the same time it is clear from the emerging experience with good practice on benefit sharing mechanisms and reparations detailed in the Knowledge Base that a continuation of the legacy of inequity associated with many large dams is not only unacceptable, but unnecessary.

This chapter uses the WCD Knowledge Base to present a number of issues of strategic importance surrounding the social impacts of large dam projects, particularly those that underpin the growing worldwide opposition to dams by affected communities. To acknowledge the costs in this regard is not to deny the benefits. However, in order to provide a foundation for the way forward, it is essential to understand the extent, diversity, and range of social impacts – particularly the adverse impacts. This understanding must extend to all manifestations and implications for various population groups; the extent to which such impacts were addressed; and the outcome of mitigation, resettlement and development measures.

The chapter begins with a brief overview of social impacts during the planning and project cycle, with an emphasis on the social costs and benefits, as well as the indirect economic benefits, that accompany dam projects and the services they provide. The impacts on people displaced from their homes and livelihoods, indigenous peoples, downstream communities, gender, human health and cultural heritage are then explored at length. Analysis of the costs and benefits of large dam projects and their distribution across different groups in society leads to conclusions on equity and the adequacy of the 'balance sheet' approach of evaluating large dams. Initiatives addressing equity concerns are then described, followed by conclusions for the chapter.

Socio-economic Impacts through the Planning and Project Cycle

Given the different types, sizes and locations of large dams in the WCD Knowledge Base, generalisations on the socio-economic impacts of dams are limited and, in many cases the nature or significance of the impacts are contested. This section briefly illustrates and characterises the social costs and benefits, and the indirect economic costs that are associated with the planning and project cycle, with particular attention to those that accompany the provision of dam-related services.

Planning and design

At the planning and design stage an important social impact is the delay between the decision to build a dam and the onset of construction. Dams are often discussed years before project development is seriously considered and once a site is identified a form of 'planning blight' can take place, making governments, businesses, farmers and others reluctant to undertake further productive investments in areas that subsequently might be flooded. Communities can live for decades starved of development and welfare investments. A related problem is the fear felt by many people living in a possible reservoir area. Such psychological stress cannot be effectively quantified in economic terms, but it is a real issue. At this stage, project beneficiaries include those sustained by the business generated by the planning process, including contractors, consultants and workers employed on the project. In the case of developing countries, particularly smaller countries with a limited 'dam industry', foreign consulting firms have often undertaken the preparatory studies.[1]

Construction

At the planning and design stage, an important social impact is the delay between the decision to build a dam and the onset of construction. This can result in communities living for decades starved of development and welfare investments.

During the construction period, dam projects require a large number of unskilled workers and smaller but significant amounts of skilled labour. New jobs are therefore created both for skilled and unskilled workers during the construction phase. The WCD Case Studies report that Kariba and Grand Coulee employed between 10 000 and 15 000 workers each. During the peak construction period, a labour force of about 15 000 people was employed at Tarbela, helping to build local capacity for subsequent national development projects. While skilled labour is typically drawn from the national labour market, international contractors are often involved at this stage as well. The construction of dams and their associated infrastructure bring significant benefits to the employees and shareholders of companies engaged in construction and the supply of equipment and materials.

The beneficial effect on local communities is often transient due to the short-lived, pulse impact of the construction economy on dam construction sites. Careful planning may, however, enhance the 'boom' phase and lead to long-lasting benefits. Roads, power lines, social services and other infrastructure installed through the building of a dam provide access to previously

inaccessible areas, allowing settlement and connecting local economies to national markets. This has both positive and negative social impacts. Exposed to the national economy, indigenous and vulnerable groups find their lands and livelihoods threatened by forces beyond their experience or control. Similarly, existing settlements at construction sites have found themselves subject to increased health problems (including malaria, sexually transmitted diseases, and HIV-AIDS) and a loss of social cohesion with the large influx of outsiders.

Irrigation

As with livelihood enhancement, the broader impacts of irrigation projects on rural and regional development were often not quantified. Dams, along with other economic investments, generate indirect economic benefits as expenditure on the project and income derived from it lead to added expenditure and income in the local or regional economy. The WCD Case Studies give examples of these 'multiplier' benefits resulting from irrigation projects. In the case of Grand Coulee, agribusiness and local communities prospered due to irrigated production yielding multiplier effects of 1.5-1.7. Similarly, the Aslantas dam spun off projects that led to a tripling in agricultural processing and machinery manufacturing for the area. Beyond these economic impacts, irrigation schemes also produce a series of indirect social benefits derived from the – typically unanticipated – multiple use of irrigation water. Estimates of project benefits usually rely simply on the expected crop output and ignore the use of water for horticulture, livestock and fish production, as well as domestic water supply.[2]

How much of a stimulus irrigation projects have provided to wider regional development is a complex issue that needs more explicit recognition during project formulation to ensure that the enabling framework is in place to encourage growth. In this regard, regional multiplier effects are useful indicators of the distribution of project benefits to a project region. However, they represent a gain in the economic welfare of the nation only when excess capacity (unemployed resources) exists. Where it does not exist they simply represent a transfer (or re-distribution) of resources from one region of the economy to another.

The role of dams in improving nutrition and food security is contested. Increases in food production from irrigated agriculture may lead to both income and price effects as farming households in irrigated areas increase their purchasing power and the price of staple foodstuffs falls for urban (and other non-farming) households. For these households irrigated agriculture and, implicitly, large irrigation dams are likely to have contributed to greater food security and improved nutrition at the household level.

At the national level, nutritional levels increased over the 25 years from 1970 by 14% in India and 30% in China.[3] These two countries are two of the largest builders of irrigation dams. However, the actual extent of the contribution of large dams to these improvements is difficult to determine. The India Case Study calculates that the share of the total increase in food production from 1950-1993 attributable to additional land brought under irrigation is

10%. The study does not assess the extent of productivity gains derived from access to irrigation water. India's Central Water Commission puts the share at 30%. In the past 50 years India achieved a marginal surplus in terms of per capita food availability. In percentage terms India also saw a decrease in the share of the rural population below the poverty line. However during this period the absolute number of people below the poverty line – that is without capacity to buy food – increased from 180 million to 300 million.[4] Again, the precise impact of dams on these numbers is not known.

National statistics mask significant local variation. Of concern is that people in food-deficit areas are disadvantaged in terms of access to basic food grains and pay higher prices than those in food-surplus areas. Contributing to these concerns is the tendency of large irrigation schemes to lead to the production of more cash crops than envisaged at planning and less food crops (as indicated in Chapter 2). Producing higher value cash crops is a rational decision for commercial farmers who may choose to purchase foodstuffs. Once poorer farmers achieve household food self-sufficiency, they will probably choose to market surplus produce and, therefore, shift to higher value crops. The concern is that those people who do not participate in the irrigation project or are otherwise marginalised due to dam construction may face higher food prices and decreased food security as a result.

Hydropower

New energy services provided by dams have benefited urban populations and others connected to power distribution systems. Typically, in countries with low levels of energy services, even small energy inputs bring significant welfare improvements. The experience in the informal settlements (*favelas*) in São Paulo, Brazil illustrates the social and environmental benefits electricity services can bring to people (see Box 4.1). As in Sao Paulo, policy measures are increasingly used to bring electricity to poorer sections of the population. The slum areas in Mumbai in India, Manila in the Philippines, and other cities and towns in the developing world are further examples of such efforts.

Employment

In terms of generating employment, the principal impact of large dam projects – aside from construction jobs – arises from the new productive enterprises allowed by the provision of water or electricity. The Case Studies provide a number of examples of anticipated and unanticipated employment generation (also see Table 4.1, p121).

Box 4.1 Bringing electricity to the *favelas* in São Paulo, Brazil

Between 1973 and 1993, São Paulo's *favelas* swelled from about 700 000 inhabitants to over 2 million. Initially, the squatters' shacks had only sporadic and illegal electrical connections, partly because the electric utility had no procedure for electrification of such structures and partly because municipal authorities thought that improvement of the *favelas* would condone illegal occupation of the land.

By 1979, the city and the electric utility, Electropaulo, came to an agreement and connected some *favelas* to the grid using simplified installation kits and no meters. Consumers were billed a flat rate, which was subsidised, for a minimal monthly consumption of 50 kWh – enough to run a couple of lamps and a radio or other domestic appliance. The cost of metering was considered too high for such low usage levels.

By 1983, some 100 000 shacks were connected, and the quality of life improved. Better lighting simplified the tasks of cleaning and maintaining the shack and caring for children and sick people. Without smoke from candles and kerosene lamps health improved. People started to use TV sets, irons and refrigerators. Where water services were provided, electric showers also became more common. For the squatters, an important benefit was receiving bills with their name and address, which gave them a certain social recognition as well as access to credit.

A decade later, electricity consumption per shack had increased to 175 kWh. Many of the dwellings had been greatly improved, and services were more reliable. Some demographers attribute the strong decline in Brazil's population growth rate – from 3.8% per year in 1970 to 1.4% today – to the adoption of new cultural values that spread partly through television, which electricity made available.

Source: Boa Nova and Goldemberg, 1999.

In the case of the Aslantas dam, an increase in employment in farming was projected. It did not materialise due to a shift to less labour-intensive crops, mechanisation and a general migration to urban areas. In the Orange River Development Project, although farm jobs decreased throughout South Africa from 1960 through the 1980s (the latest period for which data are available), they dropped less in the command area of the project. When compared to national trends the loss of at least 4 000 regular jobs was avoided and effectively some 16 000 seasonal jobs were created in downstream areas by the dams and associated irrigation development.[5] The employment impacts accrued largely to Black Africans and the 'Coloured' racial groups who represented 97% of paid farm workers.

As with any indirect economic impact, it is important to consider not just the gross number of jobs created by a project but also whether alternative uses of project resources would generate similar gains.

Employment gains are also engendered by hydropower production and other services provided by reservoirs. In the case of Tucurui and Grand Coulee, a sizeable percentage of the power produced goes to industries in the respective regions. Similarly, the creation of commercial and sport-fishing industries, as well as reservoir-based recreation and tourism, has led to job creation at Grand Coulee, Tucurui and Kariba. Inland navigation can also provide substantial employment. The Panama Canal, based on two large dams, directly employs 8 000 people on Canal operations and creates jobs in the local shipping services industry and duty-free zone.[6] As with any indirect economic impact, it is important to consider not just the gross number of jobs created by a project but also whether alternative uses of project resources would generate similar gains.

Displacement of People and Livelihoods

Many development interventions to transform natural resources, particularly large-scale infrastructure projects - involve some form of displacement of people from their livelihoods and homes. Large dams are perhaps unique amongst such projects in that they can have widespread and far-ranging ecosystem impacts due simply to the blocking of a river. The result is a series of terrestrial, aquatic and riparian impacts that not only affect ecosystems and biodiversity but also have serious consequences for people who live both near and far from the dam site. A large, multi-functional resource base like a river and its surroundings is characterised by a complex web of diverse, interconnected, implicit and explicit functional roles, dependencies and interactions. Consequently the social and cultural implications of putting a dam into such a landscape are spatially significant, locally disruptive, lasting and often irreversible.

Large dams have significantly altered many of the world's river basins, with disruptive, lasting and usually involuntary impacts on the livelihoods and socio-cultural foundations of tens of millions of people living in these regions. The impacts of dam-building on people and livelihoods – both above and below dams – have been particularly devastating in Asia, Africa and Latin America, where existing river systems supported local economies and the cultural way of life of a large population containing diverse communities.

Displacement is defined here as referring to both 'physical displacement' and 'livelihood' displacement (or deprivation). In the narrow sense displacement results in the physical displacement of people living in the reservoir or other project area. This occurs not only from the inundation of reservoirs but from the installation of project facilities and associated infrastructure. The WCD Knowledge Base records that all too often this physical displacement is involuntary and involves coercion and force – in a few cases even killing.

However, the inundation of land and alteration of riverine ecosystems – whether upstream or downstream – also affects the resources available for land- and riverine-based productive activities. In the case of communities dependent on land and the natural resources base, this often results in the loss of access to traditional means of livelihood, including agricultural production, fishing, livestock grazing, fuelwood gathering and collection of forest products, to name a few. Not only does this disrupt local economies, it effectively displaces people – in a wider sense – from access to a series of natural resource and environmental inputs into their livelihoods. This form of livelihood displacement deprives people of their means of production and dislocates them from their existing socio-cultural milieu. (See Box 4.2) The term 'affected' thus applies to people facing either type of displacement.

The timing of these social impacts varies, depending on the proximate cause. In the case of loss of home and livelihood due to the filling of a reservoir, the social impacts are quite immediate. The implications for downstream livelihoods, however, come to the fore only after completion of the dam. At this point they may set in quickly, as with changes in flow and their impact on recession agriculture or slowly, as with physical and chemical changes that are translated into degradation of ecosystem function and loss of biodiversity.

Scale of physical displacement

The WCD Knowledge Base confirms that there are many dams that have caused physical displacement – and indeed that large dam construction has physically displaced tens of millions of people worldwide in the last half century. The scale and extent of impacts will vary depending on location, size and other dam characteristics such as inundated area, and population density in the river basin. In the eight WCD Case Studies, the only one without any physical displacement was the chain of dams in the Glomma and Laagen Basin. In the Cross-Check Survey, physical displacement is reported in 68 of the 123 dams (56%). Of the dams in this sample, 52 out of 68 dams are in Latin America, Asia, and sub-Saharan Africa. Large dams on the main-stem of a

Box 4.2 Economic, socio-cultural, and health impacts of livelihood displacement

Resettlement programmes have predominantly focused on the process of physical relocation rather than the economic and social development of the displaced and other negatively affected people. The result has been the impoverishment of a majority of resettlers from most dam projects throughout the world.

According to Cernea's Impoverishment Risks and Reconstruction model, displacement epitomises social exclusion of certain groups of people. It culminates in physical exclusion from a geographic territory and economic and social exclusion from a set of functioning social networks. Thus, affected people face a broad range of impoverishment risks that include landlessness, joblessness, homelessness, marginalisation, food insecurity, increased morbidity, loss of common resources, and community disarticulation that result in a loss of socio-cultural resilience.

The key economic risks to affected people are from the loss of livelihood and income sources such as arable land, common property resources (forests, grazing land, ground and surface water, fisheries, and so on), and changed access and control of productive resources. The loss of economic power with the breakdown of complex livelihood systems results in a temporary or permanent, often irreversible decline in living standards, leading to marginalisation. Higher risks and uncertainties are introduced when diversified livelihood sources are lost. Loss of livelihood and disruption of agricultural activity can adversely affect household food security, leading to under-nourishment. Higher incidence of diseases associated with deteriorating water quality can result in increased morbidity and mortality. High mortality rates immediately following involuntary resettlement from the reservoir areas of the Kariba and Aswan High dams are cases in point. Forced displacement tears apart the existing social fabric, leading to socio-cultural disarticulation.

Source: Cernea, 1999; Cernea, 2000; Cernea and Guggenheim, 1993; McDowell, undated; Scudder, 1997a,b.

river and in densely populated regions of the world will inevitably displace people. In the Cross-Check sample, 26% of dams with a surface area less than 1 km^2 report physical displacement compared with 82% of dams over 100 km^2 in area. Yet the Cross-Check figure may understate the occurrence of physical displacement given the larger tendency towards systematic under-enumeration discussed below.

The overall global level of physical displacement could range from 40 to 80 million. According to official statistics, dams have displaced 10.2 million people in China between 1950 and 1990 (34% of all development-related displacement including that due to urban construction).[7] Independent sources estimate that the actual number of dam-displaced people in China is much higher than the official figure, with 10 million displaced in the Yangtze Valley alone.[8] Large dams in India displaced an estimated 16–38 million people.[9] Thus, in India and China together, large dams could have displaced between 26–58 million people between 1950 and 1990. The level of displacement has increased substantially after 1990 with the construction of projects such as Three Gorges in China. Among the projects involving displacement funded by the World Bank, large dams account for 63% of displacement.[10]

These figures are at best only estimates and certainly do not include the millions displaced due to other aspects of the projects such as canals, powerhouses, project infrastructure, and associated compensatory measures, such as biological reserves and so on. They also refer to physical displacement only and thus do not include communities upstream and downstream of dams that have suffered livelihood displacement.

Box 4.3 Missing numbers of affected people: Sardar Sarovar project, India, and Pak Mun dam, Thailand

For the Sardar Sarovar project, the 1979 Narmada Water Disputes Tribunal gave the number of displaced as 6 147 families, or about 39 700 people. The World Bank's 1987 mission placed the total at 12 000 families (60 000 people). In 1991, the project authorities provided an estimate of 27 000 families. According to three state governments, the current estimate of displaced families stands at 41 000 (205 000 people). This number will probably increase, since 13 years after full-scale dam construction began, resettlement surveys have still not been completed. The current estimate does not include at least 157 000 people displaced by canals. Nor does it include those moved to make space for the creation of a wildlife sanctuary and for the resettlement of people displaced by the dam, or the 900 families displaced in the early 1960s to make room for construction site infrastructure. The nature and extent of the dam's impact on downstream livelihoods were not assessed. Serious efforts to survey the affected villages and people began several years after the start of dam construction work in response to intense struggles by the affected since 1985.

In 1991, when construction started on the Pak Mun dam, 241 families were counted as displaced. By the time construction was completed it was clear that another 1 459 households had to be relocated. The true extent of the social impact only became evident when the impact of the dam on fisheries livelihoods was admitted in response to prolonged agitation by the affected people. By March 2000, the Thai government had paid interim compensation – pending a final solution to the permanent loss of fisheries livelihoods – to 6 204 households for livelihood loss during construction.

Source: Sardar Sarovar in Brody, 1999, Contributing Paper for WCD Thematic Review I.1 Social Impacts, Section 5.2; Supreme Court of India, 1999; Morse and Berger, 1992, p51, 89; WCD Pak Mun Case Study.

Under-counting of the displaced

At the planning stage, the numbers of both directly and indirectly affected people have frequently been under-estimated (see Box 4.3), and there has been inadequate understanding of the nature and extent of the negative impacts. In all the WCD Case Studies, the initial assessments of the projects failed to account for all the affected people. The level of under-enumeration ranges from 2 000 to 40 000 people. Examples from large dam projects in Africa include the tri-national Ruzizi hydroelectric project involving Zaire, Rwanda, and Burundi, the Funtua dam in Nigeria, and the Kiambere reservoir on the Tana River in Kenya, with discrepancies ranging from 1 000 to 15 000 people.[11] Similar observations emerge from other regions.[12] Among projects funded by the World Bank, the actual number of people to be resettled was 47% higher than the estimate made at the time of appraisal.[13]

The WCD Cross-Check Survey reveals a similar trend towards under-estimation insofar as 35% more people were resettled than initially planned. This figure must represent a lower bound on the error in the Cross-Check dams, given the poor reliability of the estimates. Data provided by NGOs as part of the external review of the Cross-Check Survey confirm this view, as many of the actual figures for the physically displaced were disputed.

Affected groups that are not counted or compensated

Surveys of the categories of people to be affected by dams have generally been inadequate. The scope of definition of the affected has been limited, and the totality of affected groups has not always been determined. The principal categories excluded from assessments include the landless, downstream communities and indigenous people. The WCD Case Studies show that communities situated downstream from the dam, those without land or legal title, indigenous people and those affected by project infrastructure (and not just the reservoir) were not considered as affected people at the time of design.

Among those assessed, compensation has usually gone only to those in possession of legal titles, leaving out a large number of people – often the poorest – who depend on common resources such as forests and grazing grounds for subsistence. In the WCD Case Studies on Grand Coulee, Tarbela, Aslantas and Tucurui, only those affected people with legal title were compensated for the loss of their lands and livelihoods. With such criteria for eligibility, indigenous peoples and ethnic minorities suffer disproportionately as they may lack citizenship, tenancy, or land tenure papers. One-fifth of those physically displaced by the Kao Laem dam in Thailand were from the Karen ethnic group. Because they lacked legal residence permits, they were considered ineligible for resettlement.[14]

Often, people physically displaced by canals, powerhouses, and associated compensation measures such as nature reserves are not enumerated and considered for resettlement.

Often, people physically displaced by canals, powerhouses, and associated compensation measures such as nature reserves are not enumerated and considered for resettlement. Examples of this type emerge from all parts of the world, including Sulawesi, Indonesia; the Mahaweli Development Programme, Sri Lanka; and the Sardar Sarovar project, India.[15] Further, compensation is often not paid to those affected by such additional components of a project.[16]

While not all large dams have involved physical displacement it would be much rarer to find a river whose natural function is not used or appreciated by people in some fashion. And in many cases in densely populated tropics large dams will lead to both physical and livelihood displacement. For example, the Urrá 1 dam on the Upper Sinú River in Colombia displaced 12 000 people but also affected severely more than 60 000 fishermen in the lower Sinú, where the fish population diminished drastically as a result of the dam.[17]

Physically displaced populations enumerated but not resettled

Among physically displaced people officially recognised as 'project affected,' not all are given assistance to resettle in new locations.

In India, those actually resettled range from less than 10% of the physically displaced in the case of the Bargi dam to around 90% for the Dhom dam.[18] The Yacyreta project in Argentina and Paraguay is a classic illustration of delayed and incomplete resettlement. It took the project developers 20 years to resettle just over 30% of the displaced people, leaving the remainder to be resettled in the less than two years before the reservoir would be filled. If the experience of other projects involving large displacement in a region is taken into consideration, a large proportion of the Yacyreta displaced are unlikely to be resettled.[19]

Little or no meaningful participation of affected people in the planning and implementation of dam projects – including resettlement and rehabilitation – has taken place.

The WCD Case Study on Tarbela reports that of the 96 000 physically displaced people enumerated for the Tarbela dam in Pakistan, two-thirds qualified for replacement agricultural land in Punjab and Sindh provinces. Of these, some 2 000 families or approximately 20 000 people did not receive land when the amount of land provided by Sindh fell short of that promised. In the case of Aslantas, only 75 of an estimated 1 000 displaced families asked for resettlement, with the remainder choosing cash compensation. Of these, 49 were considered eligible for resettlement and subsequently received new housing. In the case of Tucurui, of the indigenous groups physically displaced only the Parakaná people were resettled; the other indigenous group that lost land to the dam was not considered for resettlement benefits.

In the Grand Coulee project, the Colville and Spokane reservation lands, in addition to three towns, were inundated. By the summer of 1940, the rising water covered the first tracts of land and the government was clearing allotments and burning houses. But none of the owners had been paid compensation. The Colville agent reported that the Indians were growing resentful, they needed money to build new homes and improve their remaining property and they knew that white owners across the river had already been paid.[20] The Colville and Spokane tribes only received cash compensation for reservation land in 1941. Two tribal towns, Keller and Inchelium, were rebuilt and still exist today, but other smaller settlements were lost with the inundation.[21]

Experience of affected people with resettlement, mitigation and compensation

Little or no meaningful participation of affected people in the planning and implementation of dam projects – including resettlement and rehabilitation – has taken place. Involuntary, traumatic and delayed relocation, as well as the denial of development opportunities for years and often decades, has characterised the resettlement process.[22] For millions of people on all continents, displacement has essentially occurred through official coercion.[23] The starkest example from the WCD Case Studies comes from one of the earlier dam projects, Kariba, where the resistance of the Tonga people ended with the fatal shooting of eight people.[24] The displacement of people at the Sri Sailam project in India in 1981 was also achieved through force.[25] Eviction of people at the Chixoy dam site in Guatemala led to the killing of about 376 Maya Achi people from the submergence area.[26] In implementing the Miguel Aleman dam in Mexico, employees from the Papaloapan River Commission set fire to homes of 21 000 Mazatec Indians who were refusing to move.[27] In other cases – such as the submergence of 162 villages when the Bargi dam in India was filled without warning –

authorities have resorted to eviction through the filling of reservoirs prior to the departure of the displaced.[28]

Cash compensation is a principal vehicle for delivering resettlement benefits, but it has often been delayed and, even when paid on time, has usually failed to replace lost livelihoods. Compensation is understood to refer to specific measures intended to make good the losses suffered by people affected by the dam. It usually takes the form of a one-off payment either in cash or kind for land, housing and other assets.[29] The WCD Case Studies show that downstream communities affected by loss of floodplain vegetation and fisheries in Tucurui dam in Brazil and Tarbela dam in Pakistan were not compensated. The Gavaio da Montanha indigenous people, whose lands were affected by the transmission lines in the Tucurui project, were initially not considered eligible for compensation but were later given cash compensation. In the cases of Aslantas (Turkey), Tarbela (Pakistan) and Kiambere (Kenya) dams, affected people did not receive adequate compensation to buy alternate land.[30]

Further, there have been many cases illustrating inadequate compensation, unsuitable mitigation, and lack of recourse, including the Sri Sailam project in India and the Kao Laem in Thailand.[31] Delays in compensation provisions, titles to landholdings and houses, and provision of basic services have occurred. Cases illustrating inordinate delays – from 5 to 15 years – include the Aswan High dam in Egypt, the Nangbeto in Togo, the Akosombo in Ghana, the Itá in Brazil, and the Bhumibol in Thailand.[32]

Resettlement sites are often selected without reference to the availability of livelihood opportunities or the preferences of displaced persons themselves. They have often been forced to resettle in resource-depleted and environmentally degraded areas around the reservoir. Such lands rapidly lost their capacity to support the resettled population. Among the earliest instances is the Liu-Yan-Ba project on the Yellow River in China, which displaced 40 000 people from fertile valleys and relocated them to windswept uplands. Erosion and loss of fertility ultimately led to the abandonment of painstakingly reclaimed farmland, and the drastic reduction of farmland led to extreme poverty.[33] Similar experiences have been recorded from Hoa Binh in Vietnam, Sirindhorn in Thailand, Batang Ai in Sarawak Malaysia, and other rice-growing East Asian countries with large rural populations.[34]

The loss of cultivable land and inability to gain good-quality replacement land has significantly affected indigenous peoples and peasant farmers. Examples are the Chinantec and Mazatec Indians displaced by the Miguel Aleman and Cerro de Orro dams in Mexico; the Kuna and Embera people in Panama; the Parakana, Asurini, and Gavio da Montanha people in Brazil; and the Tonga in Zambia and Zimbabwe.[35]

The replacement of agricultural land, basic services and infrastructure at resettlement sites has often failed to materialise, was inadequate, or was delayed for many years. Absence of livelihood opportunities forces affected people to abandon resettlement sites and migrate. Examples include Tarbela, where allotted agricultural land was of poor quality and basic services such as electricity, health facilities and schools were not provided. Electricity was only provided after 25 years. Similar experiences are recorded from resettlement sites at Tucurui, Sirindhorn dam in Thailand, and Akosombo in

Ghana.[36] In northeast Thailand, 15 000 farming families were left without lands as a result of failed resettlement schemes between 1960 and 1970.[37] Government reports in China characterised reservoir resettlement problems as 'seven difficulties' (*qui nan*) and 'four inadequacies' (*si cha*). The seven difficulties include shortages of electricity, drinking water, schools, food, medical services and means of communication and transportation. The four inadequacies refer to the insufficient amount and poor quality of irrigation, housing, flood control and reservoir maintenance facilities.[38]

Resettlement programmes have predominantly focused on the process of physical relocation rather than on the economic and social development of the displaced and other negatively affected people.[39] Lack of accountability on the part of the state for promised entitlements has led to poor (and incomplete) implementation of resettlement measures. Finally, long delays in the onset of resettlement programmes are common and lead to great uncertainty and psychological and social anxiety for those awaiting resettlement. These and other problems have severely eroded the effectiveness of resettlement and rehabilitation programmes in creating development opportunities for the resettled and have heightened the risk of impoverishment for those being resettled.

An inverse relationship between the scale of displacement and the possibility of properly carrying out resettlement is evident.

That the livelihoods of those resettled have not been restored therefore comes as little surprise. At least 46% of the 10 million Chinese resettled as a consequence of reservoirs are still in 'extreme poverty'.[40] In the case of India, 75% of the people displaced by dams have not been rehabilitated and are impoverished.[41] A monitoring study in 1993 found that 72% of the 32 000 displaced people from the Kedung Ombo dam in Indonesia were worse off after resettlement.[42] Conditions among the 800 ethnic minority Nya Heun families displaced by the recently constructed Houay Ho dam in Laos are reported to be appalling, with people suffering from severe lack of food, shortage of arable land and insufficient clean water.[43]

Often the sheer scale and level of displacement makes adequate rehabilitation and livelihood restoration difficult. An inverse relationship between the scale of displacement and the possibility of properly carrying out resettlement is evident. For example, the inundation zone of the Danjiangkou project in the Hubei province of China, implemented in 1958, covered four rural counties and 345 villages. Although the Chinese Government tried hard in the 1980s and 1990s to improve the living standards of the physically displaced in the Danjiangkou area, many unsolved problems persist. In 1996, an estimated 35 000 of those resettled around the city of Shiyan had incomes below the official poverty line.[44] And in India, the sheer extent of displacement is making resettlement a daunting task for the Sardar Sarovar project (see Box 4.3). Since the start of the resettlement process in 1984, less than 20% of the recognised displaced people have been resettled.[45]

Elements for positive mitigation, development and resettlement outcomes

Impoverishment of affected people is increasingly seen as unacceptable but it is also unnecessary since there are a wide range of opportunities available for making not only resettlers, but all affected people project beneficiaries. This is in the interests of all stakeholders since, as beneficiaries, affected people add to the stream of project benefits, while reducing costs. The problem of making them beneficiaries lies not with affected people, who time and again have shown the capacity to respond to opportunities that are available, but with the inadequate laws, policies, plans, financing capacity and political will of governments and project authorities.

For resettlement to lead to the development of those resettled, the process has to address the complexities of resettlement itself and to effectively engage the full range of political and institutional actors. A positive outcome requires several enabling conditions such as low level of displacement, resettlement as development policy with supporting legislation, a combination of land and non-land based sustainable livelihood provisions, strong community participation and accountability and commitment from government and project developers.

Providing a legal framework that governs the process of displacement is important for protecting rights of affected people. For instance, China's Reservoir Resettlement Act specifies the rights of affected people and defines the obligations of the State and the procedures for settling conflicts and the redress of complaints.[46] Recent changes in Chinese policy serve as an interesting model for other countries. Minimising displacement is another enabling condition to effectively address resettlement needs.

In some cases, project proponents have made an effort to resettle people as communities in order to minimise socio-cultural disruptions. In the Kainji project, Nigeria measures were taken to maintain community cohesion and identity.[47] Social science input had a strong influence in determining the outcome there. Baseline demographic and socio-cultural studies played an important role in informing planners about distinct social and cultural features of people living in the proposed impact areas.

In cases where compensation packages were negotiated with project affected people and other stakeholders, the process has resulted in fewer instances of injustice and better outcomes for the resettlement process. Even where everyone may not see negotiated compensation as the most appropriate or effective option, affected people tend to feel more satisfied for having engaged in the negotiation process, as attested by the Zimapan resettlement program in Mexico.[48] In case of the Mubuku III hydropower project in Uganda, public consultation meetings with Ugandan local council system and community leaders were held for identification and valuation of land.[49] This minimised displacement by enabling adjustments to the routing of canals.

For resettlement to lead to the development of those resettled, the process has to address the complexities of resettlement itself and to effectively engage the full range of political and institutional actors.

The plan developed by the Chinese government for people affected by the Xiaolangdi dam provides an example of an integrated strategy that combines land and non-land based activities to ensure livelihoods.[50] Resettlement plans focus on building the skills of the adversely affected through substantial investments in imparting new, relevant skills that are in demand in the regional and local economy, enhancement of existing skills, and special measures to facilitate capacity building amongst women. This approach calls for resettlement plans to develop linkages between negatively affected people and other sectors of the economy, underlining the importance of incorporating the overall development of the affected economy into the resettlement programme.

An inclusive process involving all groups – including host communities – enables initiatives to promote resettlement as development to be managed jointly by the people and project and government institutions as a long-term process that can contribute to the stream of project benefits. In the case of Itá dam in Brazil, a sustained struggle by the local community for proper resettlement resulted in joint negotiation for benefit-sharing, resettlement as a community, and consultative implementation of the programme.[51] The accord between affected people and the utility Electrosul resulted in a community managed resettlement programme.

Empowering people, particularly the economically and socially marginalised, by respecting their rights and ensuring that resettlement with development becomes a process governed by negotiated agreements is critical to positive resettlement and rehabilitation.

Evolving policy frameworks in countries such as Ghana and China reflect two of the more positive attempts to learn from past resettlement experience. In the Ghanaian case, with the benefit of the administrative continuity of the Volta Resettlement Authority, planners at Kpong dam were able to avoid some of the mistakes made earlier at Akosombo.[52] While the legal framework related to land and resettlement is comprehensive and improvement was noted at Kpong, not all good intentions were successfully pursued.[53] China's resettlement experience before 1980 was in many ways inadequate, and the new policy sought to improve matters. How effectively such policy improvements are translated into successful resettlement and development outcomes remains to be seen.

Past and current experiences of affected people and the rapidly changing context reinforces the argument that displacement needs to be located in the broader perspective of the tensions between the local versus the national and international interests. Just as displacement is not an inevitable consequence of infrastructure development, resettlement need not necessarily result in impoverishment. Empowering people, particularly the economically and socially marginalised, by respecting their rights and ensuring that resettlement with development becomes a process governed by negotiated agreements is critical to positive resettlement and rehabilitation.

Indigenous Peoples

Large dams have had serious impacts on the lives, livelihoods, cultures and spiritual existence of indigenous and tribal peoples. Due to neglect and lack of capacity to secure justice because of structural inequities, cultural dissonance, discrimination and economic and political marginalisation, indigenous and tribal peoples have suffered disproportionately from the negative impacts of large dams, while often being excluded from sharing in the benefits.[54] In the Philippines, almost all the larger dam schemes that have been built or proposed were on the land of the country's 6–7 million indigenous people.[55] Similarly in India, 40–50% of those displaced by development projects were tribal people, who account for just 8% of the nation's 1 billion people.[56] These costs are not balanced by any receipt of services from dams or by access to the benefits of ancillary services or indirect economic multipliers in the formal economy.[57]

In general, development planning and implementation have inadequately addressed the special needs and vulnerabilities of indigenous and tribal peoples. In many cases, large dams have only perpetuated this disregard and exacerbated the problem – even causing multiple displacements of these peoples. The Waimiri-Atroari of northern Brazil numbered 6 000 in 1905. Eighty years later, massacres and disease left only 374 Waimiri-Atroari alive. In 1987, the Balbina dam flooded two of their villages, displacing 107 people.[58] Similarly, in the unique Biobio region in Chile, the Pehuenches were pushed higher and higher up in the valley throughout the last century. The Pangue and Ralco projects would inundate much of the remaining ancestral land of the Pehuenches.[59] For the Ibaloy indigenous people currently living in the fertile Agno river basin in the Philippines, the San Roque dams would be the third to impact their lands.[60] Similar experiences are recorded in Indonesia, Malaysia, Thailand, Brazil, Argentina, Mexico, Panama, Colombia, Guatemala, United States, Canada, and Siberia.[61]

For indigenous peoples and ethnic minorities, dam-induced displacement can trigger a spiral of events that spreads beyond the submergence area. A case in point is the situation of the 100 000 Chakma people displaced by the Kaptai hydropower dam in the Chittagong Hill Tracts, Bangladesh. The project submerged two-fifths of their cultivable land; as a consequence, 40 000 Chakma left for India and another 20 000 were supposed to have moved into Arakan in Burma. The Chakma have never gained citizenship for themselves or their children in India. The conflict triggered by land shortage between the Buddhist Chakma people and Muslim Bengali settlers has cost 10 000 lives since the project was completed in 1962.[62]

The rights of indigenous peoples and ethnic minorities are often poorly defined or enshrined in the national legal frameworks, and consequently their entitlements have lacked effective protection. The Bayano dam in Panama that forced the indigenous Kuna and Embera peoples from their traditional territories resettled them on land that was less fertile and subject to encroachment by loggers. The Panamanian government systematically failed to fulfil agreements made with the affected indigenous people at the time of construction, as well as commitments negotiated later. Among the violations was the government's failure to compensate adequately for the loss of traditional territories and provide legal titles to the new lands.[63] What happened in Panama in the 1970s is similar to what has happened in Malaysia in the 1990s.[64] In the case of the Bakun project, rights to indigenous common land in the Ulu Belaga site were not recognised or properly assessed.[65]

Industrial countries' experience with indigenous peoples in the era of building large dams was not very different from that of developing countries. Dams built during the 1950s and 1960s cost the indigenous nations of the Missouri River basin in the United States an estimated 142 000 hectares of their best land, including a number of burial and other sacred sites, leading to further impoverishment and severe cultural and emotional trauma. A guarantee used to rationalise the plan – that some 87 000 hectares of Indian land would be irrigated – was scrapped as the project neared completion.[66]

Despite changes over the years, new projects in industrial countries raise similar issues. A case in point is the second stage of the Churchill Rivers project in Labrador,

Canada, consisting of two dams and two river diversions that will flood a large area of hunting territory of the Innu people who live on both sides of the provincial boundary. The Innu have yet to be clearly recognised as the owners of their lands, and the whole area is the subject of an unresolved Innu land claim currently being negotiated with the Canadian government.[67]

In the last two decades, international and national laws progressively aimed at empowering indigenous peoples to play a decisive role in development planning and implementation. Constitutions in some countries recognise the vulnerability of indigenous people in mainstream development processes arising from their distinct culture and history and have designed safeguards to protect their rights.[68] The scope of international law has widened and currently includes a body of conventional and customary norms concerning indigenous peoples, grounded on self-determination. In a context of increasing recognition of the self-determination of indigenous peoples, the principle of free, prior, and informed consent to development projects and plans affecting these groups has emerged as the standard to be applied in protecting and promoting their rights in the development process.

Downstream Livelihoods

Downstream impacts can extend for many hundreds of kilometres and well beyond the confines of the river channel. The serious implications come to the fore only after completion of the dam and a number of the impacts only develop over time. In general, the downstream riverine communities have lacked social, economic and political power to seek mitigation, let alone development benefits.

Downstream communities throughout the tropics and subtropics face some of the most drastic impacts of large dams, particularly where the changed hydrological regime of rivers has adversely affected floodplains that supported local livelihoods through flood-recession agriculture, fishing, herding and gathering floodplain forest products. The disruption of downstream economies that results from the insertion of a dam and the subsequent reduction in natural floods can create uncertainty in livelihoods and render existing skills unproductive – leading to migration, dependence on informal wage labour in urban areas and impoverishment.

In northwestern Nigeria, the Bakolori dam on the Sokoto River reduced average flood levels by 50%, leading to a fall in cropped area of 53% and to a quarter of the households dropping dry-season cultivation as a component of their livelihood strategy.[69] Similarly, significant impacts on floodplain agriculture are seen in Niger, Chad, Nigeria, Sudan, Senegal and Mali.[70] In the case of Manantali dam on the Senegal River, between 500 000 and 800 000 people suffered from loss of access to productive floodplains that provided most or part of their means of survival.[71] The creation of the Sobradinho reservoir in Brazil affected the livelihood of 11 000 farm families downstream from the reservoir who depended on traditional floodplain agriculture.[72] The WCD Case Studies reveal that the cumulative impact of Tarbela dam and Kotri barrage has affected the grazing activities of pastoral communities in Pakistan.

Substantial losses to downstream fishery production as a result of dam construction are reported from around the world. Along with subsistence agriculture, fisheries constitute an important livelihood activity among large rural populations as well as an important low cost source of protein. Many of these households depend on fisheries either as their primary or a supplementary source of livelihood. The impact of dams on fisheries is no less in temperate regions, where salmon runs over large stretches of river in North America and Canada have been destroyed, affecting the livelihood, food security and cultural organisation of Native American communities. The Grand Coulee Case Study reports that the loss of salmon also had severe cultural and spiritual consequences integral to the First Nations' way of life. One problem faced by indigenous peoples who live downstream from dams in boreal zones is the discharge of power generating water on top of frozen rivers during winter, as in case of the Kolyma River in northern Yakutia.[73]

Adverse impacts on downstream fisheries have been severe, even in projects implemented in the 1990s. These impacts were not adequately assessed in, among others, the Urrá I project in Colombia; the Singkarak project in West Sumatra, Indonesia; the Lingjintan project in China; Theun Hinboun in Laos; and Pak Mun in Thailand.[74] Downstream communities and livelihoods are typically left out of any explicit reckoning of project impacts and subsequent efforts to manage these impacts – apart, perhaps, from installation of fish passages. One explanation for this is that the downstream communities are not only dispersed but also have typically lacked social, economic, and political power to press their case for mitigation and development. While the people affected by the flooding of the reservoir could assert their right to mitigation by refusing to move, those affected downstream have no such leverage.

Downstream impacts are not only among the most significant unassessed and unaddressed aspects of large dams, they are also indicative of the magnitude and spread of impacts associated with an altered river regime. The extent to which mitigation and development can be designed and implemented to address these complex and diverse concerns effectively is open to question. As demonstrated in a case from northern Nigeria, an examination of the economic value of downstream uses of water may also provide a convincing argument for setting aside dam projects (Box 4.4).

In the case of Manantali dam on the Senegal River, between 500 000 and 800 000 people suffered from loss of access to productive floodplains that provided most or part of their means of survival.

Box 4.4 Economic value of downstream floodplains, Hadejia-Nguru, Nigeria

In northern Nigeria, extensive floodplains exists where the Hadejia and Jama'are Rivers converge. The floodplains provide essential income and nutrition benefits in the form of agriculture, grazing resources, non-timber forest products, fuelwood and fishing for local populations, and help to recharge the regional aquifer which is an essential groundwater source. However, in recent decades the floodplains have come under increasing pressure from the construction of the Tiga and Challawa Gorge dams upstream. The maximum extent of flooding has declined from 300 000 hectares in the 1960s to around 70 000–100 000 hectares more recently with plans for a new dam at Kafin Zaki. Economic analysis of the Kano River Project, a major irrigation scheme benefiting from the upstream dams, shows returns to water of $1.73 per 1 000 m^3 and when the operational costs are included, the net benefits of the project are reduced to $0.04 per 1 000 m^3.

A combined economic and hydrological analysis was conducted to simulate the impacts of these upstream projects on the flood extent that determines the downstream floodplain area. The economic gains of the upstream water projects were then compared to the resulting economic losses to downstream agricultural, fuelwood and fishing benefits (valued at $32 per 1 000 m^3 of water in 1989 prices). Given the high productivity of the floodplains, the losses in economic benefits due to changes in flood extent for all scenarios are large, ranging from $3 million to $24 million. As expected, there is a direct trade-off between increasing irrigation upstream and impacts on the floodplains downstream. Full implementation of all the upstream dams and large-scale irrigation schemes would produce the greatest overall net losses, around $20 million.

These results suggest that the expansion of the existing irrigation schemes within the river basin is effectively 'uneconomic'. The introduction of a regulated flooding regime would reduce the scale of this negative balance substantially, to around $16 million. The overall combined value of production from irrigation and the floodplains would however still fall well below the levels experienced if the additional upstream schemes were not constructed.

Source: Acreman et al, 2000, Contributing Paper for WCD Thematic Review II.1 Ecosystems

Gender

Gender relationships and power structures are all too often detrimental to women. Extensive research has documented gender inequalities in access to, and control of, economic and natural resources. In Asia and Africa for example, women may have use rights over land and forests, but are rarely allowed to own and/or inherit the land they use.[75] Communities near the Tarbela dam in Pakistan practised the purdah system (seclusion of women) for centuries. This section employs the WCD Knowledge Base to examine whether dams aggravate, or ameliorate gender disparities.

Given the gender-blindness of the planning process large dam projects typically build on the imbalance in existing gender relations. For affected communities dams have widened gender disparities either by imposing a disproportionate share of social costs on women or through an inequitable allocation of the benefits generated. However, the WCD Knowledge Base also provides evidence of cases where dams have served as opportunities for reducing gender disparities, primarily among women in households or communities that receive access to project services.

Widened gender disparities

In spite of the fact that many countries and funding agencies have adopted specific gender policies in recent years aimed at mainstreaming gender issues in their development interventions, actual project planning and implementation continue to overlook gender aspects. An assessment by the World Bank's OED of a number of World Bank projects noted that the experiences studied '...were largely oblivious of the gender aspect of resettlement.'[76] After the Asian Development Bank approved a gender policy in 1998, a review of its dam projects observed that the impacts on gender at the project preparation and implementation stages were often not considered.[77] Where planning is insensitive to gender, project impacts can at best be neutral, and at worst aggravate existing gender disparities to the extent of radically affecting the pre-project gender balance.

Dam projects often impose the gender bias of the developer (typically the state), with negative effects for local arrangements that provide livelihood opportunities for women. When the Mahaweli dam in Sri Lanka was built the prevalent inheritance rule, which allowed women the independent right to co-own and control land, was undermined by a new arrangement that allowed the household to nominate one heir, usually a son.[78] In tribal communities in India women do not have land rights and therefore they have not been compensated for the land they have lost as users.[79] Instead women's interests are seen as linked to the household and only men and major sons are given land according to the local government's resettlement policy for the Sardar Sarovar project. Women traditionally held land rights amongst the egalitarian Gwembe Tonga community in Northern Rhodesia (now Zambia). However, the British colonial authorities that built the Kariba dam only recognised men as land owners and women lost their land without compensation during displacement and resettlement.[80]

Forests, fisheries and other common property resources, which support subsistence livelihoods, are often not replaced during resettlement with women often bearing a disproportionate share of the resulting costs. The Tarbela Case Study notes that women have suffered more than men have from the disruption of their social life resulting from involuntary displacement from their ancestral land, which severed their relationship with water, forests and other natural resources. At the Pak Mun dam the loss of local edible plants due to submergence resulted in loss of income and sources of subsistence. Again this affected women disproportionately, as they are responsible for collecting and processing these plants.

The general impoverishment of communities and the social disruption, trauma and health impacts resulting from displacement have typically had more severe impacts on women. In Ghana, the general impoverishment resulting from involuntary displacement associated with the Akosombo dam led to increased male migration to urban areas and an increase in households headed by women.[81] In South Africa, farm-workers' households headed by women suffered unduly during the displacement process caused by the Gariep and Vanderkloof dams: 30% of men-headed households followed farmers to new farms compared to 15% of women-headed households. As a result, 75% of women-headed households ended up living for more than one year in the no-man's land along the highways called the corridor, where some even gave birth. As the Kariba case illustrates, the influx of immigrants during construction and the resulting urbanisation can increase levels of sexually transmitted diseases and, more recently, HIV-AIDS prevalence rates which have negatively affected local women.

Displacement can make women's position inside and outside the family more precarious. For example, at resettlement sites for the Sardar Sarovar dam in India, Kariba in Zambia-Zimbabwe and Nangbeto dam in Togo-Benin, increased alcoholism markedly increased domestic violence.[82] As men face powerlessness, women (and children) become scapegoats.[83] Reduced fishing opportunities in coastal and mangrove areas downstream of Tarbela dam destroyed the structures of families traditionally organised around this activity and accelerated male out-migration. Women faced increased responsibilities as de facto heads of households, while household income was severely affected.

While women in affected communities bear a disproportionate share of the costs, they have often had less access to the benefits generated by dams. The employment created during the construction of large dams generally benefits men, as illustrated in the Grand Coulee Case Study where it was only at the later stages of the construction phase that the government agreed to hire women, and only for the administrative clerical work. The allocation of the irrigated land made available by dams is also often done in a manner that exacerbates gender inequalities. In the Mahaweli irrigation scheme in Sri Lanka, fully 86% of the land allocations were made to men, and only two local women-headed households were granted land. In addition, the prevalent inheritance rule, which allowed women the independent right to co-own and control land, was undermined by a new arrangement that allowed the household to nominate one heir, usually a son.[84]

While women in affected communities bear a disproportionate share of the costs, they have often had less access to the benefits generated by dams.

Dams as opportunities for addressing existing inequalities

There are also examples in the Knowledge Base where dams have provided benefits to women. As gender is a relational concept, access by women to the benefits generated by a dam is a necessary but not sufficient condition for positive gender impacts.[85] As these impacts are seldom documented, little is known regarding the ways in which benefits generated by dams affected existing gender disparities and relationships. Where dams have improved the general supply of services, (see Table 4.1) the increased availability of water for household uses, electricity and food (through irrigation) are likely to have benefited women by reducing time spent on chores and improving nutrition. Where social services are provided as part of resettlement programmes, these may represent an improvement compared to the pre-displacement situation. For example, 80 000 people resettled from the Akosombo dam benefited from services including the following: 82 school blocks, 46 markets, 146 public latrines, 52 boreholes, 6 wells and 162 water stand pipes.[86]

Large infrastructure projects like dams or their associated irrigation schemes can present unique opportunities for reforms.

Where dams achieve an improvement in living standards in impact areas, this can have a positive spill over effect on gender equity. For example, improved family income resulting from irrigated agriculture at Aslantas dam enabled farmers to give both boys and girls higher education. This, along with the eradication of illiteracy, has contributed to ending polygamy in the basin.

Because gender disparities and even severe marginalisation of women exist in many countries, large infrastructure projects like dams or their associated irrigation schemes can present unique opportunities for reforms, in areas like land tenure, that can contribute to reversing the situation. Cases in Egypt, Tunisia, and Sri Lanka demonstrate how land tenure reforms have benefited poor people (including women) recruited as new landholders in irrigation schemes.[87] A survey of 32 villages conducted in 1991-92 showed that women owned 6% percent of the almost 2 500 irrigated plots in the Middle Valley of the Senegal River. While this reflects a continued gender imbalance, in the context of the Senegal valley it represents a positive gender impact. In traditional systems of rainfed and recession agriculture women and marginal communities had only use and not property rights. In this case government agencies used the opportunity of centralised control over the allocation of irrigated land to provide women-headed households with ownership rights.[88]

Cultural Heritage

Although improvements have been noted in recent years, potential cultural heritage impacts are still largely ignored in the planning process, especially in industrialised countries.[89] Large dams have had significant adverse effects on this heritage through the loss of local cultural resources (temples, shrines, and sacred elements of the landscape, artefacts and buildings) and the submergence and degradation of archaeological resources (plant and animal remains, burial sites and architectural elements). The latter may be part of the cultural life of local communities, or they may predate the arrival of people currently inhabiting the dam site. Shoreline erosion processes can expose subsurface archaeological remains, encouraging looting and illicit digging for artefacts and valuable remains. Dams can also cause loss or damage of cultural heritage through land reclamation and irrigation projects and the construction of power lines, roads, railways and workers towns.

In most cases no measures have been taken to minimise or mitigate the loss of cultural and archaeological resources. Affected communities repeatedly raised the treatment of burial sites at the WCD Regional Consultations and other public hearings. During the construction of the Inanda dam in South Africa, remains of human bodies buried under the reservoir site were exhumed and all buried in one hole, profoundly disturbing local communities.[90] The Grand Coulee Case Study records the submergence of Native American burial sites by the dam waters. The tribes used funds provided by the authorities, and their own means, to relocate burial sites exposed by receding reservoir waters. The risk of submerging ancestral graves is one of the main reasons the Himba people in Namibia oppose the planned Epupa dam.[91]

The assessment of lost or buried cultural heritage resources not directly linked to local people has been at least equally significant, but often more difficult to estimate. The difficulty lies in the fact that no investigation of cultural and archaeological resources has taken place as part of the planning process of most dams. Given that river valleys often hosted the most ancient civilisations, the importance of losses from existing dams can be assessed by default, on the basis of the quality and quantity of finds in areas affected by dams where some cultural heritage assessment did take place. When the Madden dam in Panama dropped to its lowest historical limit in 1998, it exposed thousands of artefacts, cultural features and human burial sites.[92] In 1988 in India, reconnaissance surveys in 93 of the 254 villages to be submerged in the Narmada Sagar dam impoundment area yielded hundreds of archaeological sites ranging from Lower Palaeolithic to historic temples and iron smelting sites.[93]

In the WCD Case Studies, two dams – Pak Mun and Aslantas – were redesigned to avoid impacts on cultural and archaeological resources. The Aswan High dam, (see Box 4.5) admittedly an exceptional case, illustrates not only how important potential losses of cultural heritage can be, but also how efforts to conserve cultural resources can improve understanding of cultural heritage. A study in the United States demonstrated that although submergence may be a way of preserving archaeological resources, it is more cost-effective to excavate and manage these resources prior to reservoir inundation than to leave them for possible future underwater archaeological expeditions.[94]

Despite the established potential for significant and often irreversible losses of cultural resources due to dam construction, cultural heritage management is still not adequately considered in the planning process. In Turkey, for example, only 25 of 298 existing dam projects have been surveyed for cultural heritage, and of these only five have had systematic rescue work conducted.[95] In Argentina, despite the fact that many

Box 4.5 The Aswan High dam: a milestone in the history of archaeology

The potential adverse effects of the Aswan High dam on the monuments of ancient Nubia were recognised in 1954, one year after selection of the dam site. Thanks to an international effort driven by UNESCO, the ancient monuments of Egypt and Sudan were saved from inundation. Equally important, the international rescue operation led to decades of intensified archaeological research in the vicinity of the dam, greatly enhancing understanding of the civilisation of Nubia. This radically altered the knowledge of Egyptian archaeology, resulting in the rewriting of the prehistory of the Nile Valley. What the Director General of UNESCO called 'a task without parallel in history' subsequently led to the launch of numerous other operations supported by UNESCO to save world cultural heritage.

Source: Hassan, 2000, in Brandt and Hassan, 2000, WCD Working Paper on Cultural Heritage Management

provinces have updated their legislation on cultural heritage, cultural resources management activities related to large dams were either poorly done or not at all.[96]

The India Case Study reveals that although projects like Narmada Sagar, Tungabhadra, Bhadra, and Nagarjunsagar have paid some attention to major temples and places of worship, almost all the dams built so far suffer from lack of cultural heritage studies (let alone mitigation measures). In China, the Three Gorges Project illustrates the potential to do damage through neglect of cultural heritage. The combined problems of time constraints, under-budgeting, and a shortage of qualified personnel are seriously hampering the salvation and preservation of the impressive archaeological and cultural sites in the areas to be affected.[97]

Human Health

Environmental change and social disruption resulting from large dams and associated infrastructure developments such as irrigation schemes can have significant adverse health outcomes for local populations and downstream communities. The issue of equity – in terms of pre-existing nutritional and health conditions of the population and the capacity to resist new health problems – is at the root of the adverse health impacts of dams.[98] Among the resettled, access to drinking water, health services and ability to cope with new social and physical environment determines health conditions.

Numerous vector-borne diseases are associated with reservoir development in tropical areas. Schistosomiasis (or Bilharzia) spread through snails breeding in still or slow-moving waters was a significant public health problem that emerged from many early projects, such as Kariba, Aswan and Akosombo.[99] Rift Valley Fever has also spread due to the Aswan and Kariba dams and irrigation systems along the Blue Nile in Sudan.[100] Most reservoir and irrigation projects undertaken in malaria-endemic areas increase malaria transmission and disease.[101] The increase was more pronounced for dams below 1 900 meters of altitude and less pronounced above that altitude.[102] Similarly in India, the Sardar Sarovar and Upper Krishna projects demonstrated a high potential for malaria transmission in the short term and thereafter leading to transmission of Japanese encephalitis.[103]

In new dams in tropical, sub-tropical, and arid regions there is rapid eutrophication resulting in problems of excessive aquatic weed growth or 'blooms' of toxic cyanobacteria. This is reinforced by enhanced nutrient pollution through growth of towns, agriculture and mining operations in the catchment. In China, a high incidence of primary liver cancer has been linked to the presence of cyanobacterial toxins in drinking water.[104]

Another problem is the accumulation of high levels of mercury in reservoir fish. Mercury, naturally present in a harmless form in many soils, is transformed by bacteria feeding on the rotting biomass in reservoirs into methylmercury, a central nervous system toxin. Alternatively, effluent from human activities such as mining may lead to the accumulation of mercury in reservoirs. As methylmercury passes up the food chain it becomes increasingly concentrated in the tissue of the animals eating contaminated prey, potentially threatening human health (see Box 4.6).

Socio-cultural disruptions can be traumatic for communities. The Kariba Case Study reports on the strong emotional response of the Gwembe Valley Tonga to their involuntary resettlement. When 50 people died a mysterious and sudden death in 1959 in the Lusitu area, those relocated attributed these deaths, together with deaths from dysentery and measles that occurred earlier in the year, to 'bad spirits'. The 'good spirits' of the Tonga had been drowned by the lake and could no longer protect them from such ailments.

Destruction of community productive bases in agriculture and fisheries can give rise to food shortages, leading to hunger and malnutrition. The Kariba Case Study recalls that the serious food shortages of 1958-60 can be largely traced to resettlement-associated factors. Food shortages due to resettlement are also reported in Vietnam, China, Malaysia, Thailand and India.[106]

In recent years, the high incidence of HIV/AIDS in construction and settlement areas is a growing concern. In the Lesotho Highlands Project Area, infection rates are far higher than in surrounding areas.[107] Communities are concerned about transmission from migrant workers arriving to work in the Maguga project in Swaziland.[108]

Initial assessment and other available information is often not considered until the impacts manifest themselves in alarming proportions and mitigation measures are unprepared and inadequate. When the Diama and Manantali dams were filled in the mid-1970s, an epidemic of Rift Valley Fever occurred, schistosomiasis prevalence rates reached record levels and riverside inhabitants experienced diarrhoeal disease, malnutrition and malaria.[109] All this occurred despite the experiences with the transmission of these diseases from earlier African dams. Among the impact forecasts for the Tucurui region were the results of a National Research Institute for Amazonia study for Eletronorte underlining the association between macrophytes and the proliferation of insect vectors. After Tucurui was filled in 1984, an unusual proliferation of *Mansonia* mosquitoes in rural areas close to the reservoir forced farm families to leave their homes. Test subjects received over 500 bites per hour at the height of the infestation.

Despite several decades of precedence, documented experiences from different regions, and the availability of sophisticated assessment techniques and instruments such as Health Impact Assessment, health concerns were not integrated in the design of the dams and infrastructure to the extent possible. Mitigation suffered from lack of preparedness and commitment and health concerns were not addressed effectively. For

Box 4.6 Mercury and human health at Tucurui

Mercury can have lasting negative impacts on human health. Levels at 50 to 125 mg/kg in human hair indicate a low risk of neurological damage. Foetal damage can occur at half the lower limit. Clearly defined neurological effects appear at concentrations of over 125 mg/kg.

In the early 1990s scientists from the University of Helsinki in Finland carried out a series of studies at Tucurui to assess the origins and effects of mercury in tropical reservoirs, with co-operation of Brazilian institutions (including Electronorte – the utility operating Tucurui). The studies provide the following findings:

- Tucunaré fish caught at five locations had an average of 1.1 mg/kg net weight of mercury, more than double the maximum safety level of 0.5 mg/kg.
- the average concentrations of mercury found in hair taken from adults in the fishing community was 47 mg/kg (with a standard deviation of 10.2mg/kg);
- one individual was found with a concentration of 240 mg/kg ;
- this concentration was seven times that of non-fish eating people, adults in this group had fish at meals 14 times a week; and
- the main source of mercury is gold mining operations upstream.

The WCD Case Study revealed considerable lack of agreement with the results of the study on the part of Electronorte. Given the irreversible, accumulating, and serious nature of the health impacts of mercury poisoning further research to resolve this issue is of great importance.[105]

Sources: WCD Tucurui Case Study; Jobin, 1999, p175.

After Tucurui was filled in 1984, an unusual proliferation of Mansonia mosquitoes in rural areas close to the reservoir forced farm families to leave their homes.

those affected, this translates into an increase in pain and suffering and a decline in educational achievement and productivity. It can also put severe pressure on the capacities of public health systems in developing countries.

Equity and the Distribution of Costs and Benefits

To this point, the Global Review has identified – and in some cases quantified in physical, economic, environmental, or social terms – a series of costs and benefits that arise from the decision to build a large dam. As indicated, shortfalls in technical, financial, and economic performance led a significant number of the large dams in the WCD Knowledge Base to fail a limited post construction cost-benefit evaluation using the metric on which they were approved. The Global Review has also established that building and operating dams has serious – and largely negative impacts – on ecosystems, biodiversity and human livelihoods. This chapter has documented a wide range of adverse social impacts due to large dams, including displacement, health and cultural heritage impacts. Large dams have also provided substantial socio-economic benefits through the delivery of water, electricity and flood control, as well as various ancillary services. In many cases these benefits extend beyond the time frame proposed in the original project documents

This section shows that large dams in the WCD Knowledge Base tend to produce benefits that accrue to groups other than those who bear the social and environmental costs. Those who bear the costs are quite often poor, vulnerable (such as indigenous peoples), or unrepresented (such as future generations). Considering present societal commitments to human rights and sustainable development it is clear that dams in the Knowledge Base have led to inequitable outcomes. In the light of the range of opportunities that exist for making affected people beneficiaries and enabling them to contribute to the stream of project benefits, such outcomes are unacceptable on equity grounds.

In the past, if the expected benefits of a dam exceeded the predicted costs the project went ahead. The narrow nature of the technical and economic analyses undertaken does not necessarily mean that public authorities that chose dams as a development option were unaware of the social and environmental costs. Rather, within the context of knowledge available and the value system of those making decisions at any given time, the sacrifices were judged to be worth the benefits of pushing ahead with the project. This approach to decision-making continues largely intact today.

The emergence of equity as a critical ingredient of development underlines that this 'balance sheet' approach is unacceptable as it ignores the typical mismatch between the distribution of the gains and losses of a project across different societal groups. Large dams can be seen as an extreme example of this dilemma, as public resources – both monies and rivers – are devoted to projects that all too often result in inequitable distribution of costs and benefits.

This section uses the benefits and costs of the large dams in the WCD Case Studies to illustrate and examine these issues. First, it presents the typical approach to balancing costs and benefits along with an overview of the benefits generated by the dams in the case studies. Then it discusses the distribution of project gains and losses to different groups with the assistance of a review of the groups adversely affected by the case study dams. Using this information it draws the implications for equity and the balance-sheet approach. The chapter closes with a presentation of recent initiatives that have addressed the past legacy of inequity and provided guidance on mechanisms for benefit sharing.

Benefits, costs and the balance sheet approach to net benefits

Each WCD Case Study dam generated a number of services and benefits for people (see Table 4.1). The Kariba dam, for instance, provided inexpensive electricity to the mining sector and urban areas of Zambia and Zimbabwe and also contributed to the economy through tourism, irrigation and commercial fishing, creating thousands of jobs, directly or indirectly, in these sectors. Tucurui dam supplies electricity to industry and to two Brazilian states (Para and Maranhao) that depend almost entirely on it for their supply (97% and 100%, respectively).

The 84 000 ha of land irrigated by the Aslantas dam has led to substantial improvements in the living standards of

Table 4.1 Illustration of the services and benefits generated by large dams in the WCD Case Studies

Services and benefits	Socio-economic significance to beneficiaries
Glomma & Laagen Basin	
Electricity: 10 145 GWh/yr (1998)	■ Electricity benefited 1.3 million basin residents, landowners, and industries; Public revenues from power installations in G&L area of around $60 million in 1998 of which 80% went to G&L region.; 2,350 people are permanently employed in the G&L area power sector
Flood protection	■ Flood peaks reduced by 20%
Grand Coulee	
Electricity: 24 050 GWh/yr (1995-1998)	■ Supply to industrial sector, agriculture and urban areas.
Irrigation: 276 700 ha irrigated	■ About 1 400 farms; value of production : US$637 million (1998)
Tourism: 3 million visitors/yr	■ Thousands of jobs created to service 44 major tourist facilities and parks and recreation areas in the project area.
Kariba	
Electricity: 3 860 GWh/yr (1996)	■ Supplying mainly the mining sector, urban populations and commercial farmers
Tourism: 485 000 visitors/yr	■ 20 hotels (about 1,000 beds) established on the reservoir.
Fishing: 23 250 tons of kapenta/year from Kariba Lake	■ The fishery benefits mainly commercial kapenta fishing companies
Irrigation: 2 700 ha irrigated	■ 450 permanent and 3,000 casual jobs in irrigation schemes
Tarbela	
Irrigation: 9% of total annual water supplies (22 % in dry season) for the 18 million hectares of the Indus system	■ Benefiting millions of holders of irrigated land and associated workers
Electricity: 15 100 GWh/yr (1998)	■ Provides 28% of Pakistan' s annual power generation from the inter-connected national grid system
Tarbela employment	■ 4 000 permanent jobs in the project; secondary employment in agro-industry and irrigation
Gariep and VanderKloof	
Irrigated area: 138 000-164 000 hectares[110]	■ 40 000 job units created or saved (regular and seasonal labour combined)
Electricity: 660 GWh/yr (avg 1971 to 1998)	■ Supply to national grid
Water Supply: 151 Mm3/year	■ Supply to industrial sector and cities through inter-basin transfers
Tourism: 200 000 visitors/yr	■ 200 people employed in 18 recreational facilities identified (with at least 1 000 beds)
Aslantas	
Irrigated area: 84 000 hectares[111]	■ Over 11 000 farm families or 80 000 people (but decrease in total farm labour force); Gross value of production $160 million, Supports about 100 agro-industrial units
Electricity: 650 GWh/yr (avg 1995-99)	■ Municipalities and industries in the basin
Tucurui	
Electricity: 20 000 to 23 000 GWh/yr (1995-1998)	■ Aluminum industry consumes over half of the electricity. 8 million people supplied with electricity in northern Brazil
Reservoir Fisheries: 3 200 tons/yr	■ Commercial and subsistence fisheries
Pak Mun	
Electricity: 290 GWh/yr (avg 1995-99)	■ Supply to Northeast region of Thailand
Tourism (existed pre-project)	■ 140,000 visitors in Keng Saphue in 1999 (less than pre-project)

Source: WCD Case Studies.

thousands of farmers, while the electricity supplied by the dam has benefited municipalities and industries in the basin. Grand Coulee dam in the United States irrigates more than 276 700 ha, generates 24 050 GWh per year, and together with Lake Roosevelt and the Columbia Basin irrigation project is the basis for a flourishing tourism sector with 3 million visitors annually. Tarbela dam generates 28% of Pakistan's electricity. In addition, the dam provides over one-fifth of the water in the dry season for the massive Indus Basin irrigation system. Gariep and Vanderkloof dams in South Africa irrigate more than 160 000 ha in the most arid provinces of the country and, despite the fact that they were built during the Apartheid era, they created (or saved) jobs that mostly went to then-marginalised groups. Through inter-basin transfers, the Orange River Development Project dams have also supplied fresh water to distant cities and industrial areas.

In many cases the exact number of beneficiaries is not known, particularly where the water or power is connected to a common transmission or distribution system. Still, it is possible to roughly approximate these numbers, for example, the hydropower dams in the Glomma and Laagen system supply electricity to 1.3 million people living in the basin.

These benefits need to be viewed in the context of the costs implied, especially the economic, social and environmental costs documented earlier in this chapter. For example, in the case of Tucurui the economic costs and benefits of power production must be balanced against a series of social and environmental impacts, including those already referred to in this chapter and Chapter 3:

- loss of 285 000 hectares of tropical forest and associated populations and species of wildlife;
- the suffering of communities subject to forced displacement and the plague of *Mansonia* mosquitoes;
- the risk of mercury poisoning;
- the loss of downstream fisheries and agricultural productivity; and
- the physical and livelihood displacement of the Parakana, Gavio da Montanha, and Asurini peoples.

Furthermore, there is no agreement on whether the net greenhouse gas emissions from the reservoir, spillway, and turbines are offset by the saving in emissions from fossil fuel sources made possible by the large amount of power produced by Tucurui.

Clearly, in the case of large dams that produce a multitude of benefits and costs, the application of a balance-sheet approach is not easy to do in a comprehensive fashion. In many cases the benefits and costs (especially social and environmental costs) are not of the same currency and cannot be explicitly weighed against each other.

The Commission asked participants in the final WCD Case Study stakeholder meetings for their views on the overall development impacts of the WCD Case Study dams. Some stakeholders were able – in their own minds – to sum up the positive and negative aspects of a project. Others rejected the questions as too simplistic an approach to a very complex, multi-dimensional problem. The Global Review itself demonstrates that poor accounting in economic terms for the social and environmental costs and benefits of large dams implies that the true economic efficiency and profitability of these schemes remains largely unknown. In addition, as covered in more detail in Chapter 6, even efforts to extend traditional cost-benefit analysis to cover this task have failed to accommodate the myriad impacts and concerns that arise.

Who gained and who lost?

One of the drawbacks of the approach taken above is that it reduces the benefits and costs of large dams to abstract flows (or numbers), and hides the impacts that have had very real consequences for people and ecosystems. An analysis of the distribution of costs and benefits – that is who gained and who lost – provides another method of assessing costs and benefits of large dam projects.

Re-distribution of access to resources

The construction of a dam requires an investment of human-made capital and, as a result, generates a series of new entitlements that are then distributed – either through political-administrative means or through markets – to members of society. At the same time, the construction of a large dam will have profound effects on the natural and social landscape of the setting in which the dam is located. These changes will affect the legal, customary, or de facto entitlements to natural resources, environmental quality and socio-cultural integrity experienced by local communities and others holding rights to the resources in the area.

Dams are unique among large infrastructure projects in the scope and manner in which they affect the pattern of access to resources, and their distribution across space, time, and societal groups:

> *dams take a set of resources – a river and the lands along its banks, generating food and livelihood for local people; and transform them into another set of resources – a reservoir, hydro power and irrigation, providing benefits to people living elsewhere. There is a sense, therefore, in which large dams export rivers and lands, removing them from the productive domain of one community to make them available to another.*[112]

Grand Coulee provides a vivid example. Native Americans were physically displaced by a project that provided power to industry and households in a city some 250 km away. Furthermore, the water and land that had previously supported their livelihoods (particularly the salmon fishery) was dammed and diverted to provide white settlers with irrigated farmland.

Poor accounting in economic terms for the social and environmental costs and benefits of large dams implies that the true economic efficiency and profitability of these schemes remains largely unknown.

In other cases, the resources as transformed simply continued to benefit those who previously had access to them. The Tarbela dam illustrates this: in addition to securing existing water supply, the expansion of the cropped area by 39% and the increase in the cropping intensity have primarily benefited those owning land prior to the construction of the dam.

In the cases listed the distribution of project benefits was explicit. In other cases, however, particularly where ancillary benefits are concerned, the redistribution of resources and benefits happens in a haphazard fashion. The illustration provided earlier of the increase in fisheries in the Tucurui reservoir is an example. In this case, those living near the reservoir have benefited while those living downstream from the dam have seen their fisheries decrease substantially. Similarly, the India Case Study noted that in many irrigation dams, people lose access to

The WCD Case Studies show that the direct adverse impacts of dams have fallen disproportionately on rural dwellers, subsistence farmers, indigenous peoples, ethnic minorities, and women.

their forest and highly preserved natural resources along the river to give way for a dam that irrigates land for downstream people who have often overexploited their local natural resources. These examples show that even if no intentional redistribution is made, a dam can effectively take a resource from one group and allocate it to another.

Finally, the impacts of large dams on ecosystems and biodiversity have consequences for future generations. Dams are long-lived assets and thus are designed to provide benefits beyond the medium term. The decision to dam a river and the ensuing effects on the environment will likewise have many lasting and even irreversible environmental impacts.

Profile of adversely affected groups

The WCD Case Studies show that the direct adverse impacts of dams have fallen disproportionately on rural dwellers, subsistence farmers, indigenous peoples, ethnic minorities, and women (see Table 4.2). These groups, who are also often the poorest segments of society, tend to be over-represented in the numbers of people who are displaced from reservoirs sites or lose access to their traditional livelihoods. Within the displaced, compensation programmes tend to ignore livelihood impacts on landless groups and women. In downstream areas,

Table 4.2 Profile of groups adversely affected by large dams: illustrations from WCD Case Studies

Dam project	Displaced people	Profile of displaced people	Others adversely affected
Glomma and Laagen 1945-70	None displaced.		None documented in case study.
Grand Coulee 1934-75	5 000 to 6 500 people.	1 300 to 2 000 from Colville and Spokane tribes not compensated until 1990s; rest are settlers.	Several thousand members of First Nation groups (Colville, Spokane, Nez Perce and Canadian tribes) located throughout the upstream basin, due to flooding of fishing falls and blockage of salmon.
Kariba 1955-59	57 000 people.	Tonga subsistence farmers; ethnic minority in Zimbabwe; most were resettled in resource-depleted areas.	Thousands of downstream people lost floodplain livelihoods and lakeside inhabitants experienced increased prevalence of schistosomiasis.
Tarbela 1968-76	96 000 people.	Composed of 93% farmers; 5% artisans or semi-skilled workers; and 2% boatmen.	Pastoralists, landless people, low cast groups of fisherfolk, boatmen, basket makers, and weavers suffered from loss of wetlands, forests, and grazing.
Gariep and Vanderkloof 1963 to late 1970s	1 380 people: families of 40 white farmers and 180 black farm workers	Most white farmers felt they were fairly compensated. Black farm-workers were not eligible for compensation and eventually moved to other farms and urban areas.	Farmers living along the river, where 1 million sheep suffered from proliferation of biting blackfly leading to livestock losses.
Aslantas 1975-85	1 000 families.	Mostly former immigrants from Eastern European countries; composed of small and medium-size landowners and landless families.	None documented in case study, although it is claimed that customary users of forests that were not recognised were the most adversely affected.
Tucurui 1975-85	25-35 000 people plus indigenous peoples.	Subsistence farmers, fisherfolk, pastoralists, and riverbank cultivators (126 of 2 247 disputes are not yet settled.)	100 000 people (subsistence farmers, pastoralists, fisherfolk, and other natural-resources-dependent communities) affected by reduced water quality, loss of riverbed cultivation, and decreased downstream fish populations.
Pak Mun 1991-94	1 700 families.	Rural families dependent on rice farming and fisheries income; cash compensation failed to provide livelihood regeneration.	More than 6 000 families of subsistence farmers and fisherfolk suffered loss of livelihood from fisheries reduction.

Source: WCD Case Studies.

communities suffering from altered river flows are mainly subsistence farmers whose livelihoods are largely based on the exploitation of resources offered by the natural flow of the river (fisheries, floodplain farmlands and pastures).

Finally, to the extent that burgeoning populations are at once evermore reliant on a degraded natural resource base and demand more environmental quality (as incomes rise), future generations are also likely to be among the adversely affected. Where the performance of large dams in economic terms falls short, future generations may also bear a disproportionate share of the costs without a commensurate share of the benefits.

Profile of beneficiaries

As noted earlier, dams have benefited the general public through their contribution to food production and increased access to electricity, along with providing other direct benefits and multiplier effects. The preceding sections of this chapter have also included some specific examples of how the direct and indirect benefits of large dams have been shared with the poor, such as through the provision of water supply, electricity, employment on construction sites and jobs in dam-based irrigation, industrial and tourism industries.

Examination of the Case Study dams confirms that those who receive the benefits, usually urban dwellers, commercial farmers and industries, are typically not the same groups that bear the social costs. Immigrants generally provide most of the labour force in construction works (qualified and non-qualified jobs). They develop and operate tourism facilities and manage commercial fishing companies (Kariba). While many small to medium farmers participate in irrigation schemes (Tarbela, Aslantas, Grand Coulee) such schemes also benefit large landowners and those from privileged groups (Orange River). Electricity generation has mostly benefited the industrial and mining sector (Kariba, Tucurui) and urban areas (Grand Coulee, Pak Mun). Flood control is provided to urban areas (Grand Coulee as part of the Columbia River system) as well as rural areas (Glomma and Laagen). Fresh water supplied by dams is mainly directed to the industrial sector and urban areas.

Equity

The Case Study dams demonstrate a lack of connection between groups benefiting from dam projects and those adversely affected, confirming the general finding from the Knowledge Base. Assessment of these distributional impacts of a dam does not in and of itself provide a value system to prescribe what should be done in judging projects and confronting the distributional issues raised. Judging the social and moral acceptability or fairness of the patterns of distribution of positive and negative impacts is an equity issue. The debate on the social impacts of dams is thus about the equity dimension – whether the distribution of costs and benefits is judged to be fair or just. A strong equity constraint dictates that there be no losers, that those who bear the costs from large dams should receive a proportionate amount of benefits. However, equity often speaks more to the costs of dams than to their benefits, directing attention to those who are most at risk. This speaks to the vulnerabilities of isolated, less powerful populations for whom 'development' can all too easily mean loss. Put simply, a large dam that renders the poor and the vulnerable worse off is inequitable.

Implications for the balance-sheet approach

That large dams in the Knowledge Base have led to inequitable outcomes challenges the assumptions that underpin the 'balance sheet' approach and hence the idea that simply 'adding' up the costs, benefits, and impacts of large dams will lead – all things considered – to the best choice for society. Indeed, there are significant moral and ethical concerns that such a balance-sheet exercise does not address. It is implicitly based on the assumption that if the overall balance of impacts is positive, then those who gain would share the benefits with those who lose out. Thus, all would be better off in the end. However, examination of the distribution of gains and losses in the case studies demonstrates that such benefit sharing has seldom occurred. Those who bear the social and environmental costs and risks of large dams are frequently not the same people who receive the social and economic benefits of the water, electricity, and ancillary services that dams produce.

Where costs and benefits accrue to different groups, the standard procedures for adding up and discounting the expected costs and benefits do not provide an appropriate measure of changes in societal welfare.

The importance of the distribution analysis of impacts and the concept of equity to decision-making can be understood as follows. If the loss of access to previous sources of livelihood is offset by access to new benefits made available by the dam, former resource owners and users can have different but better living conditions than before. If loss of ecosystem function results from large dam projects the resulting costs may be included on the balance sheet. In other words, adverse social and environmental impacts of large dams do not, by themselves, invalidate the balance-sheet approach. Rather, the crucial distinction is the failure to balance the loss of entitlements that some groups experience with a corresponding gain in new entitlements.

The lack of validity of the balance-sheet approach in such a situation is confirmed by economic theory. Where costs and benefits accrue to different groups, the standard procedures for adding up and discounting the expected costs and benefits do not provide an appropriate measure of changes in societal welfare.[113] In order to apply the balance sheet approach equitably, the costs to affected groups need to be minimised and an equitable share of benefits ensured. With regard to the environment and intergenerational equity this implies the need to ensure that ecosystem needs are met in the present so that future generations can have access to a non-declining income stream, to which natural capital makes an important contribution.[114]

Initiatives for the equitable distribution of costs and benefits

The major equity issues arising from the discussion of the distributional impacts of large dams is the impoverishment of those who previously inhabited the reservoir site and those who derived their livelihood from the resource base that is transformed by dam construction and operation. There is injustice when the rights of physically displaced people are violated, including when they lose their land and access to the river and when downstream people experience reduced access to floodplains and fisheries, but are excluded from access to project benefits. Societies are increasingly rejecting these outcomes and searching for more equitable solutions.

Chapter 3 has already covered the measures that are available to avoid, minimise or reduce the ecosystem impacts of large dams. These are likely to be an important component of efforts to resolve intergenerational issues or social inequities that are linked to ecosystem impacts. This final section of the chapter describes some recent initiatives to explicitly confront past social inequities through reparations or to ensure that new projects deal with equity issues in a proactive manner through benefit sharing.

Progressive national legislation and policies

Adopting a benefit-sharing approach requires that the project design and planning process consider such mechanisms from the outset.[115] Progressive national legislation and policies provide the legal framework and standardise benefit sharing, thus having a far broader impact than project level approaches. This approach has been widely implemented in the energy sector where project proponents allocate a percentage of the electricity sales revenue to resettlers and local administrative units.[116] Examples include, the Lubuge, Yantan, Shuikou and Ertan Hydroelectric projects in China, the Rio Grande Hydroelectric Project in Colombia, as well as several projects in Brazil (see Box 4.7). Other mechanisms for benefit sharing include the supply of energy at preferential rates (as required in Norway) and payment of property or local government taxes (as required in France and Norway) which are assigned to affected areas.[117]

Japan's Act on Special Measures for Reservoir Area Development provides various measures for people who are affected by a dam project and for the development of areas around the dam/reservoir.[118] The Act provides for a combination of measures, including compensation for property and other losses, improvement of the living conditions and industrial base of the affected area, and measures for resettling people through the Fund for Reservoir Area Development. The beneficiary municipalities, affected municipalities and central government contribute to this Fund, which finances development in the reservoir area. The Fund also promotes solidarity between the downstream beneficiaries and the displaced people.[119]

In the Senegal River valley, state-led distribution of irrigated land gave lower caste groups access to land ownership, previously denied to them under traditional tenure systems.[120] The India Case Study provides examples of agriculture and homestead land being provided even to those who were landless as part of resettlement processes. In some cases, previously marginalised farmers were given more land than they had originally.

A comparison of access to electricity in Zimbabwe and South Africa documents the

Box 4.7 Royalties to communities: a Brazilian law for hydropower benefit-sharing

In Brazil, Law No. 7990, dated 28 December 1989, requires that royalties be paid to the federal government for using water for power generation purposes. The royalties paid by each power plant generating more than 10 MW represent 6% of the value of the power produced. The royalties are distributed as follows: 10% to the federal government, 45% to the state(s) where the venture is located and 45% to the municipal districts affected by the venture. The total amount paid out by the Tucurui dam in 1996 reached $19 million, with the total royalties for 1991 through 1996 topping $103 million. The Itaipu dam, in the south of Brazil, pays annually about $13 million in royalties. The royalties are among the leading sources of income for some of the municipal districts.

However, royalties by themselves will not address social injustice, as the way they are used to benefit local government units depends on broader political and social factors. In some cases the allocation of these resources is done in a non-transparent way. In others, results are visible. A municipality like Itaipulandia has paved all the roads in the city and provides agricultural supplies to the population. In addition, local young residents are funded to study at Brazilian universities on the condition that they return to the community for five years.

Source: WCD Tucurui Case Study; Itaipu dam in Ferradas, 1999, WCD Contributing Paper for Thematic Review I.1 Social Impacts

The inequitable distribution noted in the past in sharing risks and allocating benefits is not inevitable.

importance of proactive government policy and investment in confronting existing inequities in the electricity sector as a whole.[121] Almost 40 years after commissioning the Kariba dam, only about one-fifth of Zimbabwean households (mainly located in urban areas) have access to electricity. Prohibitive power pricing policies adopted by the government continue to exclude poor people from this service. In contrast, as a result of deliberate policy on the part of the post-Apartheid government in South Africa, the percentage of households there having access to electricity increased from 20 to 50% in the 1990s.

Project level benefit-sharing mechanisms

Some project proponents also initiate benefit-sharing agreements with concerned communities where national or local regulatory frameworks do not exist. A wide variety of mechanisms are developed, ranging from making affected communities primary beneficiaries of the project services – for example irrigation, electricity, water supply, fishing rights – to formal business agreements concerning equity and revenue sharing in the project itself. Several dam projects in the WCD Knowledge Base shared direct project benefits with resettlers by moving displaced people into the newly irrigated areas, for example Andhra Pradesh II and III, Brazil Ceara Water Resources, China's Daguangba Multipurpose project.[122] In the case of Hydro-Québec in Canada, the Provincial power utility proposed partnership agreements to local communities for all new hydropower projects.[123]

Reparations

There are an increasing number of examples of reparations being made for past inequities.[124] In North America, the process of paying reparation to Native Americans for dam-induced impacts has started. The Grand Coulee Case Study relates that in 1994 the United States Congress upheld the claims for damages and reparation made in 1951 by Colville Confederated Tribes who had lost homes, lands and salmon runs to the Grand Coulee Dam in the Colombia basin in the 1940s. A total of $54 million was paid in reparations in addition to an annual payment of $15 million as long as the dam continues to produce electricity. A $200-million Missouri River Trust fund proposed through legislation in early 2 000 aims to move American Indian Tribes closer to compensation for land lost to federal dams.[125]

A few precedents are emerging on reparations in developing countries as well. Approval of a World Bank loan for the Ghazi Barotha hydro project in Pakistan in December 1995 was made conditional on a process for the resolution of disputes over compensation for people displaced 20 years before by the Tarbela dam. An initiative has begun in Zambia to raise the living standards of communities suffering the consequences of displacement by the Kariba dam more than four decades ago through the Gwembe Tonga Rehabilitation and Development Programme. In a negotiated settlement resulting from prolonged agitation, more than 10 000 families affected by Bargi project on the Narmada River in India gained rights to cultivate drawdown lands, fish in the reservoir and manage forests in the catchment area jointly.[126]

China attempted to systematically address the problems faced by the reservoir-resettled people beginning with new policy and institutional initiatives in the early 1980s. In 1986, the Ministry of Water Resources and Electric Power launched a large rehabilitation programme aimed at improving living conditions of some 5 million reservoir-resettled people across 46 resettlement areas in the country.[127] Although such measures have had a significant impact, a 1994 World Bank report cited the Chinese government as saying that some 46% of the country's reservoir resettlers 'were at great risk.'[128]

An appropriate legal and policy environment, accompanied by clear political will to act, can therefore ensure that poor and vulnerable groups marginalised in the past by large dam projects can begin to share in the benefits generated by such projects. This means that the inequitable distribution noted in the past in sharing risks and allocating benefits is not inevitable. Indeed, as discussed in Chapter 6, there are a number of explanations for past failures in performance – explanations that lead to a series of recommendations for making the transition to more equitable outcomes.

Findings and Lessons

Past decision-making and planning efforts have often neither adequately assessed nor accounted for the adverse social impacts of large dams. As a result, the construction and operation of large dams has had serious and lasting effects on the lives, livelihoods and health of affected communities, and led to the loss of cultural resources and heritage. At the same time a simple accounting for the direct benefits provided by large dams – the provision of irrigation water, electricity, municipal and industrial water supply, and flood control – often fails to capture the full set of social benefits associated with these services. It also misses a set of ancillary benefits and indirect economic (or multiplier) benefits of dam projects.

The WCD Knowledge Base provides the following findings on the adverse impacts of the displacement of people from their homes and livelihood by large dams:

- 40-80 million people were physically displaced by dams worldwide;
- millions of people living downstream from dams – particularly those reliant on natural floodplain function and fisheries – have also suffered serious harm to their livelihoods and had the future productivity of their resources put at risk;
- many of the displaced were not recognised (or enumerated) as such, and therefore were not resettled or compensated;
- where compensation was provided it often proved inadequate and where the physically displaced were enumerated many were not included in resettlement programmes;
- those who were resettled have rarely had their livelihoods restored, as resettlement programmes have focused on physical relocation rather than on the economic and social development of the displaced;
- even in the 1990s, impacts on downstream livelihoods were not adequately assessed or accounted for in the planning and design of large dams; and
- there is a clear relationship between the magnitude of displacement and the ability to rehabilitate and restore livelihoods adequately - the larger the number of displaced people, the less likely it is that livelihoods can be restored.

In sum, the WCD Knowledge Base demonstrates a generalised lack of commitment or lack of capacity to cope with displacement. Large dams in the WCD Knowledge Base have also had significant adverse effects on cultural heritage through the loss of cultural resources of local communities and the submergence and degradation of plant and animal remains, burial sites and archaeological monuments.

The WCD Knowledge Base indicates that the poor, other vulnerable groups and future generations are likely to bear a disproportionate share of the social and environmental costs of large dam projects without gaining a commensurate share of the economic benefits. Specific cases include:

- Indigenous and tribal peoples and vulnerable ethnic minorities have suffered disproportionate levels of displacement and negative impacts on livelihood, culture and spiritual existence;
- affected populations living near reservoirs, displaced people and downstream communities have often faced adverse health and livelihood outcomes from environmental change and social disruption; and
- among affected communities gender gaps have widened and women have frequently borne a disproportionate share of the social costs and were often discriminated against in the sharing of benefits.

These inequitable outcomes documented in the WCD Knowledge Base invalidate the prevailing 'balance-sheet' approach to decision-making. The balancing of gains and losses as a way of judging the merits of a large dam project – or selecting the best option – is not acceptable where the mismatch between who gain from the benefits and those who pay the costs is of such a serious, pervasive, and sometimes irreversible nature.

The review also shows that the true economic profitability of large dam projects remains elusive as the environmental and social costs of large dams were poorly accounted for in economic terms. More to the point, failures to account adequately for these impacts and to fulfil commitments that were made have led to the impoverishment and suffering of millions, giving rise to growing opposition to dams by affected communities worldwide. Innovative examples of processes for making reparations and sharing project benefits are emerging that provide the basis for optimism that past injustices can be remedied and future ones avoided.

Endnotes

1 Lang et al, 2000 eco041, WCD Submission.

2 Meinzen-Dick, 1997, p50.

3 UNDP, 1999, p213, 214.

4 Saxena, pers comm, 2000; UNDP, 1998, p175, 177.

5 These figures relate only to areas along the Orange River downstream of the Gariep and Vanderkloof dams, and do not include irrigated areas along the Fish and Sundays rivers, which received water, diverted from the Orange River by ORDP as well.

6 Panama Canal Office of Public Affairs, undated.

7 ADB, 1999b, p1.

8 Jing, 1999, WCD Contributing Paper to Thematic Review I.3 Displacement.

9 Fernandes and Paranjpye, 1997, p17.

10 World Bank, 1996a, p90-92.

11 Cook, 1994, p25.

12 OED, 1993, p11.

13 World Bank, 1996a, p88.

14 WCD Thematic Review I.2 Indigenous People, section 2.1.7.

15 WCD Thematic Review I.2 Indigenous People, section 2.1.11.

16 Morse and Berger, 1992; Parasuraman, 1999, p83.

17 Correa, 1999, WCD Regional Consultation Paper.

18 Dhom dam in Parasuraman, 1999, p154; Bargi dam in Mander et al, 1999, Contributing Paper for WCD Thematic Review I.3 Displacement, p64.

19 Bartolome and Danklmaier, 1999, Contributing Paper for WCD Thematic Review I.3 Displacement.

20 Balsalm, 1940a, and 1940b.

21 WCD Grand Coulee Case Study.

22 WCD Thematic Review I.3 Displacement, section 3.1.

23 WCD Thematic Review I.3 Displacement, section 5.

24 Colson, 1971, cited in De Wet, 1999, WCD Contributing Paper for Thematic Review I.3 Displacement, p9.

25 Fact Finding Committee on the Sri Sailam Project, 1986, cited in Mander et al, op cit, p10.

26 Stewart et al, 1996; World Bank, 1996b; Chen, 1999, WCD Regional Consultation Paper; Colajacomo, 1999, Contributing Paper for WCD Thematic Review I.2 Indigenous People.

27 WCD Thematic Review I.2 Indigenous People, section 2.1.9.

28 Mander et al, op cit, p6.

29 WCD Thematic Review I.3 Displacement, section 3.2; Bartolome and Danklmaier, op cit.

30 Kiambere dam in Mburugu, 1994, p53. Note that in Turkey farmers are asked to register the value of their land for tax purposes. Farmers that record a lower value than the real asset value in order to avoid tax will have difficulty in purchasing equivalent land elsewhere as the government uses these declarations as the basis for calculating land compensation amounts.

31 Fact Finding Commission on Sri Sailam Project, 1986, cited in Mander et al, op cit; Kao Laem dam in WCD Thematic Review I.2 Indigenous People.

32 Aswan High dam in Fahim, 1981, and Fernea, 1998, cited in De Wet, op cit, p11; Nangbeto and Akosombo in De Wet, op cit, p11; Itá in Bermann, 1999, WCD Regional Consultation Paper; Bhumibol in Sluiter, 1992, p62.

33 Jing, op cit, p11.

34 Hoa Binh dam in Srettachau et al, 2000, WCD Regional Consultation Paper; Sirindhorn dam in Sluiter, op cit, pvii; Batang Ai dam in ADB, 1999a, p5.

35 Miguel Aleman and Cerro de Orro dams in Nahmad, 1999, WCD Regional Consultation Paper; Panama in Huertas and Pacheco, 1999, WCD Regional Consultation Paper; Brazil in WCD Tucurui Case Study; Zambia and Zimbabwe in WCD Kariba Case Study.

36 Sluiter, op cit, p62.

37 Sluiter, op cit, p63.

38 Jing, op cit, p7.

39 Cernea, 2000; WCD Thematic Review I.3 Displacement, section 1.3.

40 World Bank, 1994, p2-3; Jing, op cit, p5.

41 Cernea, 2000, p2.

42 OED, 1996b, p86.

43 WCD Thematic Review I.2 Indigenous People, section 2.1.10.

44 Jing, op cit, p35.

45 Supreme Court of India, 1999.

46 ADB, 1999b, p20-21.

47 De Wet, op cit, p18.

48 Robinson, 1999, Contributing Paper for WCD Thematic Review I.3 Displacement, p4.

49 Driver, 2000, Submission to WCD Thematic Review 1.3 Web Conference.

50 Jing, op cit, p18-19.

51 Bermann, op cit.

52 De Wet, 1999, op cit, p21.

53 World Bank, 1993, p18.

54 WCD Thematic Review I.2 Indigenous People; WCD Grand Coulee Case Study, Annex 9; WCD Tucurui Case Study.

55 WCD Thematic Review I.3 Displacement.

56 WCD Thematic Review I.2 Indigenous People, section 2.1.1.

57 WCD Thematic Review I.2 Indigenous People.

58 WCD Thematic Review I.2 Indigenous People, section 2.1.6.

59 Silva Orrego, 1997, p159; Opaso, 1999, WCD Regional Consultation Paper.

60 Gapuz and Shalupirip, 2000, WCD Regional Consultation Paper.

61 WCD Thematic Review I.2 Indigenous People, section 2.1.1.

62 WCD Thematic Review I.2 Indigenous People, section 2.1.6.

63 Huertas and Pacheco, op cit.

64 WCD Thematic Review I.2 Indigenous People, section 2.1.5.

65 Soong, 2000, WCD Regional Consultation Paper.

66 WCD Thematic Review I.2 Indigenous People, section 2.1.5.

67 WCD Thematic Review I.2 Indigenous People, section 1.1.4.

68 In Canada it is the Constitutional Law of 1982; in the Philippines it is the Constitution of 1987; in India it is the fifth and sixth schedules under the Indian constitution; in Brazil it is the Article 231& 232 of the 1988 constitution. National laws reflect contemporary indigenous rights norms, in Chile, Ecuador, Bolivia, Colombia, and Argentina.

69 Adams, 1985, cited in WCD Thematic Review I.1 Social Impacts.

70 WCD Thematic Review I.1 Social Impacts.

71 Horowitz et al, 1994.

72 Ferradas, 1999, Contributing Paper for WCD Thematic Review I.1 Social Impacts.

73 WCD Thematic Review I.2 Indigenous People, section 2.1.11.

74 Urrá I dam in Correa, 1999, WCD Regional Consultation Paper; Pak Mun dam in WCD Pak Mun Case Study; Other dams in ADB, 1999a, p23-24.

75 Mehta and Srinivasan, 1999, Contributing Paper for WCD Thematic Review I.1 Social Impacts.

76 OED, 1998.

77 Projects included in the studies are: Batang Ai in Malaysia, Sing Karak in Indonesia, Lingjintan in China and Theun-Hinbon in Lao PDR. ADB, 1999a.

78 Agarwal, 1996, cited in Mehta and Srinivasan, op cit.

79 While most adivasi communities in the Narmada Valley are classified as 'encroachers', they had usufructory rights and control over land. Mehta and Srinivasan, op cit

80 Colson, 1999, cited in Mehta and Srinivasan, op. cit.

81 Anane, 1999 soc210, WCD Submission.

82 Mehta and Srinivasan, op cit.

83 Colson, 1999, cited in Mehta and Srinivasan, op cit, p.12.

84 Mehta and Srinivasan, op cit, p22.

85 Gender compares men and women. Where both benefit, but men benefit more than women, the gender impact can be negative because the benefit/s in question can result in wider gender gaps.

86 Tamakloe, 1994.

87 van Koppen, 1999.

88 Niasse, 1997.

89 Brandt and Hassan, 2000, WCD Working Paper on Cultural Heritage Management.

90 Gwala, 2000.

91 Kinahan, 2000, in Brandt and Hassan, op cit, p18.

92 Norr et al, 2000, in Brandt and Hassan, op cit, p35-36.

93 Ota, 2000, in Brandt and Hassan, op cit, p52.

94 Faught, 2000, in Brandt and Hassan, op cit, p11.

95 Brandt and Hassan, op cit, p59.

96 Politis and Endere, 2000 in Brandt and Hassan, op cit.

97 Childs-Johnson, 2000, in Brandt and Hassan, op cit.

98 WHO, 1999, WCD Working Paper on Human Health, p6.

99 Kariba dam in Hira, 1969, and Mungomba et al, 1993; Akosombo and Aswan dams in Jobin, 1999, p278, 298-300.

100 Jobin, op cit, p300-303, 327-330, 425-427.

101 World Bank, 1999b, p2.

102 WHO, op cit, p21.

103 Government of India, 2000, in WCD India Country Study.

104 WHO, op cit, p19.

105 For a treatment of the value of information in the presence of uncertainty and irreversibility see WCD Thematic III.1 Economic Analysis, Chapter 7.

106 China in Jing, op cit, p10; Vietnam in Sluiter, op cit. pVII; Malaysia in ADB, 1999a, p5, 2000; Thailand in Sretthachau et al, 2000, WCD Regional Consultation Paper; India in Laxman, 1999, p208.

107 Macoun et al, 2000.

108 Mncina and Ginidza, 1999.

109 WHO, op cit, p12.

110 Area irrigated before dam is included.

111 Area irrigated before dam is included.

112 Brody, Contributing Paper for WCD Thematic Review I.1 Social Impacts, section 5.5.

113 Arrow and Lind, 1970; World Bank, 1980; Belli et al, 1998, cited in WCD Thematic Review III.1, Economic Analysis, Chapter 5 and 6.

114 The extent to which other types of capital can substitute for natural capital is debated. The degree of substitutability, as well as the degree of irreversibility of ecosystem impacts will be important determinants of the optimum balance between leaving a river in its 'natural' state or going ahead with a dam (or determining environmental flow requirements). WCD Thematic Review III.1, Economic Analysis, Chapter 7.

115 Van Wicklin, 1999 soc184, WCD Submission, p8.

116 Van Wicklin, op cit.

117 Milewski et al, 1999, soc 196,WCD Submission and Adeler and Flatby, pers comm, 2000.

118 Kuriki, pers comm, 2000.

119 Kuriki, op cit.

120 Niasse, 1991, p101.

121 Bond, 2000 eco033, WCD Submission, p6-7.

122 Van Wicklin, op cit.

123 Milewski et al, op cit.

124 Reparation is defined as action or processes that repair, make amends, or compensate for damages. In a legal sense, there are three generally recognised forms of reparation: restitution, indemnity (or compensation), and satisfaction Johnston, 2000, Contributing Paper for WCD Thematic Review I.3 Displacement, p14.

125 Johnston, op cit, p42.

126 Sinha, 1998, soc009, WCD Submission.

127 Jing, op cit, p5.

128 World Bank, 1994, p2-3 cited in Jing, op cit.

Chapter 5

Options for Water and Energy Resources Development

Part of the Commission's mandate was to look at the alternatives for energy and water services that were considered in the past when building large dams and to consider the current options. This chapter examines the current state of knowledge on existing and emerging options for meeting water and electricity needs. As part of the larger discussion of planning and decision-making processes, Chapter 6 critiques the past assessment of alternatives to large dams.

Options normally emerge in response to demand or supply. The choices available to a society at any given time also depend on factors such as natural resource endowments, technological capability, institutional capacity, finance, market conditions, cultural preferences, awareness and education. These can act either as barriers or as enabling conditions, depending on whether they impede or promote the consideration and adoption of a particular option. Creating conditions for certain options to emerge as competitive responses to demand and supply requires support. Policies, institutions and regulatory measures can either help or hinder innovation, modernisation, maintenance, continuation and sustainability of different options.

The chapter focuses on identifying the range or mix of options available today to meet water and electricity needs in different societies and in urban and rural settings. It documents the large range of generic options currently available. However, given concerns about a number of barriers that have led to limited assessment of options in the past, it is not enough simply to identify the technologies and policies that can satisfy water and energy needs. It is also necessary to identify the obstacles that prevent the more widespread adoption and use of various options. Obstacles may be generic to an option – such as the high cost of a technology – or they may be specific to a particular context – such as limited wind potential. Only a thorough and integrated examination of the options and obstacles can yield a precise list of alternatives for consideration in a given regional, country or local context. The chapter therefore indicates options that represent significant opportunities across all contexts and provides snapshots of opportunities in specific countries, regions or contexts.

The investigation of options is organised around the four 'needs' areas that are the focus of this report: agriculture, energy, water supply and flood management. Broadly, options consist of technological, policy and institutional responses. They may be categorised further based on whether they contribute to demand-side management (DSM), supply-side efficiency or represent new supply options. For example, policies and institutional options to improve management of existing systems may respond to supply-side efficiency, while a new dam represents a technological option for new supply. Previous chapters have presented and analysed the contribution of large dams to these services and the performance of large dams over time. Chapter 5 focuses on the alternatives, locating large dams in the larger mix of options.

This report confirms that selecting the most appropriate combination of options depends on giving all the options equal and appropriate consideration in any assessment process. Assessment should be based on the respective merits of available options in the given context and should include not just a set of technical, financial, and economic criteria, but also full integration of social and environmental criteria. The options listed below are not exhaustive and the Commission does not endorse particular options. Rather, the intention is to highlight the options and issues that should be considered and explored as part of the options assessment process. Much more detailed information on these and other choices can be found in the WCD Thematic Reviews on irrigation, electricity, water supply and flood management options and the related contributing papers.[1]

Agriculture and Irrigation

Efforts to promote sustainable water management practices have necessarily focused on the agricultural sector as the largest consumer of freshwater. Governments have several objectives in deciding the nature and extent of inputs in agriculture. These include achieving food security, generating employment, alleviating poverty and producing export crops to earn foreign exchange. Irrigation represents one of the inputs to enhance livelihoods and achieve economic objectives in the agricultural sector with subsequent effects for rural development.

Just as strategies and approaches to rural development are context-specific, there are numerous and diverse alternatives to agricultural development and irrigation that need to be examined. The diversity relates to scale, level of technology, performance, and appropriateness to the local cultural and socio-economic setting. Government policies and institutions play an important role in the promotion of particular water appropriation technologies and methods. Each method has different implications for food production, food security at local and national levels, and the distribution of costs and benefits.

The growth of modern 'conventional' irrigation since 1900 has been characterised by large water projects that harnessed rivers through the construction of diversion structures and canal systems. Since 1950, the spread of such technology accelerated through state-sponsored large-scale irrigation and an emphasis on large dams for water storage. Irrigated areas increased from 40 million hectares in 1900 to 100 million hectares by 1950 and to 271 million by 1998.[2] Dams support 30-40% of this area, with the remainder supplied from direct river abstraction, groundwater and traditional water harvesting systems.[3] Since the 1970s, the predominant focus has been on providing irrigation to support the green revolution package of hybrid seeds, chemical fertilisers and pesticides. Conditions for higher growth were created in such areas through subsidised infrastructure, agricultural inputs and electricity for pumping.

Irrigated agriculture has contributed to growth in agricultural production worldwide, although inefficient use of water, inadequate maintenance of physical systems and institutional and other problems have often led to poor performance. Emphasis on large-scale irrigation facilitated consolidation of land and brought prosperity for farmers with access to irrigation and markets. Chapter 4 documents the major multiplier effects produced by successful large irrigation schemes. However, the scale of support to rain-fed areas was limited, even though such systems supported more than 80% of farmers in the developing countries of Asia and Africa. As a consequence, there has been a widening income gap between irrigated and rain-fed areas. Even within large-scale irrigation systems, inequities of this nature are observed, leading to the marginalisation of smallholders.

Agriculture in Asia and Africa supports hundreds of millions of smallholder cultivators who depend on land for subsistence, livelihood and food security.

The regional economic and development context for agriculture differs markedly for industrial and developing countries. In the former, agriculture tends to be capital-intensive with large, highly mechanised holdings requiring minimal labour. In contrast, agriculture in Asia and Africa supports hundreds of millions of smallholder cultivators who depend on land for subsistence, livelihood and food security. These farmers generally do not have access to support mechanisms or capital resources to risk growing high-value crops in volatile market conditions. The low productivity of the land and labour of many subsistence cultivators is also symptomatic of absence of support and widespread neglect of their agriculture and irrigation systems.

There is considerable scope for enhancing the viability, adoption and performance of other sources of irrigation water such as groundwater, direct river abstraction and traditional water harvesting systems.

This section presents a brief overview of some of the options available for agricultural development, with an emphasis on those most likely to be considered as alternatives to irrigation, particularly irrigation supplied by large dams. The presentation is grouped according to three levels of options:

- improving performance and productivity of existing irrigation systems through improved basin and system-level management, on-farm technological options to enhance the productivity of land and water, and policy and institutional reforms to improve incentives for water efficiency and demand management;
- improving the productivity and livelihood opportunities offered by alternative supply-side measures through enhancing rain-fed agriculture, supporting local and traditional water appropriation techniques and adopting new technological options such as water recycling; and
- investing in conventional supply-side measures to develop new irrigation areas based on direct abstraction from rivers and groundwater.

A final option is to import food from other countries rather than trying to achieve either a higher degree of food self-sufficiency or security through domestic production. This may be possible in countries with a small farming population, for individual crops with a high water demand, or for countries with significant foreign exchange earnings. However, it would be extremely counterproductive in countries with a large and poor rural population if it interfered with the income-earning potential of small farmers and their incentives to produce.

Improving performance and productivity of existing irrigation systems

As noted in Chapter 2, there is considerable under-performance of large dam irrigation schemes and scope for improving the performance of existing systems. Increasing competition for water has highlighted the inefficiencies in irrigated agriculture and thus increased demands for a more effective and integrated approach to managing existing irrigation systems, particularly surface water irrigation. There is also considerable scope for enhancing the viability, adoption and performance of other sources of irrigation water such as groundwater, direct river abstraction and traditional water harvesting systems. In the past, emphasis on performance enhancement programmes has had mixed results, however, with rapid decline in gains achieved leading to the need for periodic restoration, often heavily subsidised by the state. Policy interventions and institutional reforms to support technical interventions have been inadequate.

Improved basin and system level management

As reviewed in Chapter 2, the risk and consequence of sedimentation of reservoirs is site-specific, but of considerable importance in a portion of the larger dam population. Measures to improve the sustainability of existing water resources systems through sediment flushing and catchment management can increase the contribution and longevity of irrigation systems. Enhancing infiltration and reducing surface erosion through catchment-protection initiatives may improve sustainability of reservoirs and irrigation systems, but must account for the trade-off with subsequent losses in annual water yield and the potential for lower dry season flows. A review of 94 catchment experiments from around the world suggests a loss in water yield of 10, 25 and 40 mm for a 10% increase in catchment cover respectively for scrub, deciduous hardwood and pine and eucalyptus.[4]

Further, as explained earlier in Chapter 2, increases in vegetation lead to greater evaporation without necessarily leading to greater absorption and a reduction in surface run-off. Thus, the balance between evaporation and gains from increasing absorption will determine whether catchment measures lead to increases or decreases in dry season flow.[5] The utility of these options will therefore be site-specific and depend on the techniques applied and must be developed and evaluated in the larger context of natural resources management in the catchment. In particular the contrast between the effects on soil and water conservation of vegetative cover and structural measures such as embankments, ditches and small dams needs to be considered.

Salinity affects approximately 20% of irrigated land worldwide.[6] Controlling salinity and reclaiming saline land is an urgent priority in order to increase productivity of existing land, make better use of irrigation, and demonstrate that new irrigation areas can be managed in a sustainable manner. Application of drainage technology and maintenance of existing drainage is one way of containing salinity. But reliance on physical drainage of saline effluent alone is insufficient to tackle the problem, and an integrated approach combining management of surface water, groundwater and agricultural practices is essential. Salt-tolerant crops and vegetation can form part of such strategies to remove excess surface water and lower water tables. The irrigation of crops or pastures in 'series' arranged in order of increasing salt tolerance is a further example of saline management. If practised early enough, the integrated (or conjunctive) management of surface water, groundwater and salinity can prevent the build-up of salts (see Box 5.1). Trials on this are currently under way in Australia and California.

Salinity affects approximately 20% of irrigated land worldwide. Controlling salinity and reclaiming saline land is an urgent priority in order to increase productivity of existing land, make better use of irrigation, and demonstrate that new irrigation areas can be managed in a sustainable manner.

Water quality is another important factor affecting basin productivity. For example, efforts to improve salinity problems upstream may have adverse effects down-

Box 5.1 Conjunctive management of salinity

Conjunctive management, that is groundwater pumping with reuse of the saline effluent for irrigation, is an economic and sustainable means of salinity control. In the Shepparton Irrigation Region in Australia, one project covering 600 ha with 15 groundwater pumps has been in operation since the early 1980s. The salinity of the diluted groundwater is kept below threshold level for the crops and the management system prevents salinisation of the root zone. Long-term sustainability depends on the ability to obtain a salt balance in the area protected by the drainage pumps.

Source: Heuperman, 1999, Contributing Paper for WCD Thematic Review IV.2 Irrigation Options

stream. An innovative approach to the problem of managing the disposal of saline effluent comes from the Murray Darling Basin of Australia where disposal is regulated through regional load quotas or 'salt credits.' These credits are used to ensure highly saline water is released only during periods of high flow, when disposal has the least impact on river water quality.

The review of irrigation system performance in Chapter 2 revealed a considerable lag time between commissioning of irrigation dams and full development of irrigation infrastructure. Accelerating the full development of associated infrastructure, such as canal networks, and providing an integrated package of agricultural support measures for new irrigators may be a cost-effective option for enhancing performance.

In most irrigation systems, particularly those with long conveyance lengths, a disproportionate amount of water is lost as seepage in canals and never reaches the farmlands. For example, 40% of the water diverted from the Indus basin in Pakistan is lost in conveyance. In the late 1980s it was estimated that improvements in supply efficiency could save some 14.8 billion m^3/yr of water.[7] Canal lining is one such improvement. Experience with buried plastic linings in China demonstrates good performance over 18 years.[8] The United States introduced a low-cost lining programme and since 1946 some 4 600 km of various types of linings have been installed.[9]

But in the absence of good quality control and effective maintenance the canal linings often have not achieved the predicted improvements in water savings and reliability of supply. Studies from Pakistan in the early 1990s demonstrated that a focus on targeted maintenance is more cost-effective than deferring problems to rehabilitation programmes or lining canals.[10]

Inadequate maintenance is a feature of a number of irrigation systems in developing countries. An impact evaluation of 21 irrigation projects by the World Bank concluded that a common source of poor performance was premature deterioration of water control structures.[11] Often poor maintenance reduces irrigation potential and affects the performance of systems.

The scope for improvements through better maintenance is significant. The Food and Agriculture Organisation (FAO) of the United Nations estimates that the total area that can be used with improved maintenance is 150 million hectares.[12] In India, such steps could add approximately 9 million hectares to the irrigated area.[13] In countries like Nigeria, only 52% of irrigation from large-scale schemes was actually used in 1993.[14] In the combined Gezira-Managil schemes in Sudan, 126 000 ha had to be taken out of production due to sedimentation and weed growth in canals.[15] Performance-based maintenance contracts can be used as a cost-effective management measure to maintain system integrity. Such an arrangement can be extended to cover operation of the main canal system in tandem with decentralised farmer management at the secondary and tertiary level.

Proposals to improve system efficiency should be viewed in a river basin context examining the interactive effects of surface water, drainage re-use and groundwater recharge.[16]

On-farm technologies for enhancing the productivity of land and water

Raising the efficiency of surface irrigation provides considerable scope for improvements within existing irrigation systems, whether the water comes from dams, rivers, or groundwater. There are large variations in efficiency. As cited in Chapter 2 the average water use efficiency that was reported for the Aslantas project by the WCD Case Study was 40%. However, the range of field-level efficiency for the project varied from 25% to 55%. Efficient surface irrigation, however, has produced figures of 60%.[17]

In a number of regions, irrigation has promoted cultivation of water-intensive crops such as sugarcane and rice. For example, intensive cultivation of sugarcane in India and Pakistan has led to unsustainable use of water. A number of technologies exist for improving water use efficiency and, hence, the productivity of water in irrigation systems.

Micro-irrigation methods, such as sprinkler and drip systems, provide an opportunity to obtain higher efficiency gains than those available in surface irrigation. Field application efficiencies are typically in the range of 70-90%.[18] The output produced with a given amount of water is increased by allowing for more frequent and smaller irrigation inputs, improved uniformity of watering and reduced water losses. These methods have found wide usage in water-scarce regions. Nearly all of Israel's irrigated area, 68% of Jordan's and 40% of Brazil's irrigated area uses micro-irrigation methods, as do small areas in China, India and parts of Africa.

The capital-intensive nature of micro-irrigation technology and low cost of water has prevented more widespread adoption among smallholders in developing countries, but local manufacture has brought down the cost and improved viability for high-value crops. Further development in this area may make such pressurised irrigation technology attractive to farmers in developing countries. Low technology versions, such as making portable drip line and bucket kits for small household irrigation, have recently gained support.[19] Another method for increasing water use efficiency is to change crop cultivation practices (see Box 5.2).

Micro-irrigation methods, such as sprinkler and drip systems, provide an opportunity to obtain higher efficiency gains than those available in surface irrigation. Field application efficiencies are typically in the range of 70–90%.

As with other water management initiatives, the potential for efficiency gains in water use needs to be examined within a river basin context, as in many cases losses upstream are recouped downstream.

Policy and institutional reform

Policy and management initiatives are fundamental to raising productivity per unit of land and water and increasing returns to labour. They are often interlinked and require political commitment and institutional co-ordination.

Agricultural support programmes tend to be developed and implemented in relative isolation from irrigation systems. Typically there is weak co-ordination between agencies responsible for agricultural activities

Box 5.2 Cultivation techniques can reduce irrigation water use

The Muda Irrigation Scheme in Malaysia reported a reduction in irrigation duration from 140 to 105 days and a reduction in overall water use of 28% as a result of the shift from transplanted rice to wet seeded rice. Although the impetus for this change was shortage of farm labour, it had a considerable water conservation benefit.

Source: Fuji and Cho, 1996, cited in Guerra et al, 1998

(such as extension services, land consolidation, credit and marketing) and those responsible for irrigation development. Price incentives are also inadequate to raise productivity and the outcome is a significant gap between potential and actual yields. In the absence of better opportunities from agriculture, many farmers seek off-farm employment. Incentives to enhance production are necessary and can result from a more integrated set of agricultural support measures and the involvement of joint ventures that provide capital resources and market access to smallholder farmers. Appropriate arrangements need to be introduced for such joint ventures to ensure an equitable share of benefits.

As demonstrated in Chapter 2 the extent of recovery of the costs of operations and maintenance in large irrigation dams in the WCD Knowledge Base and irrigation systems in general is often limited. In a recent study of 16 projects, annual irrigation fees varied from zero in Thailand to as high as $130/ha in Colombia.[20] Collection rates varied from 50 to 100%. One rationale for higher fees is that they will encourage more efficient irrigation practice and a shift to more water-efficient and higher-value crops.

For example, in the case of Pakistan and India reported above, farmers are attracted to the intensive cultivation of sugarcane by its profitability. This is because water charges are low, capital costs are not recovered, and the mitigation costs of attending to waterlogging or salinity problems are not borne by the farmer. Removing the subsidies inherent with supplying irrigation and drainage services may encourage adoption of technologies for increasing water use efficiency and promote a shift to less water-demanding crops, especially in arid and water-scarce regions. Ideally, pricing structures for irrigation should reflect the cost of supplying water and associated externalities, and should be designed with stepped rates to provide security for basic livelihood needs.

In many systems farmers are charged on a per hectare basis independent of the volume of water used. This removes any incentive to save water. One obstacle to volumetric charges is the practical difficulty of measuring water delivered through an open canal system to a large number of smallholders. A solution may lie in water user organisations acting as intermediary bodies. They can enter into contracts with irrigation agencies and recover the charges directly from the irrigators.

One of the major contributors to poor performance of large irrigation systems is the centralised and bureaucratic nature of system management, characterised by low levels of accountability and lack of active user participation. Agency reform and management transfer have been initiated in more than 25 countries where governments are gradually reducing their roles in irrigation management and transferring responsibility for various levels of the systems to farmers' organisations and water user associations. The major impetus for the transfer lies in the desire to cut back public expenditure on operation and maintenance costs.[21]

The structure of farmer involvement varies from transfer of assets to a range of joint-management models. As yet, there is no general evidence to suggest that irrigation performance has improved as a result of transfer alone, although there are promising examples indicating that decentralisation may be a required, but not sufficient measure to improve performance.[22] Experience has shown that in order to be effective, a strong policy framework is required, providing clear powers and responsibilities for the farmers' organisations.[23]

Water rights and trading are highly contentious issues. Win-win situations occur for farmers when they trade a part of their water to replace lost income while at the same time being able to finance water use efficiency gains from their remaining water allocation. In the United States, Colorado has one of the most advanced institutional support networks for water markets. In recent years some 30% of a district's annual water entitlements has moved through the rental market. The price at which farmers sell water is often significantly higher than their cost of supply.

Enabling conditions for water markets are clear and secure entitlements along with effective administrative systems and infrastructure to regulate the trade and to monitor compliance. There should be clear environmental limits on the extent of the trade. These prerequisites are not present in many developing countries. Chile is often cited as an example where the development of water markets has helped to avoid expensive new water infrastructure, such as dams, by allowing transfers of water rights from agricultural to urban sectors.[24] Concerns over water trading stem from the equity and livelihood implications of permanent trades of water (as opposed to yearly rentals) from small to large farmers or from rural to urban areas.

Improving alternative supply-side measures

Prior to the advent of large-scale irrigation, a number of traditional water appropriation techniques and irrigation methods were used in different contexts. Several supply options existed, complementing each other and contributing to agriculture, food production and livelihoods. In today's context, local solutions and large-scale irrigation need not be mutually exclusive. They can co-exist and complement each other, raising food production and enhancing livelihoods.

For these systems to function optimally, a number of enabling conditions are required. Water appropriation systems need repair and maintenance, desilting and weed clearing. Innovations and enhancement of traditional methods are needed to improve crop productivity. To optimise the productivity of the restored water management systems, appropriate land use, including cropping patterns, mix and rotation need to be promoted. Sustaining these location-specific systems and practices will depend on protection of sensitive catchments, floodplains and deltas. Lastly, improving community stakes will be an important factor for greater application and long-term functioning.

Enhancing rain-fed agriculture and supporting local techniques

Some 80% of agricultural land worldwide is under rain-fed cultivation, contributing to 60% of food production.[25] Given the number of low-income households that rely on rain-fed agriculture throughout the developing world, the enhancement of opportunities in this sector can have a major effect on productivity and livelihoods.

Over a period of time these farming practices and irrigation methods have been marginalised by irrigation policies, the lack of institutional support and low levels of investment and research. In recent years, increasing attention has been paid to the successes of such methods and their impor-

Box 5.3 A local approach to integrated water management, Rajasthan, India

Water-harvesting structures to store rainwater runoff were constructed in the Bhagani Tildeh river catchment in Rajasthan in response to water scarcity. Tracts of land inundated by the construction of embankments and dug retention ponds are referred to as *johads*. The production of crops and milk have increased many-fold in the Alwar district of Rajasthan due to the greater availability of water through the construction of 2 500 *johads* in 500 villages. The total expenditure incurred was in the order of $3.6 million (in 1998 prices), with 73% contributed by local people. The *johads* have recharged the groundwater table resulting in an average rise from 200 feet below ground level to 20 feet.

Sources: Thakkar, 1999, Contributing Paper for WCD Thematic Review IV.2 Irrigation Options

tance is now being recognised by policy-makers.[26] The productivity of these rain-fed systems can be enhanced through improved agricultural support, local commitment to develop reliable water supply sources and small-scale, low-cost technologies. A progressive development path, where support provided is tailored to increasing local capacity, is emerging as an important conceptual approach to the development of such systems. For example, simple technologies such as treadle pump and low-cost drip systems can be introduced as a first stage innovation. In a few years – once farmers have recovered their investment costs – the returns can be re-invested in more advanced technologies such as small motorised groundwater pumps, thus gradually building the technological and capital base of the farmer.[27]

Box 5.4 Rainwater harvesting for domestic and agricultural use in China

Gansu Province lies on the loess plateau in central China and is one of the driest and poorest areas of the country, with annual per capita incomes of around $70–80 in rural areas. In dry years, traditional rainwater collection could not always provide sufficient water, and people were forced to trek long distances to rivers or to depend on government water trucks. In response to the 1995 drought, the 1-2-1 project was launched: it provided 1 clay tiled roof catchment area, 2 upgraded cement water cellars, and plastic sheeting for concentrating rainwater runoff on 1 field. The project ensured that some 1 million people not only had sufficient water but also could grow their own food and earn limited income through cash crops. The total implementation cost for the project came to $12 per capita.

Source: Gould, 1999, Contributing Paper for WCD Thematic Review IV.3 Water Supply

Agriculture in rainfed areas is supported by a number of water appropriation methods. Many of these have been used for centuries and have been adapted to satisfy local needs.[28] Rainwater harvesting uses small dams and embankments to trap run-off, usually with the aim of recharging groundwater. Surface sources such as springs and streams are also diverted into natural depressions using bunds and check dams. In floodplains, cultivation uses natural moisture brought in by annual floods. A variety of water management techniques are used to tap the soil water in wetlands and valley bottoms. Many of these localised practises serve multiple purposes, such as checking saline ingress in coastal areas and recharging groundwater in floodplains. Effectiveness, sustainability and productivity are location-specific and depend on the extent to which they are spatially integrated. One successful example comes from the Rajasthan area of India (see Box 5.3).

In many regions rainwater is harvested where it falls. In Argentina, Brazil, and Paraguay, natural or artificial depressions are used to store rainwater for crop and live-stock production. In the semi-arid areas of Argentina, Brazil and Venezuela rainwater run off captured from roads is collected in drainage ditches or street gutters and transported to cultivation areas.[29] Rainwater harvesting initiatives in Gansu province in China have provided both domestic and irrigation water to areas previously in drought (see Box 5.4).

There are hundreds of thousands of small reservoirs or water tanks around the world, ranging from farm dams in Africa used to provide intermediate storage, to recession reservoirs in Cambodia that are used to enhance soil moisture for cultivation. In India and Sri Lanka, more than 500 000 tanks store rain water, sometimes supplemented by water from streams or small rivers.[30] Tank systems have provided irrigation and supported agriculture and livelihood for centuries in South Asia, although reliability problems, siltation, poor maintenance and degradation have reduced the area served by small tanks in India from 4.6 million hectares in 1960 to 3.3 million hectares by 1987-88.[31] In addition to helping irrigation, tanks can play several important roles, such as flood control, soil erosion control, conservation of runoff during heavy rainfall and recharge of groundwater.

The implications of the widespread replication of these small-scale storage methods on reducing demand for irrigation water and enhancing the sustainability of groundwater supplies, as well as their impact on existing surface water sources, could be far reaching. As described in Box 5.5 and Chapters 3 and 4, the floodplains and deltas of a number of the world's major rivers support wetland ecosystems of exceptional productivity that in turn support large rural communities. Floods provide natural irrigation that fertilises floodplain soils.[32] As the floodwaters recede, arable crops are grown. Some soil moisture persists in the dry season and provides grazing for migrant herds. Floodplain cultivation is among the most productive and widely practised of agricultural systems in central and Sub-Saharan Africa. Similarly, wetland and delta cultivation is widely practised in Southeast Asia. A form of flood agriculture specific to arid environments is spate irrigation, which diverts seasonal flood flows from dry riverbeds. Large areas under spate irrigation are found, for example in Morocco, Yemen and Pakistan.[33]

As noted earlier, there is a need to adequately account for the effects of structural alternatives such as dams and levees on these natural irrigation 'options'. It is also necessary to consider the option of managed flood releases from existing facilities as a means of re-establishing and supporting these productive systems.

Adopting water recycling

Reuse of irrigation drainage water can be a major supply source in areas where intensive surface irrigation is currently practised. Without formal arrangement, farmers in the Eastern Nile Delta in Egypt use approximately 3 billion cubic meters of saline drainage water every year for irrigation after suitable dilution or treatment. Similar examples exist in the North China plains, Arkansas valley in Colorado, the Pecos valley in New Mexico and Australia.[34] Drainage water reuse requires increased levels of management skills to deal with increased levels of salinity, toxic concentrations and associated health concerns.

Box 5.5 Wetland and floodplain agriculture

Cultivated wetlands and valley bottoms, with some form of traditional water management constitute 12 200 km^2 in Central Africa, East Africa and the Gulf of Guinea. Flood recession agriculture is practised over an estimated 10 400 km^2 in the Gulf of Guinea and the Sudano-Sahelian region. In Senegal, the floodplain runs 600 km downstream of Bakel and covers an area of 10 000 km^2. In good years half of it will be inundated, supporting close to a million people who depend on cultivating the floodplain. In southern African countries seasonally waterlogged lands called *dambos*, augmented by small human powered devices like shallow treadle pumps, support crop production and smallholder food security. In countries like Zambia, *dambos* cover 38 000 km^2. Similarly, wetland and delta cultivation is widely practised in East Asia. In Vietnam, the Mekong delta constitutes the principal rice producing area of the country. Rice is also grown extensively in the wetlands of Cambodia and in smaller quantities in Laos and Thailand.

Sources: Gulf of Guinea and Sudano-Sahel in FAO, 1995, p10; Senegal in FAO, 1997, p136; Zambia in Postel 1998, p213; Southeast Asia in Friedrich, 2000, WCD Regional Consultation Paper

Reuse of urban wastewater is a significant source for irrigation in a number of countries. In Israel, 275 million m^3 of wastewater are used for irrigation after treatment. This is approximately 22% of the total agricultural use of water.[35] Strict controls are needed on the level of treatment required for various classes of water, with more stringent requirements for irrigating food grains than fodder crops. Examples from Ghana and Kenya also show the potential for peri-urban irrigation based on water reuse.[36] Where regulation of water quality is weak, this practice raises significant health concerns.

Investing in conventional supply-side measures

The potential to expand irrigation into new areas has sharply declined due to increasingly constrained resources and significant increases in the unit development cost. Conventional sources of water for irrigation besides reservoir storage behind large dams include diversion from rivers and lakes and groundwater abstraction.

Diversion canals or lift irrigation pumping schemes supply irrigation systems by abstracting water from rivers. For example, river diversions and pumped irrigation serve 80% of the irrigated area in Kenya and 68% in Nigeria.[37] The lack of over-season storage implies that the capacity of the system to provide multiple crops depends on the reliability of river flow. The size of run-of-river schemes can vary from a few hectares to hundreds of thousands. Run-of-river diversions can be used to supplement storage-based systems. In Sri Lanka, for example, diversion weirs were built to capture drainage water from upstream dam-based irrigation projects for reuse.

Groundwater abstraction has played an important role in the global expansion of irrigated agriculture. The availability of electricity, centrifugal pumps and well-drilling technology gave a major boost to rapid growth, particularly by individual farmers. Countries such as China, the United States, India, Pakistan, Bangladesh, Saudi Arabia and the North Africa region have high rates of groundwater use. Currently in China, 8.8 million hectares of land are irrigated through groundwater wells, constituting 18% of the total irrigated area.[38] The Ogallala aquifer in the United States waters a fifth of that nation's irrigated land.[39] By the late 1990s, groundwater irrigated over half of all irrigated land in India, contributing to 78% of additional irrigated area created between 1984 and 1994.[40]

Groundwater – employed on its own or in conjunction with surface irrigation – is often more productive than surface irrigation per unit applied. The determining factor appears to be the higher degree of control available to farmers who are often prepared to pay considerably more for reliable sources of supply such as groundwater.[41] Improved management of surface irrigation systems, leading to greater reliability of supply, may similarly increase production and returns to water.

Continued withdrawal of groundwater at current levels is, however, becoming unsustainable in many places. The Ogallala aquifer, for instance, is being depleted at the rate of 12 billion m^3 annually. Falling water tables, increased pumping costs and historically low prices have led to a reduction in the area irrigated by the Ogallala of 20% over a 10-year period.[42] Groundwater depletion has been a growing concern in the North China plains for over three decades, with water levels falling by 30 metres since the 1960s.[43] Efficiency improvement measures are as important to groundwater systems as to surface water irrigation. Effective regulation is also necessary to curb overpumping, which leads to lowering water tables and can limit access of poorer farmers.

Sustainable use of groundwater can be achieved through controlled abstraction and associated recharge measures. Recharge can be achieved by spreading surface water over large areas, using recharge wells, and water harvesting techniques as described above. Floodplains perform a significant natural recharge function. As floodwaters soaks through, underground reservoirs are recharged, and these supply water to wells beyond the floodplain. Recharge can also be a suitable approach for controlling saline intrusion and land subsidence and for reducing pumping costs. Most of the artificial recharge systems used to date have focused on small-scale systems or municipal water supply uses.

Currently, almost half of the large dams in the world provide irrigation services. The spread and contribution of dams to irrigation and food production and the environmental and social implications of their use were described in previous chapters. As discussed in Chapter 4, irrigation projects can have significant multiplier effects on the local economy in terms of contributing to the development of agricultural processing and related industries. The WCD Knowledge Base does not elaborate on the comparative effects of the different options in terms of engendering such multiplier effects, although this will be an important consideration for options assessment.

Obstacles and enabling conditions

A number of policy, institutional, and regulatory factors hinder the emergence and widespread use of an appropriate mix of options that would respond to different development needs, sustain a viable agricultural sector, provide irrigation and offer livelihood opportunities to large populations. First, policy and institutional support for innovation, modernisation, adaptation, maintenance and extension of traditional irrigation and agricultural systems was lacking in the past. Increasing recognition of this has led a number of actors to place priority on improving rain-fed agriculture and developing small-scale irrigation capacity. An extension of this priority is the need to protect (or restore) the natural functioning of deltas, floodplains and catchments in order to sustain and enhance the productivity of traditional systems in these areas.

Sustainable use of groundwater can be achieved through controlled abstraction and associated recharge measures. Recharge can be achieved by spreading surface water over large areas, using recharge wells, and water harvesting techniques.

Second, the institutional framework needs to be redefined by transferring management to decentralised bodies, local governments and community groups (water users associations or other appropriate bodies) for recovering tariffs and maintenance. Stronger commitment is required to transform irrigation bureaucracies into more efficient, service-oriented organisations capable of managing water and land in an integrated and sustainable manner. Irrigation water needs to be appropriately priced so that charges are based on volume used, taking into account the need to support basic needs and serve both equity and conservation. Third, emphasis must be given to developing a package of agricultural support measures that are mutually reinforcing and develop intersectoral linkages in the local economy so as to spur rural development. These efforts also need to counteract the tendency of current policies and extension services to impose high transaction costs and

The priority for a sustainable and equitable global energy sector is for all societies to increase the efficiency of energy use and the use of renewable sources. High-consumption societies must also reduce their use of fossil fuels.

risks on smallholder farmers in developing countries. Measures to enhance security of tenure are also required.

Farmers also need access to international markets through reduction in barriers and supportive domestic policies. In response to structural adjustment programmes and international agreements on world trade in agriculture, many developing countries have liberalised their agricultural policies, including cutting tariffs and subsidies. However, tariff and non-tariff barriers to OECD markets – such as the large production and export subsidies for farmers in the US and European Union – limit the ability of developing countries' to diversify their agricultural sectors and gain the benefits of increased international trade, as well as increasing rural poverty where local farmers cannot compete with cheap imports.[44]

To sum up, future assessment of alternatives will need to clearly consider the following:

- improvements to the efficiency and productivity of existing irrigation systems before planning and implementing new ones;
- adaptation and expansion of local and traditional water management solutions;
- more co-ordinated management of surface and groundwater resources; and
- improvement of the productivity of rain-fed agriculture.

To make progress in this area will require concerted efforts in policy formulation and institutional reform.

As an aid to this assessment it will be useful to more fully understand the contribution of alternative irrigation and agricultural options to food production and livelihood security. For this purpose, analysis that extends beyond the immediate costs and benefits of these options to the secondary economic impacts on specific social groups is necessary.

Energy and Electricity

The range and scale of energy resources and technologies for electricity demand-side management and supply have expanded dramatically in the last quarter-century due to advances in individual technologies and greater success in adapting existing and new technologies to local settings. While countries have different energy resource endowments, there is no supply global crisis on the horizon as is anticipated for fresh water. The world's renewable sources and fossil fuels are sufficient to meet foreseeable global demands for electricity generation over the next 50-100 years using existing or near-term technologies.[45] Moreover, the range of energy systems and technologies that may be called upon to convert primary energy sources into electricity has dramatically expanded in the last few decades. The priority for a sustainable and equitable global energy sector is for all societies to increase the efficiency of energy use and the use of renewable sources. High-consumption societies must also reduce their use of fossil fuels.

Key factors in the expansion of options include the improved capacity of developing countries in design and manufacturing, growing experience in adapting new technologies to rural and decentralised settings, and enhanced cost-competitiveness of the new technologies due to volume production – ranging from wind-turbines to compact fluorescent bulbs. These have given developing countries the opportunity to 'leapfrog' over older options when extending services to rural and urban areas. Innovation and change were spurred by the oil price shocks in the 1970s and 1980s, which prompted major academic, government and industry programmes to develop alternatives. As oil prices in the late 1980s and 1990s fell back in line with historic real prices and the perceived shortages and security threat receded, alternatives were set aside. Recognition of the causes and scale of the threat of global climate change in the 1990s have refocused work on alternatives, and galvanised the thinking on sustainable development, including the role of the power sector. A tripling of oil prices in early 2000 provides a further reminder of the need for continued long-term research and development of alternatives to fossil fuel technologies.

Some observers suggest that the world has entered a period where the revolution in electricity technology – coupled with the revolution in digital technology – will pave the way for a profound transformation in the delivery of and access to electricity services early in the 21st century.[46] It is certainly broadly accepted that the long-term trend is towards a global energy system that is less carbon-intensive and less reliant on finite energy resources. Nonetheless, there is considerable debate on the means and timing of the transition, its shape in different regions of the world, and its contribution to addressing the larger global equity issues of disproportionate resource use. There is also considerable inertia and resistance to change in the existing system, but the direction implied is clear: a shift towards 'cleaner fossil fuels;' a significant, accelerated shift towards the use of renewable energy sources for electricity generation and a focus on improving efficiency in the delivery and use of electricity services.

The world's demand for electricity has doubled over the past 22 years.[47] People are using more electrical services in the technology-driven digital economy, and are using electricity more widely in post-industrial, transitional and developing economies. Yet the supply and use of electricity is highly skewed between industrial and developing countries and between the rich and the poor in developing nations. There are enormous opportunities for demand-side management in industrial economies. Government efforts to reduce greenhouse gas emissions and make a contribution to climate stabilisation targets are indications of moves in that direction.

There is also considerable scope for efficiency improvements in developing economies, where they would moderate the required investment in new supply. In addition, almost 2 billion people, both urban and rural poor, have no access to electricity at all.[48] Rural populations are often low-income and live in settings where centralised energy services are expensive, both in national or consumer terms. Decentralised, off-grid systems are an important option in these areas.

Rural populations are often low-income and live in settings where centralised energy services are expensive, both in national or consumer terms. Decentralised, off-grid systems are an important option in these areas.

Figure 5.1 provides a simple schematic representation of the electricity sector today, showing generation, transmission, distribution and end-use components. As shown, there are three general ways to improve the delivery of electricity services:

- demand-side management options, concerned with efficiency on the user side of the electricity meter;

- supply-side efficiency measures, concerned with how efficiently electricity is generated by the centralised or local supplier and transmitted and distributed to users; and
- new supply options, which replace existing generation options or supply incremental growth in demand beyond what can be achieved by options in the first two categories.

New supply options may be further divided into grid and off-grid options. The latter includes small isolated 'mini' grids and stand-alone supply to individual customers and homes. These options are outlined below beginning with demand-side management and then passing on to supply-side options (efficiency and new supply) and finally covering options for rural electrification.

Demand-side management

In the context of this report, demand-side management (DSM) represents an opportunity to reduce the need for electrical generation and consequently the need for dams. The discussion also has broader dimensions. Demand-side management is about consumers using less electricity and using it more efficiently in the residential, industrial, commercial or government sectors. The major entry point for these improvements is the replacement of energy inefficient appliances. Enabling conditions that affect DSM uptake of improved appliances include the replacement cycle (whether it is a few or many years), standards, comparative cost and availability, consumer awareness and affordability. Generally investments in DSM that promote consumers' use of efficient appliances will be more than offset by the avoidance of investments in new supply and environmental and social costs of generation.

Recent investigations as part of the UNDP's World Energy Assessment, to be finalised in 2001, indicate significant potential for electricity efficiency improvements in all countries.[49] The technical potential in

Figure 5.1 Schematic of Electricity Options

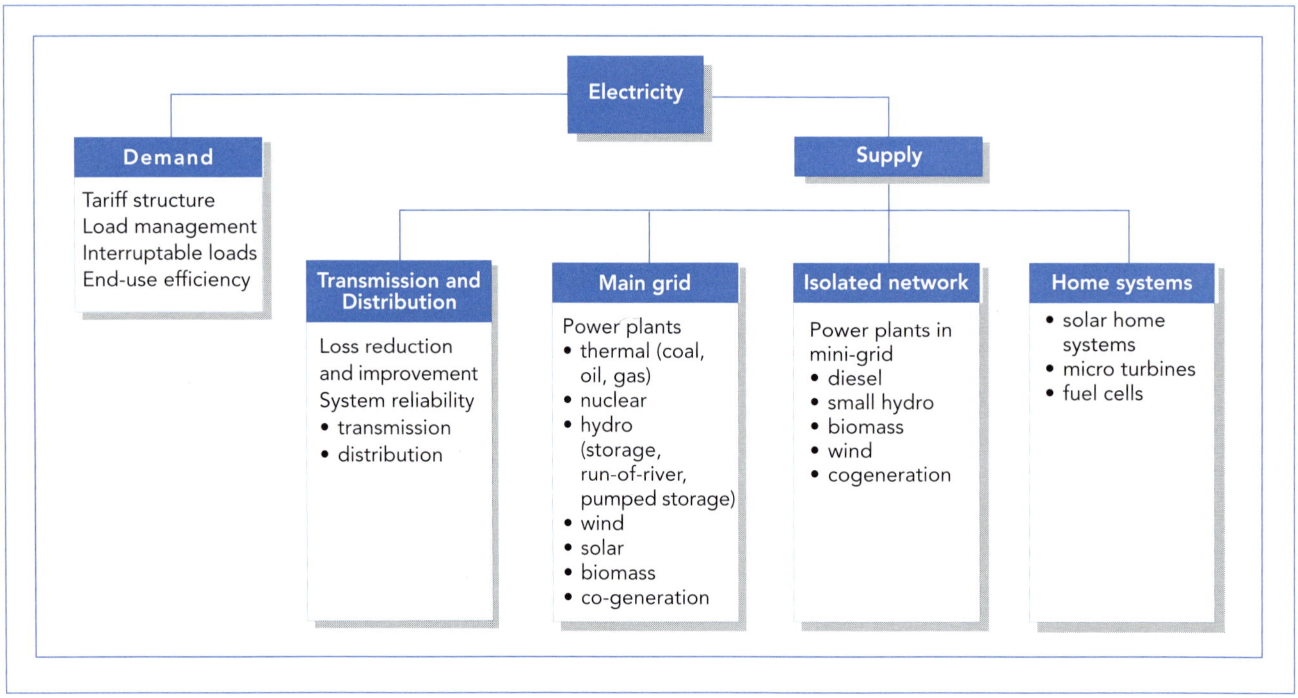

countries with a high per capita consumption, such as the United States, may be up to 50%, including modification of consumption and improved conservation behaviour. Others see less potential. Recent successful programmes in Asia and elsewhere also illustrate that there is great potential for DSM, particularly in the modern sectors of developing countries.[50] Most efficiency measures and technologies are cost-effective at today's electricity prices, and the use of full environmental and social costing of electricity supply options makes them even more so.

Despite the promise, actual global investments in energy efficiency and the savings from them continue to be small compared with the potential. And while utility-government partnerships in DSM grew strongly in the 1980s in North America and Europe, the move to open markets has led to lower electricity prices and a perverse disincentive for DSM in terms of tariffs. At the same time, many power utilities have dramatically cut their DSM programme budgets.

Several market and institutional barriers continue to obstruct the accelerated adoption of efficient end-use appliances and realisation of the potential of DSM:

- Utilities closest to the consumer often view efficiency in terms of loss of market share, and in trying to meet a growing demand tend to think at first of new supply options.
- Subsidies are still given for energy supply and consumption (especially for energy-intensive consumers), and there is a lack of credible commitment to energy efficiency from governments and international agencies.
- Governments find it easier and more politically attractive to approve new supply options than to get consumers to use energy more wisely.
- The structure of the international energy industry remains biased against DSM.

The latter bias stems in part from the diversity of producers in the industry which inhibits the development of a lobby structure similar to that of the energy supply industry which is dominated by some 50 global players.[51]

The fundamental enabling condition to improve the prospects for DSM is that governments lead the way and clearly define the support roles for utilities. Effective policies that can reduce transaction costs include continuation and expansion of information and product promotion, training programmes on energy efficiency service for suppliers and consumers, minimum efficiency standards, labelling and endorsement of high-efficiency equipment and selective government procurement. Other steps will require innovative financing and contracting mechanisms, removal of subsidies for conventional supply options and creation of power-sector regulatory frameworks and market incentives that encourage DSM. Ultimately, building capacity to design and deliver efficiency policies and programmes in government agencies and utilities is required.

Most efficiency measures and technologies are cost-effective at today's electricity prices and the use of full environmental and social costing of electricity supply options makes them even more so.

Supply-side options

Before electricity reaches the consumer, there are two broad ways to improve electrical services: investment in supply-side efficiency and new generation. The latter option involves two further issues – substituting current generation with new sources (for example replacing fossil fuels with wind power), and meeting increments of demand growth in economies that are expanding.

Supply-side efficiency

From the point where the electricity is generated to its end-use destination, numerous technical and non-technical losses may occur. The technical losses include consumption at the power station, step-up transformer losses, and transmission and distribution losses. There is considerable scope to reduce conveyance losses in many countries and these require urgent attention, especially where programmes have not been introduced.

Losses between what is sent out from the power station and what is recorded as power consumed on the customer's meter can be as high as 35-40% in some countries – and more in exceptional cases. Making existing transmission and distribution systems work better can lead to considerable savings, and often defer the need for new supply. In India, for example, 35% of electricity generated is lost in the transmission and distribution system before reaching the consumer. Yet it is considered technically feasible to reduce these losses to around 15%.[52] Moreover, the improvements are easier to implement than DSM. Investment in improving the technical efficiency of transmission and distribution efficiency systems may be undertaken in a few years and typically involve action by a single agency.

Major advances in electronic control systems and direct current transmission technology, including back-to-back AC/DC converters, are paving the way for operation and grid connection of a diverse array of options on electrical systems, particularly renewables.

Advances in transmission technology also show particular promise for improving transmission and distribution efficiency. Major advances in electronic control systems and direct current transmission technology, including back-to-back AC/DC converters, are paving the way for operation and grid connection of a diverse array of options on electrical systems, particularly renewables. These technologies enable connection of small generators with intermittent power (for example wind) to the grid and allow interconnection of grids at different voltage levels.

A recent and significant trend in the power sector is the regional interconnection of electricity grids. This ranges from bilateral arrangements for electricity sales between two neighboring countries, to cooperative power pool arrangements, to full spot-market and open pools involving several countries. Power pools and arrangements are in place in Europe and in North America. They are now emerging in Central America, parts of South America, Asia and Africa. China is in the process of transforming into five regional energy pools under the Electricity Law (1997) and early stages of power pool formation are underway in other parts of Asia.

In parallel with the trend towards linking grids, is the rapid expansion of regional pipeline networks – mainly for gas but also for oil. By linking and therefore expanding power markets these two trends are likely to increase the ability of already installed capacity to meet demand, increase system operating reliability, and reduce vulnerability to drought in grids with a mix of hydro- and thermal power. These changes will affect the future competitiveness of different electric power generation technologies.

New generation options

Excluding hydro, which represents 19% of the world's electrical supply, all renewable sources including biomass, wind, solar, geothermal, ocean energy sources and co-generation currently constitute around 1.4-1.6% of generation globally.

Every electricity-generating technology – including coal, oil, natural gas, nuclear, biomass, wind, solar, geothermal and ocean energy – has technology-specific characteristics that make it suitable or unsuitable for different roles in the system. There are also technical limits to the amount of power that any intermittent source (such as hydropower, wind or solar) can supply to the grid before reliability of supply is affected. Numerous technical criteria influence the utility preference for a particular generation source, especially in terms of the role of the option in the system (namely, peak, base, reserve or intermittent power supply). Other considerations include comparative cost, reliability, flexibility, efficiency, availability, experience and familiarity with the technology and its operation. Public policy and regulation are also important factors in deciding technology choices in the power sector.

While conventional generation technologies are mature, each is undergoing technological advances, with manufacturers and suppliers working to improve efficiency, reduce costs and construction times, and minimise emissions and environmental impacts. In the face of competition and rapid market change, power utilities and producers favour proven low-risk options with short construction times, and preferably off-the-shelf technologies. Gas-fired combined cycle systems, which combine high efficiency and flexibility with a comparatively low initial investment cost, have become the preferred choice where gas is available. When it is not available, the choice depends on the energy resource base. Typically the choice would be coal-fired plant for baseload (40% of generation globally) and oil-fired gas turbines for peaking and standby. Nuclear power, which was the fastest growing source globally between 1976 and 1996, is faced with high investment costs, limited public acceptance, and unresolved concerns on high-level waste disposal and decommissioning.[53] The outlook for nuclear power remains highly controversial, and countries such as Sweden and Germany have plans to eventually eliminate nuclear generation. The use of co-generation and combined heat and power (CHP) facilities is growing. CHP plants provide space and water heating along with electricity in cold climate settings, with very high efficiencies.

With current trends in power sector reform and financing, private developers are showing limited interest in large hydropower projects except for uniquely competitive low-cost sites and medium and small high-head hydropower in hilly and mountainous areas with few other energy resources.

Wind power is the fastest growing of the renewable energy options and is competitive with other conventional options when a back-up generation source is available and when government support (for example carbon credits for offsetting GHG emissions) is provided as an incentive. Global annual growth in installed capacity of wind turbines averaged 40% between 1994 and 1998.[54] In 2000, installed capacity stood at more than 13 000 MW in 50 countries (though this is currently small in terms of total electrical generation, equivalent to 0.4% of installed global capacity). Europe has 70% of the grid-connected wind capacity, North America about 19% and Asia about 10%. About 45% of the European wind capacity is installed in Germany.

In the last two decades, the cost of wind power under good wind conditions dropped by 75%, bringing it within reach of avoided fuel costs of modern fossil-fuelled power plants.[55] The European Wind Energy Association estimates that, fuelled by further cost reduction resulting from volume production, the installation of new wind capacity could rise rapidly at 20-30% annually so that by 2020 a total of 1.2 million MW of wind capacity could be installed world-wide, providing 10% of the world's electricity.[56] Much of the future development is expected to occur offshore.

Among advanced technologies in research and development, microturbines and fuel cells show the greatest near- and mid-term promise.

The cost of solar photovoltaics (PV) has dropped 80% in the past two decades and will need to fall by a further 50-75% in order to be fully competitive with coal-fired electricity. Photovoltaic technologies have reached a global production level of 120 MW. Although PV production is expected to keep rising, this technology will not significantly contribute to grid power requirements in the short term due to high cost. The long-term potential is considerable, however. Another renewable technology, solar thermal systems, can already almost compete with conventional thermal in settings with high solar insolation levels.

In addition to wind and solar, biomass and ocean energy systems (wave, tidal energy, and ocean thermal) have application for grid power. Biomass options are commercial where biomass fuel is readily available. But for large-scale applications, the extent to which this potential can be used will depend on numerous factors such as competition for water and land with other users, concerns over the loss of biodiversity to plantations, and technologies for biomass conversion to convenient energy services. Consequently, the greatest potential for biomass is seen in decentralised local systems.

Among advanced technologies in research and development, microturbines and fuel cells show the greatest near- and mid-term promise. Fuel cells are electrochemical devices that convert hydrogen and oxygen directly into electricity and heat. A number of companies are currently investing significant amounts in fuel cell research and development (R&D) and expect the commercialisation of the technology for use in vehicles and in grid and off-grid electricity supply by 2005. While natural gas is expected to be the main source of hydrogen initially, in the future hydrogen could be produced at remote hydropower sites, wind farms, solar stations and ocean power plants.

Electrification in rural areas

In rural areas, electricity can be supplied from existing grids under traditional rural electrification or from decentralised local grids. Perhaps the greatest expectation for change is the deployment of new and renewable technologies for rural electrification in decentralised mini grids and small simple devices for households. There is an opportunity to leapfrog the distribution stage and reach some of the 2 billion people who currently have no access to centralised electricity grids.

At the moment, policy-makers generally prefer centralised electrification options because the technologies are already proven, they retain the flexibility to cater for both urban and rural demand, investment credit (domestic and foreign) and established credit mechanisms backed by government guarantees are available, and the technologies are accepted by key institutions, such as planning bodies, utilities, local government agencies and others. There are other subjective reasons in favour of centralised electrification including the vested interests of various groups. Although grid extension will remain important, it is unlikely that centralised approaches will continue to consume the bulk of the financial resources available for rural electrification.

Rural electrification based on decentralised approaches has been taken up by utility planners for social objectives, but there is a strong preference in many areas for community-based decisions on the options. Decentralised electrification options are attractive for a number of reasons:

- They are simple and flexible ways to extend energy services to remote areas that have a low gestation period, may cost less, and involve low adverse environmental impacts. They include simple household lighting systems and mini-grids powered by diesel generating sets, small steam or gas turbines, micro-hydro units, windmills coupled to generators, modified engines using biofuels coupled with generators and photovoltaic systems.
- Community-based solar programmes are gaining momentum and can be used for rural industry, to meet needs like crop processing or water pumping and for household lighting, with a range of financing mechanisms to recover capital and operating expenses.
- Rapid improvements in technology and organisational learning have made systems based on renewable energy increasingly attractive and available.

Some of the key elements of a rural electrification policy include:

- financial support for community institutions for ownership and implementation of technologies based on local resources or strategies;
- subsidisation of most aspects of rural electrification for meeting cost of delivery or the inability of low-income consumers to pay for the services at cost recovery levels; and
- consideration of sustainability of any rural electrification programme to ensure its success, taking into consideration economic, environmental and social aspects.

Rural electrification based on decentralised approaches has been imposed on utility planners for social objectives, but there is a strong preference in many areas for community-based decisions on the options.

Obstacles and enabling conditions

Numerous barriers prevent the rapid spread of the new renewable technologies and DSM innovations including: the direct and hidden subsidies that have been built up over time for conventional energy technologies, a lack of commitment from governments, and the lobbying power of the conventional energy industry.

Over the life cycle of an energy system, tax and fiscal policies have an impact on the competitiveness of each electricity generation option. Since many developing countries had access to concessional financing for capital but not operating costs, hydropower often had a clear advantage over the longer-term financing for other options. For emerging renewable technologies, the cost of prototypes is high, but dramatic cost reductions would ensue due to mass production if the technology were widely accepted.

Regulatory change has had a major impact on utility practices, and power sector reform can create enabling conditions for the adoption of decentralised and renewable generation technologies for grid supply. The reforms will also affect rural electrification. In many countries, especially in the South, power sector reforms are extremely controversial. Concerns are raised about the lack of transparent and accountable regulatory mechanisms that contribute to corruption as well as about arrangements that favour private power producers, impacts on tariffs, and the consequent hardships for the low-income population.

Governments can take several steps to reduce the barriers to reform of the electricity sector and the adoption of DSM, supply-side efficiency measures and non-fossil fuel based generating technologies

- In strategic options assessment, it is important to identify and make transparent the level and nature of subsidies through the life-cycle cost of each option from extraction onwards.
- Fiscal incentives or subsidies can stimulate implementation of emerging technologies until prices drop enough for them to be commercially competitive.
- Tariff policies influence the ability of utilities to upgrade and improve existing assets and invest in new supply options and are currently being used to allow consumer preference to enter the equation at the bottom line (for example European and other initiatives towards 'green' power pricing).
- In developing countries, where availability and cost of electrical services are of greater concern, tariff blocks that increase as consumption increases and lifeline tariffs for low-income consumers, cross subsidised by higher income consumers, can expand access.
- Governments can also ensure that power sector reform is founded on a transparent regulatory framework in which consumers participate.

Water Supply

Domestic, municipal, and industrial consumption accounts for less than one-fifth of water use worldwide, and only about 5% in Africa, Central America and Asia. Major shortages occur in developing countries where more than 1 billion people have no access to clean water, and supplies are unreliable for many more. Distribution of water supplies is also inequitable, with the urban poor using unreliable and poor-quality sources and paying excessive amounts to water vendors. Urban demands are often given priority over rural demands.

The issues and options differ between industrial and developing countries:

- In industrial countries, coverage rates are generally high, demand pressures are moderate, and the focus is shifting to reducing consumption and increasing conservation in order to reduce or defer development of new supply sources.
- In developing countries, the low levels of service coverage coupled with high demand pressures have focused attention on new supply sources. Poor performance of existing water supply systems and weak management capacity are widespread and a large proportion of urban poor and disadvantaged communities' lack access to water supply.

Still, the exhaustion of supply sources near urban centres and deterioration of water quality are issues of concern across both developed and developing countries.

A number of opportunities exist to meet various kinds of water supply needs. In developing countries, extending services to unserved populations in rural and urban areas and revitalising sources that currently supply water to urban areas are priorities. Local resources and strategies may be more appropriate, effective and affordable in providing discrete supplies to communities. In situations where water needs cannot be met through local sources, other solutions are available. Demand-side measures, such as influencing consumption patterns, are relevant in industrial countries and among high-consumption urban water users in developing countries. New sources of supply are limited, however.

Demand-side management

DSM is gaining widespread acceptance in many industrial countries. Its potential to reduce consumption is proportional to the prevailing level of use. DSM has a high potential in the United States, where average per capita consumption is about 400 litres per capita per day. Developing countries generally have low average per capita consumption rates, yet there is considerable variation and potential for savings among high volume users. In New Delhi, for example, daily family use for those supplied with piped water ranges from 700 litres for low-income families to 2 200 litres for affluent families.[57] Tariffs are heavily subsidised, with little incentive for conservation.

Washing and sanitation are heavy users of water in domestic and commercial settings. Institutional standards, regulations and sanctions are necessary to promote water-efficient technologies, including regulatory standards for appliance and equipment manufacturers and subsidies to consumers to retrofit water-saving devices. In Denmark, per capita consumption dropped 24% over 10 years due to widespread adoption of water-efficient technologies, including toilets, showers, and washing machines.[58]

In many locations, including the United States, South Africa and Europe, block tariff structures with low lifeline tariffs and progressively rising rates for high levels of consumption have been successful in controlling or reducing water demand. For these to be effective an efficient yet cheap system of water metering is essential.

A significant proportion of high-quality domestic water is used in conventional sewerage systems to transport waste and treat it although coverage varies. However, in Japan conventional sewerage covers only 39% of the population, compared with 80% in Europe.[59] A number of low-cost and alternative sanitation systems that have low water requirements are available, such as offset pit latrines and septic tanks used in parts of South America. In India, 10 million people use a two-pit pour flush toilet, which has the potential of replication on a large scale.[60]

In many locations, including the United States, South Africa and Europe, block tariff structures with low lifeline tariffs and progressively rising rates for high levels of consumption have been successful in controling or reducing water demand.

Community management of water supply systems has a positive impact on both coverage and efficiency. This system works best when using discrete water sources in rural and urban areas, and it can reduce demand for external water resources. For example, in Bwera, Uganda, a community manages a large-scale system where a gravity-flow water supply serves over 60 000 people.[61]

Education and information are key elements in a long-term commitment to more efficient use of water. Priority areas include raising awareness about efficient technologies and practices, financial incentives and savings, metering, and the rationale underlying any new tariff structures.

Supply-side alternatives

The loss of water from leakage, illegal connections, and measurement problems is high. In Asian cities, this can amount to 35-40%, with individual cities reaching as high as 60%.[62] Stabilising and reducing losses from piped systems can help extend services. For example, figures in the United Kingdom suggest a 29 litre per capita per day decrease in system loss as a result of mandatory leakage targeting set by the regulatory authority. Proposals for a new dam in Yorkshire were deferred by introducing a leakage reduction programme and other conservation measures.[63] To succeed, leakage reduction programmes require strong institutional capacity and regulation to enforce standards. Reducing system-operating pressure may offer a relatively simple initial step to reduce leakage.

Rainwater harvesting through rooftops, tanks, and other methods are an alternative source of domestic water supply. Major rural supply initiatives in Gansu, China and in northeast Thailand are typical of the increasing number of traditional rainwater harvesting technologies that have been upgraded to provide affordable and sustainable supplies.[64] With the rising cost of conventional water supply, rooftop catchment systems are spreading in Argentina, Barbados, Brazil, Costa Rica, Dominican Republic, Chile, Mexico and Peru.[65] Rainwater harvesting is not restricted to rural areas (see Box 5.6); it is also used by millions of residents on the peripheries of new cities, like those in Tegucigalpa, Honduras. A related measure for increasing water supply in rural areas is the placement of artificial arrays in areas of high fog density to capture cloud moisture.

In places with short, intensive rainy seasons, much of the rainwater runs off already saturated surfaces. Even minor earthworks such as contour bunds and desilting existing village ponds can increase short-term above-surface storage enabling infiltration to occur over a longer period of time. Where sub-surface storage is available these methods can raise the water table and make groundwater available longer into the dry season. In Chennai, India, the metropolitan water supply utility tackled the problems of groundwater depletion and seawater intrusion through a conservation-based strategy. A series of check dams increased groundwater levels by 5-10 meters.[66] These were supplemented by controls on private water extraction and compulsory rainwater harvesting for new buildings.

As previously discussed in this chapter, maintenance of natural vegetation can have important effects in terms of improving water quality, but with variable effects on water supply. In South Africa, the Working for Water programme in the Western Cape region supports the eradication of alien

Box 5.6 Rainwater harvesting is spreading to urban areas

In Germany subsidies are available to encourage households to construct rainwater tanks and seepage wells. Due to savings in monthly water charges and other concessions, investments are paid back in 12 years. In Japan, 70% of the facilities in Tokyo's Ryogoku Kokugkan sumo wrestling arena are supplied by stored rainwater.

Sources: Gould, 1999, Contributing Paper for WCD Thematic Review IV.3 Water Supply; Down to Earth, 1998, p23

vegetation, thereby enabling the restoration of the indigenous fynbos vegetation. Fynbos provides less cover and vegetative mass and thus the programme aims to provide both employment and water yield gains.[67] Efforts by New York and New Jersey to purchase and protect the Sterling Forest from development brought renewed attention to the water quality benefits of forest management around single-purpose facilities near urban areas. Sterling Forest protects a catchment that provides a series of reservoirs that supply water to more than 2 million people.[68] In the latter case, catchment management serves to avoid the need to invest in water treatment facilities or, in the extreme case, new supply sources. Such options are highly location specific.

The recycling of wastewater can be another significant source of supply. By 1999, enough water was recycled in the Bay Area of California to meet the needs of 2 million people. The target there is to increase capacity to serve 6 million people by 2020.[69] Agriculture used 32% of the recycled water, 27% went to groundwater recharge, 17% supported landscape irrigation, 7% went to industry, and the remainder for environmental and other uses.

Desalination contributes to water supply in 120 countries, with 60% of the 11 000 desalination plants being located in the water-scarce Middle East. Though costs of desalination have been reduced dramatically, they are still high and the technique is very energy intensive.[70] Meaningful potential exists, but a major technology breakthrough is required to increase desalination's global contribution significantly.

Inter-basin transfer of water is often offered as a solution to local water scarcity. The impacts of such transfers require careful scrutiny, particularly as there will be no return flows within the basin as would occur with other abstractions. As noted above, water can be transferred from agricultural use through alternative and more proactive programmes, such as water trading, that effectively cross-finance water efficiency gains in irrigated agriculture.

Enabling conditions

As with the other sectors, a commitment to policy, institutional and management reform is essential to implement demand-side management of water supplies and alternative supply approaches. Initiatives include the following:

- In the United Kingdom, United States and Australia, regulations mandate exploration of all economically justifiable demand-management initiatives before a licence for new abstraction is granted.
- Management efficiency is fundamental to conservation and improved planning. No single public or private management approach is appropriate for all contexts. Stronger institutional capacity and accountability are required to improve the performance of public agencies.
- Effective regulatory mechanisms need to be in place to safeguard access and affordable cost of water for the urban and rural poor.

Meeting the needs of those currently not served in both urban and rural areas is a priority and requires concerted efforts. It requires an appropriate mix of demand-side management measures to enhance water use efficiency and reduce wasteful consumption, increases in supply efficiency and the development of new supply sources.

Meeting the needs of those currently not served in both urban and rural areas is a priority and requires concerted efforts. It requires an appropriate mix of demand-side management measures to enhance water use efficiency and reduce wasteful consumption, increases in supply efficiency and the development of new supply sources. In the latter case, due consideration is required to non-conventional alternatives such as reducing conveyance losses, rainwater harvesting, catchment management and water recycling.

Integrated Flood Management

Floods differ widely in their nature, in the characteristics of the affected floodplains, and in their implications. As documented in earlier chapters, in some contexts millions of people are dependent on annual floods for their livelihood, while in others similar floods threaten life and property. A number of options have emerged to predict, manage and respond to floods and at the same time obtain the most socially beneficial and economically sustainable outcomes. Institutional and policy support is needed to achieve integrated flood management, which consists of flood alleviation and utililisation, mitigation and risk management rather than a strict reliance on structural flood control based on dykes, levees and dams.

Floods are a complex phenomenon, and the success of flood intervention strategies depends on a number of factors. Chapter 2 considered the performance of dams as a structural response to flood control and highlighted the basis for concerns and the shift towards flood management as an approach in place of flood control. Vulnerability to flood events determined by characteristics of the population and land-use systems at risk and their capacity to cope and recover, are key issues in flood management strategies. Absolute flood control may be neither achievable nor desirable. The more appropriate objective is to predict, manage, and respond to the flood situation in order to prevent widespread losses and obtain the best outcome in each situation.

Following major destructive flood events in the last few decades, significant changes in flood policy have occurred around the world, redefining interventions to some extent. For example,

- coastal flooding of 1953 led to the Delta works in the Netherlands;
- the 1988-89 floods in Bangladesh led to the Flood Action Plan and the National Water Management Plan; and
- the Upper Mississippi floods of 1993, the Rhone floods in 1993, the 1997 floods in the Rhine and the 1998 flood in China drew attention to the role of non-structural catchment measures.

Intervention strategies in flood management have gradually shifted from a focus on structural responses to flood control to introducing or expanding the role of non-structural responses as part of integrated strategies for floodplain management.[71]

This section looks at the strategies and options available for integrated flood management and control. These strategies consist of three broadly complementary approaches, namely;

- reducing the scale of flood through a number of structural and non-structural means;
- isolating threats through structural, technological and policy alternatives; and
- increasing people's capacity to cope effectively with floods.

As with the role of large dams, associated structural measures and the relationship

Table 5.1: Complementary approaches to flood management

Reducing the scale of floods	Isolating the threat of floods	Increasing people's coping capacity
■ Better catchment management	■ Flood embankments	■ Emergency planning
■ Controlling runoff	■ Flood proofing	■ Forecasting
■ Detention basins	■ Limiting floodplain development	■ Warnings
■ Dams		■ Evacuation
■ Protecting wetlands		■ Compensation
		■ Insurance

between floods and natural floodplains have been discussed in earlier chapters, and therefore the focus here is on alternative measures for flood management.

Reducing the scale of floods

Reducing the scale of floods implies managing the quantity and quality of surface water runoff. Catchment management measures include:

- infiltration measures, such as infiltration trenches, detention basins, infiltration ponds, retention ponds and wetland areas to reduce runoff; and
- forest protection, reduced impact logging practices, avoidance of clear felling techniques and less intensive agricultural practices to reduce soil erosion and landslides that lead to channel siltation, raising flood levels proportionately.

Small-scale storage of runoff and improvements in drainage are other approaches to flood mitigation, particularly at the local level. Check and warping dams (built for erosion control) can lead to groundwater recharge and store initial quantities of run-off during storm events. Wire mesh dams are also used for this purpose in the highlands of Trinidad where the mesh traps water-borne debris during flash floods, blocking much of the flow.[72]

If afforestation is considered as part of a packet of measures, its effects on the full range of hydrological function and downstream uses should be considered. In assessing the likely effectiveness of source controls, pre-flood conditions (such as frozen or saturated ground) must be considered.

Flood flows can be stored in the lowlands, as detention basins are normally dry except when required for flood storage. In some cases, lakes on rivers can be used, such as the Dongting lake in Hunan province in China. Traditional tank systems in flood-prone villages of Madhubani district, in Bihar, India, are used for floodwater diversion and storage.[73] Natural wetlands are also important assets for flood storage and agricultural fields can be used for micro-storage.[74]

Isolating the threat of floods

Along with dams, earthen embankments, levees, dykes and bunds have been the dominant flood-control option in most of the world. One key characteristic of this option is that it interferes with natural drainage patterns in the area protected. This area may still be flooded from local precipitation, with minor tributaries causing local flooding, inland flooding or drainage congestion. In the absence of adequate drainage, the protected area will suffer crop loss due to waterlogging.[75] Significant morphological changes can occur following the construction of embankments, as sedimentation of channels and bank erosion lead to raised river beds and the need for even higher levels of protection (as noted in Chapter 2).

Structures such as dwellings may be modified in a variety of ways to reduce the risk of floodwater penetration: waterproofing walls; fitting openings with permanent or temporary doors, gates, or other closure devices; fitting one-way valves on sewer lines; or building boundary walls around the house structure. Other possible measures include sump-pumps that begin operating in basements when water levels rise, and contingency plans and facilities designed to be operated when a flood is anticipated. Contingent flood proofing depends upon a reliable flood warning system.[76] This approach also includes raising dwelling places. For example, in the floodplain *kampungs* (traditional villages) of Malaysia, houses are built on stilts to raise them above anticipated flood levels. Similar adaptations to floods are found throughout Southeast Asia along rivers, estuaries and coastlines.

Increasing people's coping capacities

New integrated approaches for flood alleviation, mitigation and flood risk management emphasise:

- integrated catchment and coastal zone management, and wise planning and use of floodplains and coastal zones;
- empowering local communities to make choices about land development and flood alleviation;
- reducing the impacts of humans on the environment by promoting flood disaster resilience (see Box 5.7);
- valuing and preserving the best of indigenous adaptations and improving local capacities to respond; and
- addressing problems of equity (for example alleviating poverty and lack of access to resources as a means of addressing flood vulnerability).[77]

Box 5.7 Flood resilience

'Resilience' may be taken as the opposite of vulnerability and may be enhanced by promoting access to knowledge and resources achieved through development processes and poverty reduction programmes. Alternatively, traditional social and cultural systems may enable resilience. The vulnerability of poor, rural Malaysians to floods in eastern Peninsular Malaysia is reduced by close kinship systems that exist in the floodplain villages.

Source: WCD Thematic Review IV.4 Flood Control, Section 4.2; Malaysia in Chan, 1995

Emergency planning and management has three phases: preparedness, response and recovery. The capacity of individuals, households, groups, and communities to cope with flooding depends upon their knowledge, resources, organisation and power:

- their knowledge about how to identify that a flood threatens, how to mitigate effects of floods, what to do before, during and after a flood, the causes of flooding and appropriate mitigation measures;
- the resources at their command, including their skills and physical assets, and the support of others that they can call upon; and
- the extent of their organisation, including within households, within neighbourhood groups, and within whole communities, as a way of pooling knowledge, skills, resources, and planning and co-ordinating activities to achieve optimum use and power in relation to other groups in society.

A flood management strategy will need to cover flood warnings, flood mitigation, any necessary evacuation and post-flood recovery. A clear commitment by national or federal governments to the emergency planning and management process will enhance its effectiveness.

Enabling conditions

Enabling conditions that will promote an integrated approach to flood management include:

- Promoting public involvement and devolution of decision-making to the lowest possible level enables integrated catchment management. This is essential as integrated flood management strategies at catchment level are most appropriate to local conditions.

- Funding should have a multi-functional approach. Integrated management of the catchment will increasingly result in multi-functional options being adopted.
- Institutional design is critical to the success of a flood hazard management policy, and co-ordination across various institutions has been a crucial factor.
- Flood hazard management and emergency response agencies have a key role in enhancing local communities' coping capacities by involving them in decision-making on all matters relating to floods.

The overriding message is that local ownership of flood alleviation strategies and options is necessary for long-term success.

Findings and Lessons

This chapter has examined the options for fulfilling energy, water and food needs in today's circumstances and the barriers and enabling conditions that determine choice or adoption of particular options. Many options currently exist – including demand-side management, supply efficiency and new supply options. These can all improve or expand water and energy services and meet evolving development needs across all segments of society. An overview of options for all the sectors covered suggests the following general findings and lessons:

- Demand-side management options include reduced consumption, recycling and technological and policy options that promote water and power efficiency at the point of end-use. DSM has significant untapped and universal potential and provides a major opportunity to reduce water stress as well as achieve other benefits such as the reduction of greenhouse gas emissions.
- Improving system management can defer the need for new sources of supply by enhancing supply and conveyance efficiency. Needless loss of power and water can be avoided through reductions in water leakages, improving system maintenance and upgrading control, transmission and distribution technology in the power sector.
- Basin and catchment management through vegetative and structural measures offers an opportunity across all sectors to reduce sedimentation of reservoirs and canals and manage the timing and quantity of peak, seasonal and annual flows, as well as groundwater recharge. The multi-functional nature of the hydrological system; the types and importance of downstream uses of water; and the on-site costs and benefits of the measures themselves will determine the attractiveness of different interventions.
- A number of supply options have emerged that are locally and environmentally appropriate, economically viable and acceptable to the public, including water recycling, rainwater harvesting and wind and solar (off-grid) power.

The ability of various options to meet existing and future needs or to replace conventional supplies depends on the specific context, but in general they offer significant potential individually and collectively. More specifically the different sectoral alternatives are as follows:

- In the irrigation and agriculture sector, preference is for improving the performance and productivity of existing irrigation systems and alternative supply-side measures that involve rain fed, as well as local, small-scale and traditional water management and harvesting systems, including groundwater recharge methods.

- The priority for achieving a sustainable and equitable global energy sector is for all societies to increase the efficiency of energy use and the use of renewable sources. High-consumption societies must also reduce their use of fossil fuels. Decentralised, small-scale options based on local renewable sources offer the greatest near-term and possibly long-term potential in rural areas.
- In the water supply sector, meeting the needs of those currently not served in both urban and rural areas through a range of efficient supply options is the priority. Further efforts to revitalise existing sources, introduce appropriate pricing strategies, encourage fair and sustainable water marketing and transfers, recycling and reuse, and local strategies such as rainwater harvesting also have great potential.
- In the case of floods, as absolute flood control may be neither achievable nor desirable, it is necessary to manage floods to minimise flood damage and maximise ecological benefits. An integrated approach to flood management will involve reducing a community's vulnerability to floods through structural, non-structural, technological and policy alternatives, and increasing people's capacity to cope with floods.

Numerous market, policy, institutional, intellectual and regulatory barriers hinder the emergence and widespread application of an appropriate mix of options in response to needs in the power and water sectors. The barriers to be overcome include capacity and resource constraints, the dominance of conventional approaches and interests in development planning, a lack of awareness and experience with non-conventional alternatives, inadequate access to capital and a lack of openness in the planning system. These are further analysed in the next chapter. While they are context-specific, hidden subsidies and other incentives to conventional options may limit the use and rate of adoption of even superior alternatives. To better enable the selection and use of the broader range of options will require that options are comprehensively and fairly evaluated by all stakeholders throughout the planning, decision-making, and financing process.

Endnotes

1 Sources are cited in the sections below only when they do not appear in these Thematic Reviews or where it is otherwise necessary for clarity.

2 Postel, 1999, p41; FAOSTAT, 1998.

3 WCD Thematic Review IV.2 Irrigation Options, Section 1.3.

4 Bosch and Hewlett, 1982.

5 Bruijnzeel, 1990.

6 Postel, 1999, p93.

7 WCD Thematic Review IV.2 Irrigation Options, Section 4.3.4.

8 Huasham et al, 1995 in WCD Thematic Review IV.2 Irrigation Options, Annex 8.

9 Mitchell, 1995, in WCD Thematic Review IV.2 Irrigation Options, p118.

10 Murray-Rust and Vander Velde, 1994.

11 OED, 1990, p4.

12 FAO, 1995, p280.

13 WCD India Country Study, Section 3.3.1.

14 FAO op cit, p233.

15 FAO, op cit.

16 Seckler, 1996.

17 Cornish, 1998, p20.

18 Cornish, op cit.
19 Frausto, 1999, Contributing Paper for WCD Thematic Review IV.2 Irrigation Options, p18.
20 FAO et al, 1999.
21 Vermillion, 1997.
22 WCD India Country Study, Section 3.3.6; Vermillion, op cit.
23 Bandaragoda, 1999; Vander Velde and Tirmizi, 1999.
24 Brehm and Quiroz, 1995; Hearne and Easter, 1995 cited in Hearne and Trava, 1997.
25 WCD Thematic Review IV.2 Irrigation Options, Section 3.4.1.
26 Agrawal and Narain, 1997; Thakkar, 1999, Contributing Paper for WCD Thematic Review IV.2 Irrigation Options.
27 Frausto, op cit.
28 WCD India Country Study; Agrawal and Narain, op cit; Barrow, 1999.
29 Ringler et al, 1999, p10.
30 Agrawal and Narain, op cit.
31 WCD Thematic Review IV.2 Irrigation Options, Annex 1.
32 Acreman et al, 1999, Contributing Paper for WCD Thematic Review II.1 Ecosystems.
33 WCD Thematic Review IV.2 Irrigation Options, Annex 5.
34 WCD Thematic Review IV.2 Irrigation Options, Annex 6
35 Shevah, 1999 in WCD Thematic Review IV.2 Irrigation Options, Section 4.3.2.
36 DFID, 2000.
37 FAO, op cit, p234.
38 Ministry of Water Resources and Electric Power, PRC, 1987, cited in Postel, 1999, p56.
39 National Research Council, 1996, cited in Postel, 1999, p77.
40 Thakker, op cit.
41 Dhawan, 1998, cited in WCD Thematic Review IV.2 Irrigation Options; Molden et al, 1998.
42 WCD Thematic Review IV.2 Irrigation Options, Section 3.2.3.
43 ADB, 1999c.
44 Smith, 2000, Contributing Paper for WCD Thematic Review IV.2, p17, 30.
45 UNDP et al, 2000.
46 Flavin, 1999, Contributing Paper for WCD Thematic Review IV.1 Electricity Options, Annex H; Economist, 5 August 2000.
47 IEA, 2000.
48 UNDP et al, 2000.
49 UNDP et al, 2000.
50 Rumsey and Flanigan, 1995; Worrell, 1999 working draft.
51 UNDP 2000, Chapter 6 p1.
52 Sant et al, 1999 eco013, WCD Submission.
53 IEA, 1998 and balances of non-OECD countries.
54 Kowalski and Schuster, 2000, p165.
55 WCD Thematic Review IV.1 Electricity Options, Section 3.4.
56 EWEA, 1999, Contributing Paper for WCD Thematic Review IV.1 Electricity Options, Annex H.
57 WCD Thematic Review IV.3 Water Supply Options, Section 6.5
58 White et al, 1999, eco018, WCD Submission, p9.
59 Lane, 1999, Contributing Paper for WCD Thematic Review IV.3 Water Supply Options.
60 WCD Thematic Review IV.1 Electricity Options, Section 2.4.2.
61 WCD Thematic Review IV.1 Electricity Options, Section 3.12.
62 McIntosh and Yñiguez, 1997, cited in WCD Thematic Review IV.3 Water Supply Options, Annex 1.
63 Yorkshire Water, 1997, eco082 WCD Submission A7.1.
64 Gould, 1999, op cit.
65 Ringler et al, 1999, p10.
66 SANDRP, 1999 opt080, WCD Submission, p20.
67 Preston, 1999. Contributing Paper for WCD Thematic Review IV.3 Water Supply Options.
68 Stapleton, 1996, p2-5.
69 Dickinson, 1999, Contributing Paper for WCD Thematic Review IV.3 Water Supply Options.
70 Costs range from $1.50-5.00 per cubic metre.
71 WCD Thematic Review IV.4 Flood Management Options, Section 1.2.3.
72 WCD Thematic Review IV.4 Flood Management Options, Section 4.4.2.
73 WCD India Country Study, Annex 5.
74 Delaney, 1995 in WCD Thematic Review IV.4 Flood Management Options, Section 4.4.2.
75 WCD Thematic Review IV.4 Flood Management Options, Section 4.4.5.
76 WCD Thematic Review IV.4 Flood Management Options, Section 4.4.
77 WCD Thematic Review IV.4 Flood Management Options, Section 1.2.4.

Chapter 6:

Decision-Making, Planning and Compliance

The previous chapters suggest that the main challenge for water and energy resource developers in the 21st century will be to improve options assessment and the performance of existing assets. This will require open, accountable and comprehensive planning and decision-making procedures for assessing and selecting from the available options. It also calls for monitoring programmes, evaluation procedures and incentive mechanisms that ensure compliance with project commitments, especially in the area of environmental and social performance. To do this we need a better understanding of why large dams were proposed and developed, and why failures in performance and impacts on ecosystems and affected people are still not properly accounted for, monitored or resolved. This chapter draws on the WCD Knowledge Base to characterise the critical problems encountered in the past, analyse their underlying causes and chronicle recent developments that point the way forward, this is explored in the remainder of the report.

Once a proposed dam project passed preliminary technical and economic feasibility tests and attracted interest from government or external financing agencies and political interests, the momentum behind the project often prevailed over further assessments.

As a development choice, large dams often became a focal point for the interests of politicians, dominant and centralised government agencies, international financing agencies and the dam-building industry. Involvement from civil society varied with the degree of debate and open political discourse in a country. However, there has been a generalised failure to recognise affected people and empower them to participate in the decision-making process.

Once a proposed dam project passed preliminary technical and economic feasibility tests and attracted interest from government or external financing agencies and political interests, the momentum behind the project often prevailed over further assessments. In any event project planning and appraisal for large dams was confined primarily to technical parameters and the narrow application of economic cost/benefit analyses. Historically, social and environmental impacts were left outside the assessment framework and the role of impact assessments in project selection remains marginal even in the 1990s. The influence of vested interests in the decision-making process and the narrow, technical approach to planning and evaluation have meant that many dams were not built based on an objective assessment and evaluation of the economic, social and environmental criteria that apply in today's context.

Conflicts over dams have heightened in the last two decades. This results from dissatisfaction with the social and environmental impacts of dams, and their failure to achieve targets for costs and benefits. It also stems from the failure of dam proponents and financing agencies to fulfil commitments made, observe statutory regulations and abide by internal guidelines. In some cases, the opportunity for corruption provided by dams as large-scale infrastructure projects further distorted decision-making, planning and implementation. Whereas substantial improvements in policies, legal requirements and assessment guidelines have occurred, particularly in the 1990s, it appears that business is often conducted as usual when it comes to actual planning and decision-making. Further, past conflicts remain largely unresolved due to a number of reasons, including the poor experience with appeals, dispute resolution and recourse mechanisms.

The key to improved performance in the future lies in screening out undesirable dams projects as part of a process that considers the full range of options for water and energy power services, and responds positively to changing priorities. These efforts must find ways to ensure that performance in living up to existing institutional arrangements governing the planning and project cycle is improved.

The chapter groups these topics under three headings: decision-making, planning and compliance.

Decision-making and the Political Economy of Large Dams

Large dams arise from a series of decisions taken from the beginning of the planning process through to the final approval of a project and financial closure. At each stage different actors are involved, including government agencies, public or private utilities, interested parties from the region, financing agencies, consulting and construction companies and equipment suppliers. Affected people and NGOs are increasingly involved as well, often through people's movements against dams. Each of these groups promotes its own self-interest throughout the process, ranging from profits and political power to property rights and livelihoods. This section looks at the interplay of these forces in the context of rivers, dams and the development of water and energy resources.

Similar pictures emerge for the industrialised and the developing worlds. Planning processes are controlled by single-purpose government agencies or public utilities and the decision to build is taken as the outcome of a fairly limited set of political interactions at political levels commensurate with the size and importance of the dam. In the case of developing countries, the selection of alternatives for meeting water and electric power needs was, and is, frequently constrained by preferential access to international finance and the pre-existing international expertise in large dams rather than alternatives. Recently, restructuring and reform of the energy and water sectors in many countries – both industrialised and developing – has changed the role of government in decision-making and planning, with private investors and corporations taking both financing and ownership roles in these projects.

State-led decision-making

Governments were the proponents for practically all large dams and many large dams were built by government agencies themselves.[1] Centralised agencies or utilities have traditionally managed the water and energy sector within government. Like most major development projects, decision-making processes around large dams have been centralised and technocratic in virtually all parts of the world, particularly through the 1970s. The exception may be certain large dams built as part of regional development projects where local political interests have played important roles in promoting projects – often in conjunction with their representatives in central government.

Indeed, the degree to which decision-making surrounding a dam was politicised and the level at which the decision was made varied tremendously with the project. Large and spectacular dams have often been seen as symbols of development and nation building, a potent demonstration of man's ability to harness nature's forces and a tangible 'deliverable' for politicians, usually funded from the public purse. The WCD Case Studies demonstrate that for very large dams the decision to build often was taken by heads of State, whereas smaller facilities were typically guided through the process by the relevant agencies or utilities (see Box 6.1)

Dam-building in industrialised countries

In industrialised countries, alliances between local political interests and powerful, single-interest agencies and utilities responsible for water and power development drove planning and decision-making on large dams. In the United States, the political desire to settle and develop the land and resources of the western states

encouraged the construction of large dams. At the same time, however, laws governed the planning and approval process. They required agencies and utilities to perform a long series of surveys, hold public hearings, and conduct inter-agency reviews, including cost-benefit analysis. The appropriation of funds ultimately required approval by the Congress, which further scrutinised the project plan.[2]

Box 6.1 WCD Case Studies: political decisions to build large dams

From the WCD Case Studies, the predominant role of the State can be seen throughout. In the Glomma and Laagen Basin the Norwegian government was active in licensing hydropower projects initially to promote development in isolated river valleys, then to feed power based smelting industries and other heavy industries in the period after the Second World War. Hydropower development was also promoted to support specific districts in periods of depression and high unemployment.

Similar government intervention is demonstrated at the Grand Coulee project in the United States where a presidential decision was made to proceed with the Grand Coulee project in 1932. The project formed part of the federal government's campaign to bring the country out of economic depression, provide construction jobs to eight thousand people, reclaim land for irrigation and reduce price manipulation by private power companies, thereby making publicly generated electricity more widely available at low cost.

The planning, implementation and initial operation of the Kariba project was done by the Inter-Territorial Power Commission of the then Central African Federation (the former colonies of Northern and Southern Rhodesia now Zambia and Zimbabwe) in the 1940s. The priority was to deliver power to the copper industry owned by multi-national corporations.

On the Orange River in South Africa, the proposal to build a major dam and water diversions scheme was called for by Prime Minister Hendrik Verwoerd following the Sharpeville massacre in 1960 that undermined confidence in the government and led to outflows of foreign capital. One primary motivation was to demonstrate national capacity to build major projects and to restore international confidence in the country's development and investment potential.

Source: WCD Case Studies

Outside the United States, the reconstruction of Europe after World War II led to the construction of many large dams. The Marshall Plan ushered in the era of foreign aid with the transfer of $17 billion to help rebuild Europe.[3]

During the cold war era, centralised, state-driven consolidation of resources through interventions such as the building of large dams was the hallmark of communist regimes. Most of the political and economic decision-making processes for the large dams built in eastern and central Europe were top-down and technocratic. Besides the central government, other stakeholders and the general public were not in a position to express their concerns or represent their interests in the decision-making process.[4]

Dam-building in developing countries

The success of the Marshall plan in Europe led to great optimism that the key to national development was investment in capital stock. The International Bank for Reconstruction and Development (IBRD), created to help finance the reconstruction of war-torn European countries, became a focal point for these efforts and, alongside bilateral development banks, helped export the model of centralised nation building for economic development. Dams fit well with this model of foreign aid and were often the first visible sign of IBRD (later called the World Bank) presence in a country.[5]

Role of foreign assistance[6]

Both the multilateral and bilateral development banks played a significant facilitating role in getting Asia, Africa and Latin

America started in the dam business. The World Bank began financing large dams in the 1950s, committing on average over $1 billion per year to this purpose (Figure 6.1). For the period from 1970 to 1985 this amount had risen to $2 billion per year. Adding in finance by the Asian, Inter-American, and African Development Banks, as well as bilateral funding for hydropower, suggests total financing for large dams from these sources of more than $4 billion annually at the peak of lending during 1975-84.

Bilateral and multilateral development financing agencies have helped finance studies needed for dam construction, and lent money for the construction of the dams themselves. They identified development goals through strategic sectoral planning documents, provided resources and technological capacity to conduct feasibility studies, and created basin-wide institutional frameworks to plan and implement dams. Although the proportion of investment in dams directly financed by bilaterals and multilaterals was perhaps less than 15%, these institutions played a key strategic role globally in spreading the technology, lending legitimacy to emerging dam projects, training future engineers and government agencies, and leading financing arrangements.[7]

The extent and nature of this influence varied from country to country and from region to region. The India Case Study locates the orientation of Indian planners and engineers towards dams as the principal response to water resource development in the 1950s and 1960s when large numbers of dams were first built. This predated the World Bank's major involvement in India. The Bank began lending in earnest to India in the 1970s at a time when policy reforms removed restrictions on the ability of individual states to directly access foreign assistance and provided incentives for doing so.

Since then World Bank loans to India have doubled or tripled each decade. By one estimate loans for irrigation, drainage and flood control are 14% of World Bank loans to India.[8] The India Case Study reports that, in total, foreign assistance provides about 13% of public sector outlays in the irrigation sector, with the World Bank Group accounting for almost 80% of this assistance. Thus, in India the World Bank did not provide the initial impetus behind

Figure 6.1 Development assistance for large dams, 1950-1999

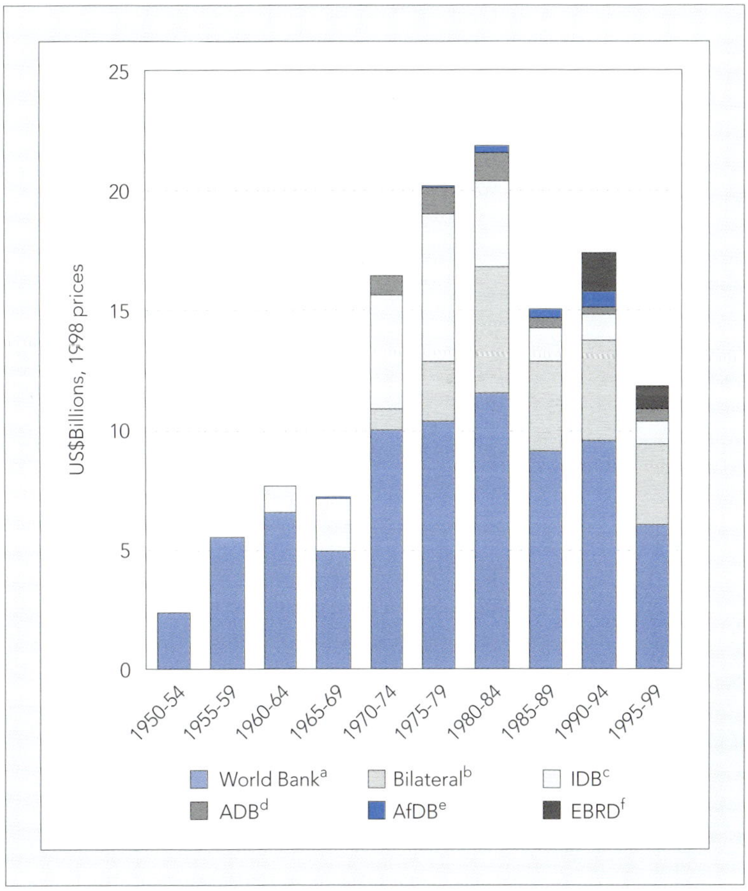

Source: [a]*Sklar and McCully, 1994 eco029, WCD Submission and World Bank, 2000;* [b]*OECD, 2000a;* [c]*IDB, 1999;* [d]*Lagman, 2000;* [e]*AfDB, 1998;* [f]*EBRD, 1996, 1999, 2000a, 2000b.*

Notes: Data for bilateral agencies also includes financing by the Commission of the European Community and includes only all hydropower investments from 1975 to 1997.

the tendency to choose dams as the response to water and energy needs, but rather provided continued and increasing external backing to the large number of dams which were built from the 1970s onwards.

As in the case of India, the WCD China Case Study shows that dam building was well advanced prior to the entry of foreign donors. Brazil also follows this pattern. Comparison of statistics on large hydropower dams commissioned in Brazil between 1950 and 1970 and the finance provided by the World Bank and the Inter-American Development Bank (IDB) show that just over 10% of the 79 large dams listed in the International Commission On Large Dams (ICOLD) database received financial assistance from these donors. However, the figure rises to over 30% of the 47 dams for the 1970-1990 period. Foreign assistance, thus, did not drive the selection of dams as an option but did provide significant finance during peak dam-building periods.

The picture differs for smaller countries. In Colombia, multilaterals helped fund the first large dam and 40% of the subsequent 50 large dams appearing in the ICOLD database. Multilaterals have played a particularly strong role in countries that have not built many dams and do not have local planning and construction expertise and capacity. In Costa Rica, which relies on hydropower for roughly 90% of its power generation, the World Bank and IDB had directly supported over half of the installed hydropower capacity by the mid-1990s.[9] In Tanzania, bilateral agencies and the World Bank have supported essentially all the large hydropower dams.[10] In these smaller countries the role of financing agencies and the firms they employ to undertake preparatory studies, design projects and build dams can be significant.

Only in the late 1980s and early 1990s has this lending activity tailed off in the face of increasing public scrutiny and criticism by civil society (see Figure 6.1). The decline followed unfavourable independent reviews of two high profile projects that were supported or considered by the World Bank – Sardar Sarovar in India and Arun III in Nepal. A number of other factors contributed to the shift away from large dam projects. They include:

- continued criticism of the pervasive 'approval culture' of the World Bank and its willingness to promote large infrastructure projects;
- internal evaluations of the Bank that documented ever-increasing 'appraisal optimism' despite evidence of poor economic and financial performance by projects in the water supply and irrigation sectors;
- failure to meet the Bank's poverty alleviation goals; and
- growing recognition of the severity of the social and environmental impacts of dams.[11]

More recently, a gradual shift towards an increased role for private sector finance in hydropower and, to a lesser extent, water supply, have also led the banks to move into a facilitation role with the emphasis on public-private partnerships and risk guarantees. Part of the financing has now been taken over by export credit guarantee agencies in donor countries that finance and underwrite risks taken by home-country engineering firms and equipment suppliers participating in projects abroad.

Role of industry and bilateral funding

Ultimately it is the country government that is responsible for taking the decision to build a dam. However, governments are naturally influenced by international expertise and financing opportunities (see Box 6.2). Once a government is politically committed and construction has begun, the nature of large construction projects makes it extremely hard to change course, even if there are cost overruns, unforeseen negative impacts, or benefits are less than predicted. The public purse generally carries the risk of poor economic performance, and there has historically been no consequence or liability for building under-performing dam projects.

For industrialised countries with a history of dam-building and expertise in related equipment, bilateral overseas aid has often become a vehicle for supporting local industry by exporting this expertise through aid programs tied to the purchase of services or equipment from the donor country.[12] Conflicts of interest have inevitably resulted between the financing agency's interest to provide contracts for home-country companies and the borrower or grant recipient's interest in providing appropriate and affordable development. In the case of bilateral agencies these conflicts of interest may be exacerbated in smaller, poorer countries where the donor plays a more central role in financial matters (see Box 6.3).

Professional associations such as ICOLD, the International Hydropower Association (IHA) and the International Commission on Irrigation and Drainage (ICID) have also played an important role in setting standards within their technical disciplines and promoting professional capacity related to the building of large dams and their associated infrastructure. These are international associations made up of members from government and industry from industrialised and developing countries alike. The associations play an important role in building capacity of member countries by collecting and disseminating technical and other information and holding annual meetings to promote formal and informal professional exchange.

Decision-making on shared rivers

The flow of water through States or provinces sharing a basin links them inextricably to a finite and common resource (see Box 6.4). Yet water resources and energy planning has frequently been undertaken at the level of administrative or political units that do not coincide with the watershed. As a

Box 6.2 WCD Case Studies and submissions: foreign involvement in dam projects

While the WCD Case Study dams built in the United States and Norway relied exclusively on national capacity, the Case Study dams in developing countries reveal the involvement of foreign firms in master plans, inventories, feasibility studies, design, construction and financing. In the case of Tarbela the World Bank even co-ordinated the Indus Water Treaty signed between India and Pakistan that gave Pakistan the opportunity to build Tarbela.

In Turkey, the comprehensive development of the water resources in Ceyhan Basin was outlined first in a 1966 study by a foreign consulting firm financed by the United States Agency for International Development (USAID). This document has guided investments in the basin for the last thirty-five years. The US Bureau of Reclamation, through USAID, undertook the initial study of the water resources of the Tocantins Basin in 1964, where Tucuruí was later built. For the Pak Mun project in Thailand, French engineering firms conducted the initial feasibility studies in the 1970's and early 1980s.

In all, the World Bank provided financing for four of the Case Study dams (Kariba, Tarbela, Aslantas and Pak Mun). Kariba was partially financed by the copper companies for which much of the power was destined. The decision of multi-national aluminium producers to invest in the Carajas region of Amazonia was subject to the decision to proceed with the Tucurui hydropower complex. Financing for the project came from internal sources and French banks.

A recent NGO report provides details on the role of 12 European companies in the design, construction and supply of equipment to 84 large dams, many of them major dams in developing countries. The report also lists the further involvement of these companies in technical studies of a larger sample of dams. Many of the projects listed are financed by home-country bilateral agencies, export credit agencies and commercial banks, as well as by the multilateral development banks. The report documents the billions of dollars that have gone to the European 'dam building' industry from projects in developing countries.

Source: WCD Case Studies and Lang et al, 2000
eco041, WCD Submission

Box 6.3 Nordic influence in the Pangani Falls Redevelopment Project, Tanzania

A 1985 Canadian study provided Tanzania with a national energy development plan that led to the decision to redevelop the old Pangani dam, raising its installed capacity from 17 MWs to 66 MWs. The Finnish International Development Agency (FINNIDA) funded the $2.5 million feasibility study in 1989-90, which was carried out by Finnish and Norwegian consultancy firms. Given the close relationship between the Finnish firm and FINNIDA, the firm not only wrote the terms of reference for the feasibility study but later was also given contracts to procure supplies and supervise construction (jointly with its Norwegian partner). In the event, the feasibility study confirmed that the dam was the best option to meet sector needs and the EIA concluded that no adverse effects existed that would prejudice the project.

As the Finnish, Swedish (SIDA) and Norwegian (NORAD) aid donors planned to finance the project, SIDA hired a Swedish firm which reviewed and confirmed the results of the feasibility study. The three Nordic donors subsequently approved grants to Tanzania to cover the costs of the project. While the aid was not 'tied', no competitive bidding was undertaken for contracts, rather, checks were made to ensure that prices offered by selected firms were competitive. A Norwegian firm supplied the turbines, a Swedish firm the generators and control equipment and a number of Finnish firms were involved in the civil works and transmission lines, including the parent company of the consultancy firm that undertook the feasibility study.

While the final 1991 project document stated that the hydrological risks to the project were small, there was sufficient concern over the availability of water supply to the project to make the Nordic donors insist on a water basin management board as a condition of the funding agreement. This decision has engendered conflicts between local, national and donor interests. The water board was to institute water fees to limit irrigation withdrawals and ensure an adequate supply of water for power generation at Pangani. With work underway in the early 1990s concern mounted as precipitation in the basin and flows at the site fell well below the 1981-92 averages. This was compounded by a lack of information on the extent of upstream withdrawals which fed traditional smallholder agriculture by the Chagga people on the slopes of Mount Kilimanjaro, as well as a series of large-scale projects financed by other international donors.

The political repercussions of charging smallholders in order to limit their water use so that electricity could be generated for consumption by industry and urban households soon manifested themselves. By 1994 resistance to the board was evident in local opposition to the tariffs. As it turns out smallholder irrigation by the Chagga is a well-studied example of a centuries-old traditional system for the management of common property – replete with a local water management 'board.' This Council of Furrow Elders is formed by elders of a specialised clan – the Wakomfongo – that plan and direct the construction of furrows as well as co-ordinate water distribution and maintenance of the furrows. Along with elders of other clans the Council administers the furrow system and resolves any problems that arise.

The potential for negative effects on the food security of traditional farmers as a result of the tariffs imposed by the official water board was acknowledged in further studies in 1995 by the original authors of the feasibility study. Yet the plan to transfer political control over water from local to centralised authority went ahead, laying the foundation for future struggles between the local people and the Tanzanian utility that operates the dam. The water board, which must mediate in this regard, is constituted by five government representatives and three representatives from each of the regions traversed by the Pangani river. No provisions were made for representation of the Chagga or other traditional water users on this board. The result is that Nordic development assistance had the paradoxical effect of undermining local resource management.

Source: Mung'ong'o, 1997; Usher, 1997a, eco026, WCD Submission

means of water storage, dams play an important role in the management of the resource and its allocation to different uses within and between countries. In the context of shared rivers, dams are a technology that allows an upstream riparian to partially 'privatise' the river by storing and using water and thereby effectively excluding downstream riparians from access to the water. In the downstream context, when faced with dwindling supplies from upstream, dams provide downstream riparians with a practical means of replacing lost dry season flows by storing wet season flows.

Relative power relationships within basins determine to a large extent how individual countries interact and whether other riparians are consulted concerning dam projects. A regional power that holds an upstream position is in a better situation to implement projects without consultation, and this has been the case in Turkey, India and China. In other cases powerful downstream neighbours whose existing resource

Box 6.4 Co-operation in shared river basins

As shown in Chapter 1, a significant proportion of the world's rivers cross international boundaries. In addition to these international basins, there are many others that cross provincial or state boundaries within a country where these states have a mandate to manage water resources. Examples include India, Australia and the United States.

Co-operation between riparian states is not new. Since AD 805 approximately 3 600 water related treaties were signed between nations. Although the majority of these relate to navigation and national boundaries, approximately 300 are non-navigational and cover issues related to water quantity, water quality and hydropower. Of these, many are limited to relatively narrow aspects and do not extend principles for integrated resource management throughout the basin. As pressure on resource use intensifies, an increase in conflicts over water may be expected and greater co-operation will be required.

Source: WCD Thematic Review V.3 River Basins

base may be affected by water resource development upstream may hold the development plans of upstream States in check. This has been the case historically, for example, with Egypt and Ethiopia.[13]

In many cases, one of the key obstacles to reaching international agreements lay in looking at water as a finite commodity and trying to allocate it on a proportional basis to different uses in different countries. In cases of water scarcity, this approach often does not give the flexibility needed to meet the multiple claims along the river course. In these circumstances it has proved helpful to extend sharing agreements to include the benefits generated by the water. The division of benefits under the 1968 Columbia Treaty between Canada and the United States on the Columbia River reported by the Grand Coulee Case Study is a case in point.

Arrangements for water sharing at provincial level are facilitated by the ability of the federal government to impose overarching regulatory frameworks, financial incentives and sanctions to ensure that provinces collaborate. A similar supra-national body is often lacking between nations and the most transparent decision-making on international rivers therefore lies within the frameworks of the many international protocols and agreements that clearly lay out the planning stages at which information should be exchanged and consultation occur. Efforts to establish accepted international principles have been negotiated through the UN for over 25 years, leading eventually to the UN Convention on the Law of Non-Navigable Use of International Watercourses. However it looks unlikely that the Convention will enter into force due to the reluctance of States to ratify it.

This situation leaves a number of key international rivers lacking a basin-wide agreement that defines a process for establishing equitable water use and therefore with no framework for good faith negotiations with other riparian States. In the absence of such agreements some States have taken unilateral action, continuing to build dams without adequate information exchange or consideration for impacts elsewhere in the basin. While this may constitute disregard for emerging international practice and the standards governing peaceful relations between riparians, it also reflects the political economy of the upstream-downstream relationship. As long as the political and economic costs of engaging in such behaviour are small relative to the economic benefits gained there is little incentive to engage in collective discussion. Clearly, as the demand for water rises and becomes ever more scarce, dams built on these international rivers are likely to increasingly affect regional relations.

Planning and Evaluation

In general project planning and evaluation for large dams has been confined primarily to technical parameters and the narrow application of economic cost-benefit analyses. Decisions of this nature were typically taken with little participation or transparency. In particular, those to be negatively affected by a dam were (and are) rarely involved in this process.

The primary concern with planning processes is that once a proposed dam project has survived preliminary technical and economic feasibility tests and attracted interest from financing agencies and political interests, the momentum behind the project and the need to meet the expectations raised often

A number of key international rivers lack a basin-wide agreement that defines a process for establishing equitable water use between riparian States.

prevail over further assessments. Environmental and social concerns are often ignored and the role of impact assessments in selecting options remains marginal. Once operations have been initiated there is a generalised lack of effort to monitor, assess and respond to operational concerns and changing values surrounding dams. Again, the political economy of large dams and the dominant power of a small number of actors often drive these planning and evaluation processes. In some cases, such as Norway, Quebec, Brazil or Nepal, a high level political choice made in favour of hydropower has driven subsequent choice of technology (large dams) and project development.

Participation and transparency

The WCD Knowledge Base shows that the most unsatisfactory social outcomes of past dam projects are linked to cases where affected people played no role in the planning process, or even in selecting the place or terms of their resettlement. In addition, governments have frequently committed themselves unquestioningly to large infrastructure projects, whose merits have not been tested by public scrutiny, without hearing alternative views on the choice of development objectives for a village, region or country. As pointed out in Chapter 4, the involvement of displaced people has the advantage of enabling them to contribute to the benefit stream of a project and thus to achieve different outcomes.

Figure 6.2 Trends in provisions for participation and information disclosure

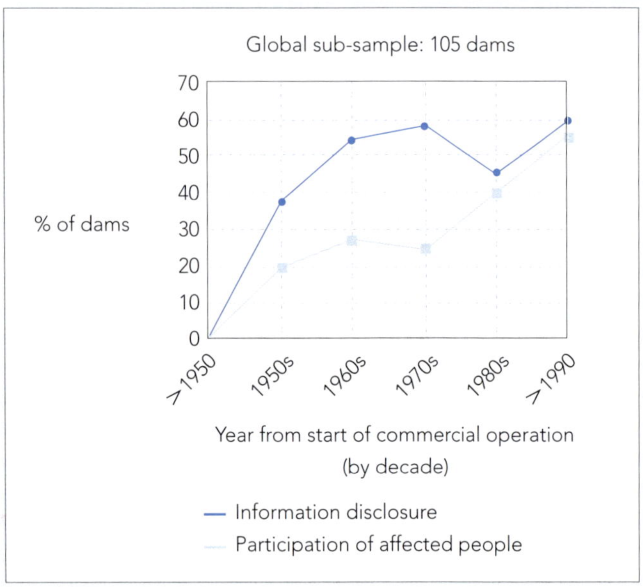

Source: WCD Cross-Check Survey.

Participation and transparency in decision-making processes involving large dams – again like most development projects – was neither open nor inclusive through the 1980s. Of the 34 dams in the Cross-Check Survey that involved resettlement of displaced people, only 7 required participation as part of the decision-making process. While there has been a growing emphasis on transparency and participation in decision-making involving large dams, especially in the 1990s, actual change in practice remains slow.

Additional results from the Cross-Check Survey illustrate that while participation has increasingly been required in the planning documents of large dams and for various activities, around 50% of projects still do not plan for the public participation of affected people. The trend for requirements for transparency through information disclosure for large dam projects is similar to that for public participation (see Figure 6.2).

The Commission's review identified the following recurring concerns and criticisms about how the public, and particularly affected people, have been involved:[14]

- insufficient time, resources and information have been made available for public consultations;
- the spectrum of participants was usually very narrow, ignoring rural communities, indigenous groups and women, and affected people's organisations whose effective participation may be constrained both culturally and linguistically;
- where opportunities for participation of affected peoples and NGOs representing affected groups have been provided, they often occur late in the process and are limited in scope. Moreover where substantial differences arise, those seeking to modify plans and decisions often must resort to legal or other action outside the normal planning process;
- there was a generalised failure to involve affected people in the design and implementation of project monitoring and follow-up; and
- the government agency staff leading the discussions had often been trained only in one sector (such as engineering) and this reduced the scope for promoting a multi-disciplinary approach.

From the experiences recorded in the WCD Knowledge Base there are recent examples that show where participation has reduced conflict and made outcomes more publicly acceptable (see Box 6.5 and Box 6.6). These contrast starkly with those where projects have been pushed through by central authorities without consultation resulting in drawn-out and acrimonious conflicts over compensation, resettlement and benefit sharing.

The failure to provide a transparent process that includes effective participation has prevented affected people from playing an active role in debating the project and its alternatives. As a result they are unable to

Box 6.5 Even late participation leads to a consensus resettlement plan: Salto Caixas dam, Brazil

The 1 240 MW Salto Caixas hydropower project in Brazil was completed in 1999 and was the fifth to be built on the Iguacu River. It was the first hydropower project in Brazil planned under the environmental regulations stipulated in the 1988 Constitution. The measures taken to comply with the new environmental legislation at Salto Caixas amounted to about one-quarter of the total project cost of approximately $1 billion. However, the EIA was only undertaken after the project was approved and land had been acquired, putting political pressure on the process. This resulted in an EIA study of 'poor quality.'

On the social side, public pressure, based in part on unsatisfactory resettlement outcomes for previous dams, led to the establishment of a 'resettlement committee'. This committee created a forum to address conflicts and meet requests of the affected people. A negotiation process involved the committee of affected people in developing an acceptable relocation programme for displaced people. Views of local people were also incorporated into the monitoring stages of the project.

Source: dos Santos, 1999, p153–154;
Verocai, 1999, Contributing Paper for WCD Thematic Review V.2
Environmental and Social Assessment, p7

Box 6.6 Public participation and project acceptance: three scenarios from Austria

In the early 1980s, nine multi-purpose dams had already been built in the Austrian section of the Danube River. Two more dams, Freudenau in Vienna, and Hainburg downstream, were planned on the main river and some low-head projects envisaged on the Mur river.

The decision to build the Hainburg hydropower and navigation dam was made explicit in 1983. Public participation was restricted to holders of property and water rights to be directly affected by the planned intervention, thus excluding environmental activist groups and other civil society organisations. With strong support from the general public, these civil society groups occupied the site of the project, and ultimately managed to stop the project. Subsequently the site was protected as a National Park.

Turning to the Freudenau hydropower dam, from 1986 to 1988, the provincial government of Vienna and the power utility promoted ideas and proposals for mitigating some of the potential impacts of the project (during construction and after completion). The public responded with great interest and as a result selected proposals and the detailed project plans were made fully accessible to the public. Information meetings on these documents were attended by more than 15 000 people. In 1991 a referendum was organised around the final project proposal. About 44% of the entitled Vienna inhabitants participated, and 75% supported the project.

The Freudenau dam was completed in 1997, however the project is not cost-effective as a consequence of the design modifications required to gain public support. The contrary was the case for the Fisching and Friesach projects on the Mur river – where following occupation of the dam site by protesters, joint-planning sessions with stakeholders led to 'significant improvements in environmental friendliness,' with associated cost savings.

Source: Hainburg and Freudenau dams
in Nachtnebel, 2000, p109-111;
Zinke, 1999, p6-9;
Fisching and Friesach dams
in Brunold and Kratochwill, 1999, p176-17

assist project planners to provide a development response that meets their needs and allows them to add to the benefits to be derived from the project. Without doubt this has magnified the negative impacts of such projects and alienated affected communities leading to active opposition to projects and considerable uncertainty for project proponents. As observed in previous chapters of this report, the outcome is often not only poor performance of the social components of projects but also schedule delays, cost overruns and poor financial and economic performance.

Options assessment

The range, scale and type of options considered in development plans in the past were limited by the boundaries of the planning and decision-making approaches of the day. Many sectoral planning studies from which projects emerged were narrow technical and economic studies, aimed at least-cost supply solutions for providing a single service such as irrigation water or electric power. When dams were contrasted with alternatives, they were typically only compared to other potential dam projects or, in the case of hydropower, with alternative large-scale thermal power generation options (see Table 6.1). In developing countries the pressure on development aid agencies to move large amounts of capital – a considerable portion of it as tied aid – argued for large-scale solutions such as large dams. Administrative efficiency is a related factor leading to a preference for financing large projects.

Table 6.1 WCD Case Studies: options assessment

Project	Alternatives considered in preliminary planning	Comparison with alternatives at appraisal	Criteria and parameters used for selection
Aslantas	Existing run of river source for irrigation was considered insufficient to support wider agriculture development. Alternative dam locations in the basin considered.	The hydropower component was compared with a thermal alternative.	Least-cost analysis for power supply.
Grand Coulee	The agreed objective was regional development through irrigated agriculture. Alternatives considered over a 15-year period related to gravity and pumped systems for the delivery of water. The 1932 Butler Report recommended the pumped option financed predominantly by hydropower revenues from the dam.		Economic analysis
Glomma and Laagen Basin	Government policy for hydropower established in early 1900s. Oil and gas discovered in early 1980s, but hydropower policy remained in place until recently.	Alternative hydropower sites considered.	Least cost and environmental and social ranking of sites in the 1980s protection and development plans.
Kariba	Kafue Gorge in North Rhodesia considered as an alternative but rejected after protracted debate with Southern Rhodesia.	A set of thermal power alternatives	Least-cost analysis for power and influence of South Rhodesia' s political interest.
Orange River	Reservoir storage was seen as the only way to achieve more reliable water supply for year round irrigation. Hydropower was a secondary benefit therefore thermal alternatives were not considered. Subsequent studies led to raising the VanderKloof dam to increase output.		Political
Pak Mun	Thermal alternatives located elsewhere considered. A higher dam option was rejected due to resettlement and environmental concerns. The revised project approved by the Cabinet included irrigation benefits.	Alternative thermal power (gas turbine).	Least cost analysis and power system studies.
Tarbela	Additional storage was considered the only option for replacing water of the eastern rivers allocated to India.	A number of alternative dams sites including Kalabagh and Gariala	Economic comparison of the sites combined with Government preference for larger storage and power potential of Tarbela
Tucurui	Tucurui responded to objectives for development of the mining-metallurgical sector and to supply power to urban areas in the Amazon Region. There was no explicit options assessment		

Source: WCD Case Studies.

Projections of demand

The needs for power, food and water are typically identified through sectoral demand forecasts, which have frequently overstated sectoral needs. The WCD Glomma and Laagen Case Study reports that in Norway gross power consumption in 1990 was 75% of that forecast in 1970. In Slovakia, water supply needs as assessed in 1985 were expected to rise to 408 and 465 litres per capita per day (lcd) in 1990 and 2000 respectively. While demand initially rose above projections (433 in 1990) it has since reversed course and had fallen to 294 lcd by 1997.[15]

Failure to adequately account for the rate of development of new supply and the effect of policy reform, when it is outside the limits of the planning exercise, may also lead to what effectively amounts to overstated demand. In the Slovakia case cited above, a doubling in the price of water and the development of alternative sources of supply by industry contributed to reducing actual demand. Projections for demand (and hence prices) for crops and other agricultural products that are widely traded can be subject to market boom and bust cycles when a series of independent decisions in different countries or provinces lead to over-production relative to demand. As documented in Chapter 2 the prices for agricultural commodities have fallen over time, instead of remaining constant or rising as assumed in many projections for irrigation projects. The same market information, technical assistance packages or consultants frequently influence these planning exercises, thus, while difficult to foresee, such over-estimates are not inevitable.

Overstating future demand has led to a perceived need for a large incremental response to meet rapidly growing needs. In many circumstances this has militated against a gradual approach of adopting smaller, non-structural options and has pushed decision-makers into adopting large-scale dam projects because they seem to be the only adequate response to the large gap between existing supply and forecast demand. A further complication is the long-lead time of large dam projects, which may take 10 years or more from initial development of a project idea to the commissioning of the structure. Changes in market conditions during construction have left proponents stranded with costs or projects that are not financially or economically viable. Of principal concern is that it is frequently the agencies that are responsible for building supply infrastructure that are also charged with undertaking demand forecasts, leading to a potential conflict of interest.

Available options

As shown in Chapter 5 there is currently a wide range of alternatives available for fulfilling water and energy needs, although the actual number available will depend on local circumstances. The number of alternatives has, however, not always been so large. For instance, alternatives to hydropower prior to the 1950s included conventional fossil fuel and biomass generation options. Nuclear power arrived in the 1960s and in the past decade the range and scale of renewable electricity supply options has dramatically expanded. Alternatives for municipal and industrial water supply have tended to be site specific and depend on whether there are groundwater aquifers, natural lakes and rivers to draw from with sufficient quantity and quality of water. On the other hand, many of the irrigation water supply and flood management options that are being considered today have been available for a long time. The principal change here is a more receptive policy context and the increasing cost of developing new water supplies.

Demand-side management options (DSM) are a more recent phenomenon. Efficiency and conservation became concepts in policy and planning in the 1970s and 1980s, but serious attention to demand management programmes has tended to depend on a perception of crisis. The oil price shocks in the early 1970s focused attention on DSM in the electricity sector in many western countries. Water scarcity and the threat of water shortages is a driving force for more efficient water use in many countries, but the response has not been universally translated into concrete action in fostering water-efficient practices.

Obstacles to consideration of options

Political economy or intellectual barriers often pre-determined what options were considered in a given context. Political economy barriers include efforts made by groups, primarily those holding economic control and political influence, to protect their own interests and to impede similar efforts by other stakeholders. In practice, these barriers were immensely varied and wide-ranging. They included soft and subtle actions such as withholding information necessary for making informed decisions from other stakeholders and from the decision-makers. At the other end of the spectrum there have been overt and even violent measures such as use of State and police power to protect favoured options.

With few exceptions, an inclusive institutional and policy structure capable of dealing with a spectrum of options has been slow to emerge in developing countries. Small-scale infrastructure alternatives often have not received integrated planning support, impeding their ability to emerge as competitive solutions. Interests promoting non-structural alternatives have rarely offered an adequate political counterbalance to interests promoting a dam option. In many cases the weight given to the infrastructure option by the key actors obstructed proper consideration of other viable alternatives. As a result, such options continue to be viewed as secondary to large projects. China provides an example of a country that has mixed both the small- and the large-scale. It has the world's largest programme for the development of small-scale rural and appropriate technology, while at the same time it has built half the world's large dams.

The hindsight provided by Chapters 2, 3 and 4, however, does suggest that in cases where dams have failed to deliver or led to large negative social and environmental impacts, a more comprehensive assessment of options may have been warranted. Whether failures to adequately assess all options implicitly lead to the selection of a dam over other equally attractive or even superior options is difficult, if not impossible, to answer. Certainly, the options currently available (as described in Chapter 5) reflect not only continued technological development over the last 50 years, but also are a product of more recent efforts to find locally appropriate, small-scale solutions that have benign social and environmental impacts. Thus, in many cases alternatives may not have been available previously and may have appeared more expensive given the methodologies employed at the time, or were excluded due to the influence of vested interests.

Parameters for project appraisal[16]

Cost-benefit analysis (CBA) emerged between the 1950s and 1970s as the dominant economic tool supporting decision-making on dam projects. Initially it was limited to a number of parameters, most of them internal to the dam owner and relatively easy to assign values to. Efforts in the last two decades to expand the scope of CBA to cover social and environmental issues have rarely led to comprehensive social and environmental valuation, and have usually been limited to incorporating the costs of resettlement and environmental mitigation.

Review of multilateral bank appraisals and the performance of CBA more generally leads to the following conclusions on the adequacy of CBA as applied to the appraisal of large dam projects:

- projections of project costs are systematically understated;
- social and environmental impacts are not valued explicitly or are only indirectly accounted for through mitigation or resettlement budgets;
- difficulty in predicting inter-annual volatility of hydrological flows, growth in demand and final design capacity (hydropower, irrigation and other benefits);
- difficulty in predicting market conditions and farmer behaviour over time (irrigation benefits);
- employing social discount rates that are too high;
- sensitivity and risk analysis is inadequate; and
- the effect of uncertainty and irreversibility of investment is ignored.

In other words, the historical and actual practice of dam project appraisal often violates the conditions under which it could, in theory, provide a reliable measure of the change in economic welfare produced by a dam project. It is worth emphasising that it is not a foregone conclusion that the net effect of fixing all of these problems would be to lower the economic profitability of dams. A number of the weaknesses of CBA may lead to understatements of the net project benefits. At the same time, it is clear that quite a number of the weaknesses can have important impacts in terms of lowering net project benefits. Improved application of CBA would assist in identifying projects that are not economically viable.

Over reliance on CBA and the implicit pursuit of economic welfare maximisation also handicap decision-making where dams have other (or additional) objectives as:

- CBA does not examine wider economic impacts – such as economic multiplier impacts; and
- CBA does not explicitly identify who gains and who loses from a project.

Although CBA is typically a prerequisite to the analysis of macroeconomic and regional impacts, as well as to distributional analysis, it is not designed to examine the potential of a project to achieve objectives in these areas. Given the continued 'partial' ability of CBA to capture even the extent to which efficiency objectives are achieved, and given that equity, macroeconomic and purely non-economic objectives are often integral objectives of water resource development projects, CBA alone is not a sufficient basis for the evaluation of large dam projects.

Influences from the larger political economy also filter through into the process of undertaking CBA. In some cases, early political or institutional commitment to a project became overriding factors, leading subsequent economic analyses to justify a decision that had in fact already been taken.

Decisions made to build dams solely on the basis of such an analysis are questionable

given the failure to undertake options assessment and to include external impacts, particularly social and environmental costs. An alternative approach to a decision support system based on CBA is to use a method that recognises that projects often have multiple objectives and not simply economic welfare maximisation. Experience to date with these multi-criteria approaches suggest that while economic criteria remain important, these decision frameworks have the benefit of allowing disaggregated information on social and environmental impacts to enter directly into the decision analysis. Such decision support systems appear particularly appropriate and useful in the case of large dams when implemented within a participatory, transparent multi-stakeholder approach.

Addressing social and environmental impacts[17]

Social and environmental issues have historically been among the least addressed concerns in dam-related decision-making. The Commission has focussed on these because they are two of the key issues that determine whether a dam proves to be an effective development project that enjoys general acceptance by the public. The environmental risks associated with large dam projects have not been generally incorporated as key factors in the decision-making process. Enforcement of existing regulations is often weak, initial assessment has not been comprehensive and it has frequently been incorrectly assumed that impacts could be effectively mitigated (see Chapter 3). Generally, monitoring of impacts and assessments of the effectiveness of environmental mitigation measures have been absent.

Similarly, the adverse social implications of large dam projects have rarely been a factor in the initial assessment and therefore have not generally influenced the decision-making process to reach a least social cost alternative. The experiences of affected people around the world as reviewed in Chapter 4 confirm the extent to which impacts remain inadequately assessed and efforts at mitigation, development and resettlement unsatisfactory.

Following the United Nations Conference on the Human Environment held in Stockholm in 1972, environmental agencies and ministries were formed at a rapid rate with approximately 60 being created by 1988 and at least another 40 by 1992. The World Bank adopted its first dam-related policy in 1977 (on dam safety). During the 1980s the Bank developed policies and guidelines that focused on the social and environmental dimensions of dams and water resources.

Environmental Impact Assessment (EIA) was adopted and formalised in many countries during the 1980s, although many developing countries only approved EIA legislation in the 1990s. EIA has become the major tool for addressing social and environmental impacts and the Commission has reviewed an extensive literature on this subject as well as hearing directly from those affected through the regional consultations. The WCD Knowledge Base demonstrates that EIA consists mostly of measures to compensate or mitigate the planned impacts and render them acceptable when the decision to proceed has already been taken. This is reflected in the tendency for EIAs in the 1990s to focus increasingly on mitigation plans. Added to this is the fragility of newly established environment ministries that may be unable to ensure compliance with many of the plans or clearance conditions.[18] There are well-documented cases, even in the 1990s, of decisions to proceed with financing or construction before an effective EIA is completed (see Box 6.7).

Political pressures and tight schedules are as relevant today as in earlier decades and EIA results often have no significant influence on the choice of a dam as the preferred option. The EIA process is also not well suited to this purpose as it was meant solely for identifying impacts and associated mitigation measures rather than as a tool for including environmental and social considerations in the final project choice and design. Many governments and financing institutions have adopted EIA in the last two decades, however the quality of assessments and their ability to genuinely influence outcomes is still under-developed. Most dam proponents see EIA as an administrative hurdle to be cleared, or a requirement to secure funding. This means that a huge political, technical and financial investment in the project has often already been made before the EIA is launched. If impacts are severe, it is often too late to change design, and project cancellation may involve loss of face and financial loss. Further, EIA operates under considerable constraints due to the political and administrative pressures imposed by project schedules as it is seen as 'delaying' the project. EIAs are also often done with inadequate baseline data on demographic trends, socio-cultural systems and ecosystem functioning. This leads to unsatisfactory outcomes.

As an impact management tool, EIA has evolved towards a tool for also setting up an ongoing environmental management system or programme when construction begins, involving appropriate experts, ministries and field activities. The transition from a planning mode, based on voluminous assessments and reports, to an implementation mode during project construction creates severe institutional and human resource challenges and in many cases the measures are either not implemented or have fallen short of the efficacy envisaged in the planning documents. The reality is that dams create huge management challenges for the implementing ministries and agencies. Where institutional capacity in the environmental area is weak the accompanying measures needed for sustainable outcomes often prove difficult to manage, particularly when compared with the physical act of designing and building the dam. This in turn may lead to public dissatisfaction with dams when affected people perceive that promises have not been kept.

Operation, monitoring and decommissioning

After large dams are commissioned there are a number of management and operational issues that require technical studies and involve either decisions at the management level or decision processes that are public in

Box 6.7 Environmental Impact Analysis (EIA): too little, too late

Even with improved environmental and social guidelines EIA still frequently fails to influence decision-making. The Theun Hinboun project in Laos was initiated in the early 1990s. The initial EIA financed by NORAD concluded that the dam would have minimal adverse impacts and significant benefits. Most of those who reviewed the document disputed these findings and NORAD undertook supplementary studies. These were completed one year after construction began, so they had no impact on the decision making process or the design of the dam.

In the WCD case studies, an EIA was conducted only for Pak Mun at the planning stage as it was a World Bank requirement. However the EIA was done ten years before the final project was approved – and examined a different project design for a different location than the one finally approved. Further, the EIA was never revised or updated. EIA's were only required in Thailand from 1992, one year after Pak Mun was approved by the Thai Government.

Source: Theun Hinboun dam in Norpower, 1993, p1-7 as cited in Usher and Ryder, 1997 eco026, p80-81, WCD Submission; WCD Pak Mun Case Study

nature. These are required for the following reasons:

- to support routine day-to-day operations such as reservoir operations and releases;
- to change operations to conform to new regulations that are introduced such as economic, technical, environmental or social regulation on dam safety, operation of reservoirs in flood events or changes to environmental flow requirements;
- to change operations when a new dam is introduced in the river basin that would impact on the operating rules of the existing dam;
- to adapt the operation to changing needs in the services provided by a dam over time, particularly when the project is multi-purpose, for example a change in power markets that increases the value of peaking power generation or a shift to recreation priorities in controlling reservoir water levels;
- for renovation, upgrading or expansion of the existing facilities; and
- for relicensing processes in some countries, or for decommissioning.

One of the most disturbing findings of Chapter 2 was the lack of monitoring of the impacts of dams and the complete failure to conduct proper ex-post evaluations of performance and impacts. That such large investments have rarely been evaluated once they have been in operation for a significant period suggests little obligation on the part of powerful centralised agencies and donors to account for the costs and benefits incurred. Perhaps more critically it signals a failure to actively engage in learning from experience in both the adaptive management of existing facilities and in the design and appraisal of new dams. The WCD Case Studies suggest that provisions for intensive monitoring of physical, social and environmental effects of projects were often weak or entirely absent. Where monitoring was present, it was often restricted to hydrology and engineering parameters related solely to the physical integrity of the dam structure.

The operation of large dams is subject to many unforeseen and unforeseeable influences over time that transform and redistribute benefits and impacts. Patterns for the release of water from reservoirs will normally change over time in response to demographic and land use changes in the river basin, shifts in water use priorities, as well as changes in the agricultural economy and the markets for electricity. Physical changes in river morphology or reservoir sedimentation as well as changes in the value that society places on ecological and social impacts of dams will influence how the dam is operated at different periods of time over its life.

As shown in Chapter 2, there are good practice examples of adaptive management to meet this changing context, drawing on sophisticated decision-support and forecasting software and in some cases accommodating stakeholder participation. Many developing countries that continue to focus on building rather than optimising operations have not yet adopted tools and policies for adaptive management and optimisation.

What happens to dams at the end of their lives? Dam decommissioning may be necessary due to safety concerns, dam owners' concerns about lower profits, or concerns about social and environmental impacts. Decommissioning can mean actions ranging from stopping electricity production to dam removal and river restoration. Several hundred dams have been deliberately removed, mostly in the United States, and most of them small. Provision is not always made in advance for who should pay for the removal or for safety or other improvement measures. As reviewed in Chapter 2, rising operations and maintenance costs may also raise questions regarding the maintenance and safety of large dams. At present, decommissioning costs are difficult to predict due to the uncertainty surrounding the various parameters affecting the costs and the limited amount of practical experience with decomissioning. One proposal is to ensure that decommissioning funds are set aside at the time of dam commissioning or during the project's license period. Such decommissioning funds are accepted practice for nuclear power plants in countries such as the United States. Decision-makers in the developed world are increasingly looking at how best to handle the end of the dam life cycle. In contrast many other countries do not yet have firm licensing periods for their dams (see Box 6.8).

Compliance[19]

Dam projects are expected to comply with the legal framework and guidelines of the country and the organisations involved in financing and constructing the dam. Where environmental and social problems have occurred in the WCD Knowledge Base, the principle cause is the lack of legal requirements for particular standards at the outset or a lack of appropriate recourse mechanisms to adequately reflect people's rights in the face of a powerful national decision. This section shows that regulatory frameworks are often weak, and the necessary provisions are not made in planning documents. Even when they are present, governments and donors alike ignore them all too frequently.

Reasons for this include:

- incompleteness, incoherence and ambiguity of national legal and regulatory frameworks;
- difficulties of accurately defining the specification of social and environmental requirements and integration of these components into the implementation agreements and schedules of projects;
- lack of transparency and accountability, frequently with opportunities for corruption at key points in the decision-making process;
- lack of meaningful participation at key points in the decision-making process
- low levels of internal and external monitoring that reduce feedback into decision-making;
- weak or non-existent legal recourse and appeals mechanisms to an independent judiciary, particularly for negatively affected and vulnerable groups; and

Box 6.8 Licensing processes and duration

There is considerable variation in the licensing procedures for dams. In some countries dam sponsors must obtain only one licence. In other countries the dam sponsors must obtain a licence for each phase in the planning and project cycle. For example in Hungary a dam project sponsor must first seek approval of the EIA, then obtain a permit to complete the activities required to prepare the project for construction. Two further licenses are required for construction and operation. Some countries exempt government operators and only require licences for private operators. Licensing procedures are often restricted to hydropower dams, with irrigation dams largely exempt from formal licensing.

There is considerable variation in the term of dam operating licences. In Spain licences are granted for 70 years, in Norway 60 years and in the USA for 30–50 years. In other cases dam licences are granted for short renewable periods of time. For example in Hungary and Vietnam licences can be granted for unspecified periods of time but they are subject to regular inspection and review. Where dams are built and operated by the private sector the duration of the licence period will need to reflect a reasonable payback period, typically set at 30 years in build, operate and transfer (BOT) agreements. Reviews conducted within the licence period have the advantage of facilitating the monitoring of operations and providing opportunities for adapting operations against the background of changing societal values and expectations. Contemporary concepts of adaptive management, transparency and accountability suggest that there should be some scope for regular review, such as every five to ten years.

Source: WCD Thematic Review V.4 Regulation

- lack of human, financial and organisational capacity.

National legal frameworks and policy provisions

There were few policy, legal and regulatory frameworks governing large dam building before the 1970s, particularly for social and environmental issues. However many countries updated their policy and regulatory frameworks in the 1980s and 1990s to give a stronger emphasis to environmental and social concerns, public participation, efficiency and cost-recovery. There is now a broad body of regulation potentially or explicitly applicable to large dams at the international and national levels, referring to both the public and private sectors.

Existing regulations in most countries tend to focus on project appraisal and implementation with insufficient focus on options assessment planning in the early stages of the decision-making process where fundamental choices are made. Few require regular assessments and evaluation of performance that could feed back to better inform decision-making based on past experience. Nor do they often provide recourse for those who may have been harmed by a particular project.

In many cases it has only been strong concerted civil society movements that have generated sufficient momentum to ensure that constructive negotiations occur, and dam projects are not imposed on displaced communities without consultation.

The Cross-Check Survey demonstrates that, since the 1950s, a growing number of projects have required dam safety, economic cost-benefit analyses and financial plans (see Figure 6.3). But economic appraisal techniques such as risk and distributional analysis were still mandated for only 20% of large dam projects even in the 1990s. Sensitivity analysis is more common and has become standard for donor-financed projects. Dam safety is a key issue for the world's ageing dams and many national bodies have taken up the challenge, assisted by the engineering networks of ICOLD.

Figure 6.3 Trends in the implementation of economic and financial analyses

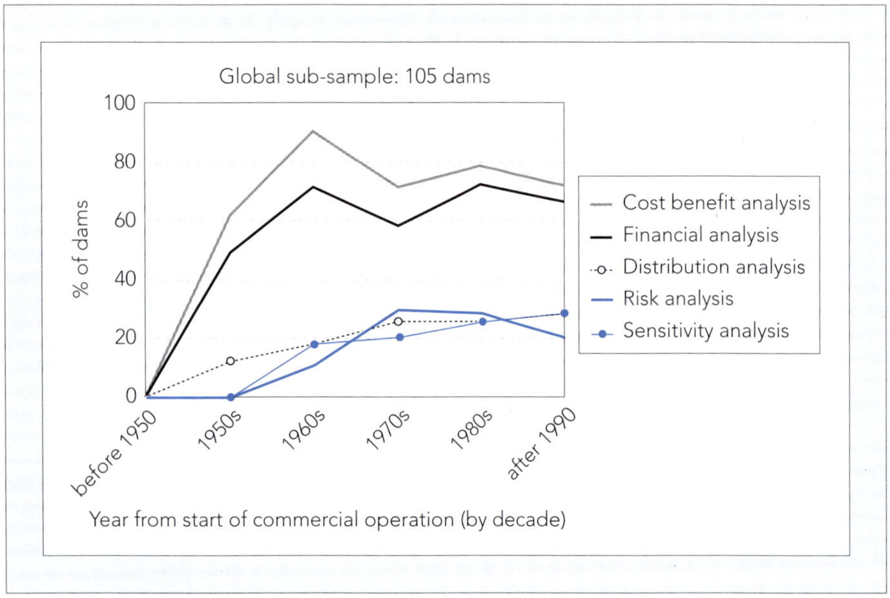

Source: WCD Cross-Check Survey.

Even requirements for large dam projects in the environmental field are far from universal although they are increasingly required. EIA is recorded for less than 40% of dams commissioned in the 1990s (see Figure 6.4). Strategic environmental assessments and baseline surveys occur at similar levels.

Corruption

Corruption is a world-wide phenomenon that affects both poor and rich countries. It may take many forms, from inducements to favour certain contractors during bidding, through to manipulation of water allocations, offsetting farmer repayments, or manipulating domestic electricity connections locally.[20] At whatever level, vested interests can distort the decision-making process, undermining development. Decision-makers may be inclined to favour large infrastructure as they provide opportunities for personal enrichment not afforded by smaller or more diffuse alternatives. The consequences frequently directly affect the poor or the environment. Allegations of corruption have tainted many large dam projects in the past but have seldom resulted in prosecution in court (see Box 6.9).

The OECD countries, and the major international financing agencies, have recognised the pervasive extent of corrupt practice and its negative consequences. Through the 1990s they have moved to assist countries in tackling corruption by making bribery payments illegal in their country of origin, debarring contractors convicted of bribery from future contracts and tightening up due diligence on bribery opportunities.[21] As of August 2000, twenty-three countries had ratified the 1997 OECD Convention on Combating Bribery of Foreign Public Officials in International Business Transactions. Its principal objective is to eliminate bribes to foreign officials, with each country taking responsibility for the activities of its companies and what happens in its own territories.[22]

Transparency International, an international NGO, has also been active in promoting workable and transparent, 'integrity pacts' for large infrastructure tenders. These have met with growing acceptance and success in Latin America. A range of legal measures and transparency processes are therefore increasingly available for ensuring that dams are built for societal good, not for personal gain.

Box 6.9 Allegations of corruption

In early 2000 the Chinese government released information that corrupt officials had embezzled $60 million (500 million yuan) from resettlement funds for the Three Gorges dam project. An official was sentenced to death for embezzling almost $1.5 million from the project.

In Lesotho a trial started in June 2000 against major international corporations involved in construction on the Lesotho Highlands Water Project (LHWP). Companies from France, Sweden, Germany, the United Kingdom and Canada have been accused of paying bribes. If the accused companies are convicted, they face debarment from future projects with the European Union.

In the United States, economists from the United States Corps of Engineers accused senior management of deliberately manipulating economic analyses to promote billion dollar investments to be managed by the Corps.

Source: China in Agence France Presse, 21 January 2000, 10 March 2000; LHWP in Sunday Independent, 11 June 2000; United States in Grunwald, 29 February 2000

Multilateral and bilateral financing agencies

Overseas development financing agencies, particularly the multilateral and bilateral

Figure 6.4 Trends in the implementation of environmental and social assessments

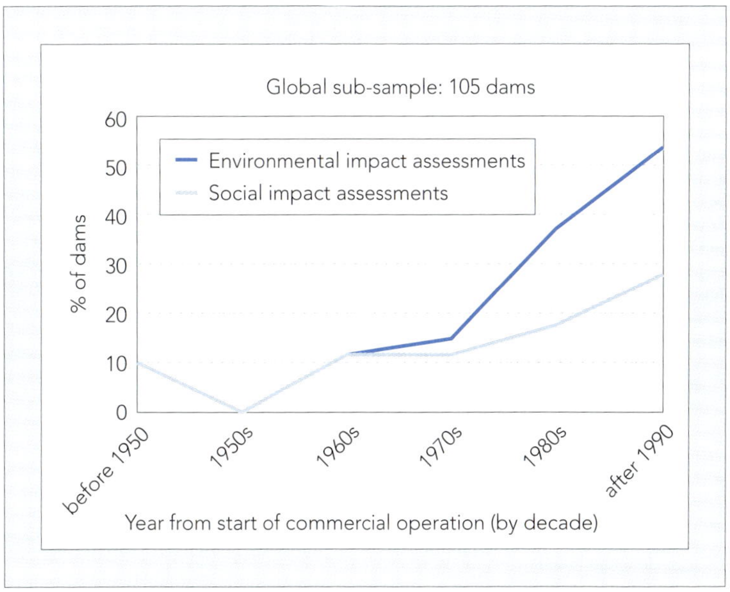

Source : WCD Cross-Check Survey.

The policies of the development banks are more concerned with project planning, design and financial management than with options assessment or with the operational phase of a large dam project.

agencies have played an important role in funding and securing large dam projects. They have adopted a broad set of policies, criteria and guidelines since the 1980s as a result of lessons learned from experience and public criticism. For example, the World Bank has adopted ten safeguard policies relating to such environmental issues as forestry, pest control and environmental assessments; and such social issues as indigenous people, cultural property and resettlement. The result of these developments is that on paper the World Bank has a comprehensive set of policies dealing with large dam projects. More recently the International Finance Corporation (IFC) and the Inter-American, Asian and African Development Banks have adopted similar guidelines.

Despite these changes, the banks' policies, like the national regulatory systems are more concerned with project planning, design and financial management than with options assessment or with the operational phase of a large dam project. In addition, they have paid more attention to monitoring the planning and construction phases than the operation of the project, which is often left to national governments. Post-implementation monitoring is generally discontinued at most five years after project commissioning. Even then, the main focus has been on comparing the project proposals with the project outcome. Weak treatment of social and environmental impacts at appraisal leads to weak assessments of outcomes at evaluation.

This approach assumes that the planning phase can anticipate and cover all future eventualities. Yet the WCD Knowledge Base shows that achieving satisfactory social and environmental outcomes requires constant adaptive management. The short term and inflexible nature of the agreement between the borrower and the bank is an obstacle to achieving this result. Further, the mitigation measures often receive less prominence in comparison with finance issues.

Numerous developed countries have bilateral aid agencies and export credit agencies which have also funded or supported the financing of dams and dam-related projects.[23] Bilateral aid agencies vary in the stringency of the requirements they have for supporting large dam projects. Yet while they are relatively small participants in the large dam sector, their funding for specific aspects of the master planning or project feasibility studies can be critical in bringing other financiers to the table.

Export credit agencies (ECAs) are increasingly financing specific portions of large-scale infrastructure projects in developing countries. ECAs provide loans, guarantees and insurance to domestic corporations and businesses for their activities overseas to support and promote export trade from their respective countries. They finance the high value, electrical and mechanical equipment components and are an increasingly important source of financing for private sector involvement in large dams.

Unlike the major development financing agencies, ECAs generally lack policies on environmental and social issues and do not necessarily adhere to internationally accepted standards and guidelines. Experiences from the Three Gorges dam in China, Ilisu dam in Turkey, Maheshwar dam in India and San Roque dam in the Philippines underline the need for ECAs to examine closely the social and environmental impacts of the projects they support. The absence of common standards among ECAs leads to ad hoc competitive decision-making (see Box 6.10).

The policies of multilateral banks have challenged the capacity of their borrower countries to actually implement their requirements. Bank staff have had to either exercise their own discretion to adapt the policies to the realities of each country or ignore cases of non-compliance by their borrowers. In either case the bank's tolerance of the staff's and the borrower's non-compliance with the policies can breed cynicism about the willingness to comply. There are no sanctions for staff members, or countries, for non-compliance. Performance criteria for staff have tended to be related to approvals and disbursement targets.

The WCD Knowledge Base offers many examples of the failure of project proponents, contractors and operators to fulfil commitments, whether explicit (project specific agreements and contracts) or implicit (applicable policies, laws, regulations and guidelines). The WCD Case Studies provide an indication of the types of breaches observed (see Box 6.11). The three basic reasons for lack of compliance have been:

- The tendency for large projects to proceed under a restricted decision-making process negotiated between governments, lenders and contractors with little public oversight, little participation by affected parties and limited disclosure and public access to information. In many cases lack of clear monitoring procedures also limited public scrutiny.
- The lack of sanctions for non-compliance, either at national or international level. In many cases local affected communities were unable to defend their interests when faced with a strong centralised government especially in countries with weak legal safeguards and recourse mechanisms.[24]
- The dependence, in many cases, on the good faith of sovereign States and public pressure to resolve disputes, adjudicate claims and ensure compensation for those who have suffered wrongs. The absence of legal sanction or, where this exists, difficulty in accessing it made it easier for developers (especially governments) to escape the consequences of non-compliance. The costs involved in seeking legal remedies were often prohibitive for those who may have been negatively affected.

The multilateral banks – and in particular the World Bank – have the most sophisticated set of policies, operational procedures and guidelines amongst the international donor community and are under regular scrutiny by civil society. In examining actual practice and compliance with standards and the realisation of the outcomes that these

Box 6.10 Export Credit Agencies: competing for business versus common standards

After the US Export-Import Bank declined support for the Three Gorges project in China, citing lack of information on environmental and social mitigation, other ECAs, with lower thresholds of social and environmental acceptability, stepped forward to issue loan guarantees to corporations. This phenomenon is especially relevant to the financing of large dam projects where ECAs are supporting projects declined by other funding agencies on environmental grounds. In June 1999, the G-8 ministerial meeting issued a statement recognising the importance of common standards among the ECAs. Later in the year the OECD Working party on Export Credits and Credit Guarantees agreed to a voluntary environmental information exchange on larger projects but fell short of agreeing on new criteria for ECA support.

Source: Udall, 2000, WCD Contributing Paper to Thematic Review V.4 Regulation, p1-3.

imply, the WCD Knowledge Base has emphasised the experience of these banks. Given that the banks have often fallen short of realising such high standards for planning and decision-making, it is legitimate to expect that the other donors and in-country agencies will have encountered similar difficulties and also fallen short of the outcomes implied by the standards set by the banks.

Box 6.11 WCD Case Studies: a compliance report card

In most of the WCD Case Studies there are examples of agreements made not being respected and commitments only partially implemented:

- For Grand Coulee dam $54 million for past losses and $15 million per year in compensation was awarded by the courts to the Colville tribe in 1994, 50 years after the dam was built. The settlement cited high level government correspondence indicating an initial intention to compensate the tribes for loss of salmon in accordance with existing treaties, this was abandoned by the late 1930s.
- Construction of Kariba dam complied with the laws of the day – however it was planned and built prior to most regulations being in place. In addition, the laws under colonial rule in Southern Rhodesia (now Zimbabwe) did not include provision for just legal redress for displaced Africans, a clear contravention of prevailing international standards.
- At Tarbela dam nearly 2 000 families had not been adequately resettled twenty years after displacement in terms of the 1967 criteria for compensating landowners.
- At Tucurui, the initiation of the second phase of the project in 1999 proceeded without an environmental impact assessment (EIA). Eletronorte, the utility that owns the project, maintains that Phase II does not require an EIA as it is the continuation of a project approved prior to the setting of EIA regulations in Brazil. Local communities, concerned about the possibility of a repeat of the social and environmental impacts of Phase I of the project, disagree with this position and have asked for a full EIA. The Case Study also points out that Eletronorte did not respect the Waters Code which stipulated that hydropower plants should not adversely affect the food and needs of river bank communities, public health, shipping, conservation and free circulation of fish, amongst others.
- In Aslantas the government agreed with the World Bank to recover a portion of the costs of the irrigation component of the scheme from farmers over 50 years. Current recovery rates are inadequate to meet this target (see Box 2.4).
- At Pak Mun, an EIA was a World Bank requirement and should have been performed on the revised project prior to construction.
- In India, a national assessment of dam projects cleared in the 1980s and 1990s shows that in 90% of cases the project authorities have not fulfilled the environmental conditions under which environmental clearance was given by the central government under the Environment Protection Act of 1986.
- In Norway, provisions for environmental flow releases from hydropower dams have allegedly dropped below the minimum established in the licensing agreements. Yet the central authorities lack legal means to monitor and sanction confirmed offenders.
- In China a review of Lingjintan dam showed that compliance with environmental clauses in construction contracts was not satisfactory due, amongst other factors, to lack of incentives, lack of accountability and poor oversight.

Source: WCD Case Studies

Findings and Lessons

Conflicts over dams have heightened in the last two decades, as awareness of their impacts and performance has grown and the debate over costs and benefits has spread. While conflict has sparked innovation in some contexts and by some stakeholders in the debate, in others it has deepened and entrenched conflict. The Global Review of large dams and their alternatives has examined the performance of large dams using a number of different lenses – technical, financial, economic, environmental and social – and explored the options that are currently available to fulfil water and energy needs.

As part of its Global Review of past experience, the Commission examined the decision-making, planning and compliance processes around large dams in the WCD Knowledge Base to better understand what factors influence these processes and the performance and results of the projects. Based on this review the findings on decision-making include:

- centralised and bureaucratic State agencies and utilities have often promoted and implemented dams as one of a small number of conventional responses to water and energy needs, a choice that, once taken, often has not been revisited even in the face of an expanding list of alternatives;
- foreign assistance has stimulated large investments in dams in developing countries, by providing financing – more than $4 billion per year during the peak

of lending in 1975-1984 – and leading financing arrangements;

- large developing countries with many large dams (including China, India and Brazil) have established internal capacity to build large dams, although in recent decades they have often used external finance and equipment to build larger projects;
- countries building fewer dams have been disproportionately influenced by foreign assistance for large dams, making them more vulnerable to conflicts between the interests of governments, donors and industry involved in foreign assistance programmes and improved development outcomes for rural people, particularly the poor.
- the multilateral banks and bilateral aid agencies, alongside the dam-building industry and international industry associations, have played a key strategic role in spreading the technology to developing countries, lending legitimacy to emerging dam projects, and fostering the technological and human resources required to build and maintain dams;
- there has been a generalised failure to include and recognise affected people and empower them to participate in decision making.
- the lack of agreements on water use within shared river basins is an increasing concern and cause for conflict, particularly as demands grow and unilateral decisions to build large dams by one country alter supply within a basin with significant consequences for other riparian States.

The end result of the influence exerted by vested interests, and the conflicts of interests that have arisen, has been that many dams were not built based on an objective assessment and evaluation of the technical, financial and economic criteria applicable at the time, much less the social and environmental criteria that apply in today's context. That many of such projects have failed to deliver by standards applicable in either context is therefore not surprising, but nonetheless cause for concern.

Focussing on the planning cycle for large dams reveals a series of limitations, risks and outright failures in the manner in which these facilities have been planned:

- participation and transparency in planning processes for large dams was neither inclusive nor open and while actual change in practice remains slow even in the 1990s there is increasing recognition of the importance of inclusive processes;
- while the number of options have increased over time, options assessment was typically limited in scope due to political and economic interests driving dam projects, lack of familiarity with other options, the perceived need to quickly proceed with large-scale projects to meet large projections in demand and the relative ease of developing new supply relative to undertaking policy or institutional reform;
- project planning and evaluation for large dams was confined primarily to technical parameters and the narrow application of economic cost/benefit analyses with many sectoral studies aimed at finding least-cost supply solutions for providing a single service such as irrigation water or electric power;
- where opportunities for the participation of affected people, and the undertaking of environmental and social impact assessment have been provided they often occur late in the process, are

limited in scope, and even in the 1990s their influence in project selection remains marginal;

- the paucity of monitoring and evaluation activity once a large dam is built has reduced the basis for learning from experience; and
- while countries that were the first to build dams are now evaluating decommissioning, removing, or re-operationalising ageing facilities that are due for re-licensing, many other countries do not yet have established licensing periods that clarify the responsibilities of the owner towards the end of the dam's effective life.

The net effect of these difficulties is that once a proposed dam project has passed preliminary technical and economic feasibility tests and attracted interest from government, external financing agencies or political interests, the momentum behind the project often prevails over further assessments. Moreover where substantial differences arise between proponents and those potentially affected, efforts to modify plans and decisions often must resort to legal or other action outside the normal planning process.

But poor outcomes and mistrust are not simply a matter of narrow and technically focussed planning and decision-making. They also stem from the failure of dam proponents and financing agencies to fulfil commitments made, observe statutory regulations and abide by internal guidelines. Among the findings on compliance are:

- in some cases, the opportunity for corruption provided by dams as large-scale infrastructure projects further distorted planning and decision-making;
- weak regulatory frameworks and lack of sanctions at the national level, particularly for options assessment and social and environmental requirements, and little enforcement of existing regulations have contributed to the poor economic, social and environmental performance of many large dams;
- large projects tend to lack public oversight of negotiations between government, lenders and contractors, including limited disclosure and public access to information;
- in many cases lack of clear monitoring procedures limits public scrutiny and accountability;
- there is a lack of sanctions at the international level for non-compliance with international norms regarding water use in shared river basins;
- within public international financial institutions, there are few, if any, sanctions for staff members, or countries, for non-compliance;
- in some countries, there is a lack of legal opportunities for affected groups to seek recourse, therefore lessening the accountability of the project developers; and
- most of the bilateral Export Credit Agencies are only beginning to develop social and environmental policies and guidelines and the lack of consistency among the agencies' guidelines has resulted in projects rejected by some on environmental and social grounds receiving funding from other sources with lower standards.

To sum up, whereas substantial improvements in policies, legal requirements and assessment guidelines have occurred, particularly in the 1990s, it appears that business is still often conducted as usual when it comes to planning and decision-making. Further, past conflicts remain largely unresolved and past impacts largely unmitigated. The WCD Global Review found that the influence of vested interests, legal and regulatory gaps, disincentives for compliance and lack of monitoring, participation and transparency amongst other things, have combined to create significant barriers to reforms that could otherwise make the planning and decision-making processes more open, responsive and accountable. Recent examples cited in this and earlier chapters are the basis of the Commission's optimism that these barriers are surmountable and these difficulties are not inevitable. The WCD Global Review indicates that there are opportunities for reducing negative impacts and conflicts, and indeed a responsibility, to:

- increase the efficiency and performance of existing assets and systems;
- better assess development needs and the full range of development options;
- avoid and minimise ecosystem impacts;
- ensure that displaced and project-affected peoples' livelihoods are improved;
- shift away from a balance sheet approach to decision-making in favour of broader, inclusive and more timely, multi-criteria approaches to planning and decision-making;
- resolve past inequities and injustices, and transform project-affected people into beneficiaries, enabling them to contribute to project benefits;
- conduct regular monitoring and periodic reviews; and
- develop, implement and enforce incentives, sanctions, and recourse mechanisms, especially in the area of environmental and social performance.

The remainder of the report builds on the findings and lessons of the WCD Global Review. It delivers a way forward that can improve planning, decision-making and compliance, capitalising on the options available – whether of a technological, policy or institutional nature – and providing economically efficient, socially equitable and environmentally sustainable solutions to meet future water and energy needs.

Endnotes

1 Many government agencies involved in water resources development in countries that build large numbers of dams maintain a construction workforce to build infrastructure: in the United States the Bureau of Reclamation and the Army Corps of Engineers and in China the Water Resources Ministry.

2 Eckstein, 1958.

3 Gillis et al, 1987, p366.

4 WCD Thematic Review V.5 Negotiation, section 3.3.

5 Sklar and McCully, 1994, eco029, WCD Submission, p12-14; Gillis et al, 1987.

6 WCD Thematic Review III.2 Financing Trends, ch. 3.

7 The total investment in dams by the multilaterals and bilaterals portrayed in Figure 6.1 is approximately $125 billion.

8 Guhan, 1995, cited in India Country Study, Section 6.1.8.

9 ICE, 1994, p15-16; ICE, 1996, table 1.

10 Usher, 1997b, p120-123.

11 Morse and Berger, 1992; Umaña, 1998, p7; Wappenhans Task Force, 1992.

12 Usher, 1997a, eco026, WCD Submission, p120-123.

13 Egypt and Ethiopia are now working towards greater collaboration through the Nile Basin Initiative.

14 WCD Thematic Review on Negotiation, section 3.4.

15 Hanusin, 1999, opt052, WCD Submission, 4-5.

16 WCD Thematic Review III.1 Economic Analysis

17 WCD Thematic Review V.2 Environmental and Social Assessment, Section 1.

18 For example see the WCD India Country Study, section 7.4.

19 WCD Thematic Review V.4 Regulation, section 3.1.

20 Lovei and McKechnie, 2000, p34-37.

21 For example the 1996 Development Assistance Committee's Rcommendation on Anti-corruption Proposals for Aid-funded Procurement.

22 OECD, 2000b; OECD, 2000c, website http://www.oecd.org/daf/nocorruption/index.htm, viewed 4 September 2000.

23 For example the United Kingdom has the Department for International Development and the Export Credits Guarantee Department.

24 For example see WCD India Country Study.

Part Two:

The Way Forward

The mandate required the World Commission on Dams to propose a framework for options assessment and decision-making processes for water and energy resources development, along with a set of criteria and guidelines for the planning, design, construction, operation and decommissioning of large dams. Part Two of the report presents a new approach to decision-making, based on the findings in the Global Review (Part One).

- Chapter 7 presents a normative framework for equitable and sustainable development and develops an approach to negotiating outcomes for water and energy development projects based on recognising rights and assessing risks.
- Chapter 8 sets out seven broad strategic priorities that should guide decision-making. Each one includes a set of principles that, if applied, will lead to more equitable and sustainable outcomes in future.
- Chapter 9 develops supporting criteria and guidelines that will help decision-makers and all interested parties implement the strategic priorities set out in Chapter 8.
- Part Two closes with Chapter 10 which stresses the need for concerted and simultaneous action and proposes entry points for the different constituencies involved in the dams debate to follow up in response to the recommendations of the Commission.

Chapter 7:

Enhancing Human Development:

Rights, Risks and Negotiated Outcomes

To improve development outcomes in the future we need to look at proposed water and energy development projects in a much wider setting – a setting that reflects full knowledge and understanding of the benefits and impacts of large dam projects and alternative options for all parties. It means that we have to bring new voices, perspectives and criteria into decision-making, and we need to develop a new approach that will build consensus around the decisions reached. This will result in fundamental changes in the way decisions are made.

The debate about dams is a debate about the very meaning, purpose and pathway of development as well as the role that the state plays.

This chapter proposes a new basis for assessing options and reaching decisions on water and energy resources development. It links our review of past experience contained in the Global Review chapters with the WCD's framework for future practice elaborated in chapters 8 and 9. In developing this framework the Commission found that applying the lessons learnt does not merely imply a change in process and procedure. The fault lines of the dams debate run far deeper and touch upon many of the fundamental norms and values that affect our lives as citizens and communities.

In moving forward the Commission recognises that the dams debate is rooted in the wider, ongoing debate on development. The emerging global vision of equitable and sustainable development provides the foundation for the Commission's findings and recommendations. This foundation relates to:

- the framework of internationally accepted norms on human rights, the right to development, and sustainability
- global trends and the emerging development paradigm; and
- a rights based approach where recognition of rights and assessment of risks provides the basis for negotiated decisions on dams and their alternatives.

From Global Review to Future Practice

Along with all development choices, decisions on dams must respond to a wide range of needs, expectations, objectives and constraints. As matters of public choice and policy they will always reflect competing interests and require negotiation. Reconciling competing needs and entitlements is the single most important factor in addressing the conflicts associated with development projects and programmes – particularly large-scale interventions such as dams.

Access to water provides a graphic illustration of such competing needs and development objectives and the reason why equity and justice considerations emerge as key issues. Riparian communities with long-standing use rights and economies that depend on local resources have an immediate interest in maintaining current use patterns and assuring fulfilment of their future needs. However, in the context of national policies, meeting development needs may require sharing water resources. To balance these needs societies will have to negotiate a framework for equitably sharing the resource. History shows that this can be done successfully provided a transparent and legitimate process is followed.

Dams have often been seen as an effective way of meeting water and energy needs. However, the Global Review has emphasised the wide range of problems associated with them. The Commission acknowledges that today's perspective on development reflects the benefit of knowledge that may not have been available to past decision-makers. Nonetheless, it is clear that the positive contribution of large dams to development has, in many cases, been marred by significant social and environmental impacts which are unacceptable when viewed from today's values.

The debate about dams is a debate about the very meaning, purpose and pathway of development as well as the role that the State plays in both protecting the rights of its citizens and responding to their needs through development policies and projects. The WCD Global Review showed clearly that large-scale infrastructure projects such as dams can have devastating impacts on the lives and livelihoods of affected communities and ecosystems, particularly in the absence of adequate assessments and provisions being agreed to address these impacts. During its regional consultations and through the WCD Case Studies the Commission was confronted with accounts by communities and individuals on the nature and extent of these impacts. These accounts give rise to fundamental concerns about the way governments and their agencies have exercised their role and responsibilities in the development process.[1]

Improving the development process and its outcomes must start with a clear understanding of the shared values, objectives and goals of development and their implications for institutional change. The Commission grouped the core values informing its understanding on these issues under five main headings:

- Equity
- Efficiency
- Participatory decision-making
- Sustainability
- Accountability

These five values run through the entire report and are the focus of concerns raised by the evidence presented in the Global Review. Applying these values to the evidence it has collected, the Commission believes that negotiated outcomes using a rights-and-risks approach will deliver the most favourable development results. Reference to these values enables all stakeholders to test decisions relating to water and energy development. If the report advances these values significantly, we will emerge at our destination – improved decision-making processes that deliver improved outcomes for all stakeholders.

In the following sections the Commission presents a new policy framework for decision-making on water and energy development options that can be applied in national and local contexts. To improve development outcomes, ensure public acceptance and reduce future controversy, this new basis for judgement needs to win the support of the full range of key stakeholders. It suggests that decision-making on water and energy management will align itself with the emerging global trends on equitable and sustainable development.

Sustainable Human Development – A Global Framework

What are these trends, and how firm is their direction and force? What do they imply for decision-making? What do they say about the rights that societies, communities and individuals within societies, are entitled to and the responsibilities that accompany these rights?

There is a globally accepted framework for setting universal goals, norms and standards. The foundations of the framework are the United Nations Charter (1945) and the

Box 7.1 Shared values and institutional practices – the UN Millennium Report

"The economic sphere cannot be separated from the more complex fabric of social and political life and sent shooting off on its own trajectory. To survive and thrive, a global economy must have a more solid foundation in shared values and institutional practices– it must advance broader and more inclusive, social purposes."

Source: Annan, 2000

Universal Declaration of Human Rights (1947) (see Annex VI for full texts).[2]

In the last two decades of the 20th century the United Nations General Assembly reinforced this framework with the UN Declaration on the Right to Development (1986) and the Rio Declaration on Environment and Development (1992) (see Annex VI for full texts). Taken with the earlier covenants and conventions on human rights, they cover a broad spectrum ranging from human rights, through social development and environment, to economic co-operation.

A rights based approach provides a principled basis for mediating development choices among competing interests.

Human rights

Reference to the framework of human rights adopted by the international community in 1948 advances the process of planning and decision-making in important ways. It articulates such rights as self determination and the right to consultation in matters that affect people's lives, the right to democratic representation of people's views on such matters, the right to remedy and the right to an adequate standard of living, freedom from arbitrary deprivation of property, freedom from violence, freedom of thought, conscience and religion and freedom of opinion and expression. More generally it includes the right to a social and international order in which these rights can be fully realised.

All people are accorded human rights without discrimination by virtue of their humanity. Reference to the human rights framework means those policies that deny the rights of some to fulfil those of others cannot be adopted. Thus any policy or law adopted must contain the intention to respect the rights and entitlements of all.

The application of a rights based approach recognises the indivisibility of civil, political, economic, cultural and social rights. It broadens the range of basic human rights beyond the socio-economic sphere of needs to include rights to life, health care, education, shelter, food, water, remedy, security, subsistence and livelihood.

Unlike needs, which are expressed as aspirations for benefits, rights and entitlements are expressed in law, allowing for their attainment or redress through the justice system. A country may use its legislative process to ensure that appropriate rights-based policies are given legal expression and to establish institutional mechanisms to uphold rights. The legal system also provides a means for resolving potential conflicts in cases where rights give rise to competing entitlements. A rights based approach thus provides a principled basis for mediating development choices among competing interests.

The right to development

In 1986 the UN General Assembly adopted the Declaration on the Right to Development (DRD).[3] It marked a significant step by the international community in developing a normative framework that specifies responsibilities in applying a human rights approach to development. It moved beyond the sphere of individual human rights to address relationships between different interest groups in society and their interaction with the state.

The Declaration on the Right to Development sets out a number of relevant concepts:

- Development is a comprehensive process aiming at the constant improvement of the well-being of the entire population;

it affects economic, civic, social, cultural and political rights.

- The promotion of, respect for and enjoyment of certain human rights and fundamental freedoms cannot justify the denial of other human rights and fundamental freedoms.
- The creation of conditions favourable to the development of peoples and individuals is the primary responsibility of their States.
- National development policies aiming at the constant improvement of the well-being of the entire population and of all individuals should be formulated on the basis of their active, free and meaningful participation and fair distribution of benefits resulting therefrom.
- The right of peoples to exercise full and complete sovereignty over all their natural wealth and resources.
- The right to self determination.
- Equal opportunity for access to food and housing.

The DRD sought to clarify the role of the State in exercising its rights, responsibilities, duties and obligations in planning and implementing national development policies and programmes. It reflects the recognition that every society acts as an organised polity in which the State is accorded powers and responsibilities. At the same time States are subject to conditions that can be summarised under the heading of good governance criteria, such as those in the Declaration on Human Rights including the rule of law, accountable bureaucracies and freedom of information. The legitimacy of the State in exercising its role is premised on the assumption that it acts in accordance with these criteria. Without good governance, the legitimacy of the State and ultimately its ability to take decisions are compromised. State authority may also be limited through adherence to the framework of international conventions that, in certain circumstances, supersedes strict sovereignty.

Sustainable Development – the Rio Principles

The Declaration of the United Nations Conference on the Human Environment (Stockholm, 1972) for the first time accepted that the environment was fundamental to human well-being, and that its management and care in the interest of advancing wider human goals was a central task of States and the international community. Articulation and codification of environmental rights took somewhat longer. The United Nations Conference on Environment and Development adopted the Rio Declaration on Environment and Development in June 1992. The Declaration contains 27 principles, usually known as the Rio Principles[4]. Several of these are of immediate relevance to water and energy resources management.

- Principle 1 states that 'Human beings are at the centre of concerns for sustainable development. They are entitled to a healthy and productive life in harmony with nature'.
- Principle 3 recognises the right to development, but insists that it be met in an equitable way that considers future generations as well as present participants in development.
- Principle 4 insists that sustainable development requires environment to be integrated with the development process and form a central feature of the aims of that process. Environment, on its own, is an insufficient goal.

- Principle 10 underlines that all concerned citizens must be involved in handling environmental issues, and must participate in the decision-making process. This participation must be accompanied by effective access to relevant information and by opportunities to seek redress and remedy in case agreements are not respected.
- Principle 13 states that States shall ensure compensation for victims of environmental damage and give priority to the further development of law regarding liability in such cases.
- Principle 15 states that the precautionary approach shall be widely applied by States according to their capabilities. Where there are threats of serious or irreversible damage, lack of full scientific certainty shall not be used as a reason for postponing cost-effective measures to prevent environmental degradation.
- Finally, Principle 22 recognises the vital role of indigenous people and other local communities in environmental management and development, and entrusts states with ensuring their effective participation in the achievement of sustainable development.

The Rio principles in conjunction with Agenda 21 thus highlighted not only the linkages between environment and development but also the importance of local communities having a significant role in shaping national development strategies.

The Global Review provided extensive evidence to illustrate that governments, in constructing dams, have often found themselves in conflict with basic principles of good governance that have been articulated in the three international instruments referred to above. This situation still prevails today. The level of conflict surrounding large dams, yesterday and today, is sufficient to illustrate that dams frequently trigger disagreements about the respective rights of governments and their citizens.

The UN Declaration of Human Rights, the Right to Development and the Rio Principles together make up an internationally accepted framework of norms empowering a concept of development that is economically viable, socially equitable, and environmentally sustainable. It is a powerful framework with a central bearing on the dams debate. Figure 7.1 illustrates how the Commission draws on these internationally accepted norms in the remainder of this

Figure 7.1 The WCD policy framework

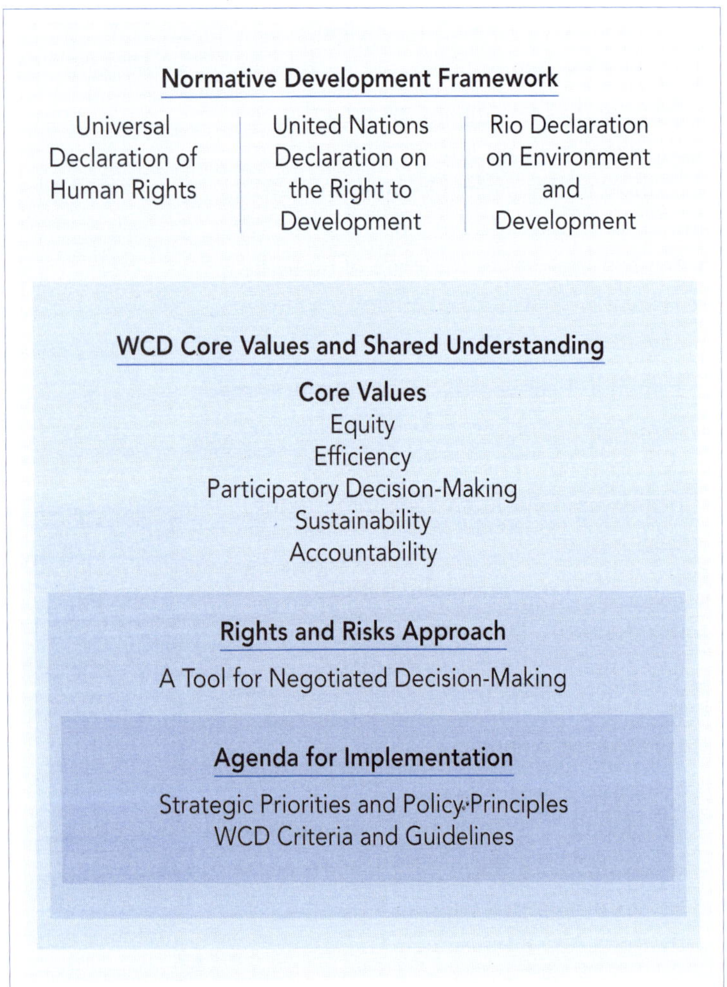

report to develop a new policy framework and corresponding guidance for water and energy resources development.

Trends and Challenges in Applying the New Development Framework

When invoking this emerging universal normative framework, one must not overstate its completeness, its complete acceptance, or the ease of applying its provisions in practice. Nevertheless, recent trends in global public policy suggest that increasing attention is being paid to the gap between aspiration and realisation. It is significant that the focus of the United Nations Development Programme's (UNDP) Human Development Report 2000 is on human rights and human development (see Box 7.2). The framework also strengthens the notion, now gaining currency in a variety of arenas, that there is a body of common concerns based on a range of international conventions and accords that transcend national sovereignty.

The adoption of a rights based approach does not on its own resolve the practical challenge of meeting human needs. During its regional consultations the Commission listened to a wide range of views and reasoning on this matter. Meeting rapidly growing needs for water and energy – particularly in the developing economies of the South – imposes difficult choices on governments. Failure to respond to these needs carries significant economic and political risks. Food security, blackouts, empty water taps and floods are among the most immediate and sensitive public service issues for which society holds government accountable. In the past, large-scale dam projects seemed to offer both apparently straightforward and highly visible options for responding to these pressures.

At the same time, the decision criteria used by governments do not always match those of organised groups of citizens. Governments are too often inclined to invoke urgent development needs as a reason for restricting rights, while civil society groups believe that full respect for rights and the search for alternatives represents the surest way of promoting equity and justice in development.

For many parts of the developing world, access to capital, technology and development opportunities determines the extent to which local and national economies are able to develop. Similarly the political economy of power, vested interests and access to resources that characterise each society have a large influence on its commitment to equitable and sustainable development.

This is not to suggest that the problems reside in the poorer countries. Pressure for water and energy development – in both

Box 7.2 Human rights and human development

The UNDP Human Development Report 2000 focuses on human rights as the fundamental framework within which human development must be pursued. It contends that societies are on the threshold of a significant advance in the recognition of, and respect for, human rights. But this will require six fundamental shifts from the thinking that dominated the 20th century:

- From state-centred approaches to pluralist, multi-actor approaches – with accountability not only for the State but also for media, corporations, schools, families, communities and individuals.
- From national to international and global accountabilities – and from the international obligations of States to the responsibilities of global actors.
- From the focus on civil and political rights to a broader concern with all rights – giving as much attention to economic, social and cultural rights.
- From a punitive to a positive ethos in international pressure and assistance – from reliance on naming and shaming to positive support.
- From a focus on multiparty elections to the participation of all through inclusive models of democracy.
- From poverty eradication as a development goal to poverty eradication as social justice, fulfilling the rights and accountabilities of all actors.

Source: UNDP, 2000

North and South – is not only caused by the imperative of meeting basic human needs, but is also driven by wasteful consumption in the richer countries or among the well-off in the poorer countries[5].

Whatever judgement is made, it is a reality that governments face very real dilemmas in trying simultaneously to satisfy urgent needs and advance the realisation of fundamental rights, even if the goal of fulfilling all people's needs and entitlements is not questioned. Notwithstanding this, the Commission believes that fulfilling development needs requires respect for fundamental rights, and not a trade off between them. We believe that an equitable and sustainable approach to development requires that a decision to build a dam or any other options must not, at the outset, sacrifice the rights of any citizen or group of affected people.

Governments face very real dilemmas in trying simultaneously to satisfy urgent needs and advance the realisation of fundamental rights, even if the goal of fulfilling all people's needs and entitlements is not questioned.

In developing its framework and recommendations the Commission has sought to draw on the broader trends and developments that reflect the changing context and international development discourse. Not all countries will recognise themselves in these statements and the trends are far more advanced in some areas than in others. Nevertheless, the Commission believes that the trends described below are not limited to any one region or group of countries, but have broad relevance. From the perspective of this shared experience we draw attention to the following elements of the evolving development paradigm:

- The world appears set to move beyond the growth paradigm, which judged progress largely in narrow economic terms, putting a strong premium on activities that offered a clear economic return. This does not mean that economic viability is no longer seen as important. If anything it has greater weight along with a greater sanction on poor economic performance at company or country level. But we are giving greater value to non-monetary or non-monetised aspects of development such as the need to conserve biological diversity, protect cultural values, or consider the needs of future generations.
- We are moving from assessing public interest in general terms to a focus on improving equity in the spread of costs and benefits from development. The growing disparity between rich and poor within and across many nations has fuelled doubts about traditional development paths.[6] The emerging consensus on the need for greater transparency and participation in development decision-making is likely to speed up this transition considerably. The focus on equity extends to recognising intergenerational equity as an important factor in dealing with resource access and use.
- An increasingly robust foundation of international covenants, charters, declarations and conventions supports the sharpening focus on equity and confirms the growing importance of equity considerations in development. In particular, a body of international instruments relating to human rights is emerging, together with institutions to oversee their further development and application. This will strengthen arguments in favour of greater transparency, participation in decision-making and accountability for compliance. Pressure from the human rights community has had an impact on governments and more recently on corporations.

- The definition of public interest is shifting from one that placed a premium on overriding interests of economic growth to one that places more weight on the rights and interests of people and communities affected by a development. The level of sacrifice that affected people are expected to endure for an often ill-defined notion of the greater public good has been increasingly challenged. Similarly, the recognition that affected communities, through their sacrifices, are in fact contributors to development projects implies a shift in focus from compensatory approaches to establishing equitable benefit sharing mechanisms.
- We have also witnessed a shift from technology-driven development choices towards a more integrated approach to managing scarce resources with technology being but one factor among others in managing demand and supply of services more effectively. Furthermore, the recognition that traditional practises and technologies can achieve great levels of efficiency in meeting local needs coupled with the advent of new technology options has increasingly challenged the notion that large and centralised systems are always the most effective and efficient way of meeting demands for water and energy.
- The emerging paradigm provides a new basis for governance and democratic decision-making. This stems from a substantial redistribution of roles and responsibilities in the public and private sectors and civil society. Many parts of the world have seen a considerable migration of national government authority, in three directions: upward to regional superstructures or international bodies, downward to provincial and local government; and outward to the private sector and civil society. This is not to say that the role of governments has become less important; but it has changed and continues to change, with implications for the way decisions are taken and implemented.

- The private sector has, by contrast, considerably expanded its role, undertaking functions that were until recently the exclusive remit of government. Apprehensions about this trend are sharp in some parts of the world and reflect growing concern about the diminishing power of citizens to control corporate activity through local and national institutions. Pressure is therefore growing on corporations to become more accountable to widely supported standards of social and environmental behaviour. Such pressure is likely to mean that corporations will face steadily rising costs and risks if they fail to comply with existing rules, regulations and standards.
- The role of civil society organisations has also expanded and their legitimacy in representing and defending interests, in participating as full actors in decision-making on development and in monitoring compliance is increasingly accepted. Civil society organisations are playing an increasingly important role in influencing public opinion and mobilising it against infringement or non-compliance with new and emerging standards of behaviour, especially on the social and environmental front.

This changing context implies a broadening range of concerns that have a legitimate right to be considered and, therefore, of

actors involved in reaching key development decisions. It will be increasingly difficult to take decisions on the narrow basis of the needs of infrastructure development. Instead, such projects need to be considered as part of the broader process of economic, social and environmental transformation.

Rights and Risks – an Improved Tool for Decision-Making

Both the findings of the WCD Global Review and the implications of the normative framework summarised in this chapter demonstrate that the traditional 'balance sheet' approach of assessing costs and benefits of a project is an inadequate tool for effective development planning and decision-making. The case of dams clearly illustrates that development choices made on the basis of such trade-offs neither capture the complexity of considerations involved, nor can they adequately reflect the values societies attach to different options in the broader context of sustainable development.

Given the significance of rights-related issues as well as the nature and magnitude of potential risks for all parties concerned, the Commission proposes that an approach based on 'recognition of rights' and 'assessment of risks' (particularly rights at risk) be developed as a tool for guiding future planning and decision making (see Figure 7.2). This will also provide a more effective framework for integrating the economic, social and environmental dimensions for options assessment and the planning and project cycles.

Rights ...

The Global Review highlighted the need for a more practical and specific approach to addressing the five values of equity, efficiency, participatory decision-making, sustainability and accountability. These values form the foundation of a rights-based approach to equitable decision-making about water and energy resources management.

Various types of rights may be relevant in the context of large dam projects. These include constitutional rights, customary rights, rights codified through legislation, property rights or the rights of developers and investors. They can be classified on the basis of their legal status, their spatial and temporal reach, or their purpose. In the spatial and temporal dimensions, one can distinguish the rights of local, basin, regional and national entities, the rights of riparian countries, or the rights of present and future generations. Regarding the purpose or subject of rights, one can distinguish rights to material resources such as land and water, and rights to spiritual, moral or cultural goods such as religion and dignity.

This approach highlights the range and complexity of relevant rights and responsibilities and the reality that rights intersect and overlap. Mechanisms for conflict resolution, adjudication and independent arbitration must begin with the assessment of these rights, entitlements and claims. This approach assumes that at the assessment stage, all claims are subject to a fair, open and transparent review. It is based on an understanding that no party's rights will extinguish another's. In fact, where rights compete or conflict, negotiations conducted in good faith, offer the only process through which various interests can be legitimately reconciled. This suggests an approach to water and energy policy that provides for negotiated processes within a legal and procedural framework, including arbitration, recourse and appeal mechanisms to ensure equitable adjudication in cases where negotiated settlements are not achievable or are contested.

Clarifying the rights context for a proposed project is an essential step in identifying those legitimate claims and entitlements that might be affected by the proposed project – or indeed its alternatives. It is also the basis for effective identification of stakeholder groups that are entitled to a formal role in the consultative process, and eventually in negotiating project-specific agreements relating, for example, to benefit sharing, resettlement or compensation.

...and Risks

The notion of risk adds an important dimension to understanding how, and to what extent, a project may impact on such rights. Traditional practice is to restrict the definition of risk to the risk of the developer or corporate investor in terms of capital invested and expected returns. These voluntary risk takers have the capacity to define the level and type of risk they wish to take and explicitly to define its boundaries and acceptability. By contrast, as the Global Review has shown, a far larger group often have risks imposed on them involuntarily and managed by others. Typically, they have no say in overall water and energy policy, the choice of specific projects or in their design and implementation. The risks they face directly affect individual well-being, livelihoods, quality of life, even their spiritual world view and very survival.

This has often led to conflict because it ignores the principle that those with a legitimate stake in a decision are best placed to assess the risks they are prepared to take to achieve a benefit. Such conflicts are exacerbated by the absence of an agreed

Box 7.3 Voluntary risk takers and involuntary risk bearers

Public and private developers of large dam projects have long understood that the sector involves managing risks of a technical, financial and even political nature. Decision-makers have not always acknowledged the differences between 'taking risk' and 'imposing risk' and between voluntary risk takers and involuntary risk bearers. The private sector regards dams as high-risk projects. As 'voluntary risk takers', private companies manage their increased exposure to risk by requiring higher financial rates of return. Their risk management procedures are relatively highly developed, using contractual agreements and sophisticated third party recourse and arbitration mechanisms.

Governments and regulators plan and manage the provision of services to the nation, and therefore also take risks. They must weigh the risks inherent in undertaking dam projects against the risks of not undertaking them. There are risks attached to other options, and to the 'do nothing' option, given growing demand for power or food, and societies are constantly balancing these different risks and opportunities.

There are those, however, on whom risk is imposed. The 'involuntary risk bearers' who are forced to bear risks include people to be displaced by the project. These people may face years of uncertainty and direct risks to livelihood even before the project is approved and before resettlement or land purchase. They may be unable to obtain finance for investments in farm infrastructure or equipment, and local government may not maintain or develop services for communities on the verge of displacement. The risks to displaced communities are compounded in cases where they have no say in the decisions but are obliged to bear the consequences. In these circumstances they often depend entirely on the capacity of government or the developer to manage the resettlement or compensation process on their behalf.

Indigenous peoples face specific cultural, social and livelihood risks. Evidence collected by the Commission illustrates that they often bear disproportionate risks associated with projects, as they were not included in decision-making processes concerning resettlement, let alone the earlier processes of assessing needs and selecting options. Downstream communities that depend on existing river flows to maintain their resource base are also often not given any say in deciding the nature of projects. Yet they face the risk of losing access to resources, or having their capacity to maintain a sustainable livelihood undermined. Often these communities do not have access to information on the nature of the risks that they face until the project is approved or completed.[7]

The case of future generations and the ecosystem is somewhat different. These 'risk bearers' cannot speak for themselves, even if the risks they face are acknowledged. Future risks can be linked to present risks. The loss of natural resources can undermine livelihood opportunities for both current and future generations. Similarly, the loss of biodiversity in the present means that it is either not available or diminished for future generations. The lower priority generally accorded to these types of risks is compounded by the absence of tangible safeguards, or the failure to implement and enforce those that do exist. In such cases, as with other involuntary risk bearers, adopting a precautionary approach is particularly relevant in order to avoid impacts. It is also essential to identify appropriate inputs by interested parties to the options assessment process and to the planning and project cycles.

The precautionary approach articulated in the Rio principles forms part of a structured approach to the analysis of risk, and is also relevant to risk management. Decision-makers faced with scientific uncertainty and public concerns have a duty to provide answers where risks and irreversibility are considered unacceptable by society.

approach and process for assessing and negotiating acceptable project outcomes – outcomes that include rejecting a dam in favour of a more acceptable alternative. Where unequal power relationships prevail and no process for good-faith adjudication among competing interests exists, the result is often protracted conflict, escalation and, eventually, 'win-lose' outcomes in which less privileged groups are further disadvantaged. This all-too-common reality points to a serious failure of the options assessment and planning process to protect and further basic human and development rights.

Dealing with risks cannot be reduced to consulting actuarial tables or applying a mathematical formula. In the end, as in the case of rights and entitlements, they must be identified, articulated and addressed explicitly. This will require the acknowledgement of risk to be extended to a wider group than governments or developers in order to include both those affected by a project and the environment as a public good (see Box 7.3). Most important, involuntary risk bearers must have the legal right to engage with risk takers in a transparent process to ensure that risks and benefits are negotiated on a more equitable basis. To that end, a framework is required that permits a transparent, balanced and participatory process of decision-making relating to key stages in the planning process. In chapters 8 and 9 we develop the practical application of this approach in greater detail to illustrate how this can be achieved at the different stages of assessment, planning and implementation.

It must, however, be stressed that not all countries possess the full range of legal and institutional structures, nor sometimes the human and financial resources, to implement such a participatory approach to decision-making effectively. It becomes, therefore, a high priority to assist those countries and communities to put the necessary structures in place and to develop the necessary capacity.

Fig 7.2 From rights and risks to negotiated agreements: a framework for options assessment and project planning

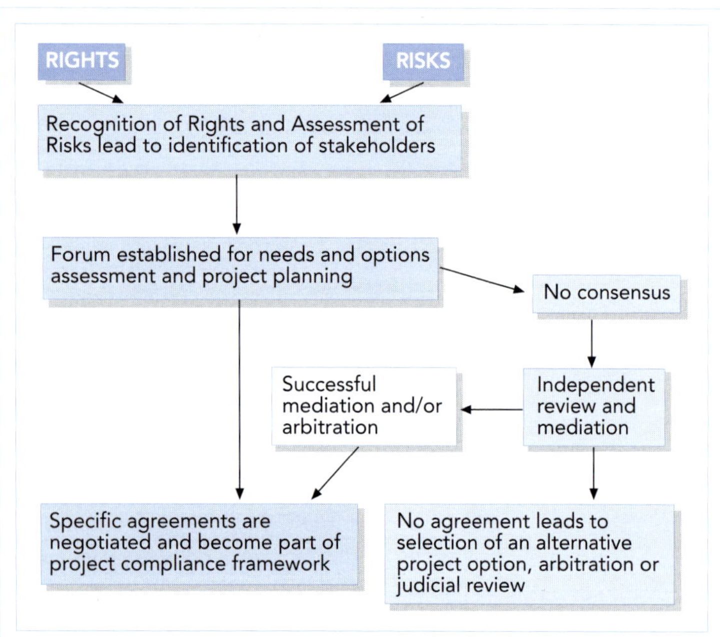

Negotiating Agreements on the Basis of Rights and Risks

In its Global Review the Commission encountered considerable experience and good practice in implementing a rights-based approach, with examples from many countries. It is also clear, however, that the approach requires a legal and procedural framework that provides for a free and informed negotiation process. This framework must provide for arbitration, recourse and appeal mechanisms to ensure equitable adjudication in cases where negotiated settlements are not achievable.

While it is easy to point to negotiated solutions as the answer to sound decision-making or to avoiding dam related conflict, a range of difficult challenges remains to be faced. Structuring a negotiation process that will lead to an optimal and widely supported decision on an option for water and energy development involves answering three questions:

- Who should participate in the decision-making process?
- What decision-making processes should be followed?
- What criteria can be applied to assess the process and its outcomes?

Who should participate?

The rights-and-risks approach is both an entry point to the options assessment process and a basis for the subsequent development of specific project options. The recognition of rights and the assessment of risk identify the interested and affected parties who possess rights or entitlements as well as risk takers and bearers. This opens the way for a negotiated approach that enables the decision-making process to assess options and reach project agreements. Those whose rights are most affected, or whose entitlements are most threatened, have the greatest stake in the decisions that are taken. The same applies to risk: those groups facing the greatest risk from the development have the greatest stake in the decisions and, therefore, must have a corresponding place at the negotiating table.

What type of process?

This chapter has suggested that global trends are increasingly leading us to decision-making based on transparent processes, full access for affected parties to relevant information, identification and empowerment of key stakeholders, and their adequate participation in the decision-making process. At the same time, the Commission recognises there is no universal formula. The most appropriate decision-making process will depend to an extent on the type of development under review, the political and cultural setting of the development, and other constraints relating to the urgency of the need and the likelihood of negative impacts.

On the other hand a process that is too complex can needlessly delay decisions and deprive potential beneficiaries of the fruits of any of the development alternatives under consideration. The goal must be a process that gives all key stakeholders a voice and a full opportunity to participate in decision-making, seeks the broadest reasonable consensus, and is transparent in the criteria used for reaching a decision. Such a process is likely to ensure the demonstrable public acceptance that projects require if they are to achieve development. However, no process will work unless all of the parties enter the negotiation in good faith. Without this there is the danger that any attempt to make the process more inclusive will end up being a recipe for stalemate, putting the achievement of needed benefits at risk.

Negotiated outcomes do not replace government decision-making: on the contrary, they rest on the State actively fulfilling its role as planner and enabler of development choices, as well as its responsiblity to provide services and safeguard entitlements. When

Box 7.4 Good governance and the UN Millennium Report

"Good governance comprises the rule of law, effective state institutions, transparency and accountability in the management of public affairs, respect for human rights, and the participation of all citizens in the decisions that affect their lives. While there may be debates about the most appropriate forms they should take, there can be no disputing the importance of these principles."

Source: Annan, 2000

a negotiation process results in a full agreement among the parties, the government (as one of the parties) need only endorse it. There will always be cases, however, where a concerted effort of all parties acting in good faith has brought agreement closer but left it unattainable. Where independent review and mediation fail to foster an agreement, alternative options should be considered or the project should go to arbitration. All decisions must be taken within the framework set by constitutions and national legislation as well as international conventions, and remain subject to a citizen's right to challenge them in the courts (see Figure 7.2).

Assessing the Process and its Outcomes

The purpose of engaging in a participatory process for decision-making is to deliver better decisions than would otherwise emerge. The proof of a decision-making process resides both in the process used, and in the outcomes delivered. A process may be deemed successful if it has been, and has been seen as:

- *Fair:* all key stakeholders perceive the process and outcomes to be fair and legitimate.
- *Wise:* the process is fully-informed, making best use of available knowledge, and continuing to make best use of knowledge over time.
- *Efficient:* the process and the solution are both cost- and time-effective, making best use of available resources.
- *Stable:* the agreement is likely to endure and can be adapted – a sign that it maintains its legitimacy.[8]

In Chapter 9, the Commission develops criteria and guidelines to illustrate how this can be achieved at the different stages of assessment, planning and implementation.

Conclusion

Large dams have increasingly been characterised by bitter conflict and deep feelings of resentment and injustice. Beginning to correct this situation will require not only new processes for taking decisions, but building confidence in these processes and their ability to deliver genuinely better outcomes for water and energy resource development. In seeking to build this confidence we do not, in many cases, begin with a clean slate, but with a difficult legacy that needs to be recognised. This legacy can only be overcome if there is a rapid investment of confidence in the legitimacy of the processes that are put in place.

This Chapter has defined the interaction between the Commission's findings and the global development debate. It has pointed out that the debate on large dams is not taking place in isolation from the broader debate on the purposes and pathways of development. It can clearly be situated within a framework relating to human rights, the right to development, and the imperative of sustainability.

Further, the Commission has sought to demonstrate that an approach based on the recognition of rights and assessment of risks can lay the basis for greatly improved and significantly more legitimate decision-making on water and energy development. This is an effective way to determine who has a legitimate place at the negotiation table and what issues need to be included on the agenda.

Finally, it has concluded that only decision-making processes based on the pursuit of negotiated outcomes, conducted in an open and transparent manner and inclusive of all legitimate actors involved in the issue, are likely to resolve the many and complex issues surrounding dams. While presenting greater demands at early stages of options assessment and project design, such processes lead to greater clarity, certainty and legitimacy for subsequent steps in decision-making and implementation.

Endnotes

1 The WCD received submissions from a wide range of interested parties, which have been listed on the WCD web site. A number of these were presented at the four regional consultations, which the Commission held in Colombo, Sao Paulo, Cairo and Hanoi.

2 UN, 1947.

3 UN, 1986.

4 UNCED, 1992.

5 Recent reports summarising the implications of high levels of consumption on the natural resource base include the Report of The World Commission on Water (World Commission on Water in the 21st Century, 2000), the World Energy Assessment (UNDP et al, 2000) and The World's Water (Gleick, 2000).

6 An overview of this discourse can be found in Sen, 1999. Current data and evolving policy frameworks for addressing poverty and equity issues have also been reviewed in the latest World Development Report (World Bank, 2000) which focuses on poverty and development.

7 Cernea, 2000, has developed a detailed assessment of these risks. The 'Impoverishment Risk Analysis' approach is referred to in greater detail in Chapter 9.

8 Susskind and Cruikshank , 1989.

do not waste
every drop
turn off taps

Chapter 8

Strategic Priorities – A New Policy Framework for the Development of Water and Energy Resources

Based on the findings of the WCD's Global Review this chapter develops the Commission's rationale and recommendations in the form of seven strategic priorities and related policy principles for future decision-making. It builds on previous chapters, starting with Chapter 1 which locates the dams debate in a broader context. This context includes the history of water resources management and large dams, the big increase in dam construction during the latter half of the 20th century, the subsequent emergence of conflict and the issues and interests that gave rise to the establishment of the Commission.

Chapter 8 draws extensively on the Knowledge Base summarised in Chapters 2 to 6 which review the performance and impacts of dams, the decision-making process and the available options for providing water and energy services. Much of this work involved reviewing existing information, but the Commission also collected important new information on all aspects of dams operation and management, especially their social and environmental impacts. The wide range of consultations with affected parties is an important contribution to knowledge about dams in development and development practice in general.

In Chapter 7 the Commission moved from the review of past experience to look at directions for the future. The chapter addresses the dams debate in the context of the wider debate on equitable and sustainable development and the corresponding framework of internationally accepted norms and standards. It introduces the rights-and-risks approach for achieving negotiated outcomes. Chapter 8 takes this forward, moving from a traditional top-down, technology focused approach to advocate significant innovations in assessing options, managing existing dams, gaining public acceptance and negotiating and sharing benefits.

Figure 8.1 The WCD's seven strategic priorities

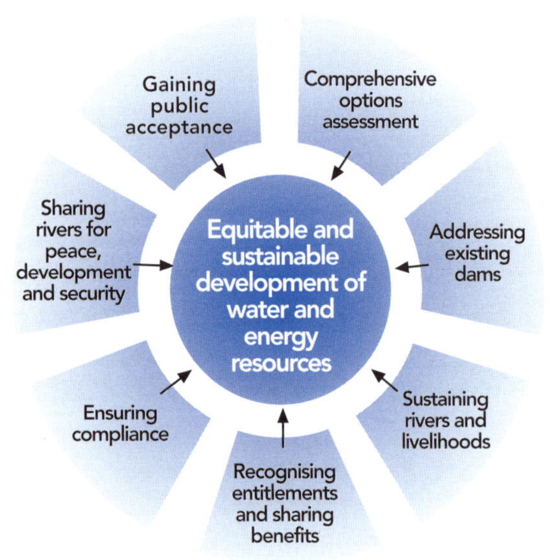

The Commission sets out this constructive and innovative way forward for decision-making in the form of the seven strategic priorities listed here and elaborated in subsequent sections of the chapter (see Figure 8.1).

The priorities are:

- Gaining Public Acceptance
- Comprehensive Options Assessment
- Addressing Existing Dams
- Sustaining Rivers and Livelihoods
- Recognising Entitlements and Sharing Benefits
- Ensuring Compliance
- Sharing Rivers for Peace, Development and Security

A key message and a set of policy principles support each of the seven strategic priorities. They are expressed in the form of achieved outcomes. A section on the rationale explains the Commission's thinking on each strategic priority and an elaboration looks at broader issues involved in achieving the strategic priority. These strategic priorities provide guidelines for all affected parties on a new way forward – one that is founded on achieving equitable and sustainable development through a process that successfully integrates social, economic and environmental considerations into decision-making on large dams and their alternatives.

Chapter 9 provides an operational approach for applying these priorities to the planning and project cycles.

Strategic Priority 1

Gaining Public Acceptance

Key Message

Public acceptance of key decisions is essential for equitable and sustainable water and energy resources development. Acceptance emerges from recognising rights, addressing risks, and safeguarding the entitlements of all groups of affected people, particularly indigenous and tribal peoples, women and other vulnerable groups. Decision-making processes and mechanisms are used that enable informed participation by all groups of people, and result in the demonstrable acceptance of key decisions. Where projects affect indigenous and tribal peoples, such processes are guided by their free, prior and informed consent.

Effective implementation of this strategic priority depends on applying these policy principles:

1.1 Recognition of rights and assessment of risks are the basis for the identification and inclusion of stakeholders in decision-making on energy and water resources development.

1.2 Access to information, legal and other support is available to all stakeholders, particularly indigenous and tribal peoples, women and other vulnerable groups, to enable their informed participation in decision-making processes.

1.3 Demonstrable public acceptance of all key decisions is achieved through agreements negotiated in an open and transparent process conducted in good faith and with the informed participation of all stakeholders.

1.4 Decisions on projects affecting indigenous and tribal peoples are guided by their free, prior and informed consent achieved through formal and informal representative bodies.

Rationale

Because of their scale and complexity, dams affect the existing rights of different groups and create a wide range of significant risks for a diverse range of interest groups. Among those affected are indigenous and tribal peoples, women and other vulnerable groups who have been shown to suffer disproportionately. This has been compounded by negligible participation of these groups in decision-making processes, with the result that planning processes for large dams have frequently overlooked gender and equity aspects. The vulnerability of these groups stems from the failure to recognise, or respect their rights, and from the significant involuntary risks imposed on them.

Failure to recognise the rights of all affected groups, whether legally sanctioned or not, coupled with the significant involuntary risk imposed on the most vulnerable, is central to the dams debate and associated conflicts.

To be socially legitimate and produce positive and lasting outcomes, development projects should provide for greater involvement of all interested parties. A fair, informed and transparent decision-making process, based on the acknowledgement and protection of existing rights and entitlements, will give all stakeholders the opportunity to fully and actively participate in the decision-making process. Instead of exacerbating existing inequalities, water and energy resources development should be opportunities for achieving a high level

A fair, informed and transparent decision-making process, based on the acknowledgement and protection of existing rights and entitlements, will give all stakeholders the opportunity to fully and actively participate in the decision-making process.

of equity. The planning process should be sensitive to, and take account of, social and economic disparities, and devise and implement mechanisms for addressing them.

Recognising indigenous and tribal people's rights

International and national policy making increasingly recognise that historical and continuing wrongs committed against indigenous and tribal peoples call for distinct measures to protect their rights. These measures include the free, prior and informed consent of indigenous and tribal peoples to developments that may affect them. To achieve this, the participation of indigenous and tribal peoples must become an integral part of the decision-making process. This is increasingly being recognised in international and national law.

International legal instruments, such as Conventions 107 and 169 of the International Labour Organisation and the evolving United Nations *Draft Declaration on the Rights of Indigenous Peoples*, recognise and support the concept of free, prior and informed consent. Other organisations reflect this trend including the Inter-American Development Bank through its operational policy, which requires informed consent of indigenous and tribal peoples to resettlement and compensation measures. Similar reforms are found at national level in a number of countries.[1]

Identifying rights and risks and recognising how they affect different parties gives planners an objective basis for identifying stakeholders.

Through acknowledging the rights of vulnerable groups, and providing for their full and active participation in the decision-making process, all the risks associated with a decision can be addressed. Requiring the free, prior and informed consent of indigenous and tribal peoples empowers them at the negotiating table.

Negotiations conducted in good faith that lead to an agreed outcome would secure wider acceptance of development policies and projects.

Elaboration of Policy Principles

1.1 Recognition of rights and assessment of risks are the basis for the identification and inclusion of stakeholders in decision-making on energy and water resources development.

Water and energy resource development projects can affect the existing rights of community groups in many different ways and can lead to a variety of risks. Legal and customary rights take many forms, including livelihood, resources, habitat, social networks and cultural heritage. Recognising this variety makes it possible to identify the risks facing communities.

Identifying rights and risks and recognising how they affect different parties gives planners an objective basis for identifying stakeholders. These stakeholders must participate fully and actively in the decision-making process and be party to all negotiated agreements throughout the process, from options assessment to final implementation, operation and monitoring. The involvement of women and other vulnerable groups in decision-making should be ensured at all stages of the planning and implementation process. There should be clear consideration for the vulnerabilities that expose women to project impacts (displacement, changes in the resource base and resulting disruptions of social and economic resources and networks) and for the specific obstacles that reduce their opportunities to share benefits generated by the project.

At the needs and options assessment stage, strategic impact assessment enables identification of stakeholders. Impoverishment risk assessment conducted at the pre-feasibility stage will enable identification of stakeholders bearing risk voluntarily and involuntarily for participation in the decision-making process.

1.2 Access to information, legal and other support is available to all stakeholders, particularly indigenous and tribal peoples, women and other vulnerable groups, to enable their informed participation in decision-making processes.

Various stakeholders have significantly different capacities to participate fully and actively in the development planning process. Rural communities, indigenous and tribal peoples, women and other vulnerable groups are at a disadvantage in accessing legal and financial resources and in their capacity to participate in negotiating decisions.

In order for these groups to participate fully and actively in negotiations, they need access to adequate resources, including legal and other professional support. Communities also need sufficient time to examine various proposals and to consult amongst themselves.

Resources committed to achieving these ends must target a continuing process of capacity building.

1.3 Demonstrable public acceptance of all key decisions is achieved through agreements negotiated in an open and transparent process conducted in good faith and with the informed participation of all stakeholders.

Participatory processes need to secure public acceptance of plans and projects for water and energy resources development. To achieve mutually agreed outcomes, stakeholders should negotiate through recognised stakeholder bodies. Public acceptance of the decision reached by stakeholders through this process should guide progress at key stages in the assessment, selection, planning and implementation of the project.

The following key principles define the nature of open and transparent decision-making processes. The process:

- is democratic, accountable and enjoys public confidence;
- safeguards the rights and entitlements of vulnerable groups by addressing imbalances in political power;
- promotes women's participation and gender equity;
- is guided by the free, prior and informed consent of indigenous and tribal peoples; and
- is based on the willing participation of all parties negotiating in good faith throughout all key stages, from options assessment to final implementation, operation and monitoring.

Negotiations should result in demonstrable public acceptance of binding formal agreements among the interested parties with clear, implementable institutional arrangements for monitoring compliance and redressing grievances.

A stakeholder forum can facilitate this process. This forum could be an existing planning institution located at the local, sub-national and national levels. Countries that already have such planning institutions must ensure representation of rural communities, indigenous and tribal peoples and other stakeholders in them. Countries without such planning institutions should consider creating a stakeholder forum for the purpose (see Chapter 9 for guidelines).

Negotiating agreements

Reaching a negotiated agreement may need assistance from an agreed independent third party from time to time. This assistance is best provided through an independent dispute resolution body that:

- is constituted with the participation and agreement of stakeholders; and
- has the necessary skills, legal and administrative capacity for this purpose.

This body should agree on a negotiating process with all stakeholders at the outset. Stakeholders should refer disagreements on any aspects of the negotiations to this body to examine them and provide assistance to the parties. This includes determining whether stakeholders are negotiating in good faith and suggesting ways of reaching a settlement.

Demonstrating public acceptance, and upholding negotiated decisions, is best achieved through binding and formal agreements.

Demonstrating public acceptance, and upholding negotiated decisions, is best achieved through binding and formal agreements. They must include mechanisms for hearing and settling subsequent grievances.

The Commission recognises that coercion and violence have been used against communities affected by dams. All project proponents – public and private – need to commit to the strict prohibition of such acts of intimidation against any stakeholders.

1.4 Decisions on projects affecting indigenous and tribal peoples are guided by their free, prior and informed consent achieved through formal and informal representative bodies.

International law includes a body of conventions and customary norms that increasingly recognise the rights of indigenous and tribal peoples. Aspects of the national laws of many countries now reflect contemporary views of indigenous rights.[2]

Some of these changes are a direct response to indigenous peoples' campaigns demanding social justice and development opportunity – including campaigns concerning dams. However, these provisions have not been very successful in protecting the rights of indigenous and tribal peoples.

To the extent that historical and present injustices continue to deny indigenous and tribal peoples the right to self-determination, countries increasingly recognise that they are entitled to distinct measures to protect their rights. This recognition has included prescriptions of non-discrimination, cultural integrity, control over land and resources, social welfare and development and self-government.

Identification of indigenous and tribal peoples

Several countries have clear laws and procedures identifying and recognising indigenous and tribal peoples. However, the situation is unclear in some other countries.

At its broadest, the adjective 'indigenous' is applied to any person, community or being that has inhabited a particular region or place prior to colonisation. However, the term 'indigenous peoples' has gained currency internationally to refer more specifically to long-resident peoples, with strong customary ties to their lands, who are dominated by other elements of the national society.

The general trend in the United Nations and other international organisations has been to accept that many of the so-called 'tribal peoples' of Africa, Asia and the Pacific are indistinguishable from indigenous peoples as far as international law and standards are concerned.[3] The International Labour Organisation's (ILO) Convention 169 applies to both indigenous and tribal peoples and thus includes many such peoples from Asia and Africa. It ascribes the same rights to both categories without discrimination. Article 1(2) of ILO Convention 169 notes: 'Self-identification as indigenous or tribal shall be regarded as a fundamental criterion for determining the groups to which the provisions of this Convention apply.'

In countries that do not explicitly define indigenous and tribal peoples, the Commission proposes an alternative approach for identifying them. Several international organisations and agencies have adopted or proposed this approach which uses the criteria listed below to recognise indigenous and tribal peoples.[4] In terms of this approach, the requirements for free, prior and informed consent should apply to groups that satisfy the following criteria:

- Historical continuity with pre-colonial societies, which is determined on the basis of the following criteria, regardless of whether they are formally recognised as indigenous or tribal peoples or not:
  - Subsistence oriented and natural resource based production systems
  - Presence of customary social and political institutions
  - An indigenous language, often different from the national language
- An experience of subjugation, exclusion or discrimination, whether or not these conditions persist.
- Vulnerability to being disadvantaged in the development process.
- Close attachment to ancestral territories and to natural resources in such areas.
- Self-identification as distinct from the dominant group or groups in societies, and identification by others as members of a distinct group.

Securing free, prior and informed consent

The requirement for free, prior and informed consent gives indigenous and tribal communities the power to consent to projects and to negotiate the conditions under which they can proceed. The effective implementation of this practice marks a significant step forward in recognising the rights of indigenous and tribal peoples, ensuring their genuine participation in decision-making processes and securing their long-term benefits.

The concept of free, prior and informed consent achieved through formal and informal representative bodies should guide decision-making on dams and their alternatives. Moreover, the Commission believes that all countries should be guided by the concept of free, prior and informed consent, regardless of whether it has already been enacted into law.

Failing that, decisions should only be made following a process of good faith negotiations that allows for the effective representation of the peoples' concerned, including genuine attempts to reconcile differences through the mutually agreed dispute resolution process, with disagreements being referred to a designated judicial body.

The customary laws and practices of the indigenous and tribal peoples, national laws and international instruments will guide the

manner of expressing consent. At the beginning of the process, the indigenous and tribal peoples will indicate to the stake-holder forum how they will express their consent to decisions. A final agreement on how to express consent will be reached before the start of the planning process.

Strategic Priority 2

Comprehensive Options Assessment

Key Message

Alternatives to dams do often exist. To explore these alternatives, needs for water, food and energy are assessed and objectives clearly defined. The appropriate development response is identified from a range of possible options. The selection is based on a comprehensive and participatory assessment of the full range of policy, institutional, and technical options. In the assessment process social and environmental aspects have the same significance as economic and financial factors. The options assessment process continues through all stages of planning, project development and operations.

Effective implementation of this strategic priority depends on applying these policy principles:

2.1 Development needs and objectives are clearly formulated through an open and participatory process before the identification and assessment of options for water and energy resource development.

2.2 Planning approaches that take into account the full range of development objectives are used to assess all policy, institutional, management, and technical options before the decision is made to proceed with any programme or project.

2.3 Social and environmental aspects are given the same significance as technical, economic and financial factors in assessing options.

2.4 Increasing the effectiveness and sustainability of existing water, irrigation, and energy systems are given priority in the options assessment process.

2.5 If a dam is selected through such a comprehensive options assessment process, social and environmental principles are applied in the review and selection of options throughout the detailed planning, design, construction, and operation phases.

Rationale

Dams have delivered benefits to society, but have also caused serious social and environmental harm. Many of the controversies over dam projects have focused attention on whether a dam was the most appropriate response to a development need or objective, and whether these were correctly identified in the first place. In some cases project objectives were not clearly stated, particularly in relation to broader national and local development goals. In others, the decision to proceed with a dam was taken before considering all options or following strong backing from specific constituencies that undermined options assessment. This failure to assess strategic options rigorously at an early stage has led to a number of disputes.

Often dams take a long time to come on stream, delaying the delivery of benefits. Because they are high cost investments they divert resources and can exclude other options that may be able to deliver benefits more quickly. These options include demand side management, alternative supply side technologies and improving and expanding the performance of existing systems. There are also some new options reaching the stage where they can compete in the market, for example renewable technologies for electricity generation such as wind and solar power.

Options assessment involves determining the relevance of individual options or a mix of options to respond to development needs in a specific location.

Options assessment involves determining the relevance of individual options or a mix of

An early focus on options assessment can reduce delays and additional costs and conflicts, benefiting all those affected by a project.

options to respond to development needs in a specific location. The challenge is to assess a wider range of alternatives earlier in the process. Experience has shown that this needs to be done in a transparent and participatory manner ensuring that human, social, environmental, technical and financial considerations get equal weight in the final decision. The increased availability of information about the expanding range of alternatives provides a sound base from which to draw.

An early focus on options assessment will exclude most questionable projects. Those that emerge will enjoy wider public support and legitimacy. It can reduce delays and additional costs and conflicts, benefiting all those affected by a project. In addition to social and environmental advantages, increased investment in options assessment can result in long term economic and financial benefits.

The outcome may not be as simple as 'build a dam' or 'do not build a dam', but could be a set of parallel and complementary interventions that together meet the defined goals. Where a large dam is selected, there are a number of options within the project that can avoid, minimise and mitigate adverse social and environmental impacts. These options relate to altering the size and location of the project and designing appropriate operating rules.

Elaboration of the Policy Principles

2.1 Development needs and objectives are clearly formulated through an open and participatory process before the identification and assessment of options for water and energy resource development.

Strategic Priority 1 presents a new perspective on identifying development needs based on recognising rights and assessing risks. It integrates the planning function of governments in the water and energy sectors with local processes to determine needs. This is consistent with a move towards a more strategic planning process that identifies options to meet expressed needs.

National policy statements on water resources, agriculture, energy and the environment should embody guiding principles that facilitate a more open process of needs assessment. Policy formulation should be a participatory process that lays the foundation for the involvement of affected groups throughout later stages of needs and options assessment.

Effective participation depends on locally appropriate processes that define the form of participation and the method for consolidating needs identified at local, sub-national and national level. Institutions or bodies representing communities should be clearly defined. Strategic Priority 5 discusses other key attributes of participation. The needs assessment will provide a framework for assessing options and linking expressed needs to development objectives for specific beneficiary groups.

2.2 Planning approaches that take into account the full range of development objectives are used to assess all policy, institutional, management, and technical options before the decision is made to proceed with any programme or project.

Once the planning process has clearly defined needs, development objectives and intended beneficiaries, it will need mechanisms to assess the appropriateness of options and for the participation of stakeholder groups. Assessing options should start early in the planning process and can be incorporated into master plans and sector plans using strategic impact assessments and other planning tools. Comprehensive options assessment must precede selection of any specific development plan, whether it includes a dam or an alternative.

The range of options being examined at the outset will be broad and go beyond technical alternatives to consider relevant policy, programme and project alternatives. It should also consider:

- institutional changes and management reforms that could influence consumption patterns, reduce demand, and affect the viability of other supply options;
- the river basin context, cumulative impacts and interactive effects, including the interaction between surface and groundwater resources;
- multipurpose functions of alternatives;
- secondary local and regional development effects of alternatives;
- subsidies that can distort comparison of alternatives;
- life cycle analysis to compare electricity generation alternatives; and
- the gestation period required before benefits are delivered.

A major consideration in selecting options is assessing institutional capacity for implementation. If capacity is weak for a particular option, and strengthening measures or external support are not viable, then the option should be rejected.

Multi-criteria analysis is a mechanism for options assessment. Selection criteria used in the analysis must explicitly reflect how each option affects the distribution of costs, benefits and impacts for each stakeholder group and how it responds to development objectives. The reasons for rejecting options should be clear to stakeholders.

2.3 Social and environmental aspects are given the same significance as technical, economic and financial factors in assessing options.

Future decision-making must increase the significance of social and environmental considerations, bringing them to the forefront of the screening process as is already the practice in some countries. The focus must shift from mitigation and compensation to make avoidance and minimisation of social and environmental impacts fundamental criteria guiding options assessment. This approach will give society a better chance to set thresholds for what is acceptable and what is not, to consider long term priorities, and to reject options that are unlikely to meet avoidance and minimisation principles. Stakeholders must agree on guiding principles to mitigate and compensate for the social and environmental consequences of options that remain on the table before taking further decisions.

Avoidance and minimisation of social and environmental impacts must become fundamental criteria guiding options assessment.

Environmental issues needing consideration include impacts on natural ecosystems and water quality and the implications of the different options for local, regional and transboundary effects. For example recent research shows that some reservoirs emit greenhouse gases. With climate change emerging as a key factor in decisions on energy options, reducing greenhouse gas emissions and maintaining climate stability requires a concerted global response.

Each case is location specific and informed decisions need an enhanced local knowledge base on social and environmental factors. Requirements include:

- social and ecosystem baseline studies at an early stage to describe existing conditions and resource endowment;
- determination of the relative weighting of environmental and social aspects in relation to technical, economic and financial aspects through an open process;
- a strategic impact assessment to determine environmental, social, health and cultural heritage impacts of alternatives and reject inappropriate alternatives at an early stage; and
- explicit assessment of future net greenhouse gas emissions of a project.

2.4 Increasing the effectiveness and sustainability of existing water, irrigation and energy systems are given priority in the options assessment process.

Planning must give priority to making existing water, irrigation, and energy systems more effective and sustainable before taking a decision on a new project. The potential is highly location specific, therefore assessment will require detailed in-country reviews that cut across sectoral boundaries and go beyond technical responses to include consideration of policy options. The management of existing water and energy systems will require a more pro-active and integrated response in order to achieve these gains. Strategic Priority 3 covers services provided by existing dam projects which are a subset of existing water and energy systems.

The energy sector can apply a range of measures to encourage more efficient production, lower distribution losses and reduce consumption. Similar opportunities exist to use alternative supply sources and conservation measures to provide water supplies for disadvantaged communities.

In the irrigation sector, enhancing existing systems by fulfilling undeveloped potential and increasing the productivity of water offers the best alternative to new construction. However, improving existing systems does not necessarily help to address the needs of the poorest sections of society. The options assessment process needs to consider alternative means to increase livelihood opportunities and local food security. This should include an objective assessment of the potential for local community based projects and other alternative or complementary measures.

2.5 If a dam is selected through such a comprehensive options assessment process, social and environmental principles are applied in the review and selection of options throughout the detailed planning, design, construction, and operation phases.

Following a decision to proceed with a dam project, decisions must be taken to determine its precise location, alignment and height, the availability and sources of construction materials, the impact of the construction process, the operational characteristics of the proposed dam, and the details of water and power distribution systems. Each of these decisions has further sets of alternatives. The process adopted for selecting alternatives requires the same multi-criteria approach proposed for the earlier stages of options assessment. It must give due prominence to social and environmental considerations and to participatory processes for decision-making. Principles agreed during the initial screening of options remain relevant when deciding on options relating to the project development and operations phases. Strategic Priorities 1, 4 and 5 provide further guidance on these matters.

Strategic Priority 3

Addressing Existing Dams

Key Message

Opportunities exist to optimise benefits from many existing dams, address outstanding social issues and strengthen environmental mitigation and restoration measures. Dams and the context in which they operate are not seen as static over time. Benefits and impacts may be transformed by changes in water use priorities, physical and land use changes in the river basin, technological developments, and changes in public policy expressed in environment, safety, economic and technical regulations. Management and operation practices must adapt continuously to changing circumstances over the project's life and must address outstanding social issues.

Effective implementation of this strategic priority depends on applying these policy principles:

3.1 A comprehensive post-project monitoring and evaluation process, and a system of longer-term periodic reviews of the performance, benefits, and impacts for all existing large dams are introduced.

3.2 Programmes to restore, improve and optimise benefits from existing large dams are identified and implemented. Options to consider include rehabilitate, modernise and upgrade equipment and facilities, optimise reservoir operations and introduce non-structural measures to improve the efficiency of delivery and use of services.

3.3 Outstanding social issues associated with existing large dams are identified and assessed; processes and mechanisms are developed with affected communities to remedy them.

3.4 The effectiveness of existing environmental mitigation measures is assessed and unanticipated impacts identified; opportunities for mitigation, restoration and enhancement are recognised, identified and acted on.

3.5 All large dams have formalised operating agreements with time-bound licence periods; where re-planning or relicensing processes indicate that major physical changes to facilities or decommissioning, may be advantageous, a full feasibility study and environmental and social impact assessment is undertaken.

Rationale

Most large dams that will operate in the 21st century already exist. A number of countries have not realised the full benefits of existing large dams because of:

- incomplete investments in delivery systems;
- lack of integration with associated systems such as local and national grids and agricultural extension services;
- lack of equity consideration in allocation of project benefits;
- poor maintenance; and
- ineffective and outdated management.

In other cases, dam owners have not made regular investments in monitoring, ongoing maintenance, modernisation and renovation due to institutional or financial limitations. In many settings owners have not done systematic assessments of opportunities for optimising or expanding facilities to improve the services existing dams provide.

Opportunities to improve the efficiency, environmental and social performance of existing dams and optimise their benefits must be taken. One of the most striking features is the persistence of social and environmental problems arising from past projects. Often promises of compensation and other benefits like local power supply and social amenities for resettled and host communities have not been kept. In many cases such promises were informal, making

Retrofitting existing dams with more efficient, modern equipment and control systems has achieved significant improvements in benefits, extending facilities and optimising operations.

compliance more difficult to achieve. Governments, industry and dam owners also recognise, often informally, that past mistakes should not be repeated; yet they remain as an unresolved legacy.

The WCD Knowledge Base provides many examples where the services provided by older dams have been restored or extended in time. In many other cases retrofitting existing dams with more efficient, modern equipment and control systems has achieved significant improvements in benefits, extending facilities and optimising operations.

The recent trend to optimise reservoir operations for new and older dams using decision support systems backed by more accurate and timely data on river flows is particularly relevant. While opportunities must be assessed on a case-by-case basis, good practice is to consider such measures as a 'new supply option' where they present significant, cost effective opportunities.

Experience shows that, conditions permitting, this approach can increase hydroelectric benefits by 5 to 10% over rule-based operating criteria without adversely affecting other water uses. This is a trend in Canada, the United States and Europe where operators are seeking full benefits from existing assets in response to power sector deregulation and competition. In some cases, optimising operations of a system of dams can postpone the need for new projects. These experiences are not confined to developed countries.[5]

Finally, it is evident that many existing dams do not have formal operating agreements, licences or concessions, particularly in the case of publicly owned irrigation and water supply dams. Wider participation in important management and operational decisions requires clear procedures and supporting legal mechanisms, especially when such decisions transform or transfer benefits and impacts. The absence of licences or formalised agreements removes the opportunity for public input and accountability. Where they do exist, licences and other agreements often lack clear performance targets, limiting public participation at re-licensing reviews.

3.1 A comprehensive post-project monitoring and evaluation process, and a system of longer-term periodic reviews of the performance, benefits, and impacts for all existing large dams are introduced.

The WCD Knowledge Base shows that historically, few comprehensive post-project evaluations have taken place after the commissioning of large dams. This applies to virtually all regions and countries. With few exceptions, there has been little or no monitoring of the physical, social and environmental effects of dams, a necessary input for such evaluations. Where post-project assessments have been undertaken, they have occurred many decades after construction usually with a narrow technical focus and little input from stakeholders.

The WCD Knowledge Base reveals that many unforeseen technical, social and environmental issues emerge during the commissioning phase and the first few years of operation. More intensive monitoring, extending from the construction phase through the first few years of operation, followed by a comprehensive post-project evaluation after 3 to 5 years involving affected stakeholders, will help to identify and resolve many early problems. The evaluation will encourage compliance with all commitments and provide a milestone to verify public acceptance. The first post-project evaluation should help confirm and strategically focus the longer-term monitoring programmes and provide 'lessons learnt' for future decisions about planning, design and operations of the dam.

Because the economic life of a dam may span many generations, it is necessary to review the project operation periodically in light of the needs it is intended to meet, and the services it can provide. These periodic evaluations at intervals of 5 to 10 years should be comprehensive, integrated, cumulative and adaptive. Where dams are part of a larger river basin and regional development scheme, the evaluations should take into account basin-level evaluations of all project and programme components linked to the dam that affect the environment and society (see Strategic Priority 4).

Enabling conditions for evaluations are context specific and measures should build on existing capacities. For many existing dams this will be the first evaluation of this nature and institutional resistance to transparency may need to be overcome. Licensed private sector operators may regard some aspects of the operation as proprietary commercial information. An essential first step is for governments, or their regulatory agencies, to clearly specify the requirements for monitoring and evaluation in the appropriate regulations, project licences and operating agreements.

Government guidelines need to clearly define roles of dam owners and operators and stakeholders who will participate in the evaluations and set out the resources and means for stakeholder input and interaction. Upgrading monitoring capacity will pose challenges in many countries due to the costs and operation of instruments and data systems, and because agencies other than the dam owners and operators may be involved. Clear responsibilities that build on existing capacities need to be defined and financial resources provided. Dam operators and the agencies involved should publish monitoring results annually, and make results freely accessible to all stakeholders.

The first post-project evaluation should provide 'lessons learnt' for future decisions about planning, design and operations of the dam.

3.2 Programmes to restore, improve and optimise benefits from existing large dams are identified and implemented. Options to consider include rehabilitate, modernise and upgrade equipment and facilities, optimise reservoir operations and include non-structural measures to improve the efficiency of delivery and use of services

Many industrialised countries are focusing on rehabilitation and modernisation to restore or extend the economic life of existing dams. In the last decade, many dam owners have implemented techniques for optimising reservoir operations, especially for dams generating electricity. They are considering other measures to improve performance and safety such as increasing spillway capacity to handle higher floods, extending reservoir storage and improving sediment flushing techniques. While new supply options may be needed in many countries, restoring or extending the life of existing dams and, where feasible, expanding and improving services from existing dams provide major opportunities to address development needs.

The WCD Knowledge Base identified three general categories of improvement:

- modernising and upgrading equipment and controls, and rehabilitating or expanding facilities associated with the dam;
- optimising operation of existing reservoirs including daily and seasonal water levels and release patterns for single or multi-purpose uses such as flood management and hydro generation. This can be done for a single dam, or in co-ordination with other reservoirs, lakes or water course diversions regulating river flow in a basin; and
- optimising the role of the dam within the larger system it services. For example optimising the use of surface and ground water inputs in agriculture where water is a limiting factor, or using load management practices to optimise the co-ordination of hydro generation with other energy sources.

The potential for increasing benefits from a particular dam, or group of dams, in a basin depends on the specific circumstances. Opportunities in all three areas noted above should be considered. Depending on the situation the potential may be considerable.

Restoring or extending the life of existing dams and, where feasible, expanding and improving services from existing dams provide major opportunities to address development needs.

Other measures have shown potential to improve the performance of existing dams and the services they provide. For example, experience is growing with flushing and sluicing practices during monsoon floods to reduce sedimentation and restore live storage in certain types of reservoirs. Accelerating the pace of investment in secondary and tertiary canal systems and drainage can significantly improve the productivity of surface irrigation systems attached to large dams. Other non-structural tariff, institutional and management practices can improve the efficiency of the irrigation and water supply services provided by existing dams but will require sectoral initiatives that may be beyond the mandate of a dam operator.

Improving performance begins with assessing each dam for potential gains from modernisation, renovation, expansion or optimisation of operations. Other gains can come from investment in necessary hydrological monitoring equipment, computer software, and the preparation of basin and system-level optimisation plans.

This must be explicitly linked with the options assessment phases of planning, clearly showing the scope for improvements to existing dams. The public should have the opportunity to comment on a survey assessing improvement opportunities for all dams. This should be followed by more detailed assessments of the specific dams which have potential for significant improvements.

3.3 Outstanding social issues associated with existing large dams are identified and assessed; processes and mechanisms are developed with affected communities to remedy them.

In all its public consultations, dam-affected communities told the Commission about the ongoing problems, broken promises, and human rights abuses associated with the involuntary resettlement and environmental impacts from dams. The WCD Knowledge Base includes significant evidence of uncompensated losses, non-fulfilment of promised rehabilitation entitlements, and non-compliance with contractual obligations and national and international laws (see Strategic Priority 6 and Chapter 9). While the Commission is not in a position to adjudicate on these issues, it has suggested ways to redress past and ongoing problems associated with existing dams.

Existing international laws have articulated a legal premise for a right to remedy, or reparations which is also reflected in the national legislative frameworks of many countries.[6] Reparation is defined as actions or processes that remedy, repair, make amends or compensate for past failures and damages. Given the nature of damages resulting from loss of land and a way of life, redress could include remedies that:

- recognise the breach of the original obligation and its consequences;
- acknowledge claims;
- assess damages;
- assign responsibility; and
- devise and implement remedial activities to repair the long-term and cumulative impact of these failures.

Assessing claims and making reparations

The responsibility for initiating the process of reparation rests with government. The affected people may also file claims with the government. In order to address reparation issues, the government should appoint an independent committee with the participation of legal experts, the dam owner, affected people and other stakeholders. The committee should:

- develop criteria for assessing meritorious claims;
- assess the situation and identify individuals, families and communities fulfilling the criteria for meritorious claims and
- enable joint negotiations involving adversely affected people for developing mutually agreed and legally enforceable reparation provisions.

States are at different stages in developing regulatory systems and institutional capacity including dispute resolution and will take different approaches to resolving this issue. However, where there are reasons to take action and alleviate hardships experienced by the people affected by dams constructed in the past, there are two practical ways of addressing pressing problems.

In order to address reparation issues, the government should appoint an independent committee with the participation of legal experts, the dam owner, affected people and other stakeholders.

- Opportunities to restore, improve and optimise benefits from existing large dams and other river basin developments should be used as an entry point to address unmitigated social problems associated with the dams in that river basin.
- In situations where no current developments are envisaged but outstanding social problems related to dams exist, meritorious claims for redress should be prioritised and assessed on the following basis.
  - affected people file genuine claims related to economic, social and cultural loses and unfulfilled promises;
  - the evidence accompanying claims filed shows that they continue to suffer harm due to unmitigated impacts, and that the impact is causally connected to the dam;
  - available mechanisms to resolve the complaint have been exhausted; and
  - the nature and extent of the harm.

Enabling conditions

To exercise their right to seek a remedy, affected people need access to political and legal systems and the means and ability to participate in prescribed ways. Affected people should receive legal, professional and financial support to participate in the assessment, negotiation and implementation stages of the reparation process.

Affected peoples must be defined according to actual experience of impacts as described in Strategic Priority 5.2, and not by the limited definition in original project documents and contracts. Further, damage from dams may require assessment on a catchment basis extending upstream and downstream. Damage assessments should include non-monetary losses. Reparations should be based on community identification and prioritisation of needs, and community participation in developing compensatory and remedial strategies.

To exercise their right to seek a remedy, affected people need access to political and legal systems and the means and ability to participate in prescribed ways.

The nature of remedies

Remedies can include restitution, indemnity (or compensation), and satisfaction. Restitution can include stopping the damaging conduct or carrying out the original obligation. Indemnity involves the payment of money for losses incurred, such as payments to compensate for loss of assets, property, and livelihoods and a variety of remedial actions, including resettlement plans and development programmes. Satisfaction includes other forms of reparation to address any non-material damage, including public acknowledgement of damage and an apology.

Responsibility for reparations

Reparations may involve multiple actors including states, financing institutions, international organisations, and private corporations. It is the State's responsibility to protect its citizens, including their right to just compensation. However, international organisations party to foreign investment agreements also have obligations and responsibilities to the rights and duties specified in the UN's declarations and instruments. The World Bank group's inspection panel and the International Finance Corporation (IFC) / Multilateral Investment Guarantee Agency (MIGA) office of the Compliance Advisor / Ombudsman acknowledge the responsibilities of the financier to comply with specific regulatory and operational policies governing its operations.

In a number of instances, efforts to assign corporate responsibility for non-compliance or transgressions related to social and environmental elements of a project have led to complaints filed in a corporation's home country.

The roles and responsibilities of all parties involved in planning, financing, building and operating the dam must be clearly established in the process of hearing and assessing a claim by an independent committee constituted by the government in consultation with the affected people and other stakeholders.

Financing reparations and compensation

While financing reparations may pose significant challenges this should not nullify legitimate claims. Priority must be given to financing a negotiated reparation plan before funding new dam projects in a specific location or river basin in a country.

Reparations can be financed with funds from national, provincial, and / or local government budgets, a percentage of loans and grants or a percentage on current income from energy and water management projects. Such funds could be allocated to a trust fund to benefit the community over the long term (see Strategic Priority 6 for more information on trust funds). Through changes in dam operations or other means, reparations can take the form of allocations of non-monetary resources, including land, water, fish and access to sacred sites.

An independent committee should be empowered to collect, manage, and award reparations. To ensure that decisions conform to the laws of the country and to international laws, such committees should include legal representatives selected by government and affected communities. Parties contributing to the fund should be represented to ensure transparent use of their funds. Accountability of the parties responsible for reparation should be ensured through contracts and legal recourse.

3.4 The effectiveness of existing environmental mitigation measures is assessed and unanticipated impacts identified; opportunities for mitigation, restoration and enhancement are recognised, identified and acted on.

The impact of large dams upon natural ecosystems and biodiversity is a major concern. In the past few decades, some countries have made considerable investments to alleviate these impacts. Widespread concern remains that dams elsewhere continue to result in significant, and even unnecessary, negative impacts on a wide range of natural ecosystems and on the people that depend on them. These ecosystems perform functions such as flood alleviation and yield products such as wildlife, fisheries and forest resources. They are also of aesthetic and cultural importance for many millions of people.[7]

A range of measures is available to enhance and restore ecosystems from their man-modified state, and many are already in use worldwide. In many cases these efforts are motivated by emergent environmental constraints and changing community priorities, as in the case of efforts to combat increasing salinity in the Murray Darling Basin in Australia, or the new South Africa Water Act which reallocates water rights. At least five countries (United States, Japan, Australia, Brazil, and France) are assessing the efficiency of existing fish passes and recommending improvements to design and operation. The design of the Mohale dam in Lesotho has been modified to allow larger flows, in anticipation of the results of the environmental flow studies currently being completed.

Countries such as South Africa, Senegal and Cameroon have introduced artificial flood releases to maintain downstream floodplains of value to local people.

Many dams in the United States have also been modified to allow for larger flows. Countries such as South Africa, Senegal and Cameroon have introduced artificial flood releases to maintain downstream floodplains of value to local people. This technique is used on the Columbia River system in the United States to reduce problems with total dissolved gases that can kill valuable fish.

There are a number of barriers and constraints to overcome. Continuous monitoring is a prerequisite to identify and assess what the actual impacts are and the possible effect of mitigation and restoration measures. Resources for implementing monitoring must be integrated in the project cost. Clear guidelines on environmental monitoring and a response to deal with impacts are needed. Other constraints will have to be addressed, particularly for privately developed hydropower projects that involve long-term supply contracts negotiated on previous release patterns. Contracts for a specified time period may not anticipate or allocate responsibility for periodic changes within the contract period, and would require renegotiation.

3.5 All large dams have formalised operating agreements with time-bound licence periods; where re-planning or re-licensing processes indicate that major physical changes to facilities or decommissioning may be advantageous, a full feasibility study and environmental and social impact assessment is undertaken.

Many dams, particularly irrigation and water supply dams, do not have operating agreements or licences. Where they do exist they are time-bound agreements. They generally set out the obligations of the public agency or private entity operating the dam, and provide a legal basis for stakeholders to participate, in an open and transparent manner, in important decisions regarding physical changes in facilities or reservoir operations.

There is a trend towards formal licensing of new and existing dams operated by public and private agencies.

Licences or operating agreements provide requirements for physical, environmental and social impact monitoring, contingency plans, operating strategies, the specific requirements for publication of monitoring and operating results, and requirements for the periodic needs and performance review of existing dams noted earlier. Given the ageing population of dams, safety issues require more attention in the form of inspections, routine monitoring, evaluations, surveillance systems, and regularly updated emergency action plans. Where practical and feasible, it is also important to update dams to contemporary standards, especially regarding spillway capacity and resistance to earthquakes.

There is a trend towards formal licensing of new and existing dams operated by public and private agencies. This applies particularly to single purpose hydropower dams and multi-purpose dams with power components. Licences for private owners are time-bound and are reviewed at periods of 20 to 40 years, or more frequently, depending on the country's policy and regulations. There are recent examples where public agencies and private owners alike have had to apply for licences for existing dams when new regulations were introduced. Such licences identify activities which the existing owners, including the government, need to comply with. They range from introducing and reporting on monitoring programmes to dam safety inspections. Harmonising the licence expiry dates for all dams in a particular river basin may have benefits in some cases, for example a cascade sequence of dams where interactive effects and cumulative impacts are a consideration.

Where re-licensing processes are in place, decommissioning may be an option. This option usually arises where a dam has exceeded its useful life, where safety reasons make it less expensive to remove the dam than to rehabilitate it, or where the costs (including environmental costs) of further operation outweigh the benefits. Experience with decommissioning is growing in North America and Europe.[8]

The effects of decommissioning, particularly on the natural environment where ecosystem restoration is a young science, have yet to be determined. In some cases environmental restoration has been well served by decommissioning, in others negative effects have been observed. Decommissioning of larger dams, particularly those with a high build-up of sediments in the reservoir may be problematic, especially if the dam is to be removed and the sediments released downstream. Removal of the dams may have many consequences on land use in upstream and downstream areas and on other socio-economic values and should be subjected to environmental, social, technical and economic assessment in the same way that new dams are.

Based on the range of issues that may surface with decommissioning, a feasibility study should be carried out to select the overall best solution, considering economic, environmental, social and political factors.

The challenge in licensing is that many countries have insufficient legal frameworks and there may be no consensus on the appropriate level of regulation and licensing. New institutional arrangements may be necessary to formalise licensing operating agreements in many countries. At present the institutional responsibility for different aspects of the operating agreement may be located in different institutions. Another barrier to overcome is that some existing dams may not have physical provisions to accommodate all the changes in regulation and meet all current standards.

Strategic Priority 4
Sustaining Rivers and Livelihoods

Key Message

Rivers, watersheds and aquatic ecosystems are the biological engines of the planet. They are the basis for life and the livelihoods of local communities. Dams transform landscapes and create risks of irreversible impacts. Understanding, protecting and restoring ecosystems at river basin level is essential to foster equitable human development and the welfare of all species. Options assessment and decision-making around river development prioritises the avoidance of impacts, followed by the minimisation and mitigation of harm to the health and integrity of the river system. Avoiding impacts through good site selection and project design is a priority. Releasing tailor-made environmental flows can help maintain downstream ecosystems and the communities that depend on them.

Effective implementation of this strategic priority depends on applying these policy principles:

4.1 A basin-wide understanding of the ecosystem's functions, values and requirements, and how community livelihoods depend on and influence them, is required before decisions on development options are made.

4.2 Decisions value ecosystems, social and health issues as an integral part of project and river basin development and prioritise avoidance of impacts in accordance with a precautionary approach.

4.3 A national policy is developed for maintaining selected rivers with high ecosystem functions and values in their natural state. When reviewing alternative locations for dams on undeveloped rivers, priority is given to locations on tributaries.

4.4 Project options are selected that avoid significant impacts on threatened and endangered species. When impacts cannot be avoided viable compensation measures are put in place that will result in a net gain for the species within the region.

4.5 Large dams provide for releasing environmental flows to help maintain downstream ecosystem integrity and community livelihoods and are designed, modified and operated accordingly.

Rationale

Since the 1970s a growing understanding of nature as the basis for long term human well being has replaced the view that it is an array of replaceable inputs to the development process. The 1992 Earth Summit in Rio de Janeiro, Brazil established the critical link for all countries between a healthy environment and economic development, refuting the idea that this is only a luxury for rich countries. Subsequently 177 countries in the world have accepted, approved, or acceded to the Biodiversity Convention and 122 to the Ramsar Convention on Wetlands.

The 1992 Earth Summit in Brazil established the critical link for all countries between a healthy environment and economic development, refuting the idea that this is only a luxury for rich countries.

Rivers and catchments

Rivers and catchments are particularly important because even relatively small changes in land-use, pollution or flows can have far-reaching effects. These often extend thousands of kilometres downstream to the deltas of major rivers and even to the coastal and marine environments. Mounting global pressure on water resources, and the fragmentation by dams of 60% of the world's rivers, makes these ecosystems an urgent global concern.[9]

Dams disrupt the existing pattern of water use and reallocate water to new uses. Where water is abundant, or where few people depend on fish, floodplains or deltas for their livelihoods, this process can often proceed smoothly. Where water is scarce, heavily used or supports economically or socially important ecosystems, this redistribution can lead to irreversible impacts, losses, inequities and conflicts. Changes to river flow, leaching of toxic elements from the reservoir bottom, and the creation of new habitats for disease vectors in slow flowing irrigation and drainage canals can all impact negatively on human and ecological health.

Equitable water use

River water is a common resource that should serve the good of all riverine inhabitants and the environment in an equitable and sustainable manner. Water reallocation through dams should explicitly take account of existing uses, and of the species and ecosystems the water supports. Many of the broader values that rivers provide are too often ignored in project planning and appraisal. The natural resources associated with rivers directly support natural habitats and the livelihoods and cultural values of millions of people worldwide. Rivers may also hold deep spiritual meaning for communities and societies.

The State must use effective mechanisms for sharing the available water between users, bearing in mind the balance between different uses – irrigation, water supply, power generation, and ecosystem – and the livelihood and quality of life needs of riverine communities.

The construction and operation of many large dams have had significant, and often irreversible, effects on many rivers, riverine ecosystems and communities. These activities threaten the sustainability of the underlying ecological processes that maintain habitat and biodiversity. Countries that have ratified the Convention on Biodiversity and the Ramsar Convention on Wetlands can use the provisions and guidance of these conventions to assist in meeting sustainability objectives.

Weighing negative and positive impacts

Dams can reduce the risks of flooding. However, especially in many developing countries, reduced flooding may also increase risks for local people by adversely affecting downstream fisheries, grazing and crop production. In all cases, negative economic and social impacts must be weighed equally against positive impacts.

In the past, dam construction has caused environmental damage due to poor assessments, vested interests, lack of knowledge, lack of enforcement of mitigation measures, lack of sufficient resources, lack of ongoing monitoring or ignorance of ecosystem functions. Research will continue to improve knowledge and understanding of ecosystem complexity. However it is difficult to mitigate all ecosystem impacts and ecosystem responses are rarely fully predictable. A multi-layered approach is needed that prioritises avoidance, especially in sensitive areas, and has in-built checks that adapt and respond to observed ecosystem changes.

> ***The natural resources associated with rivers directly support natural habitats and the livelihoods and cultural values of millions of people worldwide.***

The policy principles presented here provide a framework for the range of measures needed to ensure protection and health of ecosystems in planning, construction and operation of dams and their alternatives. No single principle can be fully effective in isolation from the others nor can a single ministry or agency be responsible for them all. Ecosystem issues are best addressed through a holistic view of the river, with all actors incorporating an ecosystem approach into their planning, operations and monitoring.

Elaboration of Policy Principles

4.1 A basin-wide understanding of the ecosystem's functions, values and requirements, and how community livelihoods depend on and influence them, is required before decisions on development options are made.

The river basin is the natural geographical unit to assess impacts on ecosystems and livelihoods.

The flow of water links riverine ecosystems, establishing a continuum from the top of the catchment to the ocean. Upstream water resource developments cannot be separated from their downstream implications.

Effective avoidance, minimisation and mitigation of negative environmental impacts from large dams and their alternatives require good baseline information and scientific knowledge of the riverine ecosystem, gathered over several years. The flow of water links riverine ecosystems, establishing a continuum from the top of the catchment to the ocean. Upstream water resource developments cannot be separated from their downstream implications.

Considering the ecosystem

Therefore, project proponents must assess the ecosystem consequences of the cumulative impact of dams, dam induced developments and other options along the full length of the river reaching as far as the delta, even where this extends into neighbouring provinces or countries. Where the resources of riparian communities could be negatively affected they should be consulted on the proposal before decisions are made (see Strategic Priorities 5 and 7).

4.2 Decisions value ecosystems, social and health issues as an integral part of project and river basin development and prioritise avoidance of impacts in accordance with a precautionary approach.

Large dam projects have frequently incorporated environmental and social considerations too late in project planning to allow their full integration into project choice and design. Environmental Impact Assessments (EIA) are often seen simply as an obstacle to be overcome in getting clearance and approval for the project. Consultants and agencies involved in planning should focus on ecosystem, social and health issues at the same time that economic and technical studies for options assessment begin. Planning teams should explicitly incorporate ecosystem, health and social findings in the final choice of project through multi-criteria analysis.

Strategic impact assessment during the options assessment stage should include independent and comprehensive assessment of ecosystem, social and health impacts and evaluation of any cumulative or inter-basin impacts. An independent panel could be used to support impact assessment.

Alternatives to large dams are frequently available, and negative impacts of large dam projects vary hugely due to site selection and the design of elements such as its height, intakes, outlets and gates. By deliberately setting out to avoid projects with the most serious and damaging negative impacts, project outcomes will prove more sustainable and acceptable in future.

The precautionary approach

The precautionary approach requires States and water development proponents to exercise caution when information is uncertain, unreliable, or inadequate and when the negative impacts of actions on the environment, human livelihoods, or health are potentially irreversible. It therefore forms part of a structured approach to the analysis of risks arising from water and other development proposals. The precautionary approach is also relevant to risk management. Determining what is an acceptable level of risk should be undertaken through a collective political process. The process should avoid unwarranted recourse to the precautionary approach when this can overly delay decision-making. However, decision-makers faced with scientific uncertainty and public concerns have a duty to find answers as long as the risks and irreversibility are considered unacceptable to society.

A precautionary approach therefore entails improving the information base, performing risk analysis, establishing precautionary thresholds of unacceptable impacts and risk, and not taking actions with severe or irreversible impacts until adequate information is available, or until the risk or irreversibility can be reduced, making outcomes more predictable. Normally the burden of proof will be on the developer.

4.3 A national policy is developed for maintaining selected rivers with high ecosystem functions and values in their natural state. When reviewing alternative locations for dams on undeveloped rivers, priority is given to locations on tributaries.

States should have a policy that excludes major intervention on selected rivers to preserve a proportion of their aquatic and riverine ecosystems in a natural state. The policy should be an integral part of the overall national water policy.

A growing number of countries have responded to increasing development pressure on rivers by setting aside certain reaches, or entire rivers, from development. For example, by 1998 the United States had designated 154 rivers under the 'Wild and Scenic Rivers Act', covering 17 200 km of a total 5.6 million kilometres of rivers in the country. Sweden has set aside four entire rivers from hydropower development, and Norway has 'protected' 35 % of its hydropower potential from development. Zimbabwe has passed similar legislation for river specific protection orders. President Kim Dae Jung of Korea cancelled a dam on the Tong River in June 2000 and declared the area protected for its natural and cultural values.

Determining what is an acceptable level of risk should be undertaken through a collective political process.

This approach helps to reconcile development of the natural resource base with the need to maintain genetic diversity as a potential resource for human development in fields like medicine. It also respects the intrinsic value of rivers. To put this policy into practice, the State, appropriate research institutions, communities and NGOs must gather the essential baseline information to inform the policy from a holistic national perspective. Gathering this information allows the country to select the least environmentally damaging options whenever dams emerge as the best option in the planning process. It enables the country to combine river development with river protection, achieving a balanced outcome that satisfies the objective of sustainable development at national and local levels.

Floodplain ecosystems and migratory fish are frequently associated with the lower stretches of the main-stems of rivers and their deltas. They are often particularly important resources for local people, especially in developing countries. In many cases main-stem dams have negatively affected the already fragile livelihoods of downstream riverine communities. Priority under the policy should therefore be given to alternative sites on the tributaries where impacts are likely to be less than on the main-stem.

4.4 Project options are selected that avoid significant impacts on threatened and endangered species. When impacts cannot be avoided viable compensation measures are put in place that will result in a net gain for the species within the region.

Too often, planned measures to mitigate impacts on threatened or endangered species have been inadequate. Failure is typically due either to insufficient knowledge and understanding of the ecosystems concerned, inadequate plans, or lack of institutional and regulatory capacity to enforce environmental mitigating measures.

The project authorities will finance compensation as an integral part of project costs for the life span of the project.

Dam proponents have often over-confidently assumed that mitigation measures will work, rather than making them work. Responsibility for their implementation may be spread across many actors. This situation has contributed to a significant increase in the rate of extinction of species sharing our planet.

Respecting international guidelines to reduce impacts

Governments have often agreed international guidelines on environmental issues, yet the record of implementation is poor. The existing international conventions contain agreed legal provisions for biodiversity protection and sustainable development issues, including measures for conserving the biodiversity of inland waters, assessing the international importance of wetlands or developing national wetland policies. Most States have ratified the UN Convention on Biological Diversity and the Ramsar Convention on Wetlands and they should urgently apply their guidelines. This will help to avoid negative impacts on rare and threatened species. States that have not yet ratified the Conventions are encouraged to do so, and in the meantime to respect their provisions.

Compensation plans

Where significant impacts on threatened or endangered species are considered unavoidable, after exhausting other water and energy options and other dam project options, project authorities should put in place a credible and monitored compensation plan. This should ensure that the population status of the species within the region shows a net gain that adequately compensates for loss of habitat to the project. Such compensation can include protecting other habitats, restoring the species in other locations and captive breeding programmes. The project authorities will finance compensation as an integral part of project costs for the life span of the project. If monitoring indicates compensation is not effective, then additional measures will be required.

4.5 Large dams provide for releasing environmental flows to help maintain downstream ecosystem integrity and community livelihoods and are designed, modified and operated accordingly.

Dam owners have often viewed releasing water from a dam for purposes other than power generation or water supply as a waste of a valuable resource. Some agencies have declared it their aim to prevent a single drop of water from reaching the sea. Yet twenty-nine countries use environmental flow releases (EFR) from dams to maintain a sustainable balance between the purpose of the dam and the needs of downstream ecosystems and resource users. Eleven countries are considering implementing EFRs.[10] This reflects the growing realisation that water and rivers support many complex processes that must be maintained to achieve sustainable development. Dams should now be specifically designed to release the necessary flow of good quality water. Targeting particular ecosystem outcomes increasingly results in flow releases that go beyond the historical notion of a 'minimum release', often arbitrarily fixed at 10% of mean annual flow. A minimum release may serve to keep the river wet but it may not be an ecologically effective measure.

Where downstream livelihoods depend on floodplains the release may take the form of a managed flood. Senegal, South Africa and Cameroon all operate dams to flood valuable downstream floodplains that benefit rural communities. These managed floods maintain hundreds of thousands of hectares of grazing land and important fisheries.

Legal measures are often required to enable implementation of environmental flows. One example is the new South African Water Act. This Act recognises that the ultimate aim of water resource management is to achieve sustainability for the benefit of all users and that the protection of the quality and quantity of water resources is necessary to ensure sustainability of the nation's water resources. Therefore the Act designates a 'reserve' that must be identified before any authorisation of water abstraction for other purposes. This basic reserve contains the minimum quantity and quality of water required to satisfy basic human needs, protect aquatic ecosystems and secure the sustainable development and use of the water resource in question. The Act opens the way to establishing the environmental flows required to maintain ecosystems.

Locally driven processes to establish the objectives of environmental flows will lead to improved and sustainable outcomes for rivers, ecosystems and the riverine communities that depend on them. Ecosystem responses to dam operating regimes are variable, so dam owners should undertake regular monitoring and a five yearly evaluation of environmental performance. This evaluation should inform modification of environmental flows where necessary (see Chapter 9).

Locally driven processes to establish the objectives of environmental flows will lead to improved and sustainable outcomes for rivers, ecosystems and the riverine communities that depend on them.

Strategic Priority 5

Recognising Entitlements and Sharing Benefits

Key Message

Joint negotiations with adversely affected people result in mutually agreed and legally enforceable mitigation and development provisions. These provisions recognise entitlements that improve livelihoods and quality of life, and affected people are beneficiaries of the project. Successful mitigation, resettlement and development are fundamental commitments and responsibilities of the State and the developer. They bear the onus to satisfy all affected people that moving from their current context and resources will improve their livelihoods. Accountability of responsible parties to agreed mitigation, resettlement and development provisions is ensured through legal means, such as contracts, and through accessible legal recourse at national and international level.

Effective implementation of this strategic priority depends on applying these policy principles:

5.1 Recognition of rights and assessment of risks is the basis for identification and inclusion of adversely affected stakeholders in joint negotiations on mitigation, resettlement and development related decision-making.

5.2 Impact assessment includes all people in the reservoir, upstream, downstream and in catchment areas whose properties, livelihoods and non-material resources are affected. It also includes those affected by dam related infrastructure such as canals, transmission lines and resettlement developments.

5.3 All recognised adversely affected people negotiate mutually agreed, formal and legally enforceable mitigation, resettlement and development entitlements.

5.4 Adversely affected people are recognised as first among the beneficiaries of the project. Mutually agreed and legally protected benefit sharing mechanisms are negotiated to ensure implementation.

Rationale

In the past, dams have displaced people from their habitats and livelihoods without giving them any control over alternatives. Besides those whose land and homes were inundated, the adversely affected people included natural resource dependent riverine communities living upstream and downstream of the dam and in other affected areas. Because some groups were not recognised as affected, the number of adversely affected people has been underestimated. Non-recognition, or partial recognition, of the entitlements of those identified as affected has resulted in inadequate restitution for losses. As a result dam projects have often impoverished adversely affected people.

Recognition of rights is an important element in establishing the existing entitlements of adversely affected people at various locations.

A negotiated process to identify those affected and develop legally enforceable mitigation and development measures depends on a number of enabling conditions. Adversely affected people need to show acceptance of the dam project by consenting to the process and to the mitigation and development measures. These measures should include a share in project benefits and redress and recourse mechanisms.

The policy principles in this strategic priority relate closely to those in Strategic Priority 1 and Strategic Priority 4 and should be read in conjunction with those strategic priorities.

Elaboration of Policy Principles

5.1 Recognition of rights and assessment of risks is the basis for identification and inclusion of adversely affected stakeholders in joint negotiations on mitigation, resettlement and development related decision-making.

Recognition of rights is an important element in establishing the existing entitlements of adversely affected people at various locations. Existing entitlements are the basis for negotiating new entitlements. The project process recognises a range of entitlements including the entitlement of affected parties to:

- participate in negotiating the outcomes of the options assessment process;
- participate in negotiating the implementation of the preferred option and
- negotiate the nature and components of mitigation and development entitlements.

Comprehensive assessment of the nature and extent of risks implied by a project allows for accurate assessment of the socio-economic conditions and the cultural context of the potentially affected people. The socio-economic, cultural, political and health impacts must be identified through a number of assessment methods such as Social Impact Assessment (SIA), Health Impact Assessment (HIA), impoverishment risk analysis and cultural heritage impact assessment with active participation of the affected people.

5.2 Impact Assessment includes all people in the reservoir, upstream, downstream and in catchment areas whose properties, livelihoods and non-material resources are affected. It also includes those affected by dam related infrastructure such as canals, transmission lines and resettlement developments.

The impact assessment studies must identify and delineate various categories of adversely affected people in terms of the nature and extent of their rights, losses and risks. Socio-economic, demographic and health benchmark surveys of all adversely affected populations must be completed and publicly reviewed prior to drafting mitigation, resettlement and development plans.

This process will facilitate assessment of the actual magnitude, spread and complexity of impacts and implications for people dependent on riverine ecosystems. This signals a departure from the way that social impacts were assessed in the past and will empower the planners and stakeholders to incorporate the full extent of social impacts and losses in the decision-making process. It will achieve a holistic approach to social impacts arising from dams and their infrastructure in river basin contexts. Such an approach is applicable to all options, dams and their alternatives, and will create a level playing field in the options assessment process.

5.3 All recognised adversely affected people negotiate mutually agreed, formal and legally enforceable mitigation, resettlement and development entitlements.

In order to enable all categories of affected people identified in 5.2 to regain and improve their livelihoods and welfare, mitigation and resettlement measures should be considered as a development opportunity focusing on a number of fundamental inputs:

- compensation for lost assets through replacement, substitution, cash and allowances;
- livelihood restoration and enhancement in the form of land-for-land options;
- sustainable non-agricultural employment and other measures;

- a share in project benefits and other development measures; and
- access to primary services such as schooling and health care.

For compensation to create effective new entitlements, customary and legal rights, and the future value of land and common property resources must all be acknowledged. To reach agreement and prevent future disputes, a transparent and legally enforceable mechanism must exist to calculate the replacement value of all affected assets. To check against under-valuation of assets, the date for calculating the value of assets must match the date of payment.

Regaining lost livelihood requires adequate lead time and preparation and therefore people must be fully compensated before relocation from their land, house or livelihood base.

In the past, cash compensation has proved ineffective in recreating lost assets and opportunities in less monetised economies and should be avoided. Where people prefer cash compensation, it must be paid with adequate safeguards that enhance long term livelihood sustainability. Regaining lost livelihood requires adequate lead time and preparation and therefore people must be fully compensated before relocation from their land, house or livelihood base. If compensation payment is delayed, interest on the compensation amount must be paid to account for inflation

Agreeing on mitigation, resettlement and development

Mutually agreed mitigation, resettlement and development provisions should be prepared jointly with the participation of all affected people, government and the developer. Sustainable mitigation, resettlement and development provisions and opportunities include:

- resettlement with land-for-land, sustainable non-agricultural employment and / or other development provisions;
- resettlement as a community or viable social unit;
- resettlement close to the original habitat for effective community recovery;
- protection against land alienation to secure negotiated entitlements; and
- other livelihood support measures including: agricultural support; access to forests, grazing lands and other common resources; access to livelihood resources for all adult members including women; and livelihood resources with the potential to sustain future generations.

Implementing a mitigation, resettlement and development plan

Stakeholders must set up a high level multistakeholder committee representing government, the developer and affected communities. The committee will be responsible for directing implementation of the mitigation, resettlement and development programme and will serve as an appeals forum to hear complaints and resolve disputes. A Mitigation, Resettlement and Development Action Plan (MRDAP) accepted by the affected people should be formalised through two legally binding contractual agreements.

- a master contract that outlines the obligations of government and the developer to carry out the actions set out in the MRDAP in time and to the full extent. The master contract will also specify penalties, incentives and remedies to facilitate compliance by government and the developer. The financial and other resources to fully comply with the MRDAP have to be secured prior to signing the master contract.

- performance contracts jointly signed by government and the developer with individual families and the community specifying entitlements (compensation, resettlement where necessary and direct benefits from the project), delivery schedule, and recourse procedures.

In order to guarantee implementation, a performance bond should be considered as part of the master contract in the case of private sector developers. See Policy Principle 6.2 under Strategic Priority 6 for a discussion of performance bonds.

While the high level multi-stakeholder committee will carry out preliminary dispute resolution and grievance redress functions, the master and performance contracts are meant to empower communities and families to seek redress from courts within the country as a final resort.

The contracts signed before construction starts remain in effect through the operation phase. Signing of master and performance contracts, and demonstration by government and the developer of their capacity to fulfil their respective commitments signifies consent by the affected people to move the project from the design stage to the construction stage.

5.4 Adversely affected people are recognised as first among the beneficiaries of the project. Mutually agreed and legally protected benefit sharing mechanisms are negotiated to ensure implementation.

People adversely affected by a dam project should be the first to benefit from the project. Appropriate mechanisms should be introduced to ensure equitable distribution of development opportunities generated by the dam.

The benefits could be related to project finance, reservoir construction, operation, downstream release and revenue sharing. The benefits under each category are elaborated in the Guideline 20 in Chapter 9. Examples of opportunities include preferential fishing rights on reservoirs, land in the irrigation command area, rights to draw down lands, equity shares, rural electrification from power generated, ownership of tourist facilities, custodian-ship over wildlife and other natural resources.[11]

The adversely affected people should participate in the identification, selection, distribution and delivery of benefits. The adversely affected people, government and the developer / financier should assess and agree on the level of benefits. As a general principle, the level of benefits should be sufficient to induce demonstrable improvements in the standard of living of the affected people.

All categories of affected people - displaced and those located upstream, surrounding the area of the reservoir, downstream of the dam and host communities for resettlement should be considered eligible in principle. They could benefit in varying degrees, or they could benefit equally, depending on the extent of risk the dam poses to their livelihoods.

Benefits could be the in the form of community assets or services; and could be individual and household focused. Once the stakeholders have agreed on the type and level of benefits, they need to decide on delivery mechanisms and timing. Commitments on benefits from the project should form part of the performance contracts with affected families and the community.

Strategic Priority 6
Ensuring Compliance

Key Message

Ensuring public trust and confidence requires that governments, developers, regulators and operators meet all commitments made for the planning, implementation and operation of dams. Compliance with applicable regulations, criteria and guidelines, and project-specific negotiated agreements is secured at all critical stages in project planning and implementation. A set of mutually reinforcing incentives and mechanisms is required for social, environmental and technical measures. These should involve an appropriate mix of regulatory and non-regulatory measures, incorporating incentives and sanctions. Regulatory and compliance frameworks use incentives and sanctions to ensure effectiveness where flexibility is needed to accommodate changing circumstances.

Effective implementation of this strategic priority depends on applying these policy principles:

6.1 A clear, consistent and common set of criteria and guidelines to ensure compliance is adopted by sponsoring, contracting and financing institutions and compliance is subject to independent and transparent review.

6.2 A Compliance Plan is prepared for each project prior to commencement, spelling out how compliance will be achieved with relevant criteria and guidelines and specifying binding arrangements for project-specific technical, social and environmental commitments.

6.3 Costs for establishing compliance mechanisms and related institutional capacity, and their effective application, are built into the project budget.

6.4 Corrupt practices are avoided through enforcement of legislation, voluntary integrity pacts, debarment and other instruments.

6.5 Incentives that reward project proponents for abiding by criteria and guidelines are developed by public and private financial institutions.

Rationale

Governments and other stakeholders need to be satisfied that once informed decisions are made, all parties will ensure that they monitor and comply with obligations throughout the life of a project. While there are good examples of the development of innovative compliance guidelines, the WCD Knowledge Base demonstrates failure on the part of developers and others to fulfil all voluntary and mandatory obligations for the assessment and implementation of approved projects. As noted in the Global Review, the failure to comply with obligations has led to impoverishment of affected peoples, under-performance and environmental degradation resulting in legitimate criticism and a serious breakdown in stakeholder trust and confidence in the process, implementation, and outcomes of decision-making.

Governments and other stakeholders need to be satisfied that once informed decisions are made, all parties will ensure that they monitor and comply with obligations throughout the life of a project.

Many States and international financing institutions have comprehensive policies, criteria and guidelines for implementing a dam project once the decision to build is taken. In the case of technical requirements, the contracting process clearly lays out the responsibilities, tasks, monitoring tools, indicators, rewards and penalties associated with contract implementation. Contracts typically provide for a penalty or bonus for late or timely completion and the posting of a performance bond by the general contractor for the site.

Resettlement, environmental mitigation and compensation are rarely, if ever, subject to such rigorous design processes and contractual terms. Often, these activities are carried out by government ministries or agencies and are not covered by contractual obligations. Where they exist, contracts for social and environmental programmes are oriented towards task completion, without any rewards or penalties related to compliance with relevant criteria and guidelines. Where resettlement numbers are underestimated, there may be no clarity on which agency or firm – aside from government – is responsible. When funds are not available for programme completion, or there is a lack of accountability, failure to resolve outstanding settlement cases has led to long-standing problems with displaced groups.

Trust and confidence in the capacity and commitment to meet obligations must be restored if new projects are to create more positive development outcomes and avoid the level of conflict that has occurred in the past. This requires the formation of new relationships and new and more effective means of ensuring compliance.

Elaboration of Policy Principles

6.1. A clear, consistent and common set of criteria and guidelines to ensure compliance is adopted by sponsoring, contracting and financing institutions and compliance is subject to independent and transparent review.

All project participants, including government agencies, multilateral, bilateral and commercial financing institutions, private sector developers and NGOs should adopt a clear set of criteria and guidelines for developing water and energy resources. This will provide a framework for assessing compliance internally and externally and demonstrating compliance in a transparent manner to stakeholders.

All project participants, including government agencies, multilateral, bilateral and commercial financing institutions, private sector developers and NGOs should adopt a clear set of criteria and guidelines for developing water and energy resources.

There are already many excellent sets of criteria and guidelines for planning, project assessment, project construction and operations. The WCD Criteria and Guidelines elaborated in Chapter 9 are not intended to be exhaustive. Rather they provide a set of conditions that must be fulfilled to restore the level of trust and confidence in the project process.

The first step is to ensure that each particular project participant makes a binding commitment to the criteria and guidelines that apply to them. This should include government agencies, bilateral and multilateral financing institutions, private sector developers and NGOs. In the case of government agencies, this may involve incorporating particular criteria into the regulatory frameworks, operational procedures and staff guidelines. For private sector developers this may involve obtaining certification, adopting internal practices and procedures, establishing codes of conduct or entering into integrity pacts, to ensure compliance with best social and environmental practice.

Ideally, participants should not only accept the WCD recommendations, but also harmonise their criteria and guidelines with those of other bodies. Not every detail needs to be the same, but all project participants should use common parameters. For example, different international financing agencies or national and provincial authorities can harmonise their criteria and guidelines.

By harmonising the criteria they employ for social and environmental management, multi-lateral and bilateral financiers, including Export Credit Agencies, will prevent developers from turning to financiers with weak guidelines to fund unacceptable projects. Project participants will reduce needless duplication of effort by employing a standard set of guidelines and tools such as strategic impact assessment and environmental impact assessment to bring projects to decision points in a more timely and less costly manner.

Ensuring compliance with criteria and guidelines

Project participants, in particular government and developers will have to show that they have lived up to their commitments in order to restore trust and confidence in the process. An external review process involving an Independent Review Panel could be the best way to achieve this (see Guideline 22 in Chapter 9). Advancing the use of such panels will require:

- establishing an accredited list of experts through a multi-stakeholder advisory group;
- giving such panels the power to invoke the 'stop lights' that the Compliance Plan will establish; and
- ensuring information from the independent panel is available to the public.

Another mechanism that can be used either with, or separate from, an Independent Review Panel is independent certification. To obtain certification, participants must show they conform with international standards for practices and procedures through regular monitoring and review by an accredited external body.

This has a number of potential practical manifestations in the case of dams. These include the development of a stewardship council on dams or the development of an International Organisation for Standardization (ISO) standard for dam management .

In recent years the Forestry Stewardship Council (FSC) has made rapid progress on certification in the forest products industry. The FSC has the capacity to influence behaviour through the certification of an internationally traded product.

Advancing the systematic management of dams-related impacts through existing recognised international standards, such as the ISO, has the distinct advantage of tapping into an existing and long-standing international structure and network. The Commission encourages industry, and particularly dam operators, to adopt ISO 14001 within a broader compliance framework that includes performance bonds, integrity pacts, and other tools.[12] ISO-based standards such as the ISO 9000 and 14000 series could be supplemented by a sector specific technical specification document for dams, incorporating the Commission's final recommendations and guidelines.

This would require agreement and effort on the part of industry, multilateral banks, NGOs and affected peoples groups, together with willingness on the part of the ISO to develop the specification (see recommendation in Chapter 10). The development of an international certification system based on the Commission's report cannot replace the integration of its guidelines into national regulatory frameworks by governments. It should be seen as a complementary approach, not as an alternative. It would encourage the private sector to promote and adapt standard voluntary codes of conduct increasingly required in today's competitive markets.

6.2. A Compliance Plan is prepared for each project prior to commencement, spelling out how compliance will be achieved with relevant criteria and guidelines and specifying binding arrangements for project-specific technical, social and environmental commitments.

Different States are at different stages in developing regulatory systems and institutional capacity and their systems will evolve to reflect their particular circumstances. The nature of existing systems will influence the range of compliance measures available. How these measures interrelate to achieve a satisfactory compliance package will need to be determined on a case by case basis.

An overarching Compliance Plan is the best way to ensure that compliance activities and measures are effectively pursued and implemented, and should be developed for each project. This plan will set out how the developer will ensure compliance with project related obligations. It may refer to regulatory and non-regulatory processes and obligations. Where the plan relies upon the State's institutional capacity to ensure compliance, the developer may have to enhance this institutional capacity before the project proceeds. The measures taken need to create trust and confidence that the Compliance Plan can and will be met.

In considering the range of tools and mechanisms available to ensure compliance, greater use should be made of existing conditions directly linked to achieving ongoing compliance. These conditions need to be built in at key stages of the project cycle to provide maximum incentive to comply. In States with well-developed regulatory systems and institutional arrangements (including judicial processes), greater reliance can be placed upon direct regulation. Where regulatory systems and supporting institutions are still evolving, the role of financiers, insurers and contractors, and the availability of a broad range of non-regulatory measures will be of particular importance in assisting States to proceed with acceptable developments.

Binding arrangements must be in place for social and environmental measures. The agreed terms of resettlement and environmental management conditions need to be incorporated into legally binding and publicly available documents, and into any relevant statutory approvals issued by the State.

An overarching Compliance Plan is the best way to ensure that compliance activities and measures are effectively pursued and implemented, and should be developed for each project.

Two mechanisms, used with significant success in other fields, can be applied to dam related projects to ensure fulfilment of commitments. They are:

Performance bonds, backed by financial assurances, to provide financial security that obligations will be met (such as guaranteeing the fulfilment of performance contracts to fulfil social and environmental commitments), and

Trust funds to hold and manage funds set aside for a particular purpose (such as resettlement or environmental measures).

Both of these measures can provide States and stakeholders with a means of achieving a new level of trust and confidence in compliance, particularly in relation to:

- the need to ensure that the full costs of resettlement and environmental performance, including monitoring and auditing, are met;

- the need to secure sufficient financial security up front to protect the State and the community from the risk of default; and
- the importance of providing a financial incentive to the contractor to complete environmental and rehabilitation works.

Social and environmental performance bonds, supported by adequate financial assurances to cover identified resettlement costs and environmental measures should be posted before project work starts. Such performance bonds are already used for the technical components of dam projects. Other types of projects carrying high environmental risk have also used them successfully.

Social and environmental performance bonds, supported by adequate financial assurances to cover identified resettlement costs and environmental measures should be posted before project work starts.

In cases where performance bonds are not appropriate, such as those where the State is the developer, trust funds should be used to achieve the same objective. Trust funds are widely employed to ensure that funds are used for a designated purpose. They can be employed to hold funds for dam related resettlement and environmental measures and for monitoring and auditing costs through the life of the project. The trust fund deed needs to be publicly available and the trustees need to be independent from the project sponsors.

Trust funds can be effectively used either alone or in conjunction with performance bonds to secure financing for ongoing monitoring and auditing obligations throughout the life of the project. These mechanisms also provide an effective means for holding and distributing royalties to fund ongoing initiatives.

As with the granting of operating licences, all contingencies cannot be anticipated. A balance must be struck on a case by case basis between providing the necessary level of certainty to stakeholders that commitments will be met, and sufficient flexibility to accommodate open and transparent adaptive management.

6.3. Costs for establishing compliance mechanisms and related institutional capacity, and their effective application, are built into the project budget.

Arrangements need to be put into place to ensure that the costs of implementing the Compliance Plan are built into the project budget. Where projects are developed by the private sector the State and affected people should be satisfied that these costs are fully financed and that appropriate arrangements are in place to link compliance to key stages in the project cycle.

When additional institutional capacity is required, the costs should be explicitly included under funding for compliance activities. Compliance activities represent a transaction cost associated with the project and should be treated as such. Competing options may impose differing compliance burdens on society. This should be reflected in analysing the acceptability of these options. Past experience shows that there is an inherent risk of failure in both the satisfactory completion of the task and the effort to ensure and enforce compliance.

Because of the high cost of compliance and the risk of failure, a project's ability to deliver on the Compliance Plan must be explicitly addressed in the multi-criteria analysis to assess options.

6.4 Corrupt practices are avoided through enforcement of legislation, voluntary integrity pacts, debarment and other instruments.

All States need to adopt and implement common and consistent anti-corruption legislation. Chapter 6 outlines progress made on this front, particularly through the ratification of the OECD Convention on Combating Bribery of Foreign Public Officials in International Business Transactions.

Implementation, monitoring and enforcement of such agreements will depend on the resources dedicated to such efforts and existing business practice. Integrity Pacts can be employed as a project-specific mechanism to complement national legislation. These pacts have the advantage of not requiring explicit legislation and can be used without waiting for lengthy legislative processes where laws do not exist.

An Integrity Pact is a voluntary undertaking that sets the contractual rights and obligations of all the parties to a procurement contract. This eliminates uncertainties about the quality, applicability and enforcement of laws in a specific country. Integrity Pacts can create greater trust and confidence in the decision-making process, a more hospitable investment climate and public support for government procurement and licensing programs.

The Integrity Pact has two principal goals, namely to enable:

- companies to abstain from bribery by providing assurances that competitors will also abstain from bribery and that government procurement agencies will follow transparent procedures and prevent corruption, including extortion by their officials; and
- the State to reduce the high costs and distortion of public procurement procedures caused by corruption.

6.5 Incentives that reward project proponents for abiding by criteria and guidelines are developed by public and private financial institutions.

Compliance with norms, regulations and practices often becomes widespread when the cost of compliance falls below the cost of non-compliance. Enforcing regulations and agreements is one way to support this process.

Compliance activities represent a transaction cost associated with the project and should be treated as such.

The level of risk and the transaction costs of constructing dams have risen steadily due to the conflict over dams. This gives project developers a financial incentive to comply with the Commission's recommendations in order to reduce conflict, leading to an improvement in market access, financing terms and profitability.

A project following the WCD Criteria and Guidelines is likely to be:

- an economically, socially and environmentally sound project; and
- a project that has achieved public acceptance through an inclusive participatory process of needs and options assessment.

This will clearly entail additional up-front costs but this investment will bear fruit later in terms of reduced risk and increased benefits.

The importance of public recognition should not be underestimated. Compliance

incentives could include awards, prizes, and other forms of public recognition in the dams community, introducing healthy competition to do better. Sanctions should include public exposure and debarment, either temporary or permanent, from participation in tenders and contracts.

Private sector financial service providers and international financial institutions (IFIs) must work together to ensure that projects and companies that comply with the WCD Criteria and Guidelines have access to one or more of the following incentives:

- enhanced access to private equity capital;
- better terms on debt finance (loans) and insurance products;
- lower rates on bond financing;
- improved accounting for environmental assets and liabilities;
- preferential and expanded access to debt financing from IFIs and their private sector windows;
- risk guarantees from IFIs; and
- interim credit mechanisms to assist pipeline projects to implement WCD recommendations.

Compliance incentives could include awards, prizes, and other forms of public recognition in the dams community, introducing healthy competition to do better.

In the case of IFIs many of the policies, procedures and instruments are already being developed. In the case of private sector instruments there is a continued need for innovation in order to convert adherence to criteria and guidelines into tangible benefits for the socially and environmentally responsible investor. A few examples of promising mechanisms to encourage investors to comply with criteria and guidelines are listed below:

Socially Responsible Investing

The current growth in socially responsible investing (SRI) in both United States and European Markets far outstrips the growth of 'traditional' investment funds and provides a window of opportunity for financing projects that emerge from proper options assessment.

Lower Cap on Insurance Liability

Premiums in the insurance industry are partly based on the extent of a project's liability. A lower cap on liability for dam projects certified as complying with the WCD Criteria and Guidelines could be arranged particularly where legally binding arrangements that reduce the risk of non-compliance are in place.

Bond Rating Systems

Bonds are increasingly used at the sovereign and corporate level for infrastructure projects. A rating system similar to the Standard & Poor credit-worthiness rating that reflects compliance could favour WCD compliant bonds.

Accounting for Environmental Costs and Liabilities

Projects that adopt the WCD Criteria and Guidelines will present a more transparent picture of the environmental costs and liabilities of projects and companies. This will lower their market risk and reduce volatility brought about by stakeholder activism in this area.

Strategic Priority 7

Sharing Rivers for Peace, Development and Security

Key Message

Storage and diversion of water on transboundary rivers[13] has been a source of considerable tension between countries and within countries. As specific interventions for diverting water, dams require constructive co-operation. Consequently, the use and management of resources increasingly becomes the subject of agreement between States to promote mutual self-interest for regional co-operation and peaceful collaboration. This leads to a shift in focus from the narrow approach of allocating a finite resource to the sharing of rivers and their associated benefits in which States are innovative in defining the scope of issues for discussion. External financing agencies support the principles of good faith negotiations between riparian States.

Effective implementation of this strategic priority depends on applying these policy principles:

7.1 National water policies make specific provision for basin agreements in shared river basins. Agreements are negotiated on the basis of good faith among riparian States[14]. They are based on principles of *equitable and reasonable utilisation, no significant harm, prior information* and the Commission's strategic priorities.

7.2 Riparian States go beyond looking at water as a finite commodity to be divided and embrace an approach that equitably allocates not the water, but the benefits that can be derived from it. Where appropriate, negotiations include benefits outside the river basin and other sectors of mutual interest.

7.3 Dams on shared rivers are not built in cases where riparian States raise an objection that is upheld by an independent panel. Intractable disputes between countries are resolved through various means of dispute resolution including, in the last instance, the International Court of Justice.

7.4 For the development of projects on rivers shared between political units within countries, the necessary legislative provision is made at national and sub-national levels to embody the Commission's strategic priorities of 'gaining public acceptance', 'recognising entitlements' and 'sustaining rivers and livelihoods'.

7.5 Where a government agency plans or facilitates the construction of a dam on a shared river in contravention of the principle of good faith negotiations between riparians, external financing bodies withdraw their support for projects and programmes promoted by that agency.

Rationale

Conflict over transboundary rivers usually results from a power imbalance amongst riparians where one State or province is sufficiently influential to exert its authority over others. Generally upstream States are considered to be in a more influential position as they can control the water source, but regional power imbalances may also make it possible for downstream riparians to exert influence over upstream States. Similar conflicts may also occur within States where rivers cross internal political borders.

Such conflicts are often caused by proposals to store or divert water by constructing dams. Experience suggests that disputes over water can be resolved and co-operation developed, even where disagreements in other spheres of international relations remain unresolved.

There are 261 international river basins worldwide. Most do not have agreements covering water allocation principles. Negotiation of such agreements between riparian States has proceeded on a case-by-case basis without any overarching globally binding legal instrument.

Experience suggests that disputes over water can be resolved and co-operation developed, even where disagreements in other spheres of international relations remain unresolved.

The 1966 Helsinki Rules on the Uses of Waters of International Rivers adopted the principle that each State had 'a reasonable and equitable share in the beneficial uses of waters in an international basin', but these rules have no status in international law. While the principles set forth in the Helsinki Rules represent what many experts contend are long accepted principles, these Rules have not achieved the level of a binding international treaty. Until it enters into force, this is also true of the UN Convention on the Law of the Non-Navigational Uses of International Watercourses. Through a resolution of the United Nations General Assembly in 1997, this Convention attracted support from 103 countries, but three countries voted against it and 27 abstained. The Convention took 27 years to develop and has yet to be ratified by enough countries to bring it into force. Among the opposing and abstaining countries are those with major dam building programmes themselves or with an interest in restricting development projects within other riparian States.

International efforts to develop a universal framework for negotiations appear to have had limited effect and some countries refuse to respect what can generally be considered as a growing body of international opinion. The Commission views the principles of the UN Convention as an emerging body of customary law and considers that States will reduce the possibility of conflict if they are prepared to endorse and adhere to them. This contrasts with a situation where some countries have followed a unilateral approach to the use of water resources and reject the need for an integrated basin-wide framework for water resources management.

In the absence of effective international agreements, other measures need to be invoked. The ability of States to implement dam projects on shared rivers is often related to financial and technical support from external agencies and the effectiveness of public opinion in influencing public policy.

In this regard, countries fall into three broad categories:

- those with the financial and technical resources to be totally independent;
- those that require financial or technical support for a significant proportion of the project itself; and
- those that may be capable of undertaking the project independently, but rely on external support for other projects and programmes in the same sector.

In addition to the application of legal principles, external financing agencies have influenced and can continue to influence countries in the second and third categories. In the first category, active networking across borders can inform public opinion and encourage moves towards a policy of co-operation.

Elaboration of Policy Principles

7.1 National water policies make specific provision for basin agreements in shared river basins. Agreements are negotiated on the basis of good faith among riparian States. They are based on principles of *equitable and reasonable utilisation, no significant harm, prior information* and the Commission's strategic priorities.

The approach adopted by a national or sub-national authority towards shared rivers may have a direct and significant influence on other water resource management strategies. The principles embodied in the 1997 UN Convention on the Law of the Non-

Navigational Uses of International Watercourses warrant support. States should make every effort to ratify the Convention and bring it into force. Where there are obstacles to endorsing the Convention, the following key principles it embodies can still provide a framework for further dialogue between riparian States:

- '*equitable and reasonable utilisation*' that promotes the optimal sustainable use of the river, taking into account the interests of other riparians;
- '*no significant harm*' to other riparians and compensation or mitigation for any harm caused; and
- '*prior information*' referring to the need to inform other riparian States on planned measures that may have a significant effect on them.

The meaning of these terms is still evolving. Particularly, the application of the principle of 'no significant harm' will often conflict at a basic level with many applications of the principle of 'equitable and reasonable utilisation'. These interactions have not fully been resolved legally or customarily, suggesting that in their application, these principles should be read alongside the Commission's strategic priorities when planning future water resources and hydropower developments.

The Commission's message is grounded on the need to obtain the consent and co-operation of riparian States in the management of shared water resources including both consumptive and non-consumptive uses. In addition to having ratified international agreements, individual States should specifically address shared river basins in their water policy or legislation, providing clarity on their intention to co-operate in water resources management. For example, this intention is reflected in the South African National Water Act. Such provisions would provide a common basis for riparian States to move towards more integrated management agreements for shared watercourses. On the basis of these agreements, the affected States can adopt a progressive approach to institutional development, starting with exchange of information, joint scientific teams to analyse data, and joint arrangements for monitoring the implementation of agreements.

The Commission's message is grounded on the need to obtain the consent and co-operation of riparian States in the management of shared water resources.

Success with a progressive approach like this will provide the confidence to embark on co-operation in other areas. Internationally there is significant interest in transboundary issues and external financing agencies have expressed interest in supporting riparian States that agree on a common approach. Such financing should be provided within a broad framework of co-operation, rather than as a specific project-related component.

7.2 Riparian States go beyond looking at water as a finite commodity to be divided and embrace an approach that equitably allocates not the water, but the benefits that can be derived from it. Where appropriate, negotiations include benefits outside the river basin and other aspects of mutual interest.

Innovative solutions are needed to solve apparently intractable problems. Often, negotiations over shared rivers have developed into disputes over allocating what may appear as an insufficient resource. A more equitable and sustainable resolution may be possible by shifting from a primary focus on the allocation of the water resource, to a focus on the benefits that derive from the use of the water, encompassing consideration of wider development objectives and the options available to meet them. This shift provides an opportunity to look more constructively at alternative programmes for meeting development objectives.

It is possible to expand the horizon of negotiations further to include other issues that optimise the comparative advantages of two or more States. Such synergies may result from differences in location, climate or resource endowment. To some extent, the Helsinki Rules began this shift. In the wider negotiation arena, the principles of sharing benefits can include an array of other resources, including co-operation in other sectors, or financial payments.

A more equitable and sustainable resolution may be possible by shifting from a primary focus on the allocation of the water resource, to a focus on the benefits that derive from the use of the water.

An approach centred on wider development objectives creates a link between discussions at transboundary level and strategic planning processes within countries that can be used to define needs more clearly and map out a wider range of alternatives. Early engagement can avoid disputes becoming polarised around a specific project proposal and entrenching negotiating positions.

7.3 Dams on shared rivers are not built in cases where riparian States raise an objection that is upheld by an independent panel. Intractable disputes between countries are resolved through various means of dispute resolution including, in the last instance, the International Court of Justice.

Openness and information sharing is a key first step in any transboundary water sharing situation. From this can follow an independent and objective assessment of the consequences and impacts of any proposed intervention. Conducted in a manner consistent with openness and information sharing, a competent, independent entity acceptable to all riparian States should conduct strategic and project-related impact assessment studies following the practice outlined in Chapter 9.

The level and intensity of impact assessments will depend on the planning stage, but in all cases should include environmental, social, health and cultural heritage assessments. Safeguards are needed to ensure independence in directing and financing the assessment team. Mechanisms such as establishing a joint trust fund for implementing the assessments should be considered. The impact assessments should be seen as part of the joint institutional strengthening activities of riparian States to provide a common, interactive approach and a sound basis for political dialogue.

Where disputes cannot be resolved, an independent panel should be established that goes beyond the remit of the impact assessment. The creation and operation of such a panel is defined in the 1997 UN Convention on the Law of the Non-Navigational Uses of International Watercourses. Good faith negotiations may of themselves lead to mutually agreeable outcomes. In the event that disputes remain, and if parties do not have recourse to dispute resolution through international, regional or bilateral agreements, the affected parties could refer the matter to the International Court of Justice (ICJ), either by mutual agreement, or directly if both parties have previously submitted to the compulsory jurisdiction of the ICJ by declaration under Article 36 of the Statute of the Court.

These policy principles apply to all watercourses including tributaries. Their application to tributary rivers is a location specific issue dependent on the significance and relationship of the tributary to the main river and the interests of riparian States. In cases where countries have already reached basin level agreements for the main stem river, such agreements should be extended to significant tributary rivers taking into consideration the Commission's strategic priorities. Where no agreements exist, riparian States should adopt an integrated approach looking at the management of the main-stem and its tributaries within the context of the entire river basin.

7.4 For the development of projects on rivers shared between political units within countries, the necessary legislative provision is made at national and sub-national levels to embody the Commission's strategic priorities of 'gaining public acceptance', 'recognising entitlements' and 'sustaining rivers and livelihoods'.

The Commission's seven strategic priorities are as relevant to rivers shared across sub-national boundaries as they are to those wholly within one province or those shared between States. Where their legal system provides for national-level control of water issues, States should embody the principles of shared waters in national water legislation and develop a locally appropriate policy framework for addressing such issues. In other cases, States at a federal level should explore mechanisms to encourage good practice and provide incentives for compliance with the strategic priorities.

Although many sub-national governments have a mandate to manage water, a national or federal government may exercise sanctions through legislative and regulatory mechanisms and provide financial incentives. In many cases dam development and operation depend on project licences issued by national and sub-national authorities, or a mix of the two. Dam projects may also be subject to clearance from national and sub-national environmental agencies. Both licences and environmental clearance can be used to ensure that accepted environmental and social provisions are met. Where federal funds are sought for project assistance, they should be made conditional on compliance with the Commission's strategic priorities. As part of their overall responsibility, States should be more pro-active earlier in the planning process to facilitate resolution of disputes between riparian provinces.

States should embody the principles of shared waters in national water legislation and develop a locally appropriate policy framework for addressing such issues.

7.5 Where a government agency plans or facilitates the construction of a dam on a shared river in contravention of the principle of good faith negotiations between riparians, external financing bodies withdraw their support for projects and programmes promoted by that agency.

The international community needs to take a strong and concerted stand in the case of shared rivers. While the decision to build a dam is often considered a sovereign decision, the decision of external agencies to support a dam depends on whether the proposed project complies with that agency's policies and guidelines. It is therefore of concern that bilateral, multilateral and export credit agencies have not yet harmonised their policies towards shared watercourses. These often-inconsistent policies make it more difficult to improve the way transboundary issues are handled. The complexity of the situation is increased by the disparate and fluid nature of financial support. This inconsistency often results in situations where, although an external agency may not be directly financing a dam

on a shared watercourse, its support for other projects in the same sector allows national resources to be allocated for this purpose.

It is essential that external agencies harmonise their policies towards shared waters and deal with the sector as a whole rather than with specific projects. Such policies should incorporate aspects of notification to riparian States, the desirability of 'consent' or 'no objection' from riparian States, and independent expert assessment of social and environmental impacts.

In the absence of agreement among riparian States, external agencies should make their involvement conditional on the findings of an independent commission as envisaged under the 1997 UN Convention on the Law of the Non-Navigational Uses of International Watercourses or other appropriate mechanisms agreeable to all parties. In cases where States proceed with projects in the absence of such a commission, or reject its findings, the external financing agency should withdraw its support from the sector concerned.

Endnotes

1 For example in many Latin American countries (Argentina, Bolivia, Brazil, Chile, Colombia and Ecuador) and in Australia, Canada, India, New Zealand, and the Philippines national laws recognise indigenous people's rights. Specific legislation on indigenous people's rights includes the Philippines Indigenous Peoples Rights Act (1997), Australian Aboriginal Land Rights (Northern Territory) Act 1976 and Chile's Indigenous Law 1993. Recent Canadian practice promotes the need for free, prior informed consent of indigenous and tribal peoples to projects in certain contexts. Goldzimer, 2000, soc013, WCD submission; Hart, pers. comm. 2000; IDS, 2000.

2 See endnote 1

3 Colchester, 1993, 1995; Tenant, 1994; Gray, 1995; Kingsbury, 1995; Fisher, 1993; Maybury-Lewis, 1996; Daes, 1996a, p72; Pritchard, 1998a, p44; Pritchard, 1998b, p61. Some Asian and African countries already accept that the term 'indigenous peoples' applies to the 'tribal peoples' and 'cultural minorities' within their borders, including Cambodia, Botswana, Nepal and Philippines. In a landmark case for the Ainu of Japan, on 28 March 1997 a local court in Sapporo, Hokkaido, recognised the Ainu as an indigenous and minority people. The Asian Development Bank has also adopted a policy on 'indigenous peoples' to guide its operations. For a more nuanced discussion of the applicability of the concept of indigenous peoples to Asia, see Kingsbury, 1998.

4 Daes1996b; World Bank, 1991.

5 Nepal: Modifiying the intake, installing an extra desander, dredging the forebay and refurbishing the generators/turbines and power house control systems at the Trushuli-Devighat hydropower station in 1995 (originally built in the 1970's) improved average annual power generation by 46%. NEA 1997.

Lao PDR: Nam Ngum is a 150 MW hydropower plant near Vientiane, the capital of Lao PDR. Surplus power is exported to Thailand and the power sales agreement between the two countries is revised every 4 years. Based on updated hydrology and other variables, studies were undertaken to optimise the monthly and day-to-day operation of reservoirs and turbines. Assuming the same average tariff, revenue from power sales to Thailand increased by about 10%. Lahmeyer International, 1990.

6 Johnston, 2000.

7 Acreman et al, 2000; WCD Thematic II.1 Ecosystems.

8 465 cases in the United States, 3 in France, 1 in Norway, a few in Canada.

9 Revenga et al, 2000.

10 Many countries are in the early stages of developing EFRs, for example adopting and implementing relevant legislation, making enquiries into available environmental flow methodologies and developing policy, but are not necessarily at the stage of formally undertaking EFRs or determining EFRs for freshwater systems. At least 29 countries have used EFRs including: Australia, Austria, Canada, Denmark, Finland, France, Germany, Hungary, Indonesia, Ireland, Italy, Japan, Korea, Lesotho, Moldavia, Namibia, New Zealand, Norway, Portugal, Puerto Rico, South Africa, Spain, Swaziland, Sweden, Switzerland, Taiwan, The Czech Republic, The Netherlands, The Ukraine, The United Kingdom, The United States. The following countries are exploring the use of EFRs: Angola, Botswana, Brazil, Chile, Israel, Mexico, Mozambique, Nepal, Thailand, Vietnam, Zimbabwe. Tharme, 2000.

11 Benefit sharing through Equity Shares-Minashtuk Project in Canada. The Minashtuk project in Canada illustrates benefit sharing through a limited partnership company. Here the Band Council of the Montagnais of Lac Saint-Jean is the majority shareholder with a 50.1% share and Hydro Quebec owns the remaining 49.9% of the shares. Minashtuk is the first project developed by Hydro-Innu. Besides guaranteeing a share in profits, it allows the Montagnais to design a project according to their priorities and in the long-term reinvest the profits in a manner that supports the economic development of their community. Milewski et al 1999, soc196, WCD Submission.

12 ISO 14001 is an internationally recognised standard that identifies the basic processes of an environmental management system (EMS). It allows an organisation to effectively identify, monitor and control its environmental impacts. It does not specify performance requirements, beyond a commitment to comply with applicable regulations and other commitments. The standard enables third-party certification for organisations that conform to its specifications. While there are many models for EMS design, the ISO 14001 standard is emerging as the dominant standard, with over 17,000 certificates granted as of June 2000. Concerns do, however, exist whether ISO standards can assure changed performance effectively. Corbett and Kirsch, 200.

13 'Rivers' is used here as a general term. The strategic priority and policy principles relate equally to all types of waters which are or might be impacted by dams.

14 The term 'riparian State' is used to mean any State through which a transboundary river flows or forms part of its boundary, or which includes part of the catchment area of a transboundary river.

U14

Chapter 9:

Criteria and Guidelines – Applying the Strategic Priorities

Many technical guidelines for the construction of dams and their associated infrastructure have been developed by professional technical networks to ensure high engineering and quality standards. What is lacking is a comprehensive and integrated framework for decision-making on the provision of water and energy services. The Commission's contribution is to provide a framework that emphasises a structured process incorporating the full range of social, environmental, technical, economic and financial criteria and standards.

The framework is built upon the seven strategic priorities described in Chapter 8 and derives its strength from recognising the rights and assessing the risks of all stakeholders in the process.

Social, environmental, governance and compliance aspects have been undervalued in the past. It is here that applying the Commission's proposals will make a difference. The framework identifies key decision points and incorporates associated criteria that translate the Commission's policy principles into a programme for implementation. Within this framework the Commission proposes a set of *guidelines* firmly anchored in examples of good practice from the Knowledge Base to describe how its policy principles can be realised. These guidelines add to existing decision-support instruments and should be incorporated by governments, professional organisations, financing agencies, civil society and others as they continue to improve their own relevant guidelines and policies over time (see Figure 9.1).

This chapter shows how implementing a decision-making process based on the Commission approach will safeguard rights, reduce the risk of conflicts emerging, and lower overall costs. The framework provides the opportunity for agencies and communities to screen out unfavourable alternatives at an early stage. It is intended to open channels of dialogue between stakeholders, increase mutual understanding and help decision-makers, practitioners and affected people assess whether needs have been adequately addressed. Proposals for dam projects that emerge from the screening process will have greater public acceptance.

Figure 9.1 WCD Criteria and Guidelines strengthen other decision support instruments

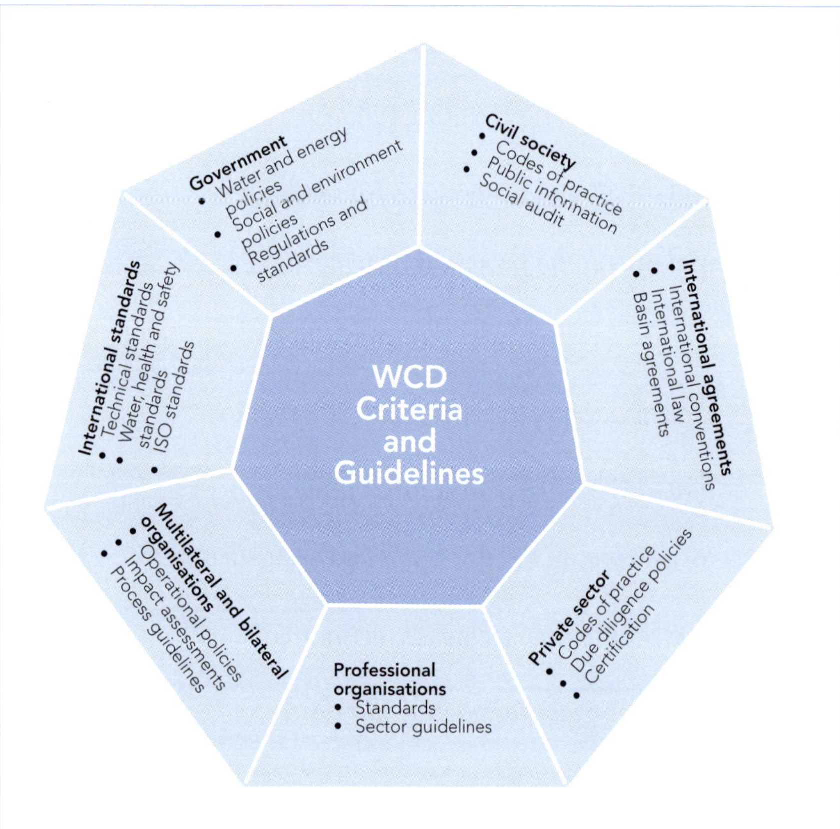

Turning the strategic priorities and their underlying policy principles into reality requires a new focus for planning and management in the water and energy sectors. This chapter concentrates on what needs to change in the way water and energy management plans are developed and projects are designed and implemented. Bringing about this change will require:

- planners to identify stakeholders through a process that recognises rights and assesses risks;
- States to invest more at an earlier stage to screen out inappropriate projects and facilitate integration across

sectors within the context of the river basin;

- consultants and agencies to ensure outcomes from feasibility studies are socially and environmentally acceptable;
- the promotion of open and meaningful participation at all stages of planning and implementation, leading to negotiated outcomes;
- developers to accept accountability through contractual commitments for effectively mitigating social and environmental impacts;
- improving compliance through independent review; and
- dam owners to apply lessons learned from past experiences through regular monitoring and adapting to changing needs and contexts.

The changes will involve reform of existing planning processes and an emphasis on the key stages where decision-makers and stakeholder groups can verify compliance. Among the multitude of decisions to be taken, five critical decision points have been identified as having a particularly strong influence on the final outcome. The first two relate to water and energy planning, leading to decisions on a preferred development plan.

1. Needs assessment: validating the needs for water and energy services.
2. Selecting alternatives: identifying the preferred development plan from among the full range of options.

Where a dam emerges from this process as a preferred development alternative, three further critical decision points occur.

3. Project preparation: verifying that agreements are in place before tender of the construction contract.
4. Project implementation: confirming compliance before commissioning.
5. Project operation: adapting to changing contexts.

The contractual steps of signing agreements and issuing licences are located within this overall framework. Although in these latter stages the Commission has focused on issues related to dam projects, the principles and general guidance presented here are also relevant to non-dam options for water and energy services.

The five decision points are supported by a set of *key criteria* that describe the processes required for compliance. The criteria are presented in the form of checklists for each decision point that provide a clear and open mechanism for determining whether the Commission's recommendations have been followed and the process can proceed to the next stage of planning or implementation.

The criteria cover the full planning and project cycles and include aspects related to existing dams. There are also many dams currently in the planning, design, or construction stage. These 'dams in the pipeline' should also be assessed to identify improvements that can be made. Practical steps are proposed to determine the extent to which such current projects comply with the Commission's recommendations, and to identify how any needed adjustments can be made.

Five critical decision points have been identified as having a particularly strong influence on the final outcome. They are supported by a set of key criteria that describe the processes required for compliance.

Recognising that guidelines are available from other sources, the Commission focused principally on what needs to be done differently. Introduction of a new decision-making framework through application of the criteria and guidelines will add significant value to existing processes and form a basis for good practice in water and energy resource development. Taken together, the criteria and guidelines will improve development outcomes and minimise problems encountered in the past.

Five Key Decision Points: The WCD Criteria

As noted, the Commission has identified five key stages and associated decision points for the energy and water sector. The most fundamental of these is selection of the preferred development plan. This determines what options will be pursued to meet needs and whether or not a dam is to be built. This decision is only taken after needs and the available options to meet those needs have been fully assessed. Each of the five stages requires a commitment to agreed procedures culminating in a decision point that governs the course of future action and allocation of resources (see Figure 9.2). At each decision point it is essential to test compliance with preceding processes before giving authority to proceed to the next stage. These points are not exhaustive, and within each stage many other decisions are taken and agreements reached. The five key stages and associated decision points are generic and need to be interpreted within the overall planning contexts of individual countries.

1. *Needs assessment: validating the needs for water and energy services*. Confirmation is required that plans for water and energy development reflect local and national needs adequately. An appropriate decentralised consultation process is used to validate the needs assessment and modify it where necessary.

2. *Selecting alternatives: identifying the preferred development plan from among the full range of options*. The preferred development plan is selected through a participatory multi-criteria assessment that gives the same significance to social and environmental aspects as to technical, economic and financial aspects and covers the full range of policy, programme, and project options. Within this process, investigations and studies are commissioned on individual options to inform decision-making as required; for example, demand-side management studies or feasibility studies.

Where a dam emerges as a preferred option, the following key decision points occur for project preparation, implementation and operation.

3. *Project preparation: verifying agreements are in place before tender of the construction contract*. The preparation stage covers detailed planning and design. Licences issued for development of a project incorporate any conditions that emerge from the options assessment process. Tendering the construction contract is conditional upon reaching negotiated agreements for benefit-sharing mechanisms and for mitigation, compensation, development and compliance measures, in addition to technical requirements.

4. *Project implementation: confirming compliance before commissioning*. The implementation stage covers procurement and construction. Issuing the licence to operate is contingent on implementation of specific benefit sharing and mitigation measures at various stages through the implementation period. Compliance with all relevant time-bound commitments is required before commissioning the project.

5. *Project operation: adapting to changing contexts*. Any decisions to modify facilities, operating rules, and licence conditions to meet changing contexts are based on a participatory review of project performance and impacts.

The five key stages and decision points provide a framework within which decision-makers and stakeholder groups can be assured of compliance with agreed procedures and commitments. The benefits of this approach include lowering risks to livelihoods and cost escalation, reducing the number of disputes, and encouraging local ownership. In the short term, additional financial resources for needs and options assessment will be required to achieve compliance with the Commission's policy principles, and efforts will be required to strengthen institutional capacity. In the longer term, the potential exists for major cost savings and increased benefits.

The remainder of this chapter describes each of the five key decision points and provides a related list of criteria for checking compliance. Stakeholder involvement is central to these processes. The composition of a forum of stakeholder groups associated with each stage is different and evolves throughout the process (see Guideline 1: *Stakeholder Analysis*).

Figure 9.2 Five key decision points in planning and project development

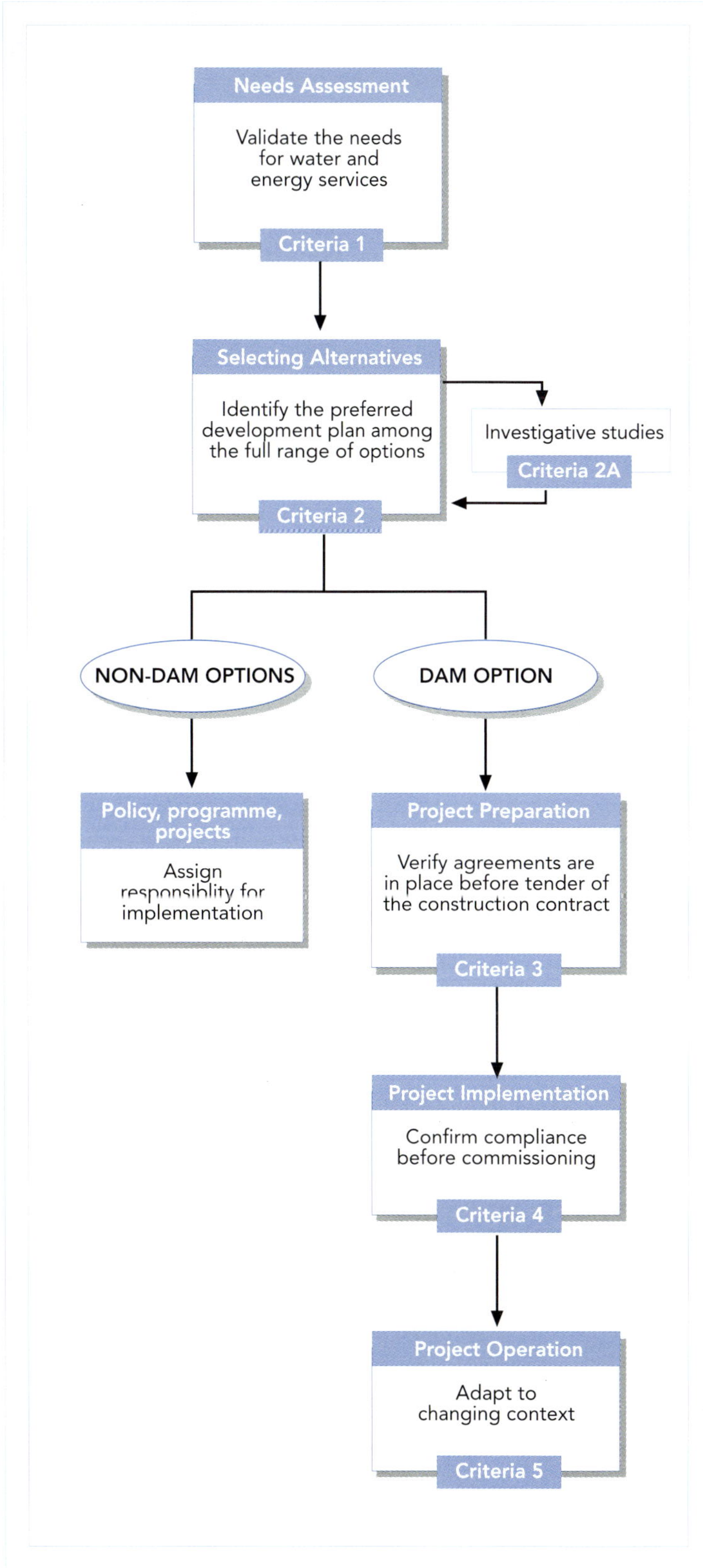

Stage 1: Needs assessment: validating the needs for water and energy services

Intended outcome

A clear statement of water and energy services needs at local, regional, and national levels that reflects decentralised assessments and broader national development goals. An assessment based on participatory methods appropriate to the local context resulting in a clear set of *development objectives* that guide the subsequent assessment of options.

Determining needs and setting priorities between and within sectors are continuous processes specific to individual countries. The Commission's policy principles describe characteristics that should be reflected in such processes and define a shift in emphasis to more open and inclusive procedures. Country-level responses are required to ensure that priority setting embodies a fair representation of basic water and energy needs and provides the appropriate balance between local and national demands.

A country's policy framework for water, energy, social issues and the environment guides the whole planning process. The open and participatory approach to needs and options assessment envisaged by the Commission requires a review of these policies to identify and address elements that may hinder its implementation.

The primary influence defining a country's portfolio of development activities, is a set of development objectives that may be embodied in a five-year plan or in a regulatory planning framework. Ensuring that the outcome of the needs assessment for water and energy services reflects stakeholder priorities requires an entry point early in the planning process. The Commission proposes a validation process to confirm the setting of priorities and the formulation of development objectives.

An open and decentralised planning process provides opportunities for public scrutiny. In situations where this has not been implemented, a programme of national and sub-national public hearings, targeted community consultations, and field surveys can assess the validity of the needs assessment. The subject of the consultations may relate to an overall development strategy, master plan, sector plan, or basin plan, and the breadth of consultation can be determined accordingly. The WCD Knowledge Base demonstrates the need for such a review, since plans are often narrowly focused, reflect social and environmental impacts inadequately, are weak in identifying affected groups, and fail to deal adequately with the distribution of costs and benefits.

In countries where a large proportion of the population does not have access to basic services, a key parameter in the validation process should be the extent to which basic human needs will be met. To ensure that these needs are given prominence, the process of validation should empower those who are least able to influence planning systems.

Responsibility for this validation process rests with the State. Independent facilitation for consultations and surveys and the presence of civil society groups will enhance confidence that the needs of disadvantaged groups are being considered. Briefing materials, records of meetings, and results of the overall process should be available in appropriate languages. If the development objectives are not confirmed by the validation exercise, they should be reviewed and updated using processes consistent with the Commission's policy principles.

Stage 1 Criteria Checklist

NEEDS ASSESSMENT	Selecting alternatives	Project Preparation	Project Implementation	Project Operation

Needs assessments may have been conducted through a range of processes including national, regional, sector-specific, or basin-wide plans. The verification process to be applied will need to be tailored to suit the particular circumstances.

Gaining Public Acceptance

- A consultation plan was developed using a stakeholder analysis to define the groups involved. The plan defines mechanisms for verifying needs at the local, sub-national and national level (Guideline 1).
- Verification of the needs for water and energy services was achieved through a process of public consultation and the results of public consultation were disseminated to stakeholders.
- Development objectives reflect a river-basin-wide understanding of relevant social, economic, and environmental values, requirements, functions, and impacts that identifies synergies and potential areas of conflict.
- An appropriate process was established to address any disparities between the needs expressed through the public consultations and the stated development objectives.

Comprehensive Options Assessment

- Legal, policy and institutional frameworks were reviewed and any bias against resource conservation, efficiency and decentralised options, and any provisions that hindered an open and participatory assessment of needs and options were addressed.

Addressing Existing Dams

- Outstanding social and environmental impacts from past projects were evaluated and incorporated into the needs assessment (see Chapter 8, policy principle 3.3).

Sustaining Rivers and Livelihoods

- Ecosystem baseline studies and maintenance needs were assessed at a strategic level (Guidelines 14, 15)

Stage 2: Selecting alternatives: identifying the preferred development plan

Intended outcome

A mix of alternatives that reflects the needs and meets the *development objectives* has been selected through a multi-criteria assessment of the full range of policy, programme, and project alternatives and included in a preferred *development plan.*

A major concern about past projects is that too limited a range of options was considered before deciding to construct a dam. To address this, the inventory of options must respond to the agreed development objectives (Stage 1) and explicitly identify the beneficiary groups. The inventory of options needs to be sufficiently diverse in terms of policy, programme, and project alternatives, project scale, and geographic coverage.

Strategic impact assessments provide an initial level of screening to remove alternatives that have unacceptable social and environmental consequences. They need to reflect the importance of avoiding adverse impacts and the precautionary approach. The assignment of relative weights to designate the importance of various parameters should be a participatory process and form the basis of a multi-criteria analysis to screen and rank alternatives. For example, the gestation period in delivering benefits, the scale of adverse impacts and costs are all key considerations.

The information available on each option will not be at the same level of detail. Decisions need to be taken during the screening process on whether to commission further investigations and studies on individual alternatives, while at the same time not jeopardising or delaying alternatives that can deliver benefits within the short-term.

Studies are required to assess the extent to which policy and programme options can meet the development objectives. The policy principles cover a number of such areas, including:

- optimising existing investments by increasing operational efficiency and improving productivity;
- demand-side management assessment;
- decentralised supply options and community-level initiatives; and
- policy and institutional reforms.

Some options will need reconnaissance, pre-feasibility, and feasibility studies appropriate to the stage in the process and incorporating social and environmental surveys and impact assessments. The decision to allocate financial resources to such studies should be taken within the overall context of the options assessment process. For example, it may have been agreed that detailed investigations of supply-side approaches should await the outcome of demand-side studies that could influence the scale of any project intervention.

Criteria relevant to project-related studies are described as a subset of the process of selecting alternatives (see Stage 2A). The findings of the studies are fed back into the screening process for consideration with all other remaining options. This approach deviates from existing practice in many countries by cutting the direct link between the feasibility study and project approval. It encourages a broader consideration of all feasible options in setting priorities.

Stage 2 Criteria Checklist

Needs Assessment	SELECTING ALTERNATIVES	Project Preparation	Project Implementation	Project Operation

Gaining Public Acceptance

- Stakeholders participated in creating the inventory of options, assessing options, and in negotiating those outcomes that may affect them (Guidelines 1, 2).
- An agreed dispute resolution mechanism for negotiated processes was established with the participation and agreement of stakeholders (Guideline 2).
- Indigenous and tribal peoples gave their free, prior and informed consent to the inclusion in the development plan of any planned option that would potentially affect them (Guideline 3).

Comprehensive Options Assessment

- Strategic impact assessments and life cycle analysis were integrated and undertaken as an initial step in the process (Guidelines 4, 7, 8, 14, 17).
- A multi-criteria assessment was used to screen and select preferred options from the full range of identified alternatives (Guideline 6).
- The screening of options:
  - covered all policy, programme, and project alternatives;
  - gave social and environmental aspects the same significance as technical, economic and financial factors;
  - gave demand-side options the same significance as supply options;
  - prioritised consideration of improving performance of existing systems;
  - considered river-basin-wide aspects and cumulative impacts;
  - took account of potential changes in climate; and
  - reflected the precautionary approach.
- Distributional and risk analyses were conducted at an appropriate level (Guidelines 9, 11) and environmental and social impacts were valued where appropriate (Guideline 10).
- Approval to proceed with any project-level investigations was informed by a comprehensive assessment of options (see Criteria Checklist 2A).
- Rejection of any options was explained in an open and timely manner.

Addressing Existing Dams

- Provisions were made for resolving outstanding social and environmental impacts (see Chapter 8, policy principle 3.3)

Sustaining Rivers and Livelihoods

- An established policy exists to maintain selected rivers with high ecosystem functions and values in their natural state.
- Consideration of options took into account: avoiding dams on the main-stem of rivers wherever possible; avoiding or minimising negative impacts on endangered species, ecosystems, livelihoods, human health and cultural resources; and respecting the provisions and guidance of relevant international treaties.

Recognising Entitlements and Sharing Benefits

- For any project option, stakeholders negotiated the guiding principles and criteria for: benefit-sharing, mitigation, resettlement, development and compensation measures (Guidelines 2, 18, 20).

Ensuring Compliance

- Sufficient institutional capacity exists, or will be enhanced, to monitor and enforce commitments for social and environmental components.

Sharing Rivers for Peace, Development and Security

- Any objections from riparian states were resolved through good faith negotiations or independent dispute resolution procedures (Guideline 26).

Stage 2A: Investigative studies

The authority to proceed with a preparatory study for a dam project should not be a signal that the project will be implemented. Rather, such a decision should be integrated into the overall options assessment process. This will provide a break in the traditional planning cycle to encourage more open decision-making. Viewing project options within the overall framework of options assessment also facilitates the rejection of projects that fail to meet social and environmental objectives in favour of better alternatives. The WCD Knowledge Base has demonstrated that more rigorous estimates of project costs are also required as a part of such studies, and the risk of cost overruns must be fully considered in the assessment process.

Meaningful participation in preparatory studies is central to the success of the investigation and the ultimate outcome. Careful analysis to recognise the rights and assess the risks of stakeholder groups is essential. A forum of stakeholder groups needs to be identified based on project boundaries. Agreement on the participatory elements of the studies should be formalised in a consultation plan.

The strategic impact assessment undertaken early in the options assessment will have outlined the key unknowns and the areas to be investigated across all sectors. The issues will be defined in more detail in the scoping stage of project-related impact assessments. On this basis, terms of reference and work plans for the diverse range of sector specialists can be integrated. Project-related impact assessments have to go beyond environmental and social impact assessments to include health and cultural impacts. To be effective, they require an improved level of baseline studies.

Preliminary negotiations with project-affected people, their community representatives, and other stakeholders are central to the preparatory studies in considering mitigation measures for any unavoidable adverse impacts and investigating benefit-sharing plans. By the time a study reaches feasibility stage, the scope of such measures should be clearly defined in order to reduce the likelihood of protracted negotiations and a breakdown of discussions later in the process. For the proposed project to be part of a preferred development plan, the acceptance of the project affected people and the prior informed consent of indigenous peoples should be obtained.

Ultimately the results of the study, including any outstanding issues, will be fed back into the screening and ranking exercise (see Criteria Checklist 2) for comparison with remaining alternatives prior to any decision to proceed further with detailed project development. The following plans, with indicative budgets, need to be developed as a minimum requirement to act as a foundation for any further project planning:

- an outline environmental management programme, including provision for an environmental flow to maintain downstream ecosystems;
- an outline social mitigation, resettlement, and development plan; and
- an outline monitoring plan, including outcome-based indicators.

A compliance plan will be required to cover these aspects and other regulatory requirements throughout subsequent stages of project planning, development, and operation.

Stage 2A Criteria Checklist

Needs Assessment	**SELECTING ALTERNATIVES**	Project Preparation	Project Implementation	Project Operation

Project-related pre-feasibility and feasibility studies need to meet the following criteria. Policy and programme related studies may also be required, and are covered in Criteria Checklist 2.

Gaining Public Acceptance

- Stakeholders participated in baseline, impact and investigative studies and the negotiation of outcomes that potentially affect them (Guidelines 1, 2, 14, 17).
- The studies and impact assessments were open and independent, and were preceded by a participatory scoping phase (Guideline 5).

Comprehensive Options Assessment

- The investigations were analysed on a river-basin-wide understanding of social, economic, and environmental values, requirements, functions, and impacts including cumulative impacts, and the precautionary approach was applied. (see Guideline 5).
- The recommendations of studies undertaken on resource conservation measures, demand-side management, local supply-side options and improvement of existing systems were reflected in the demand forecast for the sector.
- Within-project alternatives were assessed using a multi-criteria approach (Guideline 6).

Addressing Existing Dams

- Studies examined possible synergies from interactive operation of related water resource infrastructure in the basin.

Sustaining Rivers and Livelihoods

- An environmental flow requirement to maintain downstream species, ecosystems and livelihoods was defined (Guideline 15).
- Impacts on fish have been assessed and measures to avoid or minimise impacts were considered, including an effective fish pass where feasible (Guideline 16).

Recognising Entitlements and Sharing Benefits

- Stakeholders negotiated agreements for compensation, mitigation, resettlement, development and monitoring measures affecting them, including draft contracts where necessary (Guideline 19).
- Effective benefit-sharing strategies were identified and agreed with people adversely affected by the project (Guideline 20).

Ensuring Compliance

- Institutional capacity to monitor and enforce commitments for social and environmental components of the project was analysed and measures to strengthen capacity identified.
- An independent panel reviewed the assessment of impacts and the planning of social and environmental mitigation plans (Guideline 22).

Sharing Rivers for Peace Development and Security

- Riparian states were notified of options affecting them and agreed procedures for impact assessments. Objections were addressed through good faith negotiations and agreed dispute resolution procedures (Guideline 26).

Stage 3: Project preparation: verifying commitments are in place before tender of the construction contract

Intended outcome

Clearance to tender the construction contract is given by the relevant authority and includes conditions for the award of the contract and operations. Mitigation and monitoring measures are formalised into contracts between responsible parties, and compliance arrangements are in place.

The Commission considers that all large dams should have time-bound licences. Where a large dam emerges as a preferred option, a licence for project development should be issued to the developer by the appropriate regulatory agency. Project preparation continues with detailed planning and design stages, including drafting of tender documents and plans for benefit-sharing and mitigation. Adaptation of the criteria described here may be necessary where the design and construction are part of a single package.

Negotiations with all project-affected people, their community representatives, and other stakeholders will continue in good faith based on the outline agreements reached during the feasibility stage. They would cover all environmental and social plans; development programmes, including benefit-sharing plans; and construction-related issues. A clear agreement with the affected people on the sequence and stages of resettlement will be required before construction on any project preparatory works begins, such as access roads or river diversion works. In cases where these negotiations stall, an independent dispute resolution process is required. The negotiated agreements will result in signed contracts between the developer and affected communities and individuals, with clear targets for assessing compliance.

The responsibilities of the developer in relation to mitigation, development of affected communities, and benefit-sharing will be clearly reflected in the licence and the compliance plan. The operation phase should be contingent on compliance with specific commitments identified in the licence. Conditions for impoundment and commissioning should be explicitly stated.

If the tender cost estimate is substantially higher than the feasibility study estimate, the choice of project should be reviewed to see if it still meets the selection criteria. A similar check is required if needs have changed substantially since the outset, or if the project scope has changed materially. If the actual price of the lowest responsive tender exceeds agreed cost limits, procurement should be interrupted and options reviewed.

Social and environmental mitigation measures should be defined in the tender in similar detail to construction elements, namely the 'bill of quantities'. The tender should clearly identify responsibilities of the contractor, the developer, and the government in relation to:

- the environmental management plan;
- measures to mitigate adverse social impacts, including development opportunities for affected communities;
- access to and management of new resources in the reservoir;
- the construction method and schedule, and the construction camp;
- impact monitoring and reporting during the operations stage; and
- compliance instruments.

Stage 3 Criteria Checklist

Needs Assessment	Selecting Alternatives	**PROJECT PREPARATION**	Project Implementation	Project Operation

Gaining Public Acceptance

- Stakeholders participated in the project design and the negotiation of outcomes that affect them (Guidelines 1, 2).
- Indigenous and tribal peoples gave their free, prior, and informed consent to the project as designed (Guideline 3).

Comprehensive Options Assessment

- The stakeholder forum participated in assessing alternatives for the detailed layout of the dam, associated infrastructure, and its operation.

Addressing Existing Dams

- Cumulative and interactive impacts of existing infrastructure were addressed in the design of the dam and agreements reached with stakeholders and operators to modify operating rules of existing dams where needed.

Sustaining Rivers and Livelihoods

- Acceptable rules were developed for reservoir filling, commissioning and operation.
- The final design includes provisions for emergency drawdown and decommissioning and is sufficiently flexible to accommodate changing future needs and values, including ecosystem needs and ecosystem restoration (Guideline 12).
- An environmental management plan incorporating environmental flows and other mitigation and enhancement measures was agreed with stakeholders and defines monitoring and evaluation programmes.
- The developer provided sufficient evidence to demonstrate that proposed mitigation and development measures will be effective in meeting their objectives.

Recognising Entitlements and Sharing Benefits

- Mitigation, resettlement, monitoring, and development plans were agreed with affected groups, and relevant contracts signed (Guideline 19).
- Detailed benefit sharing mechanisms, and the means to deliver them, have been agreed and set in place with affected groups (Guideline 20).

Ensuring Compliance

- Independent panels reviewed and endorsed mitigation plans (Guideline 22).
- Provisional sums for mitigation are included in the tender, and their financing has been confirmed.
- A Compliance Plan was prepared, presented to the stakeholder forum and formalised. Individual compliance measures include mechanisms for dispute resolution (Guideline 21).
- The developer has allocated funds for an effective monitoring and evaluation system covering project performance, safety and impacts. Institutional capacity exists to monitor and enforce agreements effectively.
- A transparent process for short-listing contractors and selecting tenders is in place and contractors with a record of under-performance or corruption on past projects were identified and debarred where appropriate.
- Relevant performance bonds have been secured, trust funds established and integrity pacts signed (Guidelines 23, 24, 25).
- The licence for project development defines the responsibility and mechanisms for financing decommissioning costs.

Sharing Rivers for Peace, Development and Security

- Resolution was achieved where affected riparian states had outstanding objections (Guideline 26).

Stage 4: Project implementation: confirming compliance before commissioning

Intended outcome

Clearance to commission the project is given by the relevant authority after all commitments are met. Relevant elements of performance bond sureties are released. The operating licence is confirmed, including specific requirements for monitoring, periodic review and adaptive management.

Issuing the licence to operate will be contingent on compliance with mitigation measures in addition to technical requirements. The licence will contain a number of conditions for the operation stage, including compliance with operating rules, public notifications, dam safety, monitoring and periodic review. All contingencies cannot be anticipated, and a balance should be struck on a case-by-case basis between the necessary level of certainty and sufficient flexibility to accommodate open, transparent, adaptive management.

Phasing of resettlement is required where the reservoir is being impounded as the dam is constructed. Special attention is necessary to ensure that compensation and development measures are in place well in advance.

Reservoir filling, commissioning of productive capacity and the initial years of operation are critical phases that require special attention, intensive monitoring and continued dialogue with stakeholder groups. Agreements on operating conditions are required for three stages of commissioning:

- reservoir filling;
- test operation; and
- initial operation.

Full compliance with technical, environmental and social measures must be achieved before the project is commissioned and enters full operation. This includes the broad commitments of the project developer as laid down in the project licence, compliance plan, and related agreements, as well as the commitments of the contractor acting as agent of the developer. Once the developer has met specified staged commitments, the associated financial sureties of any performance bond or outstanding contractual payments may be released.

Arrangements for public safety must be in place in order to warn the downstream population that sudden releases of water may occur and may be dangerous. Compensation should be paid for any loss of livelihood, such as the loss of fishing opportunities. If test operations cause downstream inconvenience, compensation should also be paid – for example, if a test takes place during the dry season and damages recession agriculture plots.

A range of agreements on initial and long-term operation should be incorporated in the operating licence and provisions for implementation verified. Licence periods should not normally exceed 30 years. They should include:

- agreement on environmental flow releases to the downstream river;
- releases to the downstream river for other functions (navigation, water supply, downstream irrigation);
- operating rules during normal and exceptional floods;
- procedures for opening spillway gates;
- monitoring and publication of relevant operation data; and
- periodic review of operating rules.

Stage 4 Criteria Checklist

Needs Assessment	Selecting Alternatives	Project Preparation	**PROJECT IMPLEMENTATION**	Project Operation

Gaining Public Acceptance

- Stakeholders participated in monitoring mitigation measures and in negotiating outcomes that affect them (Guidelines 1, 2).
- Consultation mechanisms were agreed in advance with stakeholders for any technical, social, environmental, or other problems that may be encountered during reservoir filling and commissioning.
- Contingency plans for emergency drawdown of the reservoir were agreed with stakeholders before commissioning and were widely disseminated.

Comprehensive Options Assessment

- Affected stakeholders have reviewed any changes proposed to the tender design that substantially affect impacts, mitigation measures, benefit sharing, operational practices, or the monitoring programme.

Addressing Existing Dams

- Institutional co-ordination mechanisms that recognise interactive effects and cumulative impacts are in place to adjust operation of existing dams.

Sustaining Rivers and Livelihoods

- Required environmental mitigation measures were implemented.

Recognising Entitlements and Sharing Benefits

- The mitigation, resettlement and development action plan has been implemented and disputes resolved (Guideline 19).

Ensuring Compliance

- An independent panel reviewed and endorsed implementation of social, environmental, health and cultural heritage mitigation measures (Guideline 22).
- Preparations have been made to implement licence conditions for operations, implement continuing mitigation measures, undertake monitoring and regular evaluation, and disseminate information.
- Monitoring of social, environmental and technical aspects includes an intensive phase to cover the rapid changes that occur in the impoundment and commissioning period.
- The developer has complied with pre-commissioning commitments as defined in the Compliance Plan (Guideline 21).

Sharing Rivers for Peace, Development and Security

- Mechanisms were initiated for sharing monitoring information with riparian provinces or States (Guideline 26).

Stage 5: Project operation: adapting to changing contexts

Intended outcome

Conditions for operating under the licence are fulfilled and the licence conditions modified as necessary to adapt to changing contexts. Monitoring programmes feed back into project operation. A process is initiated to decide on reparations, if necessary.

The objectives of dam management need to be transformed from purely technical to development-oriented goals that include social and environmental considerations. This has major repercussions for operation, monitoring, and evaluation of both existing dams and dams implemented in the future. Adaptive management is needed to continuously assess and adjust operational decisions within the changing context of environmental, social, physical and market conditions. This requires a close relationship between the local community, other stakeholders and dam owners and operators in order to minimise problems and quickly resolve any that do arise.

Monitoring programmes should:

- include a full range of technical, environmental, social, and economic parameters decided openly with the stakeholders;
- have an intensive phase in the first five years, or after a major change in operation;
- feed back into project operations; and
- be available to all stakeholders, perhaps in the form of an annual report.

A comprehensive project evaluation is required three to five years after commissioning and at regular intervals thereafter (every five to ten years is suggested). The evaluation is undertaken by the dam operator but is a stakeholder-driven process and may draw on the Commission's case study methodology.

The evaluations would be:

- comprehensive across all environmental, social, economic, and institutional impacts;
- integrated to cover the interactions between impacts;
- long-term to consider impacts over several decades; and
- cumulative to reflect impacts of other structural and non-structural measures in the basin.

In addition to periodic evaluations, parties other than the dam operator, or exogenous factors may prompt re-planning studies. For example, a strategic or sectoral impact assessment may indicate the need for a re-planning exercise for a group of projects. Licence conditions should make provision for such studies and for any required changes to operating conditions through good faith negotiations that recognise the rights of all parties. Any re-planning studies should be done on the basin level.

Licences should specify transparent processes for stakeholder participation in operations and procedural requirements for monitoring and evaluation, safety inspection, contingency planning and information disclosure. The regulatory authority or responsible government line agency must ensure compliance with provisions of the licence. Non-commercial aspects of the licence should be made public. Re-licensing procedures should examine present-day alternatives and be informed by an integrated review of project performance and impacts. Impact assessments should be undertaken for all major changes including decommissioning where dams are no longer required or are too expensive to maintain.

Stage 5 Criteria Checklist

Needs Assessment	Selecting Alternatives	Project Preparation	Project Implementation	**PROJECT OPERATION**

These criteria are relevant to both existing dams (Chapter 8, Strategic Priority No.3) and the operational stages of future dams.

Gaining Public Acceptance

- Stakeholders are identified for consideration of operational issues and any proposed changes that impact on them or the environment (Guideline 1).

Comprehensive Options Assessment

- Periodic evaluations of all aspects of project operation and performance are undertaken with the involvement of the stakeholder forum every 5 to 10 years and agreements renegotiated as necessary.
- Modernisation programmes and alternative operational regimes are considered as part of periodic reviews, replanning, or relicensing exercises through a participatory multi-criteria approach (Guideline 13).
- Monitoring and evaluation programmes should explicitly consider the influence of climate change (namely increasing and decreasing rainfall and flows) on benefits and dam safety.
- A full feasibility study, including analysis of alternatives and impact assessment, is undertaken for any proposal for any major physical change, including decommissioning.

Sustaining Rivers and Livelihoods

- Operations take account of environmental flow requirements (quantity and quality) and ecosystem and social impacts are monitored (Guideline 15).

Recognising Entitlements and sharing benefits

- Detailed benefit-sharing mechanisms are modified as necessary with the agreement of affected groups (Guideline 20).

Ensuring Compliance

- Adverse social and environmental impacts and reparations issues are referred to the appropriate recourse body (Guideline 19).
- Annual reports of project monitoring programmes, including social and environmental aspects, are issued promptly and corrective measures are initiated to address issues raised in the reports.
- The requirements of remaining performance bonds or trust funds outlined in the Compliance Plan are periodically reviewed, and financial guarantees are released on satisfactory compliance with agreed milestones (Guideline 23).
- Dam safety and inspection programmes are implemented.

Sharing Rivers for Peace, Development and Security

- Mechanisms exist to share monitoring information and resolve issues as they arise.

A Special Case: Dams in the Pipeline

The strategic priorities and policy principles outlined in Chapter 8 are as relevant to projects already at an advanced stage of planning and development as they are to the selection of a project in the earlier options assessment stage. Currently a large number of dam projects are at various stages of planning and development. While acknowledging that delays in implementation can cause unacceptable delays in delivering intended benefits, the WCD Knowledge Base has demonstrated that it is never too late to improve outcomes. On this basis, the Commission proposes an open and participatory review of ongoing and planned projects to ascertain the extent to which project formulation can be adapted to accommodate the principles outlined in this report.

The Commission's Knowledge Base has demonstrated that it is never too late to improve outcomes.

The essence of the process is that stakeholder groups should have an opportunity to define the scope of the review and to propose changes in keeping with the Commission's recommendations. The extent of any additional study or changes in project configuration would depend on the stage of planning, design, or construction and be determined from a synthesis of the stakeholder consultations and, where appropriate, an inter-ministerial review. General actions to guide the review for all projects would include:

- undertake a stakeholder analysis based on recognising rights and assessing risks to identify a stakeholder forum that is consulted on all issues affecting stakeholders;
- provide support to vulnerable and disadvantaged stakeholder groups to participate in an informed manner;
- undertake a distribution analysis to assess who shares the costs and benefits of the project;
- develop agreed mitigation and resettlement measures to promote development opportunities and benefit-sharing for displaced and affected people;
- avoid, through modified design, any severe and irreversible ecosystem impacts;
- provide for an environmental flow requirement and mitigate or compensate any unavoidable ecosystem impacts; and
- design and implement recourse and compliance mechanisms.

Governments may also use the review of 'dams in the pipeline' as an opportunity to compare the existing policy framework for planning and implementation of water and energy options with the criteria and guidelines proposed by the Commission. This can serve to launch a process of internal review and modification of existing policies and legislation, and reinforcement of appropriate capacity that will facilitate implementation of the Commission's recommendations in future.

At specific stages of planning and project development, regulators, developers and, where appropriate, financing agencies should ensure that the following project-related points are addressed:

Projects at feasibility stage

- The stakeholder forum confirmed that the set of options considered was appropriate, or identified other alternatives to consider as part of the project impact assessment.

- Any bias in selection of alternatives is removed or justified in a transparent fashion (for example subsidies to particular sub-sectors or groups).
- Demonstrated public acceptance exists for the recommended options.
- The assumptions underlying the economic, financial, and risk analysis are justified and subject to sensitivity analysis.
- Mechanisms for benefit-sharing are identified.
- An environmental flow requirement is determined.

Projects at detailed design stage

- The stakeholder forum is consulted on decisions related to project layout, operation and mitigation and development measures and relevant agreements are negotiated with affected groups.
- Environmental flow requirements are determined and incorporated into the design and operation rules.
- A Compliance Plan is prepared, and recourse mechanisms are identified.
- Compliance mechanisms are provided for in tender documents.
- Benefit-sharing contracts are negotiated for displaced and project affected people.
- A process for stakeholder involvement during operation is established.

Projects under construction

- The record of compliance is reviewed and a compliance plan is developed for remaining mitigation measures.
- Existing commitments for resettlement and benefit-sharing are converted into formal contracts.

- An adequate social, environmental and technical monitoring plan is financed by the developer.
- The operating rules and commissioning plan are agreed with a stakeholder forum.
- A comprehensive post project review is agreed for two to three years after commissioning, and every five to ten years thereafter.

Governments may also use the review of 'dams in the pipeline' as an opportunity to compare the existing policy framework for planning and implementation of water and energy options with the criteria and guidelines proposed by the Commission.

This process of review implies added investigations or commitments, the re-negotiation of contracts and the incorporation of a Compliance Plan. As in the case of initial planning, the additional financial costs incurred will be recouped in lower overall costs to the operator, to government, and to society in general as a consequence of avoiding negative outcomes and conflicts.

A Set of Guidelines for Good Practice

The guidelines outlined here describe in general terms how to assess options and plan and implement dam projects to meet the Commission's criteria. The 26 guidelines add to the wider range of technical, financial, economic, social and environmental guidelines. They are advisory tools to support decision-making and need to be considered within the framework of existing international guidance and current good practice. Further information is available on many of these aspects in the WCD Knowledge Base.

The guidelines are presented under the same sub headings as the Commission's seven strategic priorities. There are clear linkages between individual guidelines and cross references to them are given in the criteria checklists for the key decision points of the planning and project cycles.

Strategic Priority 1: Gaining Public Acceptance

1 Stakeholder Analysis
2 Negotiated Decision-Making Processes
3 Free, Prior and Informed Consent

Strategic Priority 2: Comprehensive Options Assessment

4 Strategic Impact Assessment for Environmental, Social, Health and Cultural Heritage Issues
5 Project-Level Impact Assessment for Environmental, Social, Health and Cultural Heritage Issues
6 Multi-Criteria Analysis
7 Life Cycle Assessment
8 Greenhouse Gas Emissions
9 Distributional Analysis of Projects
10 Valuation of Social and Environmental Impacts
11 Improving Economic Risk Assessment

Strategic Priority 3: Addressing Existing Dams

12 Ensuring Operating Rules Reflect Social and Environmental Concerns
13 Improving Reservoir Operations

Strategic Priority 4: Sustaining Rivers and Livelihoods

14 Baseline Ecosystem Surveys
15 Environmental Flow Assessment
16 Maintaining Productive Fisheries

Strategic Priority 5: Recognising Entitlements and Sharing Benefits

17 Baseline Social Conditions
18 Impoverishment Risk Analysis
19 Implementation of the Mitigation, Resettlement and Development Action Plan
20 Project Benefit-Sharing Mechanisms

Strategic Priority 6: Ensuring Compliance

21 Compliance Plans
22 Independent Review Panels for Social and Environmental Matters
23 Performance Bonds
24 Trust Funds
25 Integrity Pacts

Strategic Priority 7: Sharing Rivers for Peace, Development, and Security

26 Procedures for Shared Rivers

Gaining Public Acceptance

1. Stakeholder Analysis

The absolute value and the relative significance of 'stakes' vary, especially in what they represent for the interested party. Stakeholders have unequal power and this can affect their ability to participate in and influence decisions. A stakeholder analysis based on recognising rights and assessing risks should be used to identify key stakeholders for planned activities. The analysis will also seek to understand and address potential factors that may hinder their involvement. The analytic approach can involve stakeholder workshops, community-level surveys, key informant surveys, and literature review.

The stakeholder analysis leads to the constitution of a temporary stakeholder forum as a basis for participation and, where relevant, negotiation processes throughout the planning and project cycles. A stakeholder forum is a dynamic construct and will need to be applied to meet changing needs through the planning and project cycles beginning with needs assessment/verification and options assessment. The composition of a stakeholder forum, the level of representation of various interests, and the means of facilitating the process changes from stage to stage.

The stakeholder analysis will:

- *Recognise existing rights and those who hold them.* Those groups whose livelihoods, human rights and property and resource rights may be affected by an intervention are major rights holders and thus core stakeholders in a stakeholder forum within which negotiated outcomes should be achieved.
- *Identify those at risk* through vulnerability or risk analysis and consider them as core stakeholders, including those who face risk to their livelihoods, human rights, and property and resource rights. Special attention should be given to indigenous and tribal peoples, women and other vulnerable groups as they may face greater risks from development interventions (Guideline 3). In the case of a dam, the analysis should include those upstream, downstream and in the proposed reservoir area. Relevant civil society groups or scientists are included in a stakeholder forum to ensure that environmental risks, for which there may be no champion, are adequately reflected and discussed.
- *Identify constraints to establishing a level playing field for stakeholder involvement.* The use of capacity building, institutional strengthening, quota systems (for example, to ensure proper representation of vulnerable groups such as women), or support mechanisms, such as NGOs or independent facilitators to correct any imbalance of influence should be explored. Financial support may be necessary to ensure adequate participation.

The government planning body sponsoring the planned interventions is responsible for initiating the stakeholder analysis leading to constitution of a forum and will participate in it. The final structure and composition of a stakeholder forum should be decided in a consultative process. The assistance of independent facilitators may be helpful in achieving this outcome.

A stakeholder forum is therefore formed of individuals representing various groups and interests. Such structures may exist already and their capacity can be strengthened or modified. Where such structures do not exist, a stakeholder forum is established as the representative body of the stakeholders. The extent to which a stakeholder forum should be formalised to enable representation for the groups identified through a rights-and-risks approach will depend on country contexts. The status of a forum, and the selection of its representatives, should however ensure effective participation for all interested and affected parties and accommodate changes over time.

Effective participation in a stakeholder forum must be facilitated through timely access to information and legal and other necessary support. This is particularly the case with indigenous and tribal peoples, women, and other vulnerable groups.

Sufficient time must also be allowed for the wider body of stakeholders to examine information and to consult amongst themselves before decisions are made. Where dispute resolution mechanisms are required for negotiated processes, see Guideline 2.

2. Negotiated Decision-Making Processes

A *negotiation process* is one in which stakeholders – identified through the Stakeholder Analysis (see Guideline 1) – have an equal opportunity to influence decisions. Negotiations should result in demonstrable public acceptance of binding and implementable agreements and in the necessary institutional arrangements for monitoring compliance and redressing grievances. All stakeholder forum members should share a genuine desire to find an equitable solution and agree to be bound by the consensus reached.

Attributes of a fair negotiation process

- *The Representation of Stakeholders* in the stakeholder forum is assured through a free process of selection, ensuring the effective and legitimate representation of all interests.
- *The Integrity of Community Processes* should be guaranteed through assurances that they will not be divided or coerced, recognising that differences and internal conflicts may arise. The process and the stakeholders should be as free as possible from external manipulation. Communities may legitimately decide to discontinue their involvement in the process if their human rights are not respected or in the event of intimidation.
- *Adequate time* is allowed for stakeholders to assess, consult and participate.
- *Special Provisions for Prior Informed Consent*. In negotiations involving indigenous and tribal peoples, mechanisms to resolve disputes should follow procedures recommended by the Commission (Chapter 8, Strategic Priority 1.4 and Guideline 3).
- *Addressing Power Imbalances*. Authorities should make available adequate financial resources to enable stakeholder groups who are politically or financially weak, or who lack technical expertise or organised representation to participate effectively in the process. These resources may include financial support to representatives for logistics, for income foregone, for capacity building and for requesting specific technical advice.

- *Transparency* is ensured by jointly defining criteria for public access to information, translation of key documents and by holding discussions in a language local people can understand.
- *Negotiations are assisted by a facilitator or mediator,* where stakeholders request it, selected with the agreement of the stakeholders.

For this to be a legitimate process, the stakeholders should:

- agree on the appropriate *structures and processes* for decision-making, the required *mechanisms for dispute resolution* (including any third party involvement), and the circumstances in which they will be initiated;
- agree that the *interests* at stake and legitimate community needs are clearly identified, in particular on the basis of relevant rights and risks;
- ensure that the available *alternatives*, their relevant consequences and uncertainties are given full consideration;
- guarantee access to all relevant *information* to the stakeholder forum in an appropriate language; and
- at the outset, agree on the *timeframe* for the key milestones within the decision-making process.

Compliance with the process outlined above will be a fundamental consideration in determining whether the negotiations process was conducted in good faith.

When a negotiated consensus cannot be achieved through good faith negotiations within the agreed-upon timeframe, the established independent dispute resolution mechanisms are initiated. These may include amicable dispute resolution, mediation, conciliation and/or arbitration. It is important that these are agreed upon by the stakeholder forum at the outset. Where a settlement does not emerge, the State will act as the final arbitrator, subject to judicial review.

3. Free, Prior and Informed Consent

Free, prior and informed consent (PIC) of indigenous and tribal peoples is conceived as more than a one-time contractual event – it involves a continuous, iterative process of communication and negotiation spanning the entire planning and project cycles (see Chapter 8, policy principle 1.4). Progress to each stage in the cycle – options assessment including priority setting and selection of preferred options, and preparation, implementation and operation of the selected option – should be guided by the agreement of the potentially affected indigenous and tribal peoples.[1]

Indigenous and tribal peoples are not homogeneous entities. PIC should be broadly representative and inclusive. The manner of expressing consent will be guided by customary laws and practices of the indigenous and tribal peoples and by national laws. Effective participation requires an appropriate choice of community representatives and a process of discussion and negotiation within the community that runs parallel to the discussion and negotiation between the community and external actors. At the beginning of the process, the indigenous and tribal peoples will tell the stakeholder forum how they will express their consent to decisions including endorsement of key decisions (Guideline 1).

An independent dispute resolution mechanism to arrive at a mutually acceptable agreement should be established with the participation and agreement of the stakeholder forum, including the indigenous and tribal peoples, at the beginning of any process. It is inappropriate to set rigid guidelines or frameworks, as these must be negotiated as the process proceeds. The outline presented in Guideline *2: Negotiated Decision Making Processes* is intended to offer some basic direction to those involved with such processes and independent dispute resolution bodies.

Comprehensive Options Assessment

4. Strategic Impact Assessment for Environmental, Social, Health and Cultural Heritage Issues

Strategic impact assessment (SA) is a relatively recent tool that can be used to provide a new direction to planning processes. It provides an entry point that defines who is involved and maps out the broad issues to be considered. The Commission proposes that the SA process starts by recognising the rights to be accommodated, assessing the nature and magnitude of risks to the environment and affected stakeholder groups, and determining the opportunities offered to these groups by different development options (Guideline 1). It should also identify where conflicts between various rights exist and require mediation.

SA takes the concept of project level impact assessment and moves it up into the initial phases of planning and options assessment. It is a broad assessment covering entire sectors, policies and programmes, and ensures that environmental, social, health and cultural implications of all options are considered at an early stage in planning. It is a generic term that includes a range of planning tools for example, sectoral environmental assessments (EA), basin-wide EAs, regional EAs, and cumulative EAs.[2]

SA should be concerned with the uses and impacts of existing water and energy projects, as well as alternatives for meeting future needs. In practice, SA may have different levels of detail, depending on where it is applied. At one level, the SA would scan and identify priority issues to be addressed subsequently in more detailed planning exercises. For example, the SA would identify whether evaluations of existing projects have been undertaken, or whether outstanding social issues on specific projects had been addressed. At this level, the SA would also assess whether a sufficient range and type of options are being considered in the different planning processes to meet future needs. In cases where the SA is more elaborate and detailed, the exercise may be extended to host a generic options assessment process using stakeholder groups and multi-criteria formulations to screen and rank options (Guideline 6).

The general goals of SA include:

- recognising the rights of stakeholders and assessing the risks;
- incorporating environmental and social criteria in the selection of demand and supply options and projects before major

funds to investigate individual projects are committed;

- screening out inappropriate or unacceptable projects at an early stage;
- reducing up-front planning and preparation costs for private investors and minimising the risk that projects encounter serious opposition due to environmental and social considerations; and
- providing an opportunity to look at the option of improving the performance of existing dams and other assets from economic, technical, social and environmental perspectives.

SA must be revisited at appropriate intervals with periodic 'state-of-the-sector' reporting. Important variables determining the frequency and intensity of this on-going process include developments in the economy, in technology, in demography and in public opinion. Review of SA reports at the highest political level (for example Parliament) is desirable.

5. Project-Level Impact Assessment for Environmental, Social, Health and Cultural Heritage Issues

Project-level impact assessment (IA) is already standard practice in many countries, and the term is used here to include environmental, social, health and cultural impacts. Deficiencies in past implementation have been identified and improved processes are needed.[3]

The following changes are proposed to the way project level IAs are implemented:

- Projects should be subject to a two-stage IA: the first is a scoping phase, including full public participation, that identifies key issues of concern and defines the terms of reference for the second, assessment, phase (Guideline 1).
- The timing of the IA should allow the results to feed into the final design of the project. There should be a total integration of technical, environmental and social studies during the design stage. Although executed by different study groups, these studies should run concurrently and interactively with regular exchange of information between all study groups.
- IAs should be carried out independently of the interests of the project developer and financing mechanisms should reflect this independence.
- IA should include an Environmental Impact Assessment, a Social Impact Assessment, a Health Impact Assessment (see Box 9.1), and Cultural Heritage Impact Assessment (see Box 9.2) as explicit components and comply with international professional standards. The assessments should be sufficiently detailed to provide a pre-project baseline against which post-project monitoring results can be compared.
- An independent panel of experts (reporting formally to the highest environmental protection authority) should be appointed to assist the government and the developer in reaching sustainable social and environmental outcomes (Guideline 22). The developer must respond to all issues raised by the panel and explain how they will be addressed. The panel's findings and the developer's response are to be made public within a

reasonable period (for example, six weeks).

- Developers should open a local liaison office to ensure adequate access to information for local affected communities in appropriate languages.
- The IA process should culminate in a series of written agreements with those departments or organisations that are required to implement mitigation, development and compensation plans, or respond to the impacts. The scope of these agreements must be fully defined prior to tendering for construction.

Box 9.1: Health impact assessment

A health impact is a change in health risk reasonably attributable to a project, programme or policy. A health risk is the likelihood of a health hazard or opportunity affecting a particular community at a particular time. Health Impact Assessment (HIA) is part of the overall risk assessment process (economic, social and environmental risks) to assess the viability of a project.

The Health Impact Assessment has the following components.

- Assessing the health condition of people in reservoir, infrastructure, downstream, resettlement, irrigation, and other impact areas. The assessment process should engage local people and resources. The parameters derived from the information collected constitute the baseline health situation of the population.
- Predicting changes in health determinants that can be reasonably attributed to the project and that could affect people during each stage of the project. The changes, taken together, produce health outcomes or changes in health states. These are expressed in a minimum of three ranks: no change, increased health risk, and health enhancement. Factors determining health outcomes in past projects involving comparable social, economic and environmental conditions can be used to enhance predictability.
- Assessing the cost of preventing and mitigating the potential health impacts in the overall cost assessment of the project.
- Developing measures to prevent, minimise and mitigate health impacts with the participation of the potentially affected people and incorporating these into contractual implementation arrangements with adequate financial provision.

Source: WHO, 1999, WCD Working Paper

- The IA process continues through and beyond project construction and adequate institutional and financial arrangements for social and environmental audit and monitoring should be included in the planned measures. Contracts with monitoring agencies (for example research institutes or NGOs) should be agreed prior to tendering for construction.
- A redress procedure should be put in place that provides mechanisms for addressing grievances during the resettlement plan and following construction.
- IAs should be public documents, posted on relevant websites, and disseminated in appropriate languages.
- IA should be guided by the precautionary approach.

The precautionary approach requires States and water development proponents to exercise caution when information is uncertain, unreliable, or inadequate and when the negative impacts of actions on the environment, human livelihoods, or health are potentially irreversible. A precautionary approach entails improving the information base, performing risk analysis, establishing precautionary thresholds of unacceptable impacts and risk, and not taking actions with severe or irreversible impacts until adequate information is available or until the risk or irreversibility can be reduced, making outcomes more predictable. Normally the burden of proof will be on the developer.

Decision-makers are faced with the dilemma of how to reconcile competing or conflicting rights and needs. The precautionary approach forms part of a structured approach to the analysis of risk, as well as being relevant to risk management. Determining what is an acceptable level of risk should be undertaken through a collective political process. The process should avoid unwarranted recourse to the precautionary approach when this can overly delay decision-making. However, decision-makers faced with scientific uncertainty and public concerns have a duty to find answers as long as the risks and irreversibility are considered unacceptable by society.

6. Multi-Criteria Analysis

Multi-criteria analysis (MCA) processes use a mix of qualitative and quantitative criteria to assess and compare options that may be policies, programmes, or projects. Stakeholder-driven MCA processes are flexible and open, based on the concept of a stakeholder forum (see Guideline 1). Their primary purpose is to provide a structured process to screen and rank alternatives and help understand and resolve differences between groups of stakeholders involved in development decisions.[4]

The multi-disciplinary planning team with a time-bounded mandate supports all phases of the MCA process. The results at each stage should be made available to the stakeholder forum and for wider public review. A public hearing should be held on the outcome at each stage if the significance of the process warrants it. If the MCA process cannot resolve all conflicts, the use of MCA may still assist in identifying policy scenarios and the way different groups and interests perceive them. There are many ways of presenting the results of the MCA. One of them is the preference matrix, which demonstrates the equal emphasis given to social, environmental, technical and economic parameters (see Figure 9.3).

Box 9.2: Cultural heritage impact assessment

Cultural heritage resources are the cultural heritage of a people, a nation or humanity as a whole, and can be on the surface, underwater or underground. They comprise:

- Cultural practices and resources of current populations - religions, languages, ideas, social, political and economic organisations, and their material expressions in the forms of sacred elements of natural sites, or artefacts and buildings;
- Landscapes resulting from cultural practices over historical and prehistoric times; and
- Archaeological resources, including artefacts, plant and animal remains associated with human activities, burial sites and architectural elements.

Cultural Heritage Impact Assessment (CHIA) requires adequate time for successful completion and should be looked at in two stages. Firstly, where regions and river valleys are known to be rich in cultural resources, landscapes, or archaeological resources, consideration of these elements should be included in Strategic Impact Assessments (Guideline 4) and used as a criterion in selecting options and avoiding impacts. Secondly, a project level mitigation plan is developed where a dam option proceeds to full feasibility phase.

The following procedural aspects need to be considered:

- financial resources should be specifically allocated to CHIA;
- the assessment team should include archaeologists and, if necessary architects and anthropologists;
- where cultural assets have significant spiritual or religious significance all activities should be planned with the consent of relevant communities;
- assessments should culminate in a mitigation plan to address the cultural heritage issues identified through minimising impacts, or through curation, preservation, relocation, collection or recording; and
- a separate report should be produced as a component of the overall IA process.

Source: Brandt and Fekri, 2000, WCD Working Paper

Figure 9.3 Preference matrix for ranking options

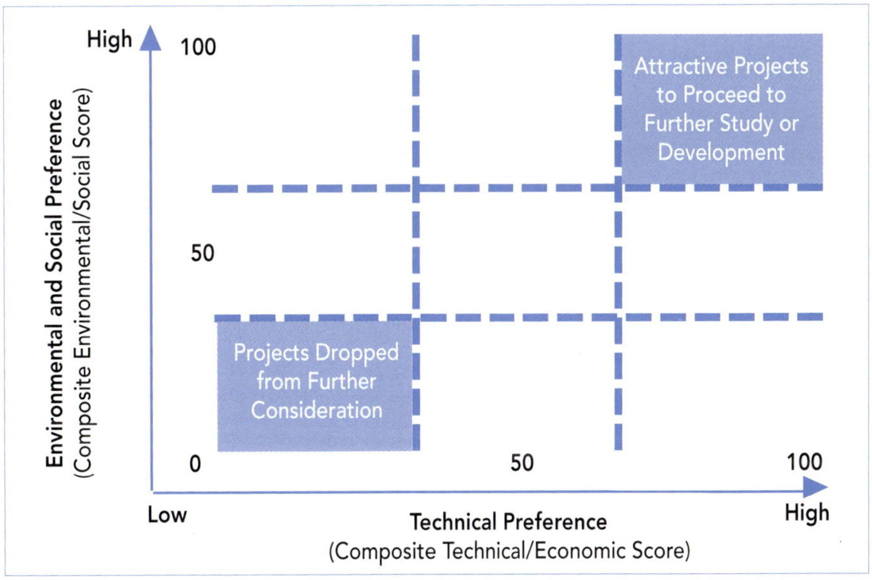

Attributes and steps of MCA processes to select the best mix of options are:

Step 1:

The sponsoring agency prepares terms of reference for the overall process and a stakeholder analysis, and establishes an information centre. Representative stakeholder groups are contacted, and the general public is informed through print and electronic media.

Step 2:

A stakeholder forum is formed and representatives of stakeholder groups identified subject to public review and comment. A multi-disciplinary planning team is formed to support the process and assembles an initial inventory of options.

Step 3:

Public comment is invited on the options inventory including proposals for additional options to be considered. The stakeholder forum confirms the comprehensiveness and adequacy of the options inventory. Where necessary, additional steps are taken to expand the inventory.

Step 4:

The stakeholder forum decides on the criteria for screening the options and criteria for coarse and fine ranking of options are established with input from the planning team.

Step 5:

Options are screened by the planning team according to the agreed criteria, results are presented to representatives of the stakeholder group for approval and subsequently announced for wider public review or comment.

Step 6:

Sequential steps of coarse and fine ranking of options (where the number of options is large) are prepared by the planning team and submitted to the representatives of the stakeholder forum at each stage. The list of options at each stage is made public and an adequate period for comment is provided between each stage. Public hearings may be held at each stage if appropriate.

Step 7:

The final selection of options that would form the basis for detailed planning is presented to agencies, communities, or groups responsible for the detailed planning.

These steps lead to preparation of a limited set of diverse development plans comprising a range of options emerging from the screening process. The multi-criteria exercise may be repeated to evaluate these alternative plans and select a preferred development plan.

7. Life-Cycle Assessment

Life-cycle assessment (LCA) is an options assessment procedure used in the energy sector to compare 'cradle-to-grave' performance, environmental impacts, and market barriers and incentives for different demand and supply options. LCA is located at the front end of the planning cycle. Its results may be fed into multi-criteria screening and ranking processes, which are a basis for deciding which options to include in subsequent stages of planning. Alternatively, the information generated by LCA is used to develop regulatory policies, for example policies addressing barriers limiting the market penetration of options otherwise considered to be in the interest of society.[5]

LCA assessments can be simple and generic or exhaustively detailed, data-rich, and elaborate. LCA procedures that quantify the potential impacts of different options on land, air and water resources, including greenhouse gas (GHG) emissions, can be transferred and adapted to different countries. The analytical framework used to assess the direct, indirect, and hidden incentives and market barriers for different options through the full chain of development is also transferable.

LCA would typically include:

- categorisation of the different stages in the life cycle of each option where the impacts and effects are relevant (for example from resource extraction through transport, manufacturing, building, operation and refurbishment to decommissioning);
- identification of the material flows and resource impacts at each stage and comparison of each option using a set of indicators (for example net efficiencies, the consumption of resources, or the impact per unit of output of the option – such as land use, water use, GHG emissions, and other gaseous, liquid or solid pollution streams); and
- identification of the range and magnitude of the direct, indirect, and hidden subsidies, external factors and incentives across each stage of the life cycle of each option.

The most advanced use of LCA is in the power sector, where it is particularly used to consider the GHG emissions of various options. These factors are becoming the prime driving force behind energy and power sector policies in many countries including Europe, Australia and Canada, and reflect the Kyoto Protocol (Guideline 8).

8. Greenhouse Gas Emissions

Recent research indicates that reservoirs can emit greenhouse gases. Precise assessments are especially important to assist in selecting climate-friendly options and if hydropower projects seek to benefit from any form of carbon credit. The emissions from the natural pre-impoundment state should be included in the comparison with other options. Good field studies with modelling predictions of emissions should be an explicit component of relevant feasibility studies.[6]

Procedures to calculate emissions for conventional and renewable options are well established and available but are continually evolving. An expert workshop convened by the Commission and held in Montreal in February 2000 decided that net emissions from reservoirs above baseline emissions are the appropriate estimates. To calculate net emissions, the planner must:

- assess the carbon (CO_2, CH_4) and nitrogen cycles (N_2O) in the pre-impoundment watershed context – this involves establishing a carbon budget, including description of flow rates, concentrations, residence time and other relevant measures;
- assess future changes to carbon inputs in the watershed from various activities, including deforestation;
- assess the characteristics of proposed reservoir(s) and inundated area(s) that will change the carbon cycle, including size, temperature, bathymetry, primary productivity and other relevant measures after dam completion; and

- assess the cumulative emissions from multiple dams on a watershed basis in cases where a dam and its operations are linked to other dams.

More baseline measurements are required on reservoirs on existing projects to extend understanding of the scale of GHG emissions to temperate and semi-arid regions of the world and to catchments with large urban populations. Such data will prove helpful in taking informed decisions on energy options and climate change.

9. Distributional Analysis of Projects

Distributional analysis provides stakeholders and decision-makers with information on who will gain and lose from a project and is an essential tool in promoting more equitable distribution of benefits and costs.[7] These gains and losses may be expressed in economic or financial terms, or they may be more simply expressed as changes in physical quantities. In some cases only the direction of a specific impact may be discernible.

Integrated distributional analysis requires assessment of the full range of project impacts including financial, social, environmental and economic aspects assessed either in a qualitative fashion, quantified in non-monetary terms, or valued in financial or economic terms. A number of methods focusing on specific aspects of distribution can be used within the overall approach at different stages of the planning cycle.

- Equity (or poverty) assessment comprises an assessment of the impacts (in economic or non-economic terms) and risks of a project on specific sub-populations or groups of concern.
- Macroeconomic or regional analysis includes an analysis of the wider economic impacts using either a simple economic or fiscal impact analysis or a formal regional or macroeconomic model.
- Economic distributional analysis includes an explicit analysis of distribution of the direct costs and benefits of the project, including those external social and environmental impacts that are to be valued (Guideline 10). This builds on the financial and economic cost-benefit analyses.

Selecting options: Integrated distributional analysis at a preliminary level should be initiated during the early stages of screening and selecting options as part of the strategic impact assessment. It can be carried forward at an increasing level of detail for projects that emerge for further consideration from this process. At the preliminary level of analysis, a matrix is prepared to identify the groups that will either receive benefits or bear the costs of the project and indicate the approximate scale of such costs or benefits. A qualitative equity assessment should also be undertaken and inform the screening process about the comparative impacts of alternatives on vulnerable groups in society.

Feasibility stage: A more detailed and integrated distributional analysis should be undertaken during the feasibility study and include both an economic distribution analysis and equity assessment. The use of a macroeconomic or regional analysis is recommended for projects with a significant irrigation component or inter-basin transfer where there are broader objectives in terms of redistributing income between regions or making a sustained contribution to the macroeconomy. The distributional analysis should be undertaken in full consultation with project stakeholders.

10. Valuation of Social and Environmental Impacts

The methodologies and applications to value environmental and social impacts of dams can be used to ensure that impacts are internalised in the economic analysis where appropriate and possible (see Table 9.1). Where it is undesirable or not possible to express such impacts in economic terms, they should be considered separately as parameters in the multi-criteria analysis (see Guideline 6).

Expertise and experience with these methods in industrialised countries are widespread, and many examples exist of their application to the impacts of dams.[8] Typically, valuation of the impacts of new dams or the decommissioning of old dams in such countries deals with recreation, tourism, fisheries and, increasingly, people's preference for healthy ecosystems.

A range of methods is available, including those based on observed market behaviour, the stated preferences of individuals, or modelling of choices made by respondents in market surveys. Their purpose is to value previously hidden costs and benefits and make them explicit in decision-making. Whatever the context, the derivation of monetary values for the unmitigated environmental and social impacts of projects is necessary when it assists the transparent, participatory, and explicit examination of project and policy alternatives. Whether these include the valuation of cultural, biodiversity or other intangible values in monetary terms will depend on the local context and on stakeholders' views. As noted, such aspects are often better addressed as an individually weighted component in a multi-criteria analysis.

Many of these valuation methods are equally applicable in the developing world and capacity to apply them increased rapidly in the 1990s. They have been adapted to the rural, developing context, particularly in combining participatory approaches with valuation methods and integrating economic valuation into multi-criteria analysis. Many of the external impacts of large dams affect household livelihoods and thus should be assessed using relatively straightforward market or revealed-preference methods. In particular, a series of relatively straightforward methods such as productivity and substitute-goods methods may be applied to estimate how changes in water quantity, quality, and flow regime affect household productivity and consumption. These methods also apply to the impact of changes in water flows on downstream communities and their natural resources, as well as impacts on major ecosystem functions and services where these provide an economic good. For example, sediment flows and deposition along the coast, which if interrupted, may lead to a need for erosion control measures.

Studies of this nature should involve at least three steps:

- a scoping exercise to identify and select impacts to be valued;
- valuation studies; and

Table 9.1: Valuation methods

	Observed Behaviour	Hypothetical Behaviour
Direct	**Market Prices**	**Stated Preferences**
	Competetive market prices Shadow pricing	Contingent valuation (dichotomous) choice, willingness-to-pay, bidding games
Indirect	**Revealed Preferences**	**Choice Modelling**
	Productivity methods Avertive (defensive) expenditure Travel cost Hedonic pricing Substitute goods	Contingent referendum Contingent ranking Contingent behaviour Contingent rating Pairwise comparisons

Source: Freeman, 1993; Pearce, pers. comm. 2000; Barbier, pers. comm. 2000.

- public meetings to report back to the stakeholder forum on the results of the studies.

The scoping exercise may be incorporated into the initial stage of project impact assessment (see Guideline 5). Finally, the information generated through valuation studies should have an explicit role in informing not only applicable cost-benefit and distributional analyses, but also the negotiations between stakeholders and decision-makers.

11. Improving Economic Risk Assessment

All infrastructure projects and commercial undertakings involve risk, uncertainty and irreversibility. Project risk assessments generally take into account technical, economic and financial aspects.[9] The Commission recognises the nature of social and environmental risks and that these can be addressed through other mechanisms (Guidelines 4, 5, 18).

The following are recommended as a general approach for technical, financial and economic risk assessment:

- the assessment of risks should be included in all steps of the planning cycle;
- identification and selection of risks for assessment should be undertaken as part of the larger stakeholder and multi-criteria processes;
- past performance of large dams should be used to identify likely ranges for the variables and values to be included in risk and sensitivity analysis; and
- sensitivity analysis should be complemented by a full probabilistic risk analysis.

Good practice involves the use of probabilistic risk analysis, a quantitative technique that employs the probability distributions of individual variables to produce a consolidated single probability distribution for the criteria of interest.

For example, in determining economic risk, the probabilities of different values for inflows and power generation can be combined with probabilities of cost overruns in a cost-benefit analysis to result in a probability distribution of net returns. This provides a robust assessment of the risk of different outcomes (see Box 9.3). It introduces a more effective approach than the simple sensitivity analysis used to assess the effect of potential changes in important variables where the cost-benefit analysis may have been re-run for a number of individual scenarios. For example, the sensitivity analysis is used to see whether the project is still profitable when the planned project costs increase by 20%.

In implementing these general recommendations on economic risk analysis across the planning cycle, a number of specific suggestions should be considered.

At all stages:

- improved prediction of project costs by using a frequency distribution of the cost overruns for similar projects.

At options assessment stage:

- a simple sensitivity analysis using agreed value ranges for key variables; and
- a qualitative comparison of options under consideration in terms of the uncertainty associated with the cost and benefit streams of each project.

At the feasibility stage

- a full probabilistic risk analysis of economic profitability;
- modelling of changes and variability in hydrological estimates that may result from climate change and their effect on delivery of services and benefit flows; and
- investigation of the likely benefits of risk reduction measures and the costs this entails.

Box 9.3: Ghazi-Barotha, Pakistan

The World Bank appraisal of the Ghazi-Barotha hydroelectric project in Pakistan used a probabilistic risk assessment of the economic rate of return (EIRR) of the power expansion programme, with the risks summarised under four scenarios: demand uncertainties, cost profiles, schedule delay, and amount of additional capacity provided by private projects. Each scenario has three alternative states. Probabilities were assigned to each scenario so that a weighted average EIRR could be obtained. This yielded 54 total possible outcomes. For each one the expected value of the EIRR, calculated as probability times its own EIRR, is then summed over all outcomes to give the expected EIRR. A probability distribution of EIRR was then calculated for the overall power sector programme and for the project alone.

The results indicate that the risk-weighted EIRR on the overall investment programme is 18.5%. This is lower than the Base Case estimate, but considerably higher than the opportunity cost of capital at 12%. The probability of the EIRR falling below the opportunity cost of capital is estimated at 8%. The risk-weighted EIRR proved quite robust to changes in the basic probabilities.

Source: World Bank, 1995.

Addressing Existing Dams

12. Ensuring Operating Rules Reflect Social and Environmental Concerns

Agreements on operating conditions should reflect commitments to social and environmental objectives in addition to the commercial interests. At all times the safety and well being of the people affected must be guaranteed throughout the project cycle. All operating agreements should be available to stakeholder groups.

River diversion during construction

Emergency warning and evacuation plans are needed in the event of overtopping of temporary diversion works. Licence conditions should assign responsibility for compensation to the downstream population for any damages that occur during such events. The compensation would be limited to those impacts caused by the breach, over and above the natural flood event.

Releases to the downstream river to satisfy drinking water and environmental requirements should be maintained during river diversion. If, for technical reasons, flow is interrupted, the operator must guarantee that alternative supplies of drinking water will be made available to the downstream population.

Reservoir filling

During the reservoir filling period, there should be releases of good quality water to the downstream river to satisfy drinking water, irrigation, and environmental requirements. If the water quality is expected to be poor, then – as with the agreement on river diversion – alternative supplies of drinking water must be made available for the downstream population.

Test operation

Test operation of the spillway (if gated), of other outlet works, and of the turbines can lead to major sudden releases to the downstream river, endangering human and animal life. The operator will be responsible for warning the downstream population that sudden releases may occur and may be dangerous. Local fishers will be compensated for days when fishing is impossible. If test operation takes place during the dry season, people experiencing damage to recession agriculture must be compensated.

Operation

A range of agreements on the operating phase should be covered in the licence:

- environmental flow releases to the downstream river;
- minimum technical releases to the downstream river (for navigation, water supply, downstream irrigation and so on);
- maximum ramp rates for downstream releases (to avoid problems with navigation and damage to the river banks);
- water allocations during normal operation;
- operation during normal and exceptional floods;
- warning of people potentially affected and rules for evacuation of people and animals;
- opening of spillway gates;
- periodic safety inspection by independent parties;
- drawdown procedure if dam safety is in doubt;
- monitoring of relevant operation data and dissemination of data to stakeholders; and
- periodic review of operating rules.

13. Improving Reservoir Operations

A range of project specific non-structural and structural methods to adapt, modify, improve, or expand operations of dams and associated facilities may be considered at different periods in time. Structural measures may include modernising equipment and control systems and improving civil structures such as spillways, intakes and canals. Non-structural measures generally involve a change in reservoir operation practices to optimise benefits, cater to changing water use priorities, enhance conjunctive operation, or improve sediment management. Dam safety improvement and updating contingency plans for operation of reservoirs in extreme flood events are other aspects of adaptive management.[10]

Detailed technical guidelines are available on ways to change reservoir operations either by adapting existing rule curves or introducing more modern computerised decision support systems, including real-time data inputs, simulations and forecasting. In adapting reservoir operations owners/operators should:

- work with stakeholders to collect views on current reservoir operations and views on the need, concerns, and limitations of potential future changes in water release patterns, including downstream impacts;
- confirm any change in the priority of water uses (such as environmental flows) and evaluate the scope to use flow forecasting to optimise reservoir operation (Guideline 15);
- use simulation models where feasible, to assess the scope for optimising the supply of water and energy (for example timing,

quantity) into the system (for example irrigation canal system and conjunctive use of groundwater, power grid or water distribution system) to improve the overall value of the services in the system;

- assess the ability to operate the reservoir to optimise delivery of services using computerised models;
- assess the scope to further optimise interactive operation of the reservoir with other reservoirs, diversions or facilities using basin-level decision support systems;
- provide clear responsibilities and procedures for emergency warning and improved preparedness of downstream countries, operator training and downstream evacuation in extreme flood events; and
- ensure monitoring systems are in place and feed into operational decision making.

Sediment management is one area where increased attention is needed. A sediment management plan would consist of:

- monitoring sediment in the reservoir, including quantitative and qualitative analysis of sediment to verify properties and pollution levels;
- minimising sediment deposition in reservoirs where possible by sluicing or density current venting;
- removing accumulated deposits where possible by drawdown flushing (drawing the water level down during high-flow seasons), and excavation of sediments; and
- catchment management programmes to reduce sediment inflow to the reservoir where possible as part of a basin-wide plan.

Sustaining Rivers and Livelihoods

14. Baseline Ecosystem Surveys

The effectiveness of mitigation, enhancement, compensation and monitoring measures require better baseline knowledge and understanding of ecosystems. Baseline assessments inform both the national policy on maintaining rivers and requirements for environmental flows and other compensation and mitigation measures. They are not restricted simply to an 'impact statement', but instead gather the necessary baseline information prior to alternatives being assessed.[11]

The baseline surveys aim to establish the link between the hydrological regime of the river and its associated ecosystems. Baseline surveys should gather relevant information on:

- the life cycle of important fish species (especially migratory species);
- the distribution of habitat for threatened or endangered species;
- important areas for biodiversity; and
- key natural resources for riverine communities.

The studies should explicitly identify where modifications to flow or water quality will have significant impacts on biodiversity, habitats, or riverine communities and provide the scientific basis for testing flow and quality scenarios against ecosystem responses (Guideline 15). Such studies would normally be undertaken over several seasonal cycles.

Appropriate research agencies staffed with specialised scientists should undertake baseline surveys, assisted where necessary by international networks. Enhanced local and regional capacity will help identify, understand and manage environmental impacts, hence improving environmental outcomes for current and future dams.

15. Environmental Flow Assessments

Dams should provide for an environmental flow release to meet specific downstream ecosystem and livelihood objectives identified through scientific and participatory processes. In some cases managed floods may be necessary to maintain downstream floodplains and deltas. Several approaches are available for assessing environmental flow requirements (EFR), ranging from 'instream flows', which refer to within-bank flows, to 'managed flood releases' designed to overtop and supply floodplains and deltas. 'Environmental flow' includes all of these and stresses the need to meet clear downstream social and ecosystem objectives rather than simply releasing a quantity of water.[12]

Environmental flow assessments (EFA) can be done at several levels of detail, from a simple statement of water depth to provide wetted habitat for a particular fish species to a comprehensive description of a flow regime with intra-annual and inter-annual variability of low flows and floods in order to maintain complex river ecosystems. Confidence in the suitability of an EFA to meet its objective is linked to the level of investment in appropriate specialist inputs.

Holistic methodologies contribute to a detailed understanding of the merits and drawbacks of a series of competing water resource options in terms of required river flow, water available for off-channel use, and the social and economic implications. Sophisticated habitat-modelling techniques provide additional detailed information on the flows required for specific valued river species or features, where the targeted rivers are of high conservation importance or have a high likelihood of conflicts over water.

EFRs are an integral part of the impact assessment process (see Box 9.4.). Continual interaction with the design team is essential to ensure that the least damaging and most flexible options are retained and that the dam design reflects the structural and operational needs of the flow release.

The Knowledge Base provides guidance on the following steps to informed decision-making leading from baseline surveys to environmental flows[13]:

Step 1: Situation assessment

Identify the extent of the targeted river system likely to be affected by a dam – upstream, downstream, and in the reservoir basin – and alert decision-makers to the likely ecological and social issues that will need to be addressed. This draws on data in the baseline surveys, where these exist.

Step 2: Specialist surveys and identification of ecosystem components

A range of specialists (ecologist, geomorphologist, sociologist, and resource economist) undertakes field surveys to provide a comprehensive description of the affected river. The studies link flow-related information with ecological and social values.

Step 3: Developing predictive capacity on biophysical responses to dam-related flow changes

The team develops data sets, models and various analytical tools that can be used in scenario creation to assist decision-making (Step 5). These may include, for instance, specifying conditions needed for a certain fish species to spawn, or how water quality differs between the rising and falling arms of a flood hydrograph, or how downstream fisheries and pastures will be affected.

Step 4: Predicting social impacts of the biophysical responses

The present river use, exploitation of river-related natural resources, and health profiles of the affected people and their livestock are quantified, and possible flow-related health risks are identified.

Step 5: Creating scenarios

Scenarios are created that include social, biophysical, and economic parameters and present a series of future options for decision-makers. Scenarios may be defined by:

- the volume of water required as yield from the dam – the rest is allocated to the river;
- protection of a valued species, community, or river feature, in which case a flow regime to achieve this would be described;
- a definition of the priorities of the competing users, and a description of the resulting flow regime and its effect on river condition; and
- river rehabilitation downstream of an existing dam, in which case the best that can be achieved within the design limitations of the dam is described.

In addition, the 'no development' scenario should always be included.

Step 6: Selection and implementation of one scenario

This requires:

- reflection of the chosen scenario in the dam design and the Environmental Management Plan; and
- monitoring of implementation to ensure that objectives are met

Box 9.4: Design and cost of environmental flows - Pollan Dam, Ireland

The EIA results for the Pollan Dam showed that migratory salmon were present upstream of the dam site, and that the dam would act as a barrier to salmon movements, affecting the fishery. The environmental water releases were designed to meet the seasonal needs of the migratory fish. Design modifications had major implications for structures such as the concrete dam, spillway, and downstream channel. The capital cost of all environmental protection measures is estimated to have increased the total cost of the project by 30% (from $6 million to $8 million). The flows have been effective in maintaining the salmon population and the recreational fishery.

Source: Smith, 1996; Bridle, pers. comm. 2000

16. Maintaining Productive Fisheries

The impact of dams on fish and fisheries is of major concern in many parts of the world. Several issues need attention in order to maintain productive fisheries (see also Guideline 15).[14]

Fish passes should be tested and shown to be efficient mitigation tools. Fish pass design has focused on the needs of leaping salmonids that usually dominate fish communities in fast-flowing rivers in the industrialised northern countries. Yet many fish species in slow-flowing tropical rivers are unable to use this kind of fish pass as they do not leap.

Box 9.5: Benefits of improving fish passes

In 1976 a pool-and-weir type fish pass was incorporated into a tidal barrage on the Burnett River in SE Queensland, Australia. Assessment of the fish pass in 1984 and 1994 showed it to be ineffective, with only 2 000 fish of 18 species ascending over a 32 month period. The fish pass was modified to a vertical-slot design with low water velocity and turbulence. Over 17 months 52 000 fish of 34 species used the redesigned fish pass. Non-leaping fish are now able to use the fish pass, benefiting the entire fishing community on the river.

Source: Flanders, 1999; env219, WCD Submission

Where fish passes exist, their effectiveness should be measured and their design improved where efficiency is low. For new dams, proposed fish pass designs should be tested hydraulically and their appropriateness for the target species explicitly assessed. (See Box 9.5.)

Where the reservoir fishery will be assessed as a project benefit, the proposal should explicitly include regional experience of similar reservoir fisheries, rigorous assessments of potential reservoir productivity, and proposals for the institutional mechanisms to manage the new fishery. Relevant contracts should be established between the project proponent, the agencies responsible for developing or managing the reservoir fishery, and the fishers, with priority given to affected people. Fisheries management objectives for dams include:

- preventing the loss of endangered and/or commercially important fish biodiversity;
- maintaining fish stock abundance;
- ensuring the long-term sustainability of the catch, employment and income; and
- producing fish for local consumption and exportable fish products.

Reservoir fisheries management concerns focus on protecting spawning grounds in affluent inflow areas, stocking to increase production (for example, of a small pelagic fishery) and advice on management of the water level to reduce impacts that harm fish stocks. Downstream river fisheries management focuses on aeration of anoxic discharge water from the dam, provision of effective fish passes, reduction of turbulence in the stilling pool, and mitigation of fish losses on the downstream floodplain through flow releases.

Recognising Entitlements and Sharing Benefits

17. Baseline Social Conditions

Constructing a social baseline is central to the planning and implementation process. It provides key milestones against which project performance and positive and negative impacts on people can be assessed through periodic monitoring and evaluation. It is also a key input to strategic social impact assessment (Guideline 4).[15]

Social baseline assessments should be done at two stages in the planning process:

- a low-intensity appraisal during options assessment, linked to Strategic Impact Assessment; and
- a more comprehensive baseline during Project-level Impact Assessment once an intervention emerges from the options assessment process.

In light of the significance of the impacts that can occur between the time a decision is made to develop a project and its actual implementation, the second baseline study may need to be updated at the tender stage of a project. Subsequent monitoring activities can follow standard practice.

Baseline studies should be undertaken for all impact areas, and in particular the areas and communities likely to be positively and negatively affected by the project. Groups to be considered include communities to be displaced, prospective host communities, downstream and upstream communities, and command area inhabitants (irrigation schemes, transmission line corridors, and other infrastructure areas). The social baseline study should be participatory and involve discussion and feedback through stakeholders (Guideline 1). It should also be accessible to the general public.

Some of the common techniques used to assess baseline social conditions combine gender-sensitive household surveys, community-level participatory appraisals, and other methods such as key informants, oral testimonies and preference assessment, direct observations and literature review. For the assessment of social processes, some of the participatory appraisal methods (such as the Venn diagram of institutional processes) can be useful. Aerial photos, satellite imagery, and geographic information systems can be combined with participatory community resource mapping exercises.

The State is responsible for ensuring that social baseline information is collected. The task should be carried out by independent institutions selected in consultation with the stakeholder forum. Large projects should be considered as an opportunity for building local capacity (in relevant government agencies, academic and research institutions, and civil society organisations) to undertake social assessment and monitoring.

18. Impoverishment Risk Analysis

The impoverishment risks and reconstruction analysis model for resettling affected and displaced populations adds substantially to the tools used for explaining, diagnosing, predicting, and planning for development. This guideline should be read in conjunction with Guideline 4: *Strategic Impact Assessment*, Guideline 5: *Project-related Impact Assessment* and Guideline 17: *Baseline Social Conditions*. At the core of the model are three fundamental concepts: risks, impoverishment and reconstruction. Impoverishment risks are analysed by separating out the components of the displacement process. They are landlessness; joblessness; homelessness; social, economic, and political marginalisation; food insecurity; increased morbidity and mortality; loss of access to common property resources; and loss of socio-cultural resilience through a community's inability to secure its interests.[16]

The internal logic of the model suggests that:

- preventing or overcoming the pattern of impoverishment requires risk reversal;
- explicit identification of risks in advance is crucial for planning counter-risk measures; and
- the transparent recognition of risks in advance will allow planners and affected people to search for alternatives to avoid displacement or to respond with mitigation and development measures or strategies and coping approaches.

The strategy to implement the impoverishment risk model includes the following:

- the baseline study covering such aspects as numbers of affected people, availability and access to resources, sources of livelihood and social, cultural, demographic, economic and political conditions and processes (Guideline 17) – these studies must incorporate variables to construct key elements of the risk model, in addition to collecting data on other aspects;
- the baseline study providing information to understand how social, economic and cultural networks, physical environment and resources support the well-being of individuals, households and communities; and
- mitigation, development and benefit-sharing measures to improve the livelihoods and well-being of affected people, and to provide the social and physical environment that would enable individuals, households and communities to successfully overcome impoverishment risks.

A four stage, two-generation model that would enable affected communities to reach full development includes[17]:

- developing benefit sharing, mitigation, and development plans with the participation of the affected people;
- enabling resettled people to cope and adapt following displacement, with continued support from the government and civil society groups;
- supporting economic development and community building within resettlement areas; and
- hand over of resettlement sites and incorporation within broader social and political institutions at a stage when resettlement and development plans are fully realised and capable of sustaining the gains for future generations.

19. Implementation of the Mitigation, Resettlement and Development Action Plan

A mitigation, resettlement, and development action plan (MRDAP) is negotiated between all affected peoples, the government and the developer. It generally has two elements – a master contract and a performance contract.[18] The affected people were identified through an Impoverishment Risk Analysis (Guideline 18). One component of the MRDAP may be a Project Benefit-Sharing Mechanism (Guideline 20). The overall obligations and responsibilities of the government and the developer will be included in the Compliance Plan (Guideline 21).

A master contract ensures that the MRDAP provisions and responsibilities are clearly understood and assigned, while a performance contract formalises provisions and commitments with affected families and communities. These two legally binding contracts are found at the government and developer level and the affected persons and community level. Where government ministries or departments act as the developer, there may not be a need for a master contract, but their obligations in this regard should be clearly stated in the MRDAP. In such cases, they will enter into agreements directly with affected people through performance contracts.

At the government and developer level

The MRDAP should have legal status. Countries with resettlement and rehabilitation acts or policies should make suitable amendments to ensure that contracts apply to all affected communities including downstream communities.

- The master contract is concluded between the developer (public corporations, private, or joint) and the government. In the contract, the developer agrees to carry out all the actions set out in the MRDAP in a timely manner. It specifies government responsibilities for providing support to acquire land, staff, schools and so on.
- A private sector developer should sign a performance bond supported by financial security (Guideline 23).
- Where the government undertakes to provide other services (including land acquisition, road building and health care), the responsible line ministry enters into agreements with other appropriate ministries to provide them.

The master contract:

- specifies penalties, incentives, remedies, and other measures to facilitate compliance by the government and the developer;
- provides for the establishment of a mitigation and development office for implementation purposes. This is usually staffed by government officials drawn from various ministries supported by staff from the developer;
- confirms the role of a multi-stakeholder committee as a subgroup designated by the stakeholder forum (including senior government officials, the developer, NGOs and affected peoples groups) to deal with grievances and supervise the work of the mitigation and development office;
- empowers the mitigation and development office to monitor the implementation of the MRDAP;
- provides for continuous monitoring of implementation by an independent field monitoring team, selected with the consent of the affected people and reporting to the multi-stakeholder committee;
- confirms the composition and role of a panel of experts for the implementation phase (Guideline 22), appointed by and reporting to the multi-stakeholder committee to assess whether the MRDAP is being implemented correctly, rehabilitation objectives are being achieved and project benefits are being provided to affected people; and
- establishes a mechanism for dispute resolution – the multi-stakeholder committee is responsible for hearing disputes and grievances related to execution of the performance contract where the mitigation and development office has been unable to resolve issues. If the committee cannot resolve the dispute or grievance, the matter will be referred to the appropriate judicial body.

At the community and affected persons level

Based on the provisions of the master contract, performance contracts are agreed with the community and affected persons detailing:

- compensation, resettlement, and development entitlements;
- schedule and method of delivery;
- institutional arrangements to deliver the commitments;
- obligations and responsibilities of the parties in the contract, namely affected peoples, community, government and developer; and
- recourse procedures.

The master and performance contracts have to be agreed at the project feasibility stage and signed prior to tendering the construction contract. The signing of the performance contracts by the affected persons and communities signals their consent for project implementation. The multi-stakeholder committee addresses all disputes related to performance contracts.

20. Project Benefit-Sharing Mechanisms

Adversely affected people are entitled to share in project benefits. Beneficiaries and benefits need to be identified and will form part of the Mitigation, Resettlement and Development Action Plan (see Guideline 19). The nature of agreed benefits can take many different forms.

Type of project benefits

Project Revenues-Related: A percentage share of project revenues/royalties, the construction budget and other profits. A joint enterprise with affected people having a share of equity.

Project Benefit-Related: Provision of irrigated land or an opportunity to purchase irrigated land, access to irrigation water, provision of electricity supply, domestic water supply from the project as appropriate. Right to reservoir fisheries, cultivation in the drawdown area of the reservoir, and contract to manage recreational/water transport facilities.

Project Construction and Operation-Related: Employment in construction, plant operation, and service sector of the project. Financial and training support for self-employment contract to provide goods and services.

Resource-Related: Preferential access to, or custodianship of, catchment resources for defined exploitation and management purposes, catchment development such as planting fruit trees or reforestation, access to pumped irrigation from the reservoir, and benefits from managed flows and floods.

Community Services-Related: Provision of better and higher levels of service including health, education, roads and public transport, and drainage; income support for vulnerable or needy households; agricultural support services including preferential planting materials and other inputs; community forests and grazing areas; market and meeting spaces.

Household-Related: Skills training and interim family support; interest-free loans for economic activities, housing improvements, provision of start-up livestock, access to public works or work for wages, free or subsidised labour-saving devices or productive machinery, access to preferential electricity rates, tax rates, water and service charges.

Identification, assessment and delivery of benefits

Definition of Beneficiaries: Beneficiaries include all people in the reservoir, upstream, downstream, and in catchment areas whose properties, livelihoods, and non-material resources are affected; and also those affected by dam-related infrastructure such as canals, transmission lines, resettlement, and other factors.

Identification of Beneficiaries: Baseline surveys must establish the nature and extent of loss to livelihoods and enumerate all categories of adversely affected and displaced individuals, families, and communities. This will be done with the participation of the affected people and reflect a rights-and-risks approach (Guideline 17).

Eligibility and Level of Benefits: All adversely affected people are entitled to benefits. The level of benefits must be assessed, agreed upon by the parties involved (affected people, government, and developer/financier) and included in the performance contract.

Benefit Delivery and Redress Mechanisms: The mitigation and development office is responsible for the delivery of benefits to the affected people (Guideline 19). The multi-stakeholder committee will hear all representations relating to identification of beneficiaries, apportionment of benefits, performance contracts, and delivery of benefits.

Ensuring Compliance

21. Compliance Plans

The preparation of an overarching Compliance Plan by the developer will address technical, financial, social and environmental obligations and commitments and provide the means for the developer to describe clearly how compliance will be ensured for a particular project. The stakeholder forum will be able to monitor compliance against the plan, which will be a publicly available document.[19]

States are at different stages in the development of regulatory systems and institutional capacity. The range of tools selected to ensure compliance for any particular project would vary from case to case. While variations in systems and capacity will result in project-specific Compliance Plans, the level of compliance should be consistent.

In using Compliance Plans in connection with the construction of dams, a number of issues will need to be addressed on a case-by-case basis, including the following:

- *The laws applicable to the construction of dams.* These will vary from country to country, and the Compliance Plan will need to be consistent with local laws.
- *The use of voluntary measures.* These will include tools such as comprehensive compliance criteria and guidelines, ISO certification, integrity pacts (see Guideline 25) and the independent review of internal processes and commitments.
- *The level of in-country institutional capacity.* Where it is insufficient to meet the requirements of the plan, provision must be made for training and other technical assistance, as required, to ensure sufficient capacity is put in place.
- *The use of performance bonds, supported by financial guarantees and trust funds.* The use of one or both of these measures will be needed to ensure sufficient funds have been set aside to secure performance. They will need to be developed and applied in a manner that best suits the particular circumstances (see Guideline 23: *Performance Bonds* and Guideline 24: *Trust Funds*).
- *The cost of compliance.* The cost of compliance will need to be built into the plan, the project budget and the evaluation process.
- *Performance indicators* and benchmarks need to be established against which compliance can be assessed.

22. Independent Review Panels for Social and Environmental Matters

Independent review panels (IRP) should be established for all dam projects. They differ from tribunals, commissions, judicial reviews or other recourse mechanisms as their principal task is reviewing assessment of impacts and the planning, design and implementation of social and environmental mitigation plans. In some countries their recommendations can be binding on all parties. In others they are only advisory. The scope of the IRP powers is laid out in its terms of reference. They report to the regulator, developer, consultants, affected peoples and financing agency to help ensure the best possible social and environmental outcomes. The IRP is not a dispute resolution mechanism, but may assist in bringing issues to the attention of the relevant body for resolution.

IRPs offer independent assessments of the issues that should be dealt with in project level impact assessments and project implementation, while also providing a mechanism to transfer best practice from one project to another, both nationally and internationally. IRPs further provide a quality control function to assure the developer, regulator, financing agency and affected groups that the necessary standards are being met and that laws or guidelines are complied with, as laid out in the Compliance Plan. They usually perform functions in the social and environmental domain similar to independent engineering inspectors for technical issues.

The composition and tasks of IRPs can be adapted to different stages of the project cycle, although it will be useful to maintain core members (normally one ecologist and one social scientist) to ensure continuity through the different project phases. Planning and appraisal may call for different skills and composition to those needed for monitoring implementation of an environmental management plan or the resettlement and development programme.

In establishing an independent panel, States and financing agencies should consider the following:

- Project level IRPs should be established by the State (as developer or regulator or the Ministry of Environment), in agreement with the stakeholder forum, as soon as the options assessment has decided on a dam as a possible option, and prior to project-level impact assessment beginning.
- IRPs are funded by the State, the developer or a financing agency according to local circumstances. The IRP forms an integral part of project costs.
- The primary reporting responsibility of IRPs should be to the national government involved and more specifically the responsible project agency and regulator. The IRP should include members able to effectively address the major issues that are addressed by state-of-the-art ecosystem, demographic, social and health assessments. They have the prerogative to add additional members to deal with issues for which the IRP has insufficient expertise.

Panels should include at least one host country national and at least one member supported by any affected people. The IRP is independent of all parties and its terms of reference should allow the panel to look into any issues deemed important without the need to justify such examination.

- The developer ensures systematic information distribution to the IRP, which has access to all project-related documentation.

- All reports following panel missions must become public documents once the developer or appropriate agency has had a reasonable time to comment (usually one month). Should the developer or other agencies request assistance on a confidential basis then that is an issue for them and the IRP to negotiate.
- The developer has the obligation to show how they are responding, or intend to respond, to the issues raised by the IRP.
- Frequency of IRP visits to the project area should be flexible. In some phases one per year may be sufficient, though once construction starts, six monthly intervals would be more appropriate.

23. Performance Bonds

Performance bonds supported by financial guarantees provide a secure way of ensuring compliance with commitments and obligations (see Box 9.6). They are used by mining and environment protection agencies and in the construction industry in many different countries. The bond is called upon, either in whole or in part, to meet unfulfilled obligations and commitments or is released when commitments are met, either in whole or in part, depending upon the circumstances.

Performance bonds have been used widely in the construction industry to ensure that work is completed within the specified time period and to specified standards (including during the construction of dams). They are also used in relation to activities that carry a high risk to the environment; for example ensuring that mine sites will be rehabilitated.

In applying the use of performance bonds to the social and environmental mitigation measures related to the construction of dams, a number of issues will need to be addressed on a case-by-case basis, including the following:

- *The activities the bond will apply to*. The bond could apply to a wide range of activities, such as physical resettlement and provision of benefits, environmental mitigation works, monitoring, auditing and decommissioning, or to aspects of each of these activities. Bonds should be carefully targeted to activities identified in an approved management plan and, preferably, should apply to the developer who is ultimately responsible for the entire project. The developer may in turn enter into performance bonds with contractors.
- *The form of security, including insurance cover, to be provided*. A package of measures can be used that collectively results in providing sufficient financial assurance. The use of bank guarantees is a cost-efficient method of providing financial assurance, but there are many others, including insurance cover.
- *Who will hold the bond and hence determine whether to release or use the security*. The relevant government agency (the environmental protection agency or mines department) has been used in most countries where performance bonds have been used to date. However, a well-structured trust fund can also be used, particularly where the government is also the developer (Guideline 24).
- *The appropriate level of financial assurance*. Considerations include the higher cost to government to do the works, a contingency sum for high-risk activities,

Box 9.6 Financial assurances and the Environment Protection Agency, Victoria, Australia

The use of performance bonds supported by financial assurances has now been successfully applied in a number of different areas. For example, following a chemical waste management company becoming bankrupt and leaving the government with a potential liability to dispose of abandoned waste, the Victorian Environment Protection Agency (EPA) required financial assurances from 38 companies involved in the waste industry. The level of the financial assurance was arrived at based upon the extent of the environmental risk – which has since been reduced where improved environmental management systems have been put into place. While the EPA has not called on any financial assurances to date, the programme has been successful in raising the performance of the industry overall and has protected the government from financial risk. The programme is being extended to cover landfills and major petroleum product storage sites.

Source: Robinson, pers. comm. 2000

making provision for staged assurance, and providing a discount for quality management, good past practice, and/or a lower risk activity.

- *The stages of the development when the security will be released.* Partial release provides a form of financial incentive to the developer to discharge its responsibilities.
- *Regular review of the level of security to reflect the actual costs.* A fall-back provision is needed that allows the State to extract the difference from the developer where the level of security proves to be insufficient.

24. Trust Funds

Trust funds have been used over a long period of time, and in a wide variety of situations, to ensure that funds set aside for a particular purpose are used for that purpose (see Box 9.7). In recent years they have been applied to the establishment and ongoing management of government-protected areas, through initiatives funded by the Global Environment Facility amongst others. There is scope to extend these funds to other areas where there is a need to set aside monies to be applied for a particular purpose, such as benefit sharing and mitigation measures associated with the construction of dams. They could also be used for decentralising responsibility to affected communities for planning and implementing their own mitigation, development and resettlement programmes.

Trust funds could be effectively used, either alone or in conjunction with bonds, to secure the financing of ongoing obligations in relation to monitoring and auditing – activities that must continue for the life of the project. This could include providing an effective means for the collection and distribution of royalties from dam-related activities to fund ongoing initiatives.

The use of trust funds would be most appropriate where the proponent is the State. In such cases, the concern is no longer one of ensuring that the risk is not passed from the developer to the government, rather it is ensuring that the risk is not passed from the State to the affected communities and to the environment.

In using trust funds in connection with the construction of dams, a number of issues will need to be addressed on a case-by-case basis, including the following:

- *The laws applicable to the establishment of trust funds.* These will vary from country to country, and the trust deed will need to be consistent with local laws.
- *The content of the trust deed.* This will need to include an open and transparent process for appointing trustees and administering the fund and for setting

out the activities the funds will be applied to, such as resettlement, environmental mitigation, monitoring and auditing. The deed must be publicly available.

- *The trustees of the fund.* The trustees will need to be sufficiently independent from the developer and have the confidence of stakeholders.
- *The role of affected people.* Their role in managing trust funds in relation to mitigation, resettlement and development needs to be defined.

Box 9.7 Suriname Central Nature Reserve

In 1998, the Government of Suriname announced the creation of the Central Suriname Nature Reserve, a protected area that covers 10% of the country. The ongoing maintenance of this protected area is secured through the Suriname Conservation Foundation Trust Fund announced in April 2000, which will ultimately administer a $15-million endowment. This sum will be invested in the fund through contributions from numerous donors. The fund will be used for long-term management support, ecological surveys, conservation awareness and education, and ecotourism as a conservation enterprise. The fund's trustees are drawn from government, sponsoring institutions, the private sector, and indigenous peoples.

Source: Famalore, pers. comm. 2000

25. Integrity Pacts

Integrity pacts relate to the procurement process, namely the supply of goods and services. (See Box 9.8 and Chapter 8 Strategic Priority 6: Ensuring Compliance) They are voluntary undertakings aimed at reducing corruption and founded on contractual rights and obligations. They can be used as one component of a Compliance Plan. Integrity pacts are of particular use in situations where regulatory systems and institutional capacity are weak, but they have universal application.[20]

Integrity pacts in various forms have now been tried and tested in many countries.

In applying the use of integrity pacts to the construction of dams, a number of issues will need to be addressed on a case-by-case basis, including the following:

- *The form and content of the pact.* The form and content of the pact must comply with accepted international models and past applications.
- *The level of in-country institutional capacity.* Where this is insufficient to meet the requirements of the integrity pact, provision must be made for training and other technical assistance, as required, to ensure sufficient capacity is put into place. This should be covered in the Compliance Plan (Guideline 21).

Box 9.8 Mendoza Province, Argentina

The Provincial Governor of Mendoza Province, Argentina decided in 1997 to amend procurement rules to include an Integrity Agreement between the Government of the Province and companies interested in bidding for government contracts. Government commitments under this agreement include providing full transparency in relationships with suppliers, ensuring that employees will not accept or demand any bribes, informing the State Prosecutor of any violations, requiring a bid bond, excluding violators from future contracts, and having the State Prosecutor oversee implementation of the policy.

Source: Wiehen, 1999

Sharing Rivers for Peace, Development, and Security

26. Procedures for Shared Rivers

A basin-wide perspective is promoted for open discussion of the issues, negotiation on sharing the benefits, and the mitigation of any adverse impacts. The procedures for equitable and reasonable utilisation, no significant harm, prior notification, impact assessment, and dispute resolution will build on provisions of the UN Convention on the Law of the Non-Navigational Uses of International Watercourses and other international agreements. Such provisions are also relevant to rivers within a country shared between a number of sub-national entities.

Prior notification

States considering options that may have a significant impact on other riparian States should notify those States at various stages and establish an effective channel of communication between all potentially affected parties. Notification should occur:

- at an early stage of planning, as part of the strategic impact assessment, and should allow potentially affected riparian States at least three months to identify relevant issues for inclusion in subsequent preparatory studies and impact assessments;
- during the scoping stage of impact assessments, to allow agreement on mechanisms for sharing technical data and information, and for participation in project-related impact assessments – potentially affected riparian States should respond within three months of the notification;
- prior to selecting an option on a shared river as part of a preferred development plan – potentially affected riparian states should receive adequate technical information about the proposed project and the results of any impact assessments, and should respond in writing within six months of the notification with their findings and response to the proposed project; and
- as required to cover any additional data and information that is available and necessary for an accurate evaluation by any potentially affected riparian States.

In the event that properly notified riparian States do not respond in a reasonable and timely manner, the notifying State would proceed with planning and development, subject to its observance of the relevant international law principles and the Commission's strategic priorities and policy principles.

In the event that a State fails to notify another riparian State which could potentially suffer significant harm by the proposed action, the potentially affected State should be able to request and receive information, make their views known, including proposing modifications, and be part of a negotiated settlement before any action to construct the dam is taken. If this opportunity is denied, remedies should be available through the International Court of Justice (ICJ), or other appropriate mechanisms. Any external financing for the dam should be conditional upon resolution of the issue as described in Chapter 8 (see policy principle 7.5).

Basin-wide impact assessment

The Commission's Guidelines on *Strategic Impact Assessment* and *Project-Level Impact Assessment*, taken together with prevailing regulatory requirements, provide the framework for a basin-wide assessment of impacts.

Impact assessments designed to suit context-specific situations need to:

- include a participatory basin-wide scoping phase;
- take into consideration the submissions of riparian states and affected communities; and
- be subjected to review by an independent panel agreed upon by all potentially affected riparian states.

All states should give the independent panel access to all necessary information.

Dispute resolution

In the event a dispute cannot be resolved within six months, either through good faith negotiations or independent dispute resolution, it should be referred to a fact-finding commission as detailed in Article 33 of the UN Convention on the Law of the Non-Navigational Uses of International Watercourses. Failing resolution through this body, the dispute should be heard by the ICJ either through case specific agreement or through compulsory jurisdiction in Article 36 of its statute.

Endnotes

1 WCD Thematic Review I.2 Indigenous People.

2 WCD Thematic Review V.2 Environmental and Social Assessment; WHO, 1999, WCD Working Paper on Human Health; Brandt and Hassan, 2000, WCD Working Paper on Cultural Heritage Management.

3 WCD Thematic Review V.2 Environmental and Social Assessment; WHO, op cit; Brandt and Hassan, op cit.

4 WCD Thematic Review V.1 Planning.

5 Ibid.

6 WCD Thematic Review II.2 Global Change.

7 WCD Thematic Review III.1 Economic Analysis, Chapter 9.

8 WCD Thematic Review III.1 Economic Analysis, Chapter 4.

9 WCD Thematic Review III.1 Economic Analysis, Chapters 2, 6, 7, 8.

10 WCD Thematic Review IV.5 Operations.

11 Brown and King, 1999; Brown et al, 1999, Contributing paper for WCD Thematic Review II.1 Ecosystems.

12 WCD Thematic Review II.1 Ecosystems.

13 Brown and King, op cit; Brown et al, 1999, Contributing paper for WCD Thematic Review II.1 Ecosystems.

14 Bernacsek, 2000, Contributing paper for WCD Thematic Review II.1 Ecosystems.

15 WCD Thematic Review V.2 Environmental and Social Assessment.

16 WCD Thematic Review I.3 Displacement; Cernea, 2000.

17 Scudder, 1997c.

18 WCD Thematic Review I.3 Displacement.

19 WCD Thematic Review V.4 Regulation.

20 Ibid; Wiehen, 1999.

WCD'S
KNOWLEDGE
BASE
Mr. Lakshmi Lal

Chapter 10

Beyond the Commission –

An Agenda for Change

Our report has distilled more than two years of intense study, dialogue, and reflection by the World Commission on Dams (WCD), the WCD Secretariat, the WCD Forum, and hundreds of individual experts on every facet of the dams debate. It contains all the significant findings that result from this work and expresses everything that we, the Commission, feel is important to communicate to governments, the private sector, civil society actors, and affected peoples – in short, to the entire spectrum of participants in the dams debate.

The debate about dams began well before the Commission, and it will continue well beyond it. We hope that one of the lasting results of the WCD process will have been to change the tenor of that debate from one of lack of trust and destructive confrontation to co-operation, shared goals and more equitable development outcomes.

The Commission alone is ultimately responsible for the conclusions and recommendations presented here.

Dams and Development: A New Framework for Decision-Making provides a solid basis for assessing options for energy and water development, and for planning and implementing projects that can achieve the desired benefits without exacting an unacceptable cost for anyone affected, or for our environment. If all parties now adopt, adapt, and implement our recommendations in good faith, much of the energy currently focused on the controversy surrounding large dams can be channelled into improving development outcomes on a co-operative and sustainable basis. But this will not happen unless the wider dams constituencies – those who entrusted us with the mandate two years ago and who created the Commission as a platform for dialogue – become the heirs of our work, go forth with it, and multiply its impact.

The evidence we present is compelling. We feel confident that the WCD Knowledge Base provides overwhelming support for the main messages in the report. We believe there can no longer be any justifiable doubt about the following:

- Dams have made an important and significant contribution to human development, and the benefits derived from them have been considerable.
- In too many cases an unacceptable, and often unnecessary and high price has been paid to secure those benefits, especially in social and environmental terms, by people displaced, by communities downstream, by taxpayers and by the natural environment.
- Lack of equity in the distribution of benefits has called into question the value of many dams in meeting water and energy development needs when compared with the alternatives.
- By bringing to the table all those whose rights are involved, and who bear the risks associated with different options for water and energy resources development, the conditions for a positive resolution of conflicts and competing interests are created.
- Negotiating outcomes will greatly improve the development effectiveness of water and energy projects by eliminating unfavourable projects at an early stage, and by offering as a choice only those options that key stakeholders agree represent the best ones to meet the needs in question.

The directions are clear. It is one thing, however, to see this. It is another to actively break through traditional boundaries of thinking, step into a different frame of mind and look at familiar issues from another perspective. This is what the Commission has had to do, and we have shown over two years that it works. We have seen similar constructive processes at work among many of the constituencies that have participated so actively in our work as they have come to understand what motivates other interest groups.

It is time now to bring the debate home. The controversy over dams has appropriately been raised to the international stage. A dissipation of that controversy, however, should allow decisions about fundamental water and energy development choices to be made at the most appropriate level. This level is one where the voices of powerful international players and interests do not drown the many voices of those with a direct stake in the decisions to be taken. For this to work, all the actors have to make that commitment to step out of their familiar frame of reference. We recommend that all parties begin by using the report as the starting point for discussions, debates, internal reviews, and reassessments of existing procedures, and for an assessment of how they can address a changed reality.

Strategic Entry Points for Follow-up

Nobody can, of course, simply pick up the report and implement it in full. It is not a blueprint. This section proposes a number of entry points to help organisations identify immediate actions they might take in response to the Commission's report. Engaging through these entry points would initiate permanent changes to advance the principles, criteria, and guidelines in the report.

These entry points don't aim to be comprehensive. Instead, they illustrate the sorts of actions different constituencies can take that would, collectively, bring about a permanent shift in the debate over our water and energy future.

The Commission calls on all interested parties to recognise that its recommendations are in the enlightened self-interest of all concerned, and form a solid base for good faith negotiations around water and energy sector planning based on accepted norms and on the evidence in our Knowledge Base. We are aware that many organisations involved with dams, water and energy will have to review existing criteria and guidelines and adapt them in light of our report.

We urge all groups to study this report and discuss how to adopt or adapt its recommendations, bearing in mind that it results from consultations that, in terms of inclusiveness and breadth of scope, are beyond the reach of any individual interest group. In this way the report will serve as a common platform for all parties to develop forms of implementation appropriate to their context and status.

This section provides an illustrative list of short- and medium-term actions specific to individual stakeholder groups. The list is addressed to all stakeholders, including governments, the private sector, bilateral and multilateral funding agencies, professional associations, and civil society. By applying it they can immediately begin incorporating the content and spirit of the report in their own professional practices and help maintain the momentum for change generated by the work of the Commission. All groups should consider the following:

We urge all groups to study this report and discuss how to adopt or adapt its recommendations, bearing in mind that it results from consultations that, in terms of inclusiveness and breadth of scope, are beyond the reach of any individual interest group.

- **Active dissemination of the report**

Water and energy development interacts with many sectors and disciplines and involves a variety of actors. Assist, where you can, in translating the report into different languages, teaching materials, checklists, and other tools that will help shift gears from prescription to implementation. Help ensure that the report reaches as many of those concerned about the dams debate as possible. NGO and professional networks especially can help ensure that people affected by dams or those planning and managing dams worldwide have access to the Commission's report and recommendations in appropriate languages.

We call on local media and publications of professional bodies to write and comment on the report and on its proposals for the future.

■ Review the report

The Commission recommends that all interested groups put in place an appropriate consultative process to review the report and propose a fitting response that will lead to effective implementation and incorporation of its recommendations in their respective policies. This may take place at the national level by government initiative, at the regional level, or at the level of individual organisations. In some cases, such as the harmonisation of the social and environmental provisions of the Export Credit Guarantee Agencies or professional associations, extensive international consultation will be essential.

Following such reviews, organisations and governments are urged to issue a public statement of support for the direction that the report takes.

Here are some questions that need answers.

- What are the reactions to the Commission's key proposals?
- How will these affect ongoing activities, immediately and in the long term?
- Which recommendations can be adopted? Why not all of the recommendations?
- What practices can be changed in response to the report?
- What national or international networks or organisations might help implement the recommendations?

The Commission welcomes a response from all parties concerning the proposals arising from the review processes. These will be posted on the WCD website (send to info@dams.org) and on any websites that follow uptake and implementation of the Commission's report.

■ Public pledges

Following such reviews, organisations and governments are urged to issue a public statement of support for the direction that the report takes. Beyond that general endorsement, make such pledges as specific as possible – for example, by endorsing the seven strategic priorities set out in Chapter 8. Governments and other organisations are further invited to report on what actions they have taken as a result of such reviews, and how their policies and actions have changed. These reports can also be sent to the WCD website (info@dams.org) to facilitate sharing of lessons and information.

■ Evaluation, monitoring, learning

The Commission has noted with dismay the absence of formal processes for evaluating the long-term performance and outcomes of large dams projects worldwide, despite the billions of dollars spent on them. This remains a huge gap in the process of learning from past good and bad practices, and has severely constrained the capacity of the sector to learn rapidly from experience, both nationally and internationally, and to promote adaptive management.

The Commission therefore urges all parties – national governments, aid agencies, financiers, professional associations, and private-sector consultants – to invest more resources in evaluating past performance through open, participatory processes that build on the Commission's case study methodology and adapt it to national circumstances. Evaluation should normally be undertaken every five to ten years as some impacts only emerge over time.

■ Review dams currently under development

The Commission recommends that public and private developers, financiers, and consultants involved in dams projects in all stages of the planning and development process give high priority to reviewing these proposals against the framework laid out in Chapter 9, and to adjusting procedures and adapting projects where necessary.

■ Promote capacity building

Adopting the recommendations of the Commission has implications for institutional capacity and financing to manage the transition in water and energy management that the Commission is recommending. Lack of capacity should not be an argument for not adopting the Commission's report. Capacity must be built if good outcomes are to be achieved, including strengthening civil society and particularly empowering women to make their voices heard. Bilateral and multilateral donors, national and international NGOs, are urged to support this transition in developing countries and wherever possible to offer support to international networks assisting in this process. Investing in the capacity and process for options assessment and decision-making should be seen as an investment in a long-term strategy of lowering the costs of future projects.

Taking the Initiative – Institutional Responses

This section contains recommendations addressed at specific stakeholders in the dams debate. They are not intended to provide a comprehensive list of what we expect, but instead to illustrate some of the salient actions that we believe each group should consider as it moves from debate over the report's thrust into actions to implement its provisions. They are entry points for follow-up. These recommendations result not only from the Commission's review of experience with past dams, but also from two years of analysis and dialogue with many partners. They complement what lies in the body of the report.

National governments

- Establish an independent, multi-stakeholder committee to address the unresolved legacy of past dams.
- Require a review of existing procedures and regulations concerning large dam projects.
- Develop a specific policy statement governing stakeholder participation in options assessment and planning, setting out the range of considerations that will be incorporated.
- Review legal, policy and institutional frameworks to assess and remove any bias against resource conservation, efficiency and decentralised options, and any hindrance to open participatory processes.
- Introduce and support a UN General Assembly resolution that welcomes the publication of the Commission's report, invites governments to accept and implement its recommendations and transmits the report to the Rio+10 process as a positive example of multi-stakeholder co-operation that can result in a substantial advance towards sustainable development.

Capacity must be built if good outcomes are to be achieved, including strengthening civil society and particularly empowering women to make their voices heard.

Line ministries

- Issue criteria and guidelines for promoting third party review and dispute resolution around large dam projects.
- Adopt the practice of time-bound licences for all dams, whether public or privately owned.

> *The WCD is a model that can be, if not replicated, then at least adapted to other similar controversial issues that are also at the interface of environment and development. These are polarised issues such as the future of biotechnology in development, or the role of large mines and extractive industry in development.*
>
> Frances Seymour, World Resources Institute

Civil society groups

National NGOs and international NGO networks

- Participate in cross-cutting partnerships with other actors on key issues.
- Gather, analyse and disseminate information widely to promote transparency and openness.
- Monitor compliance with agreements and assist any aggrieved party to seek resolution of outstanding disagreements or to seek recourse.
- Actively assist in identifying the relevant stakeholders for water and energy projects using the rights-and-risks approach.
- Contribute to the establishment of appropriate forums for stakeholders to enable them to identify, articulate and represent their legitimate rights.

Affected peoples' organisations

- Develop proposals for follow-up dialogue on the Commission's report with government authorities and project developers.
- Identify unresolved social and environmental impacts and convince the relevant authorities to take effective steps to address them.
- Develop support networks and partnerships to strengthen technical and legal capacity for needs and options assessment processes.

Professional associations and agencies

(International Commission on Large Dams, International Commission on Irrigation and Drainage, International Hydropower Association, International Energy Agency, International Association for Impact Assessment.)

- Promote a culture of evaluation and self-reflection to ensure continuous learning from all aspects of large dam projects through adopting appropriate procedures.
- Extend national committees to include a consultative group of NGOs, environmental scientists and affected peoples' groups.
- Set up joint work programmes with these groups at the national and regional levels to learn from past experience.
- Develop processes for certifying compliance with WCD guidelines.
- Extend national and international databases, such as the ICOLD World Register of Dams, to include social and environmental parameters.

International Organisation for Standardisation (ISO)

- Explore the codification of the Commission's guidelines in a sector-specific guidance document or standard that incorporates social impact management as well as public reporting and negotiated decision-making. A high degree of developing country and civil society participation in the standards-setting process is needed to ensure legitimacy.

The private sector

Suppliers, contractors, developers and consultants

- Publicise the acceptance of the Commission's principles, criteria and guidelines in corporate policy and company literature.
- Abide by the provisions of the anti-bribery convention of the Organisation

for Economic Co-operation and Development.

- Adopt integrity pacts for all contracts and procurement, as developed by Transparency International.
- Develop and adopt voluntary codes of conduct, management systems and certification procedures for best ensuring and demonstrating compliance with the Commission's guidelines, including, for example, through the ISO 14001 management system standard.
- For consulting companies, refine the use of the tools proposed by the Commission so they become standard industry practice. These include distributional analysis, multi-criteria analysis, risk and sensitivity analysis, rights-and-risks approach, and environmental flow assessments.
- Put in place mechanisms to ensure that designers of dams either participate in or at least receive evaluations of predicted social, environmental, financial, and economic performance five years after construction in order to learn from their experience. Make these evaluations available to the public.

Private financiers

- Develop criteria for innovative bond-rating systems for use in financing all options, including large dams, in the water resources and electric power sectors.
- Recognise the risk reduction opportunities inherent in proper options assessment and develop legally binding arrangements on environmental and social matters in assessing insurance, equity, and bond premiums.
- Incorporate the principles, criteria, and guidelines of the Commission in corporate social responsibility policies and statements.
- Use the Commission's guidelines as social and environmental screens for evaluating support for, and investment in, individual projects.

The example of the WCD shows that establishing a basic measure of trust among actors in a conflict-ridden environment is time-consuming and costly, but launching a sustainable mechanism for consensus building and standards setting requires no less.

Wolfgang Reinicke and Francis Deng, *Critical Choices*

Bilateral aid agencies and multilateral development banks

- Develop programmes to help countries, especially those with a significant existing or potential dam population, formulate a response to the Commission's report and find ways to implement its recommendations.
- Ensure that any dam options for which financing is approved emerge from an agreed process of ranking alternatives and respect the Commission's guidelines.
- Accelerate the shift from project- to sector-based finance, especially through increasing financial and technical support for effective, transparent, and participatory needs and options assessment, and the financing of non-structural alternatives.
- Review the portfolio of past projects to identify those that may have under-performed or present unresolved issues and share in addressing the financial burden of such projects for borrower countries. This may include, for example, cancelling the outstanding debt related to them, converting debt repayment into development assistance targeting affected areas, or providing new support to help borrower countries address unresolved economic, social, and environmental problems.

At the beginning many doubted the Commision's independence but if you've followed it closely, you will realise the Commission has done excellent work and its independence has been respected. There was a spirit of openness, of understanding, a dialogue between partners.

Aboubacry Mbodji, CODESEN, a coalition of NGOs concerned with development

- Review internal processes and operational policies in relation to the Commission's recommendations to determine changes needed in the selection of projects for lending portfolios; the appraisal process; and implementation, monitoring, and evaluation.

Export credit agencies

- Introduce and adopt common environmental, social and transboundary criteria for financial guarantees and strengthen institutional capacity to appraise projects against such criteria.
- Improve co-ordination among agencies at the international level to ensure that dam projects refused by one agency are not accepted by others.
- Require private-sector applicants for dam projects to meet due diligence criteria or voluntary codes of conduct that conform to the Commission's recommendations.
- Promote consultation and information disclosure as normal procedure.

Intergovernmental organisations

The United Nations

- Facilitate the adoption of a General Assembly resolution on the Commission's findings and encourage the inclusion of the report in the Rio+10 process.

UN technical agencies

(World Health Organisation, Food and Agriculture Organisation, United Nations Development Programme, United Nations Educational, Scientific and Cultural Organisation, etc.)

- Review technical guidelines, norms, and practices regarding water and energy resources development to integrate the principles, guidelines, and criteria from the Commission's report.
- Provide support to improve national capacity for options assessment in developing countries.

United Nations Environment Programme

- Actively promote the results of the Commission in its co-ordinating and catalysing environmental work within the UN family of organisations.
- Examine how the Multilateral Environmental Agreements that it administers might serve to strengthen the spirit and advance the proposals of the Commission.

Academic and research bodies

- Assist in the evaluation of further dam case studies following WCD methodology.
- Undertake research on alternatives to dams such as demand side management and ensure these are available to decision-makers through the options assessment process.
- Assist in improving the WCD Knowledge Base as outlined in Box 10.1.

Continuing the Dialogue

The previous section identified a small selection of specific recommendations aimed at the major groups of participants in the dams debate. It focused on actions that could be taken immediately and that, in many cases, are specific to the target group they are directed at.

We choose not to go beyond this for two principal reasons. First, the Commission has found that most problems associated with dams result from faults in the process of options assessment and decision-making on energy and water development choices. It follows that the solution lies in making improvements to the process – improvements that will make it more even-handed, equitable, transparent, and inclusive. How the process is designed and conducted will depend, to a large extent, on country or setting. There is no universally applicable model – only basic principles.

The second reason is more significant. Unlike other Commissions, whose reports were essentially aimed at governments or the international community, our report has a broad and diverse target. It is aimed equally at governments, international organisations, multinational companies, financiers, consultants, NGO networks, indigenous communities, and locally organised groups of people affected by dams. There is no natural heir to our work other than the complex and multi-faceted community of those concerned with dams issues.

It is appropriate that we seek to build no lasting international edifice but instead prefer to send our report for implementation where it belongs – to the regional, country, basin, community, and dam-specific level. We hope and expect that the report will lead to regional and national processes of dialogue, to discussions concerning specific projects or issues, to new research and training programmes, to new networks – in short, to a multiplicity of results and developments that are decentralised, dispersed, and highly diverse in character. We hope that they will include many new partnerships across sectors where earlier divides existed, connecting to understand differing interests and to identify shared objectives. This would be in keeping with the spirit of the Commission.

In fact, this is already happening. Whether or not connected to the WCD process, the debate around dams continues. For example, policies and institutional responses are evolving, due to improved awareness, in Brazil, Sri Lanka, United Kingdom, Thailand and Nepal. Actors continue to respond with tailor-made solutions to local issues such as:

- proposals for a national commission on dams, modelled on the WCD;

Box 10.1 Priorities for strengthening the knowledge base

A major constraint the Commission faced was a lack of comparative data on the development effectiveness of large dams and the actual direct and indirect consequences of such dams on local, regional, and national development, and more specifically on affected people and environments. Much information is either not available or not shared by those who hold it. Detailed studies are needed that are comparative, comprehensive, integrated, long-term, cumulative, and adaptive. There are numerous areas about which the Commission found insufficient information on which to base strategic recommendations. The priorities for more information and understanding through specific research, data collection, monitoring, and evaluation include:

- Studies to compare the direct and indirect benefits and costs of multipurpose dams with alternative options.
- Research to inform strategic thinking on the cumulative impacts of a cascade of dams or interbasin transfers on river basin environments and populations and on local, regional and national development.
- Information on the potential for multiplier effects relating to food production, water use efficiency, poverty alleviation, and non-farm linkages of dam-supplied major irrigation projects with other irrigation and agriculture options.
- The impacts of main-stem and tributary dams on downstream ecosystems, including deltas and the livelihoods of people using those ecosystems.
- A better understanding of the extent to which managed floods can offset the impacts of dams on downstream ecosystems and livelihoods.
- Improved understanding of how dams impact on women and gender relationships.
- Improved understanding of how to influence rural-urban migration and the requirements of meeting needs through decentralised rural development compared to the challenge of servicing mega-cities for water and energy.
- Baseline studies of greenhouse gas emissions from pre-project rivers for comparison with emissions from post-project reservoirs on those rivers, to facilitate study of how the change from a natural to a human-modified system influences greenhouse gas emissions.
- In anticipation of global warming, research and strategic thinking on the impact of past periods of drought on the ability of large dams to deliver expected services in arid, semi-arid, and drought-prone areas.
- Identification of techniques for recharging aquifers on a larger scale as a storage technique for avoiding water loss through evaporation, and for improving conjunctive and sustainable use of surface and ground water.
- Development of improved policies and practices to ensure that cultural heritage is dealt with in the planning and implementation of dam projects, given the importance of river basins for the origins of human civilisation.

- multi-stakeholder workshops convened to receive a debriefing following the last meeting of the WCD Forum and to develop a national dialogue;
- independent mechanisms to review and resolve escalating conflicts on dam projects;
- reviews of export credit policies following controversy on particular dams; and
- reviews of dam safety and compensation policy for dam-affected people.

We expect that such examples will multiply as our report is disseminated and becomes part of the ongoing discourse about dams and development.

Unlike other Commissions, whose reports were essentially aimed at governments or the international community, our report has a broad and diverse target. It is aimed equally at governments, international organisations, multinational companies, financiers, consultants, NGO networks, indigenous communities, and locally organised groups of people affected by dams.

The debate about dams began well before the Commission, and it will continue well beyond it. We hope that one of the lasting results of the WCD process will have been to change the tenor of that debate from one of lack of trust and destructive confrontation to co-operation, shared goals and more equitable development outcomes. We must not, however, give the impression that the Commission has laid the dams debate to rest. Several important elements of that debate must continue in order to carry the discussion on dams forward in the context of meeting development goals.

For one thing, the Knowledge Base is not complete. Extensive as the information gathered, structured, and analysed by the Commission may be, there are still gaps in our knowledge, experience, and understanding. Many issues cannot finally be settled because the information needed to resolve them is still not available. The process of information gathering, analysis, learning, consultation and review must go on (see Box 10.1).

Moreover the Commission, in the course of its work, generated a wealth of ideas for the future. These ranged from policy, regulatory and institutional tools to best practice experience at the community level. Many of these ideas inspired the principles and guidelines set out in our report. But there are also many that proved too detailed, too context-specific, or too untested to include here. They should not be lost. For example, the WCD Knowledge Base can be used to develop a series of practical tools in as many language versions as possible. Publishing a regular State of World Dams Report is another possible project. Ideas for national or international recourse mechanisms and co-ordinated monitoring of compliance include creating national dams inspection panels, developing a register of accredited independent experts for review panels, and establishing a system for assuring that individual dams or stages in dam projects conform with the Commission's guidance, through an independent and iterative verification process along the lines of the ISO 14001 third-party certification mechanism.

None of these follow-up ideas, nor much of what lies in the body of the report, will advance very far without a concerted and large-scale effort to build the necessary expertise and institutional capacity for implementing them. This point cannot be stressed enough. It is one thing to propose a model or set of actions that will sweep away the problems encountered in pursuing energy and water development, especially when dams are involved. It is quite another to put the solutions patiently in place. Calling upon developing countries to slot proposals into a framework that does not exist, or that is incapable of accommodating them, is a recipe for frustration. The Commission is convinced that helping to build the capacity to consider fully all options for water and energy development is as important as implementing any choices that are made. Often the latter cannot take place without the former.

We hope that the momentum generated during the past three years – from the Gland meeting, through all the papers and reviews written for the Commission, to the actions taken to launch and disseminate the report – will be sufficient to carry the process forward. The establishment of the Commission opened an important space for debate and dialogue, and that space has been fully used. Further spaces must now be created and filled.

A Call to Action

The work of the World Commission on Dams is over. But the concerns that led to its establishment are still with us. Dams have too often left a legacy of social injustice and environmental damage, and that legacy does not disappear because we have identified a better way of doing things in future. Early and resolute action to address some of the issues arising from the past will go a long way to building the trust required to enable the different actors to work together. So, too, would an assurance to countries still at an early stage of economic development that the dams option will not be foreclosed before they have had a chance to examine their water and energy development choices within the context of their own development process.

The experience of the Commission demonstrates that common ground can be found without compromising individual values or losing a sense of purpose. But it also demonstrates that all concerned parties must stay together if we are to resolve the issues surrounding water and energy resources development. It is a process with multiple heirs and no clear arbiter. We must move forward together or we will fail.

We do not assume, of course, that there will be no further disagreement. Dynamic debate leads frequently to better outcomes, as controversy can often be empowering. We do, however, believe that we will not turn the corner and put the principal conflicts around water and energy development behind us unless the participants in the debate decide to work together locally, nationally, and globally, in the spirit of the Commission.

> *Sri Lanka and other countries have already benefited from the WCD process... For example, Sri Lanka has appointed a committee to examine the grievances of dam-affected people and to compensate victims, as well as to take action on dam safety issues not undertaken earlier.*
>
> Tilak Ranaviraja, Ministry of Mahaweli Development

The world never stands still. The context in which decisions on energy options and water development must be made is in constant evolution. Technology is advancing, democracy and governance reform is spreading, the market is changing and the demand for greater equity is growing more resolute and persistent. As noted in Chapter 1, this century will see increasing stress on water resources worldwide. The assessment of water and energy options and of the role dams play, increasingly takes place against a background of competing sectors and interests. Growing needs and a dwindling resource base, in both quantity and quality, will require many countries to fundamentally reassess their water management policies. We are convinced that business as usual will not prove to be a viable strategy.

The work of the World Commission on Dams is over. But the concerns that led to its establishment are still with us. Early and resolute action to address some of the issues arising from the past will go a long way to building the trust required to enable the different actors to work together.

The Commission has not sought to divert these global trends or to predict their future scale and direction. We have, instead, tried to bring them together, in some form of harmony, on a score card that the range of actors in the debate can use. We believe this report is a milestone in the evolution of dams as a development option. We have conducted the first comprehensive and global review of the performance of dams and their contribution to development. We have done this through an inclusive process that has brought all significant players into the debate. And we believe we have shifted the centre of gravity in the dams debate to one focused on options assessment and participatory decision-making. The rights-and-risks approach we propose will raise the importance of social and environmental dimensions of dams to a level once reserved for the economic dimension.

We have told our story. What happens next is up to you.

A Comment – Medha Patkar

1. The process of the World Commission on Dams was unprecedented in bringing together so many of those involved in debates and conflicts over large dams. Many peoples' movements and NGOs – the constituency that first proposed a comprehensive and independent review – actively participated. Our final Report has synthesised enormous amounts of information and diverse opinions and delivered many important findings and recommendations. I hope it will be a reference for all those concerned about large dams. While signing the Report because of its many positive aspects, I still feel I must put forth this opinion on some fundamental issues that are missing or not given the central place they deserve.

2. The problems of dams are a symptom of the larger failure of the unjust and destructive dominant development model. It is beyond the scope of our report or the brief of the Commission to resolve all the underlying problems of global development. But addressing these issues is essential in any attempt to reach an adequate analysis of the basic systemic changes needed to achieve equitable and sustainable development and to give a pointer towards challenging the forces that lead to the marginalisation of a majority through the imposition of unjust technologies like large dams.

3. The frequent failure of large dams to provide their claimed benefits and this poor performance needs to be recognised and accepted. There is no reason for optimism on the feasibility of improving the poor performance of dams and mitigating their impacts. A major question is the feasibility of just rehabilitation with land for land lost by agriculturists and alternative, appropriate sources of livelihood for other displaced people. In large scale displacement, the experience shows a clear failure. Within the value framework the Commission propagates – equity, sustainability, transparency, accountability, participatory decision-making, and efficiency – large dams have not helped attain, but rather hindered, "human development".

4. An inclusive, transparent process of decision-making with equal status to all the stakeholders, equal place for local and national needs and plans, equal significance to social, environmental, technical and financial aspects of planning, would be a great advance, but does not go far enough. Even with rights recognised, risks assessed and stakeholders identified, existing iniquitous power relations would too easily allow developers to dominate and distort such processes. These developers include multilateral institutions like the World Bank that have pushed many large dams despite

non-compliance with their own policies. The State, controlled by powerful vested interests, may do the same. Understanding this takes us beyond a faith in negotiations to emphasize certain priorities and primacies.

Communities, especially those who live on and seek livelihood from their natural resource base, such as forest produce gatherers, farmers or fisherpeople, should have the first right to planning, development and management of those resources. Inequities within communities also need to be recognised and addressed. Social and environmental parameters must have a higher weightage than the technical and financial aspects in decisions concerning human development. It is necessary to stress the 'principle of subsidiarity', according to which development planning would be based on micro-catchments, working from ridge to river, and from origin to sea.

5. A full assessment of the options for meeting water and energy needs as the first part of project planning needs to be supported. But only creating a level playing field for options cannot suffice. We should instead give priority to more equitable, sustainable and effective options to satisfy basic human needs and livelihoods for all before supporting the additional luxuries of the few, unjustified in the face of the many who remain deprived.

6. The wider context of national and global political and economic trends obviously affect decisions in the water and power sector. These trends include the diminishing role of the State, the growing marginalisation of national laws and institutions, and the trampling of human rights due to the expanding role of private capital and free trade. While there may be a few welcome instances of progress towards enhanced human rights and equity, to say there is a global trend towards these goals would indeed be erroneous.

7. No undue legitimacy should be granted to corporations and international financing agencies. The sovereignty of both people and the nation-state must not be compromised for anything but the basic values and goals of humankind. It is necessary to give a serious critique of the privatisation of the water and power sectors and the resulting marginalisation of local people and corporate domination over natural resource-based communities.

8. The issues above are those raised by the peoples' movements whose role and perspectives should be given their due place. Not just with stories of eviction, repression and confrontation, but with their ideologies, strategies, and vision.

9. Over and above all this, I recognise and share the Commission's achievement of local to global consultation and, more so, a humane, well-intentioned, open and frank dialogue under an able leadership, which needs to be kept alive beyond the short life of this forum. To endorse the process and many of our findings and recommendations, I have signed the Report. To reject the underlying assumptions of a development model which has palpably failed and to caution against the massive gulf between a statement of good intent and a change in practice by entrenched vested interests I have asked for this note to be attached.

Whatever is missing or could not be addressed in the Report should form an agenda for further dialogue and research; but also for struggles for justice with people at the forefront, people more empowered by this report and otherwise to say NO to the perverted development vision, process and projects.

Medha Patkar

Annex I
Bibliography

Abdel Megeed A, Aly Makky E, 1993, 'Shore Protection of the Nile Delta After the Construction of High Aswan Dam', in Egyptian Committee on Large Dams (ed), *High Aswan Dam Vital Achievement Fully Controlled*, Cairo, ENCOLD.

Acreman M, Farquharson F, McCartney M, Sullivan C, Campbell K, Hodgson N, Morton J, Smith D, Birley M, Lazenby J, Wingfield R, Barbier E, 2000, *Managed Flood Releases from Reservoirs: Issues and Guidance*, Report to DFID and the World Commission on Dams, Wallingford, UK, Centre for Ecology and Hydrology, Contributing Paper to WCD Thematic Review II.1 Ecosystems.

Adams WM, 1992, *Wasting the Rain: Rivers, People and Planning in Africa*, Earthscan, London.

Adams WM, 1985, 'The Downstream Impacts of Dam Construction: A Case Study from Nigeria,' in *Transactions of the Institute of British Geographers NS*, Vol. 10: 292-302.

ADB (Asian Development Bank), 1984, *Chashma Command Area Development - Appraisal Report*, Manila, Asian Development Bank.

ADB, 1994, Sector *Synthesis of Post-Evaluation Findings in the Water Supply and Sanitation Sector*, Manila, Asian Development Bank.

ADB, 1995, *Sector Synthesis of Post Evaluation Findings on the Irrigation and Rural Development Sector*, Manila, Asian Development Bank.

ADB, 1997, *Guidelines for the Economic Analysis of Projects*, Manila, Asian Development Bank.

ADB, 1999a, *Special Evaluation Study on the Social and Environmental Impact of Selected Hydropower Projects*, Manila, Asian Development Bank.

ADB, 1999b, *China Resettlement Policy and Practice- Review and Recommendations*, Draft for Reviewing, Regional Technical Assistance Project, Manila, Asian Development Bank.

ADB, 1999c, *Strategic Options in the Water Sector*, Peoples Republic of China, Canada, Hydrosult.

ADB, 2000, *Study of Large Dams and Recommended Practices*, RETA 5828, Manila, Asian Development Bank.

Adeler A, Flatby, R, 2000, Chief Engineer, Section Manager, Norwegian Water Resources and Energy Directorate, 'Norwegian Legislation on Local Benefits,' personal communication with authors, 14 June.

AfDB (African Development Bank), 1998, *Review of the Bank's Experience in the*

Financing of Dam Projects, African Development Bank, Operations Evaluation Department.

Agarwal B, 1996, *A Field of One's Own: Gender and Land Rights in South Africa*, Cambridge, Cambridge University Press.

Agarwal A, Narain S, (eds), 1997, *Dying Wisdom*, New Delhi, Centre for Science and the Environment.

Agence France Presse, 21 January 2000, *Three Gorges Dam Hit by 600 Million-Dollar Graft Scam*, Agence France Presse.

Agence France Presse, 10 March 2000, *Official Sentenced to Death for 3G Corruption*, Agence France Presse.

Aleem AA, 1972, 'Effect of river outflow management on marine life', in *Marine Biology*, Vol. 15: 200-208.

Anane M, 1999, *Gender and Dams: A Case Study of the Akosombo Dam*, WCD Submission soc210.

Anderson RC, 1997, *New Independent States Issue Paper No. 1: Environmental Damage Assessment of the Aral Sea Disaster*, Central Asia Mission, US Agency for International Development.

Annan KA, 2000, *We the Peoples: the Role of the United Nations in the 21st Century*, New York, United Nations.

Arrow KJ, Lind RC, 1970, 'Uncertainty and the Evaluation of Public Investment Decisions,' in *American Economic Review*, Vol. 60: 364-378.

ASCE (American Society of Civil Engineers), 1998, "1998 Report Card for America's Infrastructure", in *Issue Brief - Dams*, American Society of Civil Engineers, 5 March.

ASDSO (Association of State Dam Safety Officials), 2000, *Dam Safety Facts*, http://www.damsafety.org/facts/html, viewed 28 June 2000.

Atakpu L, 2000, *Dams, Food Security and Livelihoods: Understanding the Nigerian Experience*, African Network for Environment and Economic Justice, Nigeria, WCD Regional Consultation Paper.

Bacon RW, Besant-Jones JE, Heirarian J, 1996, *Estimating Construction Costs and Schedules: Experience with Power Generation Projects in Developing Countries*, World Bank Technical Paper 325, Washington DC, World Bank.

Bacon RW, Besant-Jones JE, 1998, 'Estimating Construction Costs and Schedules: Experience with Power Generation Projects in Developing Countries', in *Energy Policy*, Vol. 26, No. 4: 317-333.

Balland P, 1991, *Le Littoral Méditerranéen Français: Evolution Physique. Qualité Génerale*, Agence de l'Eau Rhône-Méditerranée-Corse.

Balsam L, 1940a, Telegram to Collier J, 15 March, 311.1(a) Acquisition of Lands - USBR Correspondence, Colville Realty Office, Nespelem, Washington DC, cited in WCD Case Study Grand Coulee Dam, Annex 9.

Balsam L, 1940b, Letter to Collier J, 14 July, 311.1(a) Acquisition of Lands - USBR Correspondence, Colville Realty Office, Nespelem, Washington DC, cited in WCD Case Study Grand Coulee Dam, Annex 9.

Bandaragoda DJ, 1999, *Institutional Change and Shared Management of Water Resources in Large Canal Systems: Results of an Action Research Programme in Pakistan*, Research Report 36, Colombo, International Water Management Institute.

Barbier E, 2000, Professor, Environment Department, University of York, personal communication with authors, 14 March.

Barker D, Dave D, 2000, *The Asian Rice Economy in Transition*, unpublished draft, 5 May 2000.

Barrow CJ, 1999, *Alternative Irrigation: The Promise of Runoff Agriculture*, London, Earthscan.

Bartolome LJ, Danklmaier CM, 1999, *The Experience with Dams and Resettlement in Argentina*, Contributing Paper for WCD Thematic Review 1.3 Displacement.

Belli P, Anderson J, Barnum H, Dixon J, Tan P, 1998, *Handbook on Economic Analysis of Investment Operations*, Operational Core Services Network Learning and Leadership Center, Washington DC, World Bank.

Benade B, 1999, Eco-Impact: Environmental Consultants, 'Fishways', personal communication with authors, 23 November.

Benech V, 1992, 'The Northern Cameroon Floodplain: influence of hydrology on fish production,' in Maltby E, Dugan P, Lefeuve JC, (eds), *Conservation and Development: The Sustainable Use of Wetland Resources*, Gland, Switzerland, IUCN

Berga L, (ed), 2000, *Dams and Floods*, draft paper, Paris, International Commission On Large Dams, Contributing Paper for WCD Thematic Review IV.4 Flood Management Options.

Berga L, Yague J, Cajete J, Giron F, Mendiluce JM, 2000, *Benefits and Concerns about Dams in Spain*, papers for the Beijing 2000 Congress of ICOLD, draft.

Bermann C, 1999, *Community Managed Resettlement: The Case of Ita Dam*, WCD Regional Consultation Paper.

Bernacsek G, 2000, *Capacity and Information Base Requirements for Effective Management of Fish Biodiversity, Fish Stocks, and Fisheries Threatened or Affected by Dams during the Project Cycle*, Contributing Paper for WCD Thematic Review II.1 Ecosystems.

Berz G, 2000 "Flood Disasters: Lessons from the Past - Worries for the Future," *in Proceedings of the Institute of Civil Engineers, Water and Marine Engineering*, London, Vol. 142, March: 3-8.

Bhatia R, Cesti R, and Winpenny J, 1995, *Water Conservation and Reallocation: Best Practice Cases in Improving Economic Efficiency and Environmental Quality*, A World Bank- Overseas Development Institute Joint Study, Washington DC, World Bank.

Bizer JR, 2000, *International Mechanisms for Avoiding, Mitigating and Compensating the Impacts of Large Dams on Aquatic and Related Ecosystems and Species*, prepared for IUCN, Contributing Paper for WCD Thematic Review II.1 Ecosystems.

Blackmore D, 2000, Chief Executive, Murray Darling Basin Commission, personal communication with authors, 28 August.

Boa Nova AC, Goldemberg J, 1999, *Electrification of Shanty Towns in Sao Paolo*, International Urban Development Association, 23rd Congress.

Bond P, 2000, *Paying for Southern African Dams: Socio-Environmental Financing Gaps*, Cape Town, EMG-GEM-IRN, WCD Submission eco033.

Bosch JM, Hewlett JD, 1982, 'A Review of Catchment Experiments to Determine the Effect of Vegetation Changes on Water Yield and Evapotranspiration,' in *Journal of Hydrology 55:3-23*

Bosi M, 2000, *An Initial View on Methodologies for Emission Baselines: Electricity Generation Case Study*, IEA Information Paper, Paris, International Energy Agency.

Bourke G, 1988, 'Subduing the Sea's Onslaught' in *South*, July: 117.

Bowman M, Cantrell S, Johnson S, 1999, *Dam Removal Success Stories*, Report from

American Rivers, Friends of the Earth, Trout Unlimited.

Brandt S, Hassan F, 2000, *Dams and Cultural Heritage Management*, Newcastle upon Tyne, World Archaeological Congress, WCD Working Paper.

Brehm MR, Quiroz J, 1995, *The Water Market for Water Rights in Chile*, Washington DC, The World Bank.

Bridle R, 2000, British Dam Society, personal communication with authors, 10 June.

Bridle RC, Sims GP, 1999, 'Benefits of Dams to British Society', in Turfan M, (ed), *Benefits of and Concerns about Dams: Case Studies*, Ankara, Turkish National Committee on Large Dams.

Brody H, 1999, *Assessing the Project: Social Impacts and Larger Dams*, Contributing Paper for WCD Thematic I.1 Social Impacts.

Brown C, King J, 1999, *Information Needs for Appraisal and Monitoring of Ecosystems Impacts*, Contributing Paper for WCD Thematic Review II.1 Ecosystems.

Brown C, King J, Tharme R, 1999, *Definition and Implementation of Instream Flows*, Contributing Paper for WCD Thematic Review II.1 Ecosystems.

Brown LW, Halweil B, 1998, 'China's Water Shortage Could Shake World Food Security', in *World Watch*, July/August.

Brunold H, Kratochwill, 1999, 'Conflict Management as Part of a Successful Implementation Strategy', Conference Proceedings, *Hydropower into the Next Century, International Journal on Hydropower and Dams*, Gmunden, Austria, October 1999.

Bruijnzeel LA, 1990, *Hydrology of Moist Tropical Forests and Effects of Conversion: A State of Knowledge Review*, Paris, International Hydrological Programme of the United Nations Educational, Scientific and Cultural Organisation.

Cernea MM, 2000, 'Risks, Safeguards, and Reconstruction: A Model for Population Displacement and Resettlement', in Cernea MM, McDowell C, (eds), *Risks and Reconstruction - Experiences of Resettlers and Refugees*, Washington DC, The World Bank.

Cernea MM (ed), 1999, *The Economics of Involuntary Resettlement: Questions and Challenges*, Washington DC, The World Bank.

Cernea MM, Guggenheim SE (eds), 1993, *Anthropological Approaches to Involuntary Resettlement: Policy, Practice and Theory*, Boulder, Colorado, Westview Press.

Chan NW, 1995, *A Contextual Analysis of Flood Hazard Management in Peninsula Malaysia*, unpublished PhD, Enfield, Middlesex University.

Chen C, 1999, *The Chixoy Dam Case*, Rio Negro Community Representative, Guatemala, WCD Regional Consultation Paper.

Childs-Johnson E, 2000, 'The Three Gorges Project: There is No Dragon', in Brandt S, Hassan F (eds), *Dams and Cultural Heritage Management*, World Archaeological Congress, WCD Working Paper.

Colajacomo J, 1999, *The Chixoy Dam in Guatamala: The Maya Achi Genocide - The Story of Forced Resettlement*, Contributing Paper for WCD Thematic Review I.2 Indigenous People.

Colchester M, 1993, *Slave and Enclave: Towards a Political Ecology of Equatorial Africa*, Penang, World Rainforest Movement.

Colchester M, 1995, *Venezuela: Violations of Indigenous Rights*, Chadlington, World Rainforest Movement.

Collier M, Webb RH, Schmidt JC, 1996, *Dams and Rivers: A Primer on the Downstream Effects of Dams*, US

Geological Survey Circular 1126, Tuscon, US Geological Survey.

Colson, E, 1971, *The Social Consequences of Resettlement*, Manchester, Manchester University Press.

Colson E, 1999, 'Engendering Those Uprooted by 'Development'', in Indra D (ed), *Engendering Forced Migration: Theory and Practice*, Oxford, Oxford Refuge Studies Program, Oxford University.

Cook CC, (ed) 1994, *Involuntary Resettlement in Africa: Selected Papers from a Conference on Environment and Settlement Issues in Africa*, World Bank Technical Paper Number 227, Washington DC, The World Bank.

Corbett CJ, Kirsch DA, 2000, 'ISO 14000: An Agnostic's Report from the Frontline' in *ISO9000 + ISO140000 News*, No.2, February: 4-17.

Cornish G, 1998, *Modern Irrigation Technologies for Smallholders in Developing Countries*, SRP Exeter.

Correa I, 1999, *Urra Dam Project, Colombia*, Association of Producers for Communal Development of La Cienaga Grande de Lorica - ASPROCIG, WCD Regional Consultation Paper.

Cosgrove WJ, Rijsberman FR, 1999, *World Water Vision: Making Water Everybody's Business*, World Commission on Water for the 21st Century, draft report, November.

Crabb P, 1997, *Murray Darling Basin Resources*, Canberra, Murray Darling Basin Commission.

CWC (Central Water Commission), 1994, *National Register of Large Dams*, New Delhi, Central Water Commission, Government of India.

Davidson N, Delany S, 2000, *Biodiversity Impacts of Large Dams: Waterbirds*, funded by UNEP, commissioned by IUCN from Wetlands International, Wageningen, Netherlands, Contributing Paper for WCD Thematic Review II.1 Ecosystems.

Daes I, 1996a, *Pacific Island Workshop on the United Nations Draft Declaration on the Rights of Indigenous Peoples*, paper presented at Suva, Fiji, September.

Daes I, 1996b, *Supplementary Report of the Special Rapporteur on the Protection of the Heritage of Indigenous Peoples*, United Nations Sub-Commission on Prevention of Discrimination and Protection of Minorities, forty-eighth Session, E/CN.4.Sub.2/1996/22.

Delaney TA, 1995, 'Benefits to Downstream Flood Attenuation and Water Quality as a Result of Constructed Wetlands in Agricultural Landscapes, in *American Farmland Trust*, Center for Agriculture in the Environment, http://www.farmlandinfo.org/cae/caepubs/delaney.html.

De Wet C, 1999, *The Africa Experience*, Contributing Paper for WCD Thematic Review 1.3 Displacement.

DFID (Department for International Development), 2000, *Water*, Knowledge and Research Newsletter, Department for International Development, Issue 10, May.

Dhawan BD, 1988, *Irrigation in India's Agricultural Development: Productivity, Stability, Equity*, New Delhi, Sage Publications, second edition, New Delhi, Commonwealth Publishers.

Dickinson MA, 1999, *Water Efficiency Case studies*, California Urban Water Conservation Council, cited in WCD Thematic Review IV.3 Water Supply Options.

Dietrich W, 1999, *Impacts of Dams on River Geomorphology*, Berkeley, University of California at Berkeley, WCD Submission env082.

Dixon DA, 2000, "A Growing Problem", in *International Water Power and Dam Construction*, May: 23-25.

Dos Santos A, 1999, 'Salto Caxias Hydroelectric Power Plant: Meeting Environmental Needs', in *International Journal of Hydropower & Dams: Hydropower into the Next Century - III Conference Proceedings*, Gmunden, Austria.

Down to Earth, 1998, "The Value of a Raindrop", 15 October: 22-33.

Drinkwater KF, Frank KT, 1994, 'Effects of river regulation and diversion on marine fish and invertebrates', in *Aquatic Conservation: Freshwater and Marine Ecosystems*, No. 4: 135-151.

Driver P, 2000, *Mubuku III Hydropower Project in Western Uganda*, Contribution to WCD Thematic Review 1.3 Web Conference, 28 January.

Dynesius M, Nilsson C, 1994, 'Fragmentation And Flow Regulation Of River Systems In The Northern Third Of The World', in *Science* No. 266, 4 November 1994:753-762.

EBRD (European Bank for Reconstruction and Development), 1996, *Latvian Hydropower Plants Receive EBRD Financing*, press release, 29 April, London, European Bank for Reconstruction and Development, http://www.ebrd.com/english/opera/PRESSREL/PR1996/30APR29.HTM.

EBRD, 1999, *EBRD Signed and Approved Projects in Slovenia: as at 31 December 1999*, London, European Bank for Reconstruction and Development, http://www.ebrd.com/english/opera/COUNTRY/Slovproj.htm.

EBRD, 2000a, *EBRD Activities in Azerbaijan: EBRD Signed Projects in Azerbaijan*, London, European Bank for Reconstruction and Development, http://www.ebrd.com/english/opera/COUNTRY/AZERACHT.HTM.

EBRD, 2000b, *EBRD Signed Projects in Georgia: as at 30 June 2000*, London, European Bank for Reconstruction and Development, http://www.ebrd.com/english/opera/COUNTRY/Georproj.htm.

Eckstein O, 1958, *Water Resource Development: The Economics of Project Evaluation*, Cambridge, Massachusetts, Harvard University Press.

EEA (European Environment Agency), 1999, *Lakes and Reservoirs in the EEA Area*, Topic Report 1/1999, Copenhagen.

Eley TJ, Watkins TH, 1991, 'In a Sea of Trouble - The Uncertain Fate of the Pacific Salmon', in *Wilderness*, Fall: 20-26.

Emerton L, 2000, *Environmental Economic Tools for Project Analysis: Tana Hydropower Scheme*, IUCN working paper.

Epple R, 2000, *Dam Decommissioning in Europe*, European Rivers Network, WCD Submission opt136.

EWEA (European Wind Energy Association), 1999, *Submission from the EWEA: Greenpeace and Wind Industry Unveil Global Energy Blueprint*, Contributing Paper for WCD Thematic Review IV.1 ESDM Annex H.

Fact Finding Committee on the Sri Sailam Project, 1986, 'The Srisailam Resettlement Experience: The Untold Story', cited in Mander H et al, 1999, *Dams, Displacement, Policy and Law in India*, Contributing Paper for WCD Thematic Review 1.3 Displacement.

Fahim HM, 1981, *Dams, People and Development: The Aswan High Dam Case*, New York, Pergamon Press.

Famolare L, 2000, Director, Guianas Regional Program, Conservation International, personal communication with authors, August.

FAO (UN Food and Agriculture Organization), IPTRID, World Bank, 1995,

Modern Water Control and Management Practices in Irrigation, Water Reports 19, Rome, UN Food and Agriculture Organization.

FAO, 1995, *Irrigation in Africa in Figures*, Water Reports, Rome, UN Food and Agricultural Organization.

FAO, 1997, 'Irrigation Potential in Africa: A Basin Approach', in *FAO Land and Water Bulletin* 4, Rome, UN Food and Agriculture Organization.

FAOSTAT, 1998, *FAO Statistical Databases*, Rome, UN Food and Agricultural Organization, http://apps.fao.org/, viewed 29 September 2000.

Faught M, 2000, 'National Reservoir Inundation Study 1975-1980', in Brandt S, Hassan F (eds), *Dams and Cultural Heritage Management*, Newcastle-upon-Tyne, World Archaeological Congress, WCD Working Paper.

Fearnside P, 1995 'Hydroelectric Dams in the Brazilian Amazon as Sources of 'Greenhouse' Gases' in *Environmental Conservation*, Vol. 22, No.1, Spring: 7-19.

Fearnside P, 2000, 'Environmental Impacts of Brazil's Tucurui Dam : Unlearned Lessons for Hydroelectric Development in Amazonia', in *Environmental Management*, Vol. 26.

Fernandes W, Paranjpye V, 1997, *Rehabilitation Policy and Law in India: A Right to Livelihood*, Pune, New Dehli, ECONET, Indian Social Institute.

Fernea RA, 1998, *Including Minorities in Development: the Nubian Case*, unpublished report for the World Bank, Washington DC, World Bank.

Ferradas CA, 1999, *The Latin American Region*, Contributing Paper for WCD Thematic Review I.1 Social Impacts.

Fisher RJ, 1993, 'Creating Space: Development Agencies and Local Institutions in Natural Resource Management', in *Forest Trees and People Newsletter*, No. 22: 4-11.

Flanders D, 1999, *Effectiveness of Fishways and Fish Lifts in Queensland: Passage of Native Fish in a Modified Vertical-slot Fishway on the Burnett River Barrage, South-eastern Queensland*, by Stuart IG, Berghuis AP, Bundaberg, Queensland Department of Primary Industries, WCD Submission env219.

Flavin C, 1999, *Energy in the 21st Century*, International Cogeneration Alliance, Contributing Paper to WCD Thematic Review IV.1 ESDM Annex H.

Fox JA, Brown LD, 1998a, 'Assessing the Impact of NGO Advocacy Campaigns' in Fox JA, Brown LD (eds), *The Struggle for Accountability: The World Bank, NGOs, and Grassroots Movements*, Cambridge, Massachusetts, MIT Press.

Fox JA, Brown DL, (eds), 1998b, *The Struggle for Accountability: The World Bank, NGOs, and Grassroots Movements*, Cambridge, Massachusetts, MIT Press.

Francfort JE, Cada GF, Dauble DD, Hunt RT, Jones DW, Rhinehart RN, Sommers GL, Costello RJ, 1994, *Environmental Mitigation at Hydropower Projects; Volume II Benefits and Costs of Fish Passage and Protection*, US Department of Energy, Idaho Operations Office.

Frausto K, 1999, *Irrigation Study: Technological – Potential for Improvement*, Contributing Paper for WCD Thematic Review IV.2 Irrigation Options.

Frazier S, 1999, *Ramsar Sites Overview*, Wetlands International, Wageningen, Netherlands.

Freeman AM, 1993, *The Measurement of Environmental and Resource Values*, Washington DC, Resources for the Future.

Friedrich H, 2000, *The Biodiversity of the Wetlands in the Lower Mekong Basin*,

IUCN, WCD Regional Consultation Paper.

Fuji H, Cho MC, 1996, 'Water Management under Direct Seeding,' in Morooka Y, Jegatheesan S, Yasunobi K, (eds), *Recent Advances in Malaysian Rice Production: Direct Seeding Culture in the Muda Area*, MADA and JIRCAS.

Furness HD, 1978, *Ecological Studies on the Pongola River Floodplain*, Working Document IV, Workshop on Man and the Pongolo Floodplain, No. 14/106/7C, Pietermaritzburg, South Africa CSIR.

Gammelsrød T, 1996, 'Effect of Zambezi Management on the Prawn Fishery of the Sofala Bank', in Acreman MC, Hollis GE, (eds), *Water Management and Wetlands in Sub-Saharan Africa*, Gland, Switzerland, IUCN.

Gapuz AT, Shalupirip S, 2000, *What My People Stand to Lose with the Construction of San Roque Dam*, Philippines, Indigenous Peoples' Movement, WCD Regional Consultation Paper.

Gillis M, Perkins DH, Roemer M, Snodgrass DR, 1987, *Economics of Development*, 2nd Edition, W W Norton & Company, Inc, New York

Gleick PH, 1998, *The World's Water: The Biennial Report on Freshwater Resources*, Washington DC, Island Press.

Gleick PH, 2000, *The World's Water: The Biennial Report on Freshwater Resources*, Washington DC, Island Press.

Goldsmith E, Hildyard N, 1984, *The Social and Environmental Impacts of Large Dams*, Cornwall, UK, Wadebridge Ecological Centre.

Goldzimer AM, 2000, *Prior Informed Consent of Project-affected Indigenous Peoples: An Analysis of Case Studies*, John F. Kennedy School of Government, Harvard University, WCD Submission soc013.

Goodland R, 1997, 'Environmental Sustainability in the Hydro Industry, Desegregating the Debate', in Dorsey T (ed), *Large Dams: Learning from the Past, Looking at the Future*, Gland, Switzerland, IUCN.

Goodland, R, 2000, *Social and Environmental Assessment to Promote Sustainability*, An Informal View from the World Bank. Environment Management Series, Paper No. 74, Washington DC, World Bank.

Gould J, 1999, *Rainwater Harvesting*, Contributing Paper for WCD Thematic Review IV. 3 Water Supply.

Government of India, 2000, *Comments*, Ministry of Health and Family Welfare, 30th May, in WCD India Country Study.

Gray A, 1995, 'The Indigenous Movement in Asia', in Barnes RH, Gray A, Kingsbury B (eds), *Indigenous Peoples of Asia*, Michigan, The Association for Asian Studies Inc.

Gray A, 1998, 'Development Policy - Development Protests: The World Bank, Indigenous Peoples, and NGOs', in Fox JA, Brown LD (eds), *The Struggle for Accountability: The World Bank, NGOs, and Grassroots Movements*, Cambridge, Massachusetts, MIT Press.

Grunwald M, 2000, 'Agency says Engineers Likely Broke Rules: Corps Economist's Allegations of Rigged Lock Expansion Study Forwarded to Cohen', in *Washington Post*, 29 February.

Guerra LC, Bhuiyan SI, Tuong TP, Barker R, 1998, *Producing More Rice with Less Water*, SWIM Paper 5, Colombo, International Water Management Institute.

Guhan S, 1995, *The World Bank's Lending in South Asia*, Brookings Occasional Papers, Washington DC, Brookings Institute.

Gwala P, 2000, 'South Africa: Inanda Dam. How the Dam Affected the Mapheph-eteni Tribe', in Stott N, Sack K, Greeff L, (eds), *Once There Was A Community*, Southern African Hearings For Communities Affected By Large Dams, Environmental Monitoring Group.

Hanusin J, 1999, *Water Supply in Slovakia*, WCD Submission opt052.

Hart J, 2000, Assistant Crown Solicitor, Native Title Section, Australia, 'Northern Territories Native Title Act 1993', personal communication with authors, 13 June.

Hassan F, 2000, 'The Aswan High Dam and the International Rescue Campaign', in Brandt S, Hassan F (eds), *Dams and Cultural Heritage Management*, Newcastle-upon-Tyne, World Archaeological Congress, WCD Working Paper.

Hearne R, Easter KW, 1995, *Water Allocation and Water Markets: An Analysis of Gains-From-Trade in Chile*, World Bank Technical Paper Number 315, Washington DC, World Bank.

Hearne RR, Trava JL, 1997, *Water Markets in Mexico: Opportunities and Constraints*, Discussion Paper 97-01, Environmental Economics Programme, London, International Institute for Environment and Development.

Heuperman A, 1999, *Potential for Improvement - Some Drainage Options*, Contributing Paper for WCD Thematic Review IV.2 Irrigation Options.

Hira P, 1969. 'Transmission of Schistosomiasis in Lake Kariba, Zambia', in *Nature*, Vol. 224.

Holden PB, Stalnaker CB, 1975, 'Distribution and Abundance of Mainstream Fishes of the Middle and Upper Colorado River Basins', in *Transactions of the American Fisheries Society*, Vol. 104: 217-231.

Hollis GE, Adams WM, Aminu-Kano M, (eds) 1993, *The Hadejia-Nguru Wetlands: Environment, Economy and Sustainable Development of a Sahelian Floodplain Wetland*, Gland, Switzerland and Cambridge, IUCN.

Horowitz M, Salem-Murdock M, Niasse M, Magistro J, Nuttal C, Kane O, Grimm C, Sella M, 1994, *Les Barrages de la Controverse. Le Cas de Flueve Senegal*, Paris, Harmattan.

Huashan L, Yongtang J, 1995, 'Canal Lining Experience in China,' in *Proceedings of the Workshop on Canal Lining and Seepage*, October 1993, Lahore, Pakistan, H R Wallingford, UK

Hubbs C, Pigg J, 1976, 'The Effects of Impoundments on Threatened Fishes of Oklahoma', in *Annals of the Oklahoma Academy of Science*, Vol 5:133-77.

Huertas H, Pacheco B, 1999, *The Bayano Hydroelectric Dam in Panama*, WCD Regional Consultation Paper.

Hynes HBN, 1970, 'The Ecology Of Flowing Waters In Relation To Management', in *Journal of The Water Pollution Control Federation*, Vol. 42, No. 3: 418-424.

ICE (Instituto Costarricense de Electricidad), 1994, *Electricity and Sustainable Development in Costa Rica*, San Jose, Instituto Costarricense de Electricidad.

ICE, 1996, *Planes de Expansion de la Generacion (Escenario Base)*, San Jose, Instituto Costarricense de Electricidad.

ICID (International Commission on Irrigation and Drainage), 2000, *Role of Dams for Irrigation, Drainage and Flood Control. ICID Position Paper*, New Delhi, International Commission on Irrigation and Drainage.

ICOLD (International Commission on Large Dams), 1981, 'Dam Projects and Environmental Success' in *ICOLD Bulletin*, No. 37, Paris, International Commission on Large Dams.

ICOLD, 1988, 'Dams and Environment. Case Histories', in *ICOLD Bulletin*, No. 65, Paris, International Commission on Large Dams.

ICOLD, 1995, 'Dam Failures Statistical Analysis', in *ICOLD Bulletin*, No. 99, Paris, International Commission Large Dams.

ICOLD, 1997, *Position Paper on Dams and the Environment*, Paris, International Commission on Large Dams.

ICOLD, 1998, *ICOLD World Register of Dams*, Computer Database, Paris, International Commission on Large Dams.

ICOLD, 2000, *World Register of Dams 1998: Updating 2000 for China*, Paris, International Commission on Large Dams.

IDB (Inter-American Development Bank), 1999, *IDB's Dam-Related Projects (1960-1999)*, unpublished, Washington DC, Inter-American Development Bank.

IDS (Institute of Development Studies), 2000, *Operationalisation of Prior Informed Consent*, Sussex, Contributing Paper for WCD Thematic Review 1.2 Indigenous People.

IEA (International Energy Agency) 1998, *Key World Energy Statistics*, International Energy Agency, Paris, http://www.iea.org/stats/files/keystats/stats_98.htm.

IEA, 2000, *Implementing Agreement for Hydropower Technologies and Programmes. Annex III Hydropower and the Environment: Present Context and Guidelines for Future Action*, Paris, International Energy Agency.

IFRCRCS (International Federation of Red Cross and Red Crescent Societies) 1998, *World Disasters Report 1998*, Oxford, Oxford University Press.

IJHD (International Journal of Hydropower and Dams), 1999, *World Atlas and Industry Guide 1999*, International Journal of Hydropower and Dams, UK, Aqua-Media International.

IJHD (International Journal of Hydropower and Dams), 2000, *World Atlas and Industry Guide 2000*, International Journal of Hydropower and Dams, UK, Aqua-Media International.

IPCC (Intergovernmental Panel on Climate Change), 1992, *1992 IPCC Supplement*, Geneva, World Meteorological Organisation.

IPCC, 1996, *Climate Change 1995: The Science of Climate Change*, (eds) Houghton JT, Meiro Filho LG, Callendar BA, Harris N, Kattenburg A, Maskell K, Cambridge, Cambridge University Press.

IPCC, 1999, *Special Report on Emission Scenarios*, draft version, Geneva, World Meteorological Organization and United Nations Environment Program.

IUCN (World Conservation Union), 2000, *Vision for Water for Nature - A World Strategy for Conservation and Sustainable Management of Water Resources in the 21st Century*, Gland, Switzerland, and Cambridge, IUCN.

Jackson D, Marmulla G, 2000, *The Influence of Dams on River Fisheries*, Contributing Paper for WCD Thematic Review II.1 Ecosystems.

Jing J, 1999, *Displacement, Resettlement, Rehabilitation, Repatriation and Development - The China Report*, Contributing Paper for WCD Thematic Review 1.3 Displacement.

Jobin, WR, 1999, *Dams and Disease*, London, Routledge.

Johnson III SH, 1997, *Management Transfer in Mexico: A Strategy to Achieve Irrigation District Sustainability*, Research Report No. 16, Colombo, International Irrigation Management Institute.

Johnston, BR, 2000, *Reparations and the Right to Remedy*, Center for Political Ecology,

University of California, Contributing Paper for WCD Thematic Review I.3 Displacement.

Jubb, RA, 1972, 'The J.G. Strydom Dam, Pongolo River, Northern Zululand: The Importance of Floodplain Pans Below It', in *Piscator*, No.86: 104-9.

Kammen DM, 1999, 'Bringing Power to the People: Promoting Appropriate Energy Technologies in the Developing World', in *Environment*, Vol. 41, No.5: 10-15, 34-41.

Kassas M, 1973, 'Impact of River Control Schemes on the Shoreline of the Nile Delta', in Farvar MT, Milton JP, (eds) *The Careless Technology: Ecology and International Development*, London, Stacey.

Keller A, Sakthivadivel R, Seckler D, 2000, *Water Scarcity and the Role of Storage in Development. Research Report 39*, Colombo, International Water Management Institute

Kijne J, Prathapar SA, Wopereis MCS, Sahrawat KL, 1998, *How to Manage Salinity in Irrigated Lands: A Selective Review with Particular Reference to Irrigation in Developing Countries*, Colombo, Sector-Wide Initiative for Water Management (SWIM).

Kinahan J, 2000, 'Quaternary Surveys: Lessons from the Joint Angolan-Namibian Lower Cunene Hydropower Scheme', in Brandt S, Hassan F (eds), *Dams and Cultural Heritage Management*, Newcastle-upon-Tyne, World Archaeological Congress, WCD Working Paper.

Kingsbury B, 1995, 'Indigenous Peoples as an International Legal Concept', in Barnes RH, Gray A, Kingsbury B (eds), *Indigenous Peoples of Asia*, Michigan, The Association for Asian Studies Inc.

Knudson T, Vogel N, 1997, 'Again Dams are Facing New Pressures', in *The Sacramento Bee*, 23-27 November.

Kowalski M, Schuster J, 2000, 'Die Windmacher', in *Focus*, No. 29: 162-166.

Kudlavicz, M, 1999, *Reao: Barragen de Porto Primavera no Rio Paraná*, WCD Submission env129.

Kudlavicz, M, 2000, *Porto Primavera Dam in Rio Paraná*, WCD Submission env063.

Kuriki M, 2000, General Manager, Chugoku Regional Construction Bureau, Ministry of Construction, personal communication with authors, 14 March.

Lagman A, 2000, *Database of ADB Large Dams*, unpublished, Manila, Asian Development Bank.

Lahmeyer International, 1990, *Nam Ngum Operations Study*, prepared for Asian Development Bank, Vientiane.

Lamb D, Gilmour D, 2000, *A Succinct Overview of the Issues from the Scientific Sessions as a Basis for Interfacing with Policy*, paper presented at UNESCO Symposium on Forest-Water-People in the Humid Tropics, Bangi, Malaysia, 30 July-4 August.

Lane J, 1999, *Assessment of Water Supply Options*, Contributing Paper for WCD Thematic Review IV.3 Water Supply Options.

Lang C, Hildyard N, Geary K, Grainger M, 2000, *Dams Incorporated: The Record of Twelve European Dam Building Companies: A Report by Corner House*, Sturminster Newton, UK, Swedish Society for Nature Conservation, WCD Submission eco041.

Larinier M, 2000, *Dams and Fish Migration*, Contributing Paper for WCD Thematic Review II.1 Ecosystems.

Laxman MK, 1999, 'Testing the Risks and Reconstruction Model on India's Resettlement Experience', in Cernea MM (ed), *The Economics of Involuntary Resettlement: Questions and Challenges*, Washington DC, World Bank.

Lecornu J, 1998, *Dams and Water Management*, Report of the Secretary General, International Commission on Large Dams to the Conference Internationale Eau et Developpement Durable, 19-21 March, Paris, France, at http://genepi.louis-jean.com/cigb/article-barrages-an.html.

Ligon FK, Dietrich WE, Trush WJ, 1995, 'Downstream Ecological Effects of Dams', in *Bioscience*, Vol 45: 183-192.

Lovgren L, 1999, *Moratorium in Sweden: An Account of the Dams Debate*, WCD Submission env136.

Lovei L, McKechnie A, 2000, 'The Costs of Corruption for the Poor - the Energy Sector', in *Public Policy for the Private Sector*, June, No. 21: 34-42.

Lowe-McConnell RH, 1985, 'The Biology of the River Systems with Particular Reference to the Fish', in Grove AT (ed), *The Niger and its Neighbours: Environment, History and Hydrobiology, Human Use and Health Hazards of the Major West African Rivers*, Rotterdam, Balkema.

Lu Y, 2000, *Three Gorges Project: A Project Improving Ecological Environment of the Yangtze*, papers for the Beijing 2000 Congress of ICOLD, draft.

MacDonald A, McInally GA, 1998, "Reservoirs and Flood Control: A Northern Perspective", in Tedd P (ed), *The Prospect for Reservoirs in the 21st Century*, proceedings of the tenth conference of the British Dams Society, University of Wales, Bangor, 9-12 September, London, Thomas Telford Publishing.

Macoun A, Horta K, Tricario A, 2000, *Brief Report from Meeting Between Andrew Macoun, the World Bank Task Manager on Lesotho Highlands Project with Korina Horta, EDF and Antonio Tricario of Reform the World*, June.

Mahmood K, 1987, *Reservoir Sedimentation – Impact, Extent, and Mitigation*, World Bank Technical Paper No 71, Washington DC, World Bank.

Maltby E, 1986, *Waterlogged Wealth*, London, Earthscan.

Mander H, Hemadri R, Nagraj V, 1999, *Displacement, Policy and Law in India*, Contributing Paper for WCD Thematic Review I.3 Displacement.

Manibeli Declaration, 1994, *Call for a Moratorium on World Bank Funding of Large Dams*, annexed in McCully P, 1996, *Silenced Rivers: The Ecology and Politics of Large Dams*, London, Zed Books.

Maybury-Lewis D, 1997, *Indigenous People, Ethnic Groups and the State*, Allen and Bacon.

Mburugu E, 1994, 'Dislocation of Settled Communities in the Development Process: The case of Kiambere Hydroelectric Project', in Cook CC (ed), *Involuntary Resettlement in Africa*, The World Bank Technical Paper Number 227, Washington DC, World Bank.

McCully P, 1996, *Silenced Rivers: The Ecology and Politics of Large Dams*, London, Zed Books.

McCully P, 1997a, *A Critique of "The World Bank's Experience with Large Dams: A Preliminary Review of Impacts,"* Berkeley, International Rivers Network.

McCully P, 1997b, 'Taking Down Bad Dams', in *World Rivers Review*, Vol. 12, No. 4, August.

McCully P, 1999, *Cost and Time Overruns for Dam Projects*, WCD Submission eco061.

McDowell C (ed), undated, *Resisting Impoverishment*, Oxford, Berghahn Books for Refugee Studies Programme, University of Oxford.

McIntosh AC, Yñiguez CE, (eds)1997, *Second Water Utilities Data Book: Asia and Pacific Region*, Manila, Asian Development Bank.

Mehta L, and Srinivasan B, 1999, *Balancing Pains and Gains. A Perspective Paper on Gender and Large Dams*, Contributing Paper for WCD Thematic Review I.1 Social Impacts.

Meinzen-Dick R, 1997, 'Valuing the Multiple Uses of Irrigation Water', in Kay M, Franks T, Smith L (eds), *Water: Economics, Management and Demand*, Proceedings of the 18th European Regional Conference, Oxford, September, International Commission on Irrigation and Drainage.

Milewski J, Egre D, Roquet V, 1999, *Dams and Benefit Sharing*, Montreal, Hydro-Quebec, Direction Environment, WCD Submission soc196.

Ministry of Construction, Japan, 1999, *Japan Dam Almanac*, Tokyo, Ministry of Construction.

Ministry of Water Resources and Electric Power, PRC, 1987, *People's Republic of China, Irrigation and Drainage in China*, Beijing, China Water Resources and Electric Power Press.

Mitchell TE, 1995, 'Report on Canal Linings Used by the Bureau of Reclamation', in *Proceedings of the Workshop on Canal Lining and Seepage, October 1993, Lahore, Pakistan*, UK, HR Wallingford.

Mncina J, Ginindza N, 1999, 'Swaziland: Community Participation in the Construction of the Maguga Dam', Ekuvinjelweni Resettlement Committee Maguga Dam, in Stott N, Sack K, Greeff L (eds), 2000, *Once There Was A Community*, Southern African Hearings For Communities Affected By Large Dams, Environmental Monitoring Group.

Molden D, Sakthivadivel R, Perry CJ, de Frainture C, Kloezen WH, 1998, *Indicators for Comparing Performance of Irrigated Agricultural Systems*, Research Report No. 20, Colombo, International Water Management Institute.

Moore D, Sklar L, 1998, 'Reforming the World Bank's Lending for Water', WCD Submission eco048 in Fox JA, Brown LD (eds), *The Struggle for Accountability: The World Bank, NGOs, and Grassroots Movements*, Cambridge, Massachusetts, MIT Press.

Morse B, Berger T, 1992, *Sardar Sarovar: The Report of the Independent Review*, Ottawa, Resource Futures International Inc.

Mungomba L, Chandiwana S, Madsen H, 1993, 'Schistosomiasis Around Siavonga on the Shores of Lake Kariba, Zambia', in *Annals of Tropical Medicine and Parasitology*, Vol. 87, No.4.

Mung'ong'o C, 1997, 'Pangani Dam Versus the People,' in Usher AD (ed), *Dams as Aid: A Political Anatomy of Nordic Development Thinking*, London, Routledge, WCD Submission eco026.

Murray-Rust DH, Vander Velde EJ, 1994, 'Changes in Hydraulic Performance and Comparative Costs of Lining and Desilting of Secondary Canals in Punjab, Pakistan', in *Irrigation and Drainage Systems*, Vol. 8, Issue 3:137-158.

Nachtnebel HP, 2000, 'From Hainburg to Freudenau: An Austrian Experience With Seeking Solutions With Public Participation', in Gayer J (ed) *Participatory Processes in Water Management (PPWM)*, Proceedings of the Satellite Conference to the World Conference on Science (Budapest, Hungary 28-30 June 1999), UNESCO/ IHP-V Technical Documents in Hydrology No. 30, Paris: 105-117.

Nahmad S, 1999, *The Impact of hydro-electric dams on Indigenous People, Chinantecos, Otomies, Huicholes: A Case Study from Mexico*, WCD Regional Consultation Paper.

National Research Council, 1996, *A New Era for Irrigation*, Washington DC, National Academy Press.

NEA (Nepal Electricity Authority), 1997, *Power System Simulation Model*, Medium Hydropower Study Project, Nepal, Nepal Electricity Authority.

Niasse M, 1991, 'Les Périmetres Irrigués Villageois Vieillissent Mal: Les paysans se désengagent-ils en même temps que la SAED?', in Crousse M, Seck SM, Mathieu P (eds), *La Vallée du Fleuve Sénégal, Evaluations et perspectives d'une décennie d'aménagements (1980-1990)*, Paris, Karthala.

Niasse M, 1997, *Réforme Foncière et Equité. La Loi sur le Domaine National à l'épreuve dans les périmètres de la vallée du fleuve Sénégal*, Proceedings of the Regional Conference on Décentralisation et Réformes Foncières au Sahel, Djamena, Chad, CILSS/PADLOS, 28 July-1 August.

Norpower, 1993, *Nam Theun 1/2 Hydropower Project. Feasibility Study*, Vol.3, Environmental Impact Assessment Report, May, cited in Usher AD, Ryder G, 1997, 'Vattenfall Abroad: Damming the Theun River', in Usher AD (ed) *Dams as Aid: A Political Anatomy of Nordic Development Thinking*, London, Routledge, WCD Submission eco026.

Norr L, Fraught M, 2000, 'Archaeological Site Location and Assessment on Lake Alajuela, Panama', in Brandt S, Hassan F (eds) *Dams and Cultural Heritage Management*, Newcastle-upon-Tyne, World Archaeological Congress, WCD Working Paper.

OECD (Organisation for Economic Co-operation and Development), 2000a, *Development Assistance Committee International Development Statistics: Online Databases*, Paris, Development Assistance Committee, Organisation for Economic Co-operation and Development, http://www.oecd.org/dac/htm/online.htm.

OECD (Organisation for Economic Co-operation and Development), 2000b, *Report by the CIME: Implementation of the Convention on Bribery in International Business Transactions and the 1997 Revised Recommendations*, Council at Ministerial Level, 26-27 June, Paris, Organisation for Economic Co-operation and Development.

OECD, 2000c, *OECD Outline: Anti-Corruption Unit*, website http://www.oecd.org/daf/nocorruption/index.htm, viewed 4 September 2000.

OED (Operations Evaluation Department), 1990, *Annual Review of Evaluation Results: 1989*, Washington DC, World Bank.

OED, 1992, *Water Supply and Sanitation Projects: The Bank's Experience: 1967-1989*, Washington DC, World Bank.

OED, 1993, *Early Experiences with Involuntary Resettlement*, Washington DC, World Bank.

OED, 1996a, *The World Bank's Experience with Large Dams: A Preliminary Review of Impacts*, Washington DC, World Bank.

OED, 1996b, *The World Bank's Experience with Large Dams: A Preliminary Review of Impacts: Profiles of Large Dams*, background document, Washington DC, World Bank.

OED, 1998, *Recent Experience With Involuntary Resettlement. Overview*, Report No. 17538, Washington DC, World Bank.

Opaso C, GABB - Action Group for Bio Bio, Chile, 1999, *The Bio Bio project: A Lesson Not Fully Learned*, WCD Regional Consultation Paper.

Ota SB, 2000, 'Cultural Heritage Management Vis-à-vis Dams: The Narmada Issue', in Brandt S, Hassan F (eds) *Dams and Cultural Heritage Management*, Newcastle-upon-Tyne, World Archaeological Congress, WCD Working Paper.

Oud B, 2000, Lahmeyer International, 'Operation of Nam Ngum during 1995 Flood', personal communication with authors, 2 June.

Oud E, Muir TC, 1997, 'Engineering and Economic Aspects of Planning, Design, Construction and Operation of Large Dam Projects', in Dorsey T (ed), *Large Dams: Learning from the Past, Looking at the Future*, Gland, Switzerland, IUCN

Panama Canal Office of Public Affairs, (undated), 'Growing Participation of Panamanians', in *The Panama Canal Commission*, Balboa Heights, Republic of Panama.

Parasuraman S, 1999, *The Development Dilemma: Displacement in India*, London, MacMillan.

Pearce D, 2000, Professor, Department of Economics, University College London, personal communication with authors, 16 March.

Pillay P, 1999, 'Water Project's Former CEO in Court', *Business Day*, Johannesburg' 30 November.

Pinstrup-Anderson P, Pandya-Lorch R, Rosegrant MW, 1997, *The World Food Situation – Recent Developments, Emerging Issues, and Long-Term Prospects*, Food Policy Statement Number 26, Washington DC, International Food Policy Research Institute.

Politis G, Endere ML, 2000, 'Archaeological Heritage Management and Dams in Argentina: A Brief Review of the Current Situation,' in Brandt S, Hassan F (eds), *Dams and Cultural Heritage Management*, Newcastle-upon-Tyne, World Archaeological Congress, WCD Working Paper.

Postel, 1998, 'Water for Food Production: Will there be enough on 2025?' in *Bio-Science*, August.

Postel S, 1999, *Pillar of Sand: Can the Irrigation Miracle Last?*, New York, WW Norton & Co.

Preston G, 1999, *Example of Demand Management from South Africa*, Contributing Paper for WCD Thematic Review IV.3 Water Supply Options.

Prinz D, et al, 1998, *Rainwater Harvesting for Dry Land Agriculture - Developing a Methodology Based on Remote Sensing and GIS*, Proceedings, XIII International Congress Agricultural Engineering, Rabat, Morocco, 2-6 February.

Prinz D, Singh AK, 1999, *Potential for Improvements of Water Harvesting*, Contributing Paper for WCD Thematic Review IV.2 Irrigation Options.

Pritchard S, 1998a, 'The International Covenant on Civil and Political Rights and Indigenous Peoples', in Pritchard S (ed), *Indigenous Peoples, the United Nations and Human Rights*, London, Zed Books.

Pritchard S, 1998b, *Setting International Standards: An Analysis of the United Nations Draft Declaration on the Rights of Indigenous Peoples*, 2nd edition, Aboriginal and Torres Strait Islander Commission.

Ramsar Convention Database, 1999, maintained by Wetlands International, Gland, Switzerland. http://www.wetlands.agro.nl/ramsar_database/ramsar_quick.html

Raskin P, Hansen E, Margolis R, 1995, *Water and Sustainability: A Global Outlook*, Stockholm, Stockholm Environmental Institute.

Raskin P, Gallapin G, Gutman P, Hammond A, Swart R, 1998, *Bending the Curve: Toward Global Sustainability*, Global Scenario Group, Stockholm, Stockholm Environment Institute.

Reinicke WF, Deng F, 2000, *Critical Choices: The United Nations, Networks and*

Future of Global Governance, International Development Research Centre, Canada.

Revenga C, Murray J, Abramowitz J, Hammond A, 1998, *Watersheds of the World: Ecological Value and Vulnerability*, Washington DC, The World Resources Institute and Worldwatch Institute.

Revenga C, Brunner J, Henninger N, Kassem K, Payne R, 2000, *Pilot Analysis of Global Ecosystems: Freshwater Systems*, Washington DC, World Resources Institute.

Ringler C, Rosegrant MW, Paisner MS, 1999, *Irrigation and Water Resources in Latin America and the Caribbean: Challenges and Strategies*, Report prepared for the Inter-American Development Bank, Washington DC, International Food Policy Research Institute.

Robinson B, 2000, Chairman, Environmental Protection Authority, Victoria, Australia, personal communication with authors, July.

Robinson S, 1999, *Displacement, Resettlement, Rehabilitation, Reparation and Development, The Mexico Case*, Contributing paper for WCD Thematic Review I.3 Displacement.

Roo, H, 2000, *Benefits and Concerns About Dams, General Report on the Papers for Question 77*, 20th ICOLD Congress, Beijing, September.

Rosa LP, Matvienko B, Santos MA, Sikar E, 1999, *Relatório Eletrobrás/Fundação Coppetec - Inventário das Emissões de Gases de Efeito Estufa Derivadas de Hidrelétricas*, COPPE Report to Eletrobras.

Rosegrant MW, Hazell PBR, 1999, *The Transformation of the Rural Economy. Rural Asia Transformed: The Quiet Revolution*, Theme Paper No. 1 for the Asian Development Bank, Washington DC, International Food Policy Research Institute.

Rumsey P, Flanigan T, 1995, *Compendium of Asian Energy Efficiency Success Stories*, Washington, DC, International Institute for Energy Conservation.

SANDRP, 1999, *Assessment of Water Supply Options for Urban India - Large Dams have No Case*, WCD Submission opt080.

Sant G, Dixit S, Wagle S, 1998, *Reassessing the Role of Large Dams in Meeting Power Demand*, Pune, India, PRAYAS Energy Group, WCD Submission eco013.

Saxena NC, 2000, 'Presentation to the Prime Minister of India from the Secretary, Planning Commission', personal communication to the authors, 18 July.

Schildgen B, 1999, 'Unnatural Disasters', in *Sierra*, May/June: 49-57.

Schmid RE, 2000, 'Americans Urged to Be Aware of Dams', Associated Press, 6 June, http://dailynews.yahoo.com/h/ap/20000606/us/dangerous_dams_1.html.

Schnitter NJ, 1994, *A History of Dams: The Useful Pyramids*, Oregon, Books News Inc.

Scudder T, 2000, *Dialogue Note with WCD Commissioners on the Multiplier Effects of Irrigation Schemes*, August.

Scudder T, 1997a, 'Social Impacts', in Biswas AK (ed), *Water Resources: Environmental Planning, Management and Development*, New York, McGraw Hill.

Scudder T, 1997b, 'Resettlement', in Biswas AK (ed), *Water Resources: Environmental Planning, Management and Development*, New York, McGraw Hill.

Scudder T, 1997c, 'Social Impacts of Large Dams', in Dorsey T (ed), *Large Dams: Learning from the Past, Looking at the Future*, Gland, Switzerland, IUCN.

Seckler D, 1996, *The New Era of Water Resources Management: From 'Dry' to 'Wet' Water Savings*, Research Report 1, Colombo, International Irrigation Management Institute.

Seckler D, Molden D, Barker R, 1998, *Water Scarcity in the Twenty-first Century*, IWMI Water Brief 1, Colombo, International Water Management Institute.

Seckler D, Amarasinghe U, Mollen D, de Silva R, Barker R, 1998, *World Water Demand and Supply, 1990 to 2025: Scenarios and Issues*, Research Report 19, Colombo, International Water Management Institute.

Sen A, 1999, *Development As Freedom*, Oxford, Oxford University Press.

Shalaby A, 1999, *Egypt's Experience in Managing the Nile River System*, WCD Regional Consultation Paper.

Shevah Y, 1999, *Irrigation and Agriculture Experience and Options in Israel*, Contributing Paper for WCD Thematic Review IV.2 Irrigation Options

Shiklomanov IA, 1998, *Assessment of Water Resources and Water Availability in the World. Report for the Comprehensive Assessment of the Freshwater Resources of the World*, Stockholm, Stockholm Environment Institute.

Silva Orrego JP, 1997, 'In Defence of the Biobio River', in Usher DA (ed) *Dams as Aid: A Political Anatomy of Nordic Development Thinking*, Routledge, London, WCD Submission eco026.

Sinha R, 1998, *The Bargi Experience: Lessons Learned the Hard Way*, WCD Submission soc009

Sklar L, McCully P, 1994, *Damming the Rivers: The World Bank's Lending for Large Dams*, Working Paper 5, Berkeley, International Rivers Network, WCD Submission eco029.

Sluiter L, 1992, *Mekong Currency*, Bangkok, Project for Ecological Recovery/ TERRA.

Smith DA, 1996, *Proceedings of the Institution of Civil Engineers of Ireland*, November 1996.

Smith L, 2000, *Options in Agricultural Policy*, Contributing Paper for WCD Thematic Review IV.2 Irrigation Options.

Smith R, 1999, 'Reduction in Useful Capacity of Maghreb Dams', in *ICID British Section - News and Views*, No. 29, Winter: 13.

Soong KK, 2000, *Empty Promises, Damned Lives*, Coalition of concerned NGOs on Bakun, Malayasia, WCD Regional Consultation Paper.

Srettachau C, Nungren K, Olsson A, 2000, *Social Impact of Rasi Salai Dam*, WCD Regional Consultation Paper.

St Louis V, Kelly CA, Duchemin E, Rudd EWM, Rosenberg DM, in press, 'Reservoir Surfaces as Sources of Greenhouse Gases to the Atmosphere: a Global Estimate', in *BioScience*.

Stanley DG, Warne AG, 1993, 'Nile Delta: Recent Geological Evolution and Human Impact', in *Science*, No. 260: 628-634.

Stapleton RM, 1996, 'Deep Woods and Clear Water: What Price Sterling Forest?', in *Land and People, The Public Trust for Public Land*, Volume 8, Number 2, Fall

Stewart J, O'Connell K, Ciborski M, Pacenza M, 1996, *A People Dammed: A Witness for Peace Publication*, Witness for Peace.

Sunday Independent, 2000, 'Bigger Fish in the Dock in Lesotho Trial', in *Sunday Independent*, South Africa, 11 June.

Supreme Court of India, 1999, *Affidavit on Resettlement and Rehabilitation by the Governments of Gujarat, Maharashtra and Madhya Pradesh*, unpublished, India.

Susskind L, Cruikshank J, 1989, *Breaking The Impasse: Consensual Approaches to Resolving Public Disputes*, BasicBooks.

Takeuchi K, Harada J, 1999, *Operation, Monitoring and Rehabilitation of Dams/ Reservoirs in Japan - Institutional Framework and Empirical Studies*, Department of Civil and Environmental Engineering, Yamanachi University, Japan.

Tamakloe MA, 1994, 'Long-Term Impacts of Resettlement: The Akosombo Dam Experience', in Cook CC (ed), *Involuntary Resettlement in Africa*, Africa Technical Department Series, Washington DC, World Bank.

Tedd P (ed), 2000, *Dams 2000*, proceedings of the biennial conference of the British Dams Society, University of Bath, 14-17 June, London, Thomas Telford Publishing.

Tenant C, 1994, 'Indigenous Peoples, International Institutions, and the International Legal Literature from 1945-1993', in *Human Rights Quarterly* Vol. 16, No. 1:1-57.

Thakkar H, 1999, *Assessment of Irrigation Options in India*, Contributing Paper to WCD Thematic IV.2 Irrigation Options.

Tharme RE, 2000, *International Trends in the Development and Application of Environmental Flow Methodologies: A Review*, Water Research Commission Technology Transfer Report, draft paper, Pretoria, Water Research Commission.

The Economist, 2000, 'The Electric Revolution', 5 August 2000: 17-18.

Thomas S, 1998, 'Columbia River and Salmon: A Brief History', in *Salmon Passage Notes*, US Army Corps of Engineers, North Pacific Division, Summer.

Townshend PD, 2000, 'Towards Total Acceptance of Fully Automated Gates', in Tedd P (ed), *Dams 2000*, proceedings of the biennial conference of the British Dams Society, University of Bath, 14-17 June, London, Thomas Telford Publishing.

Udall L, 1998, 'The World Bank and Public Accountability. Has Anything Changed?', in Fox JA, Brown LD (eds) *The Struggle for Accountability: The World Bank, NGOs, and Grassroots Movements*, Cambridge, Massachusetts, MIT Press.

Udall L, 2000, *Review of Environmental Guidelines of Selected Export Credit Agencies*, International Department of First Nations Development Institute, Contributing Paper for WCD Thematic Review V.4 Regulation.

Umaña A, (ed) 1998, *The World Bank Inspection Panel: The First Four Years (1994-1998)*. Washington DC, The World Bank.

UN (United Nations), 1947, *Universal Declaration on Human Rights*, New York, United Nations.

UN, 1986, *Declaration on the Right to Development*, New York, United Nations.

UN, 1998, *World Urbanization Prospects, The 1996 Revision, Estimates and Projections of Urban and Rural Populations and of Urban Agglomerations*, New York, United Nations Population Division.

UN, 1999, *World Population Prospects, The 1998 Revision*, Volume 1, Comprehensive Tables, New York, United Nations.

UNCED (United Nations Conference on Environment and Development), 1992, *Agenda 21, The Rio Declaration on Environment and Development*, United Nations Conference on Environment and Development, Rio de Janerio, Brazil, 3-14 June.

UNDP (United Nations Development Program), 1998, *Human Development Report 1998*, New York and Oxford, Oxford University Press.

UNDP, 1999, *Human Development Report 1999*, New York and Oxford, Oxford University Press.

UNDP, 2000, *Human Development Report 2000*, New York and Oxford, Oxford University Press.

UNDP, United Nations Department of Economic and Social Affairs, and

World Energy Council, 2000, *World Energy Assessment Report*, New York, United Nations Development Program.

UNEP (United Nations Environment Program), 1999, *Global Environment Outlook 2000*, Nairobi, United Nations Environment Program.

US Bureau of Reclamation, 2000a, *Here's a Warm Thought: Temperature Control Modifications at Glen Canyon Dam*, http://www.usbr.gov/amp/wr/gctc_feature.html, viewed 7 July 2000.

US Bureau of Reclamation, 2000b, *Daily Variations in the Water Discharge of the Colorado River at Lee's Ferry-Hourly Fluctuations in Streamflow (cubic feet per second) During September 1982* - data obtained from US Bureau of Reclamation, Upper Colorado Region, February.

US Geological Survey, 2000, *Real Time Water Data*, web page, http://water.usgs.gov/realtime.html, viewed March 2000.

USACE (US Army Corps of Engineers), 2000, *National Inventory of Dams*, http://crunch.tec.army.mil/nid/webpages/nid.cfm

Usher AD, 1997a, 'Pangani Power Struggle: Nordic Dam Builders on a Tanzanian River,' in Usher AD (ed), *Dams as Aid: A Political Anatomy of Nordic Development Thinking*, London, Routledge, WCD Submission eco026.

Usher AD, (ed) 1997b, *Dams as Aid: A Political Anatomy of Nordic Development Thinking*, London, Routledge, WCD Submission eco026.

Usher AD, Ryder, G, 1997, 'Vattenfall Abroad: Damming the Theun River', in Usher AD (ed), *Dams as Aid: A Political Anatomy of Nordic Development Thinking*, London, Routledge, WCD Submission eco026.

Van Hofwegen P and Svendsen M, 2000, *A Vision of Water for Food and Rural Development*, Delft, Institute for Hydraulic Studies (IHE), draft.

Van Koppen B, 1999, *More Crops and Jobs per drop. Managing Water for Gendered Poverty Alleviation and Agricultural Growth*, Colombo, International Water Management Institute.

Van Wicklin III WA, 1999, *Sharing Benefits for Improving Resettlers' Livelihood*, Washington DC, World Bank, WCD Submission soc184.

Vander Velde EJ, Tirmizi J, 1999, *Irrigation Policy Reforms in Pakistan: Who's Getting the Process Right?*, International Researchers Conference on Participatory Irrigation Management, Hyderabad, Administrative Staff College of India.

Vermillion DL, 1997, *Impacts of Irrigation Management Transfer: A Review of the Evidence*, Research Report No. 11, Colombo, International Water Management Institute.

Verocai I, 1999, *Environmental and Social Impact Assessment for Large Dams - Thematic Review from the Viewpoint of Developing Countries*, Contributing Paper for WCD Thematic Review V.2 Environmental and Social Assessment.

Walker KF, 1979, 'Regulated Streams in Australia: the Murray-Darling River System', in Ward JV, Stanford JA (eds) *The Ecology of Regulated Streams*, New York, Pelnum Press.

Wang J, (ed) no date, *Comprehensive Dictionary of The Yangtze River (Changjiang Dacidian)*, Wuhan, Wuhan Publishing House.

Wapenhans Task Force, 1992, *Effective Implementation: Key to Development Impact*, Report of the World Bank's Portfolio Management Task Force, Washington DC, World Bank.

Welcomme RL, 1976, 'Some General and Theoretical Considerations on the Fish

Yield of African Rivers', in *Journal of Fisheries Biology*, Vol.8: 351-364.

Wescoat J, Halvorsen S, 2000, *Ex Post Evaluations of Dams and Related Water Projects: Patterns, Problems and Potential*, Boulder, Colorado, University of Colorado at Boulder, Contributing Paper for WCD Thematic Review IV.5 Operations.

White S, Dupont P, Robinson D, 1999, *International Report- Demand Management*, Report to the 1999 International Water Services Association Conference, Sydney, Institute for Sustainable Futures, WCD Submission eco018.

Whittington D, Lauria DT, Mu X, 1991, 'A Study of Water Vending and Willingness to Pay for Water in Onitsha, Nigeria', in *World Development*, Vol. 9, No. 2/3: 179-198.

WHO, 1999, *Human Health and Dams*, prepared by Birley M, Bos R, Diop M, Jobin W, Unnikrishnan P, Geneva, WCD Working Paper.

Wiehen M, 1999, *The Integrity Pact (TI-IP): The Concept, the Model and the Present Applications: A Status Report As of November 1, 1999*, Transparency International, http://www.transparency.de/activities/ip_status-report.html.

Wolf A, 2000, *Development of Transboundary Waters: Obstacles and Opportunities*, Department of Geosciences, Oregon State University, Contributing Paper for WCD Thematic Review V.3 River Basins.

Wolf A, Natharius J, Danielson J, Ward B, Pender J, 1999, 'International River Basins of the World', in *International Journal of Water Resources Development*, Vol. 15, No. 4:387-427.

Wong T, 1994, 'Determining O&M Costs Over the Life of a Hydro Station', *in Hydro in the 90's*, Kansas City, Hydro Review Worldwide.

World Bank, 1973, *Staff Appraisal Report: Ceyhan Aslantas Multipurpose Project*, Report No. 16a-TU, Washington DC, World Bank.

World Bank, 1980, *Operational Manual Statement: Economic Analysis of Projects*, No. 2.21, issued May, Washington DC, World Bank.

World Bank, 1985, *Project Completion Report: Ceyhan Aslantas Multipurpose Project*, Report No. 6756, Washington DC, World Bank.

World Bank, 1987, *Project Performance Audit Report: Ceyhan Aslantas*, Report No. 6756, Washington DC, World Bank.

World Bank, 1991, *Operational Directive 4.20 on Indigenous Peoples*, Washington DC, World Bank.

World Bank, 1993, *Early Experience With Involuntary Resettlement: Impact, Evaluation, Change*, Washington DC, World Bank.

World Bank, 1994, *China: Xiaolangdi Resettlement Project*, Report No. 12557, Washington D.C, World Bank.

World Bank, 1995, *Staff Appraisal Report: Ghazi Barotha Hydropower Project*, Report No. 14587, Washington DC, World Bank.

World Bank, 1996a, *Resettlement and Development. The Bankwide Review of Projects Involving Involuntary Resettlement. 1986-1993*, Paper No.032, Environment Department Papers, Washington DC, The World Bank, Environment Department.

World Bank, 1996b, *World Bank Background Note, Chixoy Hydroelectric Project, Guatemala*, 27 September 1996, Washington DC, World Bank.

World Bank, 1997, *Staff Appraisal Report: Pakistan National Drainage Project*, Report No. 15310-PAK, Washington DC, World Bank.

World Bank, 1999a, *World Development Report 1998/99: Knowledge for Development*, New York, Oxford University Press.

World Bank, 1999b, *Identifying Opportunities to Address Malaria Through Infrastructure Projects*, Workshop Report, June, Washington DC, World Bank.

World Bank, 2000, *World Development Report 2000/2001: Attacking Poverty*, New York, Oxford University Press.

World Bank, ADB, FAO, UNDP, NGO Water Resources Group, 1996, *Vietnam - Water Resources Sector Review*, in cooperation with the Institute of Water Resources Planning, Vietnam, World Bank, Asian Development Bank.

World Bank Water Demand Research Team, 1993, 'The Demand for Water in Rural Areas: Determinants and Policy Implications', in *The World Bank Observer*, Vol. 8, No.1, January: 47-70.

World Commission on Water in the 21st Century, 2000, *A Water Secure World: Vision for Water, Life, and the Environment*, Marseilles, World Water Council.

Worrell E, 1999, *Sustainable Energy Strategies: Materials for Decision Makers*, working draft, UNDP.

WRI (World Resources Institute), UNEP, UNDP and World Bank, 1998, *World Resources 1998-99, A Guide to the Global Environment*, World Resources Institute, New York, Oxford University Press.

WWF (World Wide Fund for Nature), 2000, *Dams: Impacts on Life in River Ecosystems*, prepared by Gujja B, Hunziker DO, Gland, Switzerland, WCD Submission env230.

Yorkshire Water, 1997, *Establishing the Economic Level of Leakage*, Bradford, UK, Yorkshire Water Services Ltd, WCD Submission eco082.

Young R, 2000, *On the Limited Economic Benefits of Dam and Reservoir Projects*, Fort Collins, Colorado, Colorado State University, WCD Submission eco066.

Zhang L, 2000, *China Social Impacts of Large Dams*, Institute for Agricultural Economics, China, WCD Regional Consultation Paper.

Zinke A, 1999, *Dams and the Danube: Lessons from Environmental Impacts*, presentation to WCD Forum, Prague, 25-26 March.

Annex II

Glossary

Active (or live) storage. Volume or cubic capacity of a lake or reservoir between the maximum and minimum operating levels.

Adversely affected people. Populations who suffer negative effects during water and energy development interventions. In the case of dam projects, this includes people whose economic, social and cultural lives are negatively affected by construction works, impoundment, alteration of river flows and any ecological consequences. The term includes displaced people, host communities, and downstream and upstream populations. It may also include groups affected by the construction of transmission lines or the development of irrigation schemes, water transfer canals, sanctuaries, and so on.

Aquifer. An underground water-bearing layer of permeable rock, sand or gravel that is capable of yielding exploitable quantities of water.

Barrage (gate-structure dam). A structure built across a river consisting of a series of gates that when fully open allow the flood to pass without appreciably increasing the water level upstream of the barrage, and that when closed raise water levels upstream to facilitate diversion of water to a canal for irrigation or to a powerhouse for the generation of electricity.

Baseline assessment. The collection and analysis of data that describe prevailing social and environmental conditions and are used in the design of project activities and as a benchmark for future monitoring studies.

Benefit sharing. Transfer of a share of the benefits generated by a project, such as a dam, to local communities or authorities. Mechanisms for benefit sharing include preferential rates (for example, of electricity generated), revenue sharing or royalties, and equity sharing (through which local populations or authorities own all or part of the project).

Catchment. The area that drains into a river system; in relation to a dam, the area upstream from the dam from which the reservoir receives water. (The term 'watershed' has been used in this document to convey the same meaning.)

Civil society. Non-governmental organisations, community based organisations, professional associations from all disciplines and other sectors of society that are neither government bodies nor the private sector.

Compensation measures. Alternative resources (land, property or money) provided to displaced people or others adversely affected by a project as mitigation for losses suffered.

Conjunctive water use. The co-ordinated use of surface water and groundwater resources.

Cropping intensity. The extent of land use in a year, which reflects the degree of multiple cropping. It is the ratio of the total area cropped per year to the irrigation command area.

Cultural heritage. The cultural practices and resources of current populations (religions; languages; ideas; social; political and economic organisations) and their material expressions in the forms of sacred elements of natural sites or artefacts and buildings; landscapes resulting from cultural practices over historical and prehistoric times; and archaeological resources; including artefacts, plant and animal remains associated with human activities, burial sites and architectural elements.

Dead storage. Storage below the lowest outlet that cannot be released under normal conditions.

Decommissioning. Removing a dam from service and, where appropriate, physically dismantling it.

Demand-side management. Reducing use of water or electricity by improving the efficiency of use by the transmission system or the consumer, whether in the residential, industrial, commercial, agricultural or government sector.

Developer. The organisation (private or public sector) responsible for promoting and implementing a project, as distinct from the contractor who constructs the project.

Discounting. The process of applying a rate of interest to cost and benefit flows that is used to find the equivalent value today of sums receivable or payable in the future.

Displaced people. Communities required (often involuntarily) to abandon their settlements (homes, agricultural land, commons, forests and so on) or suffering loss of livelihood due to construction of a dam, submergence of the reservoir area, downstream impacts, building of dam-related infrastructure such as roads, and so on.

Ecosystem. An interacting system formed by living organisms and their abiotic environment regulating itself to a certain degree and explicitly includes the human social system.

Environmental flow. The specific release of water from a dam to ensure the maintenance of downstream aquatic ecosystems and key species. The flows may include seasonal or annual flows and/or regular or irregular pulses to meet ecosystem needs. They may also be linked to livelihood needs of downstream affected people.

Environmental management system. The processes by which an organisation identifies and assesses environmental problems, sets goals to address the problems, and measures and verifies progress in solving the problems.

Ethnic minorities. Social groups with a social and cultural identity distinct from the dominant society. They have been historically disadvantaged; come from non-dominant sectors of society; have low social, economic and political status; and are determined to preserve, develop and transmit to future generations their ethnic identity as the basis of their continued existence as people.

Export credit agency. A government agency that helps finance the overseas sales of a nation's goods and services, generally by providing guarantees of working capital loans for exporters, guaranteeing the repayment of loans, or making loans to foreign purchasers of the nation's goods and services. The agency may also provide credit insurance that protects exporters against the risks of non-payment by foreign buyers for political or commercial reasons.

Externalities or External Impacts. Costs and benefits that are external to the financial aspect of decision-making, and thus do not accrue to project developers and operators.

Flood control. In relation to dams, the intention to reduce flood peaks in the river and to minimise the impact of flood events on human activities, including loss of life, social disruption, health impacts, and property and economic losses.

Flood management. A broad concept that focuses on reducing flood hazards through a combination of policy, institutional, regulatory and project measures (such as replanting catchment areas), while recognising that they can never be fully controlled. This takes into account the beneficial uses of natural floods, which are more difficult to quantify in human and economic terms but which sustain natural systems that also have economic, social, cultural and ecosystem values and functions.

Greenhouse gases. Gases that accumulate in Earth's atmosphere and trap heat. Some are naturally occurring gases, like carbon dioxide and methane; others are made by humans, such as halocarbons.

Groundwater. Water that flows or seeps downward and saturates oil or rock, is stored underground and supplies springs and wells. The upper level of the saturated zone is called the water table. Generally, all subsurface water, as distinct from surface water.

Impoundment. Body of water formed by collecting water, as by dam.

Indigenous and tribal peoples. At its broadest, the adjective 'indigenous' is applied to any person, community or being that has inhabited a particular region or place prior to colonisation. However, the term 'indigenous peoples' has gained currency internationally to refer more specifically to long-resident peoples, with strong customary ties to their lands, who are dominated by other elements of the national society.

Integrity pacts. Voluntary undertakings related to the procurement of goods and services that are used to reduce corruption, and that are of particular use in situations where regulatory systems and institutional capacity are weak, although they have universal application. The concept was first developed by Transparency International.

Large dam. A dam with a height of 15m or more from the foundation. If dams are between 5-15m high and have a reservoir volume of more than 3 million m^3, they are also classified as large by the International Commission on Large Dams. In this report, everything else is considered a small dam.

Life-cycle assessment. An options assessment procedure at the front end of the planning cycle used in the energy sector to compare 'cradle-to-grave' performance, environmental impacts, and market barriers and incentives for different demand and supply options.

Main-stem. The main course of a river, characterised by its middle and lower reaches.

Major dams. The *World Atlas & Industry Guide* of the International Journal on Hydropower & Dams defines a 'major dam' as a project meeting one of the following criteria: dam height over 150 meters; dam volume over 15 million cubic meters; reservoir volume over 25 billion cubic meters; installed capacity over 1 000 megawatts.

Mitigation measures. The reduction of potentially significant adverse impacts.

Multi-criteria analysis. An analytical process that uses a mix of qualitative and quantitative criteria to assess and compare options, which may be policies, programmes or projects.

Multipliers. The amount by which equilibrium output of the economy changes when aggregate demand - as caused for example by the expenditure by a development project - increases by one unit. As those receiving the initial round of income generated are likely to consume a portion of the additional income, this subsequent expenditure will lead to additional ripple effects of rounds of income and consumption through the economy. The net effect of these increases in output is the multiplier effect of the initial expenditure, measured as a proportion of the initial expenditure.

Multi-purpose dam. A dam that meets two or more objectives (such as irrigation, flood control, water supply, power generation, recreation, navigation or fish and wildlife enhancement).

Performance bonds. Bonds supported by financial guarantees to provide a secure way of ensuring compliance with commitments and obligations. The bond is called upon in part, or in full, to meet unfulfilled obligations and commitments or is repaid when commitments are met, either in whole or in part, depending upon the circumstances.

Precautionary approach. According to the Rio Declaration on Environment and Development, signed in 1992, where there are threats of serious or irreversible damage, lack of full scientific certainty shall not be used as a reason for postponing cost-effective measures to prevent environmental degradation.

Political Economy. Political economy is the analysis and explanation of the ways in which governments affect the allocation of scarce resources in society through their laws and policies as well as the ways in which the nature of the economic system and the behaviour of people acting on their economic interests affects the form of government and the kinds of laws and policies that get made.

Recession agriculture. A system of agriculture that depends on the moisture of the soil as the flood recedes. Recession agriculture takes place in the floodplain, which is the area subject to seasonal flooding by the river.

Rehabilitation. The physical or social restoration of an ecosystem or a community after a dam construction project has been completed, or the process of renovating a facility or system that has deteriorated and whose performance is failing to meet the original criteria and needs of the project.

Reservoir. Any natural or artificial holding area used to store, regulate or control water.

Reservoir drawdown. The extent to which the water level in the reservoir changes on a daily or seasonal basis due to release of water from the reservoir for operations (such as irrigation or daily peaking for power generation). Emergency drawdown may be for safety reasons, or in anticipation of a major flood event.

Resettlement. Physical relocation of people whose homes, land or common property resources are affected by a development, such as dam building.

Retention. Temporary storage provided by a dam. Even when a reservoir is full, the outflow may be smaller than the inflow as a result of the retention effect.

Riparian. Lying on or adjacent to a river or lake. Used to denote people, plants or wildlife living along the water's edge.

Riparian State. Any State through which a transboundary river flows or forms part of its boundary, or that includes part of the catchment basin of a transboundary river.

River. Large stream that serves as the natural drainage channel for a drainage basin. In terms of transboundary rivers, the term relates equally to all types of waters that are or might be affected by dams.

River basin. The area from which the river system under consideration naturally receives its drainage water; may encompass a series of tributary rivers and their sub-basins.

Riverine. Features or habitats relating to, formed by, or lying within a river; living along the banks of a river.

Run-of-river dams. Dams that create an hydraulic head in the river to divert some portion of the river flows. They have no storage reservoir or limited daily pondage. Included in this category are weirs and barrages.

Surface water. Water that flows or lies on the ground surface.

Tailwater. The water in the natural stream immediately downstream from a dam. Applied irrigation water that runs off the lower end of a field.

Water table. The level of groundwater; the boundary between ground that is saturated with water (the zone of saturation) and ground that is unsaturated or filled with water and air (the zone of aeration).

WCD Forum. A body with some 68 members affiliated to the broad range of stakeholders and interest groups involved in the dams debate. The Forum is partly composed of members of the Reference Group from the 1997 meeting in Gland that recommended the establishment of WCD. It also has new members subsequently invited to participate by WCD. The Forum is a consultative body.

WCD Global Review. An assessment of the performance and impacts of large dams and of alternatives for water resources and energy development, based on the WCD Knowledge Base.

WCD Knowledge Base. Materials commissioned, organised or accepted by the WCD to inform its work: in-depth Case Studies of eight large dams on four continents, together with two country review studies; a Cross-Check Survey of large dams located in 52 countries across the globe; 17 Thematic Reviews grouped along five dimensions of the debate; four regional consultations; and 947 submissions from interested individuals, groups and institutions. These materials are available at www.dams.org.

Weir. A structure built across an open channel to raise the upstream water level or to measure the flow of water. Weirs tend to be smaller than barrages and are not generally gated.

Annex III

The WCD Work Programme – Approach and Methodology

This annex provides additional details on the method used in the WCD's four main work programme tasks. In response to its mandate, the Commission began by assembling a consolidated knowledge base on the worldwide experience with large dams. To give its analysis and conclusions a solid foundation, it commissioned, organised or accepted:

- in-depth case studies of large dams in five continents, together with two country papers;
- a Cross-Check Survey targeted at 150 large dams in 56 countries;
- 17 thematic reviews grouped along five dimensions of the debate;
- four regional consultations; and
- inputs submitted by interested individuals, groups and institutions.

The work programme elements were mutually reinforcing. And there were different levels of analysis and review. For example, different types and purposes for dams were assessed (at the basin, country and regional levels); different methods were employed to look at cross-cutting issues; different timeframes were used to assess issues, including retrospective and forward-looking perspectives; and, finally, different stakeholder groups and review processes were involved. In terms of participation, the case studies involved interested and affected stakeholder groups looking at dams as a whole from a historical perspective and in the context of a specific country or basin; the thematic studies engaged review groups from all disciplines, regions and constituencies to look at past, present and future trends; the regional consultations included all constituencies in a debate of cross-cutting issues at a regional level. Finally, the WCD Forum provided the Commission with a multi-stakeholder, international-level review of the knowledge base products as a whole, as they were finalised. The following sections describe the main work programme elements in more detail.

Case Studies

The case studies were used primarily in developing the global review presented in Part 1 and addressed the question of development effectiveness. In all, the WCD produced 11 sets of reports and papers in the case study programme:

- seven full case studies of selected large dams (five in developing countries – Brazil, Thailand, Pakistan, Turkey and Zambia-Zimbabwe, and two in industrial ones – the United States and Norway);
- country-level review papers on the experience with large dams in India and China;
- a pilot case study of two dams in South Africa; and
- a briefing paper on the Russian Federation and the Newly Independent States.

Each case study had its own stakeholder review process. A common framework was used to identify the stakeholders to:

- identify and set priorities for the study team's coverage of the issues;
- review the information and analysis provided by the study team on these issues; and
- consider the findings, conclusions and lessons drawn.

The Commission took a principled decision to employ national teams of experts rather than using international consultants for the case studies. While creating greater challenges in terms of independence and neutrality it provided the Commission with a deeper insight into the political, historical and cultural contexts for water and energy resources management.

Purpose and selection of the case studies

Given that there are nearly 45 000 large dams worldwide, the case studies were not meant to be statistically representative. Case studies set out to provide an independent and in-depth review of the performance and development effectiveness of a number of large dams around the world and to draw lessons from this review. For the WCD Case Studies, a focal dam was studied intensively in the context of its river basin. Other large dams in the basin were examined to assess and illustrate the interactive and cumulative effects of these dams with the focal dam in areas such as storage and river regulation, sediment impacts, fisheries impacts, operation practices and cumulative social impacts.The WCD took a broad approach to the concept of 'development effectiveness'. This included the relevance and appropriateness of the dam as a response to the needs that motivated its construction (such as irrigation, power, flood control, water supply, navigation or multi-purpose benefits) or other goals such as using public infrastructure for regional development. The studies also looked at projected versus actual benefits, at costs and impacts, at the distribution of gains and losses among groups, and at the general conditions under which the dam was built and is now operated. The latter aspect includes decision-making and consultative processes, and the ex-post validity of the key assumptions upon which the project was originally developed. The stakeholder processes formed around the study were the basis for considering development effectiveness from different perspectives, as it is clear there is no one view on this issue.

Selection process and criteria

The selection process started with the Secretariat, under the Commission's direction, compiling an initial list of possible countries, river basins and specific dams. This list was presented and discussed at the Second Commission Meeting in Cape Town, in September 1998. A number of criteria guided the effort to refine the list of candidates. For example, because developing countries have the largest number of existing dams, and are currently most active in dam building, the selection of case studies in developing countries was emphasised.

The Secretariat contacted a number of governments and dam authorities to assess the degree of co-operation the WCD could expect in implementing the programme. This co-operation was considered essential to ensure adequate access for travel to the site, field visits, interviews with project-affected families and data collection from government authorities and other institutions – and more broadly to maximise stakeholder involvement throughout the process. In parallel with the identification of potential dams and river basins for the work programme, the Secretariat initiated a pilot case study of two dams on the Orange River in South Africa. The pilot study was done to develop, pre-test and obtain wider consensus among the WCD stakeholders on the approach and method for the full case study programme. Once the pilot methodology was agreed on, terms of reference for the case studies were developed in two stages: the first for the scoping phase, and the second for full implementation of the studies.

In December 1998, at its third meeting, the Commission discussed a further short list of candidate dams. The Secretariat then proceeded to make formal contacts with governments to enter into agreements to undertake the scoping phase of an initial group of case studies.

Diversity was the main criteria applied to select case studies from across the list of potential countries, basins and dams. These criteria generally included:

- regional diversity (with the intention of covering all continents);
- functional diversity (to cover as broadly as possible hydroelectric, irrigation, flood-control and multi-purpose dams);
- age diversity (to capture experience from different decades and long-term impacts);
- diversity in size and type (to differentiate between large and major dams, as well as between storage and run-of-river dams); and
- diversity of the catchment area (tropical, sub-tropical and temperate zones, covering a range of variables – climate, biodiversity, river morphology, sediment characteristics and so on).

The start of the case studies was also influenced by budget considerations. Because the WCD was continuously fundraising, the budget and consequently the number of case studies was not finalised until mid-1999.

After the initial contacts with dam operators and governments, there were a number of developments. The Governments of Brazil, Pakistan, Thailand, Zambia, Zimbabwe, United States and Norway agreed to fully participate in the case studies. In Brazil, the Commission decided to study the Tucuruí project. The Governments of India and China indicated initially that they were not prepared to participate in full case studies. Based on a meeting in Beijing in June 1999, China agreed to participate in a country-level review. After a change of institutional responsibilities within the Ministry of Water Resources in China, however, the government withdrew the agreement to participate actively in the country review (October 1999). The WCD then undertook an external review of dams in China (January 2000). The Government of

Table III.1 Basic data on WCD pilot and case studies

Focal dam	River basin	Basin country	Year commissioned
Grand Coulee	Columbia	United States	1941
Kariba	Zambezi	Zambia/Zimbabwe	1959
Tarbela	Indus	Pakistan	1968
Four Dams	Glomma-Laagen	Norway	1970, 1955, 1952 and 1910
Aslantas	Ceyhan	Turkey	1985
Tucurui	Tocantins	Brazil	1986
Pak Mun	Mun	Thailand	1994
Pilot study			
Gariep/Vanderkloof	Orange River	South Africa	1971

India, after meetings in New Delhi (June 1999), declined full participation; it subsequently agreed to co-operate fully with a country review paper on dams in India (February 2000).

Table III.1 lists the dams included in the pilot study and case studies, along with the year they were commissioned.

Figure III.1 Procedure for the WCD case studies

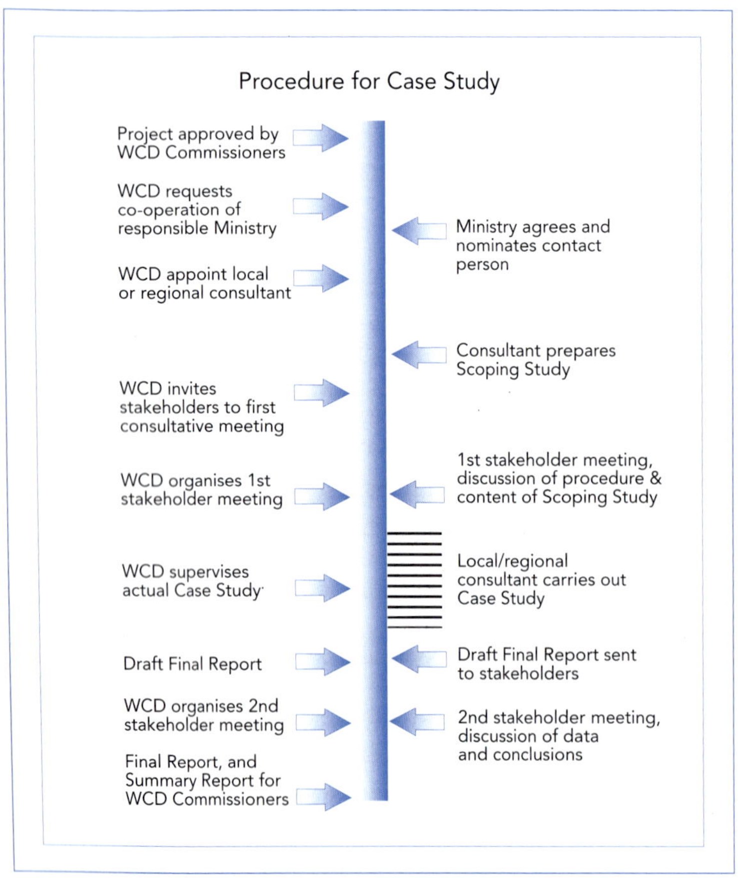

Case study method

The WCD established a stakeholder group and an inter-disciplinary study team for each case study and country study. This involved varying degrees of negotiation before the study teams and stakeholder groups were finalised. The selection of the study team and the identification of the stakeholders was an important aspect of the study to maximise participation and acceptance of the process and the study results. The standard method was then followed for each case study, based on the model developed by the WCD for the Orange River pilot. The main steps are represented in Figure III.1.

A scoping report and consultative meeting with stakeholders launched each study. The scoping phase required two to five months, depending on the study. The purpose of this phase was to identify and set priorities for the issues and to discuss with stakeholders how these would be covered within the study framework provided by the WCD. After a scoping phase meeting with the stakeholder group, the study team was revised and the terms of reference for the full study were finalised.

The study team then started parallel steps of data collection and verification, structured interviews, enquiries and invitation of submissions from interested parties.

After the data collection, field programmes, collection of inputs and analysis, the study team assembled a draft report. This was circulated and discussed by the stakeholders at a second meeting. The study team incorporated comments arising from the meeting and subsequent written follow-up, and then produced the final report. This captured the experience, lessons learned and the convergent and divergent views of the stakeholders. At times the process sparked considerable controversy and aspects of reports were contested in some countries.

At the core of the case studies were six questions:

- What were the projected versus actual benefits, costs and impacts?
- What were the unexpected benefits, costs and impacts?
- What was the distribution of costs and benefits – who gained and who lost?
- How were decisions made?
- Did the project comply with the criteria and guidelines of the day?
- How would this project be viewed in today's context in terms of lessons learned?

The study response to these particular questions, combined with additional views gathered through the questionnaires and discussions with stakeholders, were used to assess and illustrate the 'development effectiveness' of the dam, and to draw out and establish priorities on the lessons learned.

Media releases were provided for the stakeholder meetings, and at each stage the reports and results circulated to the stakeholder group were placed on the WCD website for international comment.

Each report presents the lessons learned from the case study that the stakeholders and study teams developed. These lessons were developed in one- or two-day meetings with stakeholders, with sometimes 60-70 participants. The Secretariat also undertook to brief stakeholders on the incorporation of their comments in the final report. Where full agreement with all stakeholders was not possible, divergent views were also captured in the report and written comments in the annex of the case study. The WCD website includes a list of lessons across the case studies along with the individual reports. The reader may also refer to the pilot study, which illustrates the procedures for verifying and collecting opinions from stakeholders at the meetings.

Cross-Check Survey

The WCD Cross-Check Survey provides a link between the topical issues emerging at a global and regional level as provided by the submissions, consultations and thematic reviews and the more in-depth insights derived from the WCD case studies. The survey does not provide assessments of individual dam projects nor does it aim to be statistically representative of the nearly 45 000 large dams. Rather it was designed to indicate broader patterns and trends.

A multi-stakeholder approach with inclusiveness and participation at all levels and with access to multiple information sources remained the key focus for the information-gathering process. As such, the Cross-Check represents an international collaborative survey in which legitimacy, transparency and integrity of data from contributors in more than 50 countries was the overriding objective.

Range of dams selected

The dams included in the Cross-Check Survey were drawn from a number of sources, including WCD Case Studies, existing databases and additional dams that contributed to the overall diversity of the total sample. The mix of large dams that make up the Cross-Check Survey sample includes dams:

- from different regional locations;
- of differing ages – dating from the 1930s through to the1990s;
- with a range of different heights and sizes; and
- with different purposes such as water supply, irrigation, flood management, power and recreation.

The survey sampling technique set out to construct a Cross-Check sample that would inform the dams debate and provide guidance for similar studies in the future, within the context of the WCD Work Programme. Given the emphasis on the Case Studies in the work programme, and the interest in 'cross-checking' the result of these in-depth studies against the wider experience in the selected river basins, the Case Study dams and a selection of other dams from the basin formed the first two strata in the sample. Inevitably, the selection of these two sub-samples involved some bias due to the criteria used in choosing the Case Study dams and further sampling from specific basins.

The remainder of the sample aimed to eliminate many aspects of this selection bias – that is, to correct for the over-representation of specific attributes such as size, location, purpose and age. The third sub-sample aimed to build on dams from existing surveys by selecting a sub-sample of the dams examined in the 1996 World Bank study of large dams. Specifically, dams were chosen to represent regions, countries and purposes not reflected in the first two sub-samples. Representation of dams from additional countries was introduced by incorporating dams from WCD country studies of India, China and Russia into the fourth sub-sample.

The final step in the sampling process was an attempt to correct for remaining differences in representation of dams by location, age, size and purpose in the first four sub-samples as compared with the best available information on the global population of large dams. To meet this requirement, the WCD used the ICOLD *World Register* to select 73 complementary dams (this excludes Chinese dams, as noted elsewhere). The full details of the selection methodology and criteria are found in the Cross-Check Report.

Table III.2 shows the change from the original targeted sample to the final status of the dams received for each sub-sample in the Cross-Check Survey. Numerous challenges were faced in applying a global survey of this nature as part of a time-limited process. Due to many factors (either controversy, denial of permission or logistical difficulties), information for some large dams was inaccessible. At the end of the process, contributions were received for over 80% of the original sample. During the process, some dams were substituted to ensure compliance with the original criteria for the target sample. A number of additional contributions were received for dams not in the original targeted sample.

Table III.2 Cross-Check Survey Database

Strata (sub-samples)	Selected dams	Updated sample	Submissions received
Case Study Dams	13	13	13
Non-focal dams in Case Study basins	24	24	22
Dams from existing databases	24	23*	21
Country studies	19	9*	9
Complementary dams	73	73	60
Sub total	**154**	**143**	**125**
Incomplete questionnaires			2
Final total analysed			**123**
Additional submissions	–	–	

*number changes due to exclusion of the 11 Chinese dams from the sample

Of the total questionnaire submissions received (125 selected dams and 9 additional dams), only the 125 originally selected as part of the Cross-Check methodology were considered for final analysis. Two of these questionnaires were excluded due to incomplete information. The final analysis was therefore performed on 123 questionnaires about large dams.

Who provided the evidence?

The large dam projects analysed in the Cross-Check span 52 countries in six major regions of the world:

- Over 70 contributors were contracted, consisting of 40% government departments/utilities, 40% private consultants/companies and 20% NGO/academic/research institutions.
- An additional 30 contributors were commissioned to review a select sample of 17 randomly chosen and 18 controversial projects for data verification. The constituencies chosen for review were predominantly local NGOs, who provided a second opinion on the government and private contributions.

Safety nets and data confidence

To ensure credibility and confidence in the Cross-Check Survey, it was necessary to implement a series of iterative steps and 'safety nets' to enhance the quality and quantity of submitted data. They consisted of:

- *Range of Data Sources* used by contributors, including interviews.
- *Internal Review* of all questionnaires, requesting minimum threshold data and clarifications.
- *Review by WCD staff of in-house literature* on dams in the Survey.
- *External review* of controversial and randomly selected dams. Of the 50 targeted, 31 reviews were received and analysed, with the following insights:
  - Reviewers commented on approximately one third of the original data.
  - Most reviews corroborated the original data, with 20-40% contested figures (minimum 10% discrepancies).
  - Of the figures contested, the predominant focus covered discrepancies in number of actual displaced people and, in a few instances, actual capital costs.
- *Transparency* through sharing 'raw' data from submitted questionnaires on request. Around 50% of respondents gave permission for independent review of the questionnaires they submitted. A network of civil society organisations conducted a general review, supplementing and contesting information in this sub-sample.

The questionnaires

The main questionnaire for the Cross-Check Survey was developed, tested and revised by the Commission based on its experience with the Orange River Pilot Study in South Africa. Ten categories of questions covering a wide range of concerns – including questions about technical, economic, social, environmental and decision-making aspects – form the basis of the survey. It was necessary during the process to introduce a short supplementary questionnaire for completeness.

Data analysis

Information submitted in the original questionnaires was stored in a relational database. The WCD derived 32 indicators from the analysis of the final dam sample to measure projected versus actual performance parameters. The graphs and statistics required to analyse these 32 indicators in the Cross-Check Survey were produced using the *Statistica* statistical software package. For most of the 32 indicators programs were written to draw the required graphs and calculate the appropriate statistics. This was automated to facilitate the smooth and fast production of the required results and to ensure that the analysis could be easily regenerated in the future.

The Cross-Check dams were classified according to the following 11 characteristics: region, sub-region, height, decade commissioned, reservoir area, purpose, sub-samples, reach of river, climate, income and sub-income. (See Table III.3.) Each indicator was first analysed from a global perspective – ie all valid observations and the overall trend or pattern were identified. Then the observations were split between the categories of the chosen set of 11 specified characteristics, and the emerging trends or patterns were compared between the various categories of the particular characteristic (see list of cross-classifications).

The analysis of the survey questionnaires has been mostly graphical in conjunction with some simple statistics, regression modelling and clustering techniques used for hydropower energy output. For each of the 32 indicators, different types of graphs and corresponding statistics were prepared. All data was represented at the case level to ascertain the distribution and variance. Trends and patterns were inferred from the prepared graphs in the Cross-Check Report. Where possible, appropriate scientific or other explanations were given to justify emergence of particular trends or patterns. Key findings from 24 indicators were interpreted and integrated into the performance, impacts and decision-making aspects of Part One of this report.

All observations classified by *Statistica* as extremes or outliers were investigated. In

Table III.3 Cross-classification of dams in Cross-Check sample

Classification by	Category names	Number of dams
Region	United States and Canada	10
	Latin America	17
	Africa & Middle East	20
	Europe and Central Asia	34
	Rest of Asia	42
Sub-region	North America (United States & Canada only)	10
	Central America, Caribbean & Mexico	12
	South America	5
	Middle East & North Africa	5
	Sub-Saharan Africa	15
	Eastern Europe and Central Asia	9
	Rest of Europe	25
	East Asia & Pacific	26
	South Asia	16
Height of dam (m)	15–30 m	31
	31–100 m	63
	> 101 m	29
Reservoir area (km^2)	< 1 sqr km	19
	1–10 sqr km	25
	10–100 sqr km	40
	> 100 sqr km	39
Decade commissioned	< 1950	3
	1950–59	13
	1960–69	25
	1970–79	39
	1980–89	26
	>=1990	17
Purpose	Single Purpose: Water Supply Hydropower Irrigation Flood Control Other	 9 35 13 1 1
	Multipurpose	64
Sub-samples	OED	23
	Basin Study/ Non focal	21
	Basin Study/ Focal	12
	Country Study	
	Other	58
Reach of river	Upper	50
	Middle	32
	Lower	18
	Unknown	23
Climate	Temperate	77
	Tropical/Sub-Tropical	46
Income	High Income	46
	Middle Income	46
	Low Income	31
Sub-income	OECD	46
	Middle-Upper	22
	Middle-Lower	24
	Low	31

most circumstances these observations turned out to be valid in terms of the original questionnaire submissions and hence legitimate extremes. Any spurious data points were corrected where possible. Where the validity of the extremes or outliers was suspect and unsubstantiated, the observations were removed from the analyses. This was only necessary in a few isolated instances.

These are not the only analyses that the WCD has done on the fully verified Cross-Check sample, but they illustrate the final performance, impact and decision-making indicators that can be derived from the data. The WCD has also conducted bivariate and multivariate analyses to highlight trends and patterns for regional and other cross-classifications. They are reported in the WCD Cross-Check Report.

Thematic Reviews

The WCD commissioned 17 thematic reviews and some 130 papers addressing five major areas of concern identified in the strategy and objectives paper:

- social and distributional issues,
- environmental issues,
- economic and financial issues,
- options assessment and
- governance and institutional processes.

The thematic reviews provided the baseline information, analysis and recommendations on issues that cut across the central elements in the large dams debate. The reviews considered past and current experience, as well as the forward-looking context, by synthesising the state-of-the-art knowledge, practices and key viewpoints on each topic. Within limits set by available resources and the Commission's schedule, the level of effort involved in preparing these review papers varied according to the complexity of the issue and the level of controversy surrounding it. Preparation of the review papers included setting up panels and procedures for broader peer review. This helped to bring together a wide spectrum of perspectives and approaches on the topic and to clarify the areas of potential agreement (and persistent disagreement) on highly controversial issues. The thematic review process is illustrated in Figure III.2.

Consultations and Submissions

For most of its two-year mandate the WCD followed the old adage of being 'quick to listen, and slow to speak'. Rather than plant-

Figure III.2: Thematic Review process

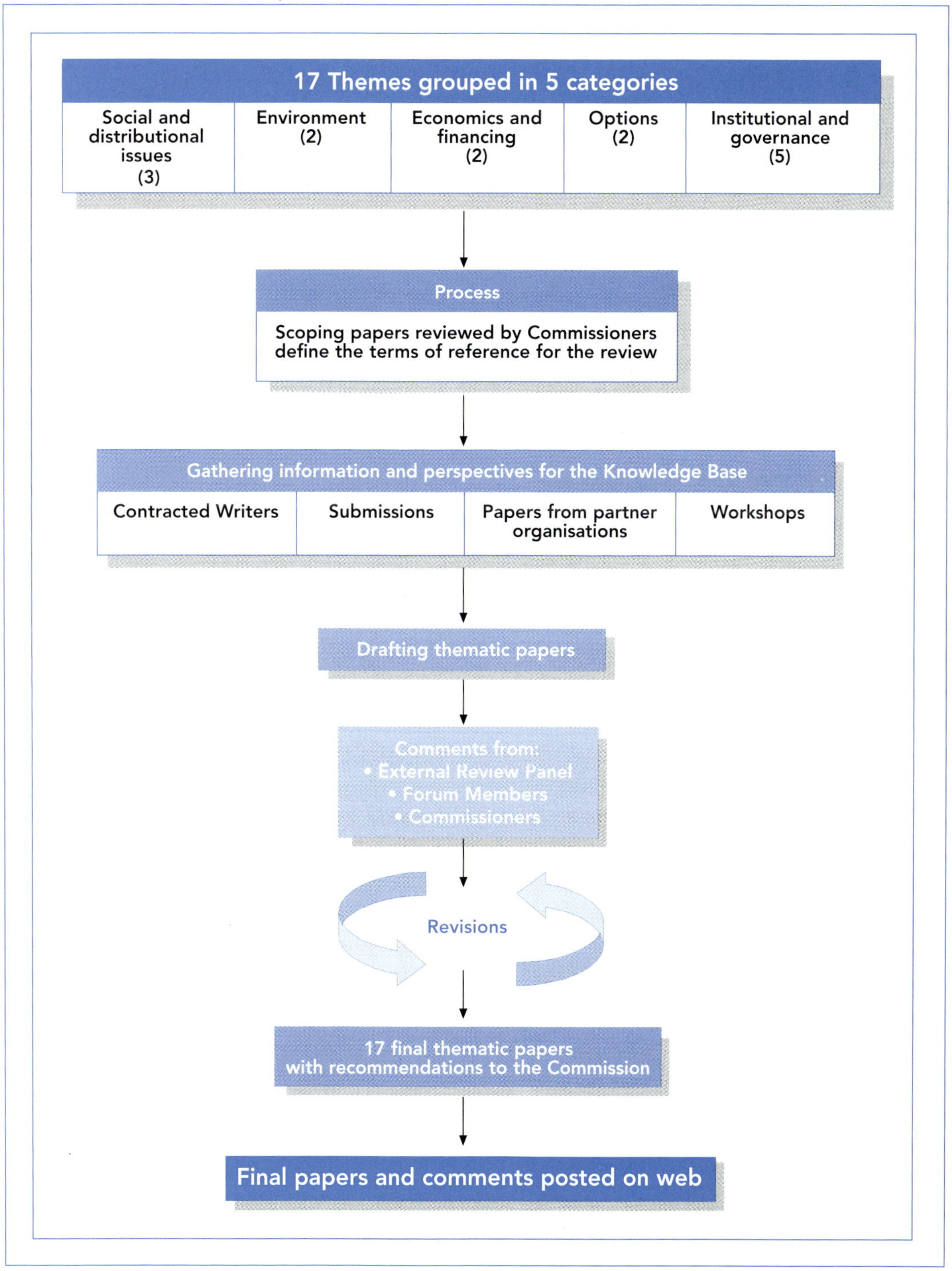

ing its opinions before it had harvested its research, the Commission canvassed views and knowledge on the costs and benefits of dams from a vast array of stakeholders. In addition to its work programme studies, the WCD established global listening posts through its submissions programme and its regional consultations. In addition to the submissions and the four regional consultations, the Commission consulted with numerous professional associations and provided input for many non-WCD stakeholder events.

A full list of submissions and reports on the regional consultations is available at www.dams.org.

Regional consultations

In total 1 400 individuals from 59 countries attended the WCD regional consultations held for South Asia, East and South-East Asia, Latin America, and Africa and the Middle East. The meetings were carefully organised to ensure broad-based participation by NGOs, governments, industries and utilities, irrigation interests, academics, financiers, and other interested parties. The WCD also participated in two hearings organised for its benefit by NGOs in Southern Africa and Europe.

Figure III.3 Topics in Submissions

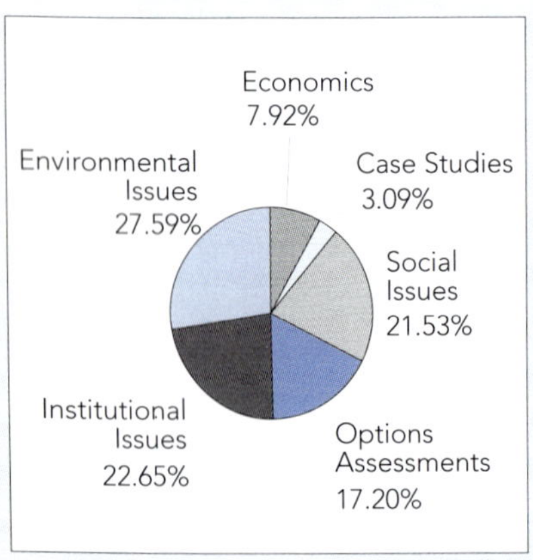

The consultation process began with the choice of themes for each meeting. The themes were based on a broad range of submissions requested through a 'Call' announced and mailed out to a large list of dams' constituents some months before the regional consultation. After selecting the themes, the Secretariat invited presenters based on the relevance of their contribution to the WCD Knowledge Base and the effort to achieve balanced regional representation.

The key issues that emerged through the consultations were:

- participation and transparency in planning and decision-making on dams;
- displacement, resettlement and reparation for those negatively affected by dams;
- dams as a means to satisfy demands for food, energy, drinking water and flood control;
- assessing the costs and benefits of alternatives in providing irrigation, flood control, electricity and water supply;
- the importance of environmental sustainability;
- regional/transboundary approaches to sharing water resources, as well as conflict resolution; and
- methods of ensuring that those engaged in dam building comply with regulations, laws and policies.

Submissions

By September 2000, the WCD had received 947 submissions from 79 countries. Of those, 400 submissions were related to the regional consultations. All submissions were entered into a central database. They were catalogued by subject along the lines of the thematic reviews: social, environmental, economic and institutional issues, and options assessment. The cataloguing system allows sorting by issue, region and country. The submissions were reviewed by the Secretariat and, where relevant, were shared with lead writers and external reviewers involved in compiling the WCD Case Studies, Thematic Reviews and the Cross-Check Survey.

Annex IV

Reports in the WCD Knowledge Base

The WCD Knowledge Base served to inform the Commission on the main issues surrounding dams and their alternatives, and complemented the regional consultations, where Commissioners heard firsthand about relevant regional experience from governments, members of civil society and the private sector.

The reports listed in this Annex are available on the website at www.dams.org. They were used to synthesise and structure the wealth of knowledge that lies within the Knowledge Base, and to distill key issues, lessons learned and proposals for the future. They remain inputs to the Commission rather than products of its deliberations and thus do not represent expressions of the Commission's views and conclusions, which are contained solely in this final report.

The Case Studies and Thematic Reviews were contracted by WCD to lead authors, who were selected for their ability to respect the Commission's guiding principles of professional expertise, independence, transparency and openness in synthesising information and perspectives from the contributing papers and the submissions. All terms of reference, and final draft reports, were peer-reviewed by local stakeholder groups (for the Case Studies) or by review panels (for the Thematic Reviews) composed of 8–10 people with varying backgrounds, regions of origin and perspectives. Working Papers were developed with partner organisations to include perspectives on particular issues.

In total, 201 people commented on the draft Thematic Reviews, and the lead writers were asked to address their comments, wherever possible, in the final report. Reviewers' comments have been included as annexes to enable the reader to better comprehend the debate. Draft and final reports for Case Studies and Thematic Reviews were available on the WCD website throughout the review process.

The Contributing Papers are listed for each Thematic Review, and the synthesis writers of the Reviews drew on them as appropriate. Some were solicited by WCD through key institutional partners; others were directly contracted to WCD terms of reference; and still others were the result of initiatives by partner organisations as direct contributions to the WCD process and knowledge base.

Case Studies

Case Study	Team Leaders
Grand Coulee Dam, Columbia River Basin, United States	Leonard Ortolano, Katherine Kao Cushing
Tarbela Dam, Indus River Basin, Pakistan	Amir Muhammed
Aslantas Dam, Ceyhan River Basin, Turkey	Refik Çölasan
Kariba Dam, Zambezi River, Zambia/Zimbabwe	Alois Hungwe
Tucurui Dam, Tocantins River, Brazil	Emilio Lèbre La Rovere, Francisco Eduardo Mendes
Pak Mun Dam, Mun-Mekong River Basin, Thailand	Songkram Grachangnetara et al.
Glomma and Laagen Basin, Norway	Jostein Skurdal
Orange River (Pilot Study), South Africa	WCD Secretariat

Country Studies

Country Study	Team of Writers
India	R. Rangachari, Nirmal Sengupta, Ramaswamy R. Iyer, Pranab Banerji, Shekhar Singh
China	Ismail Najjar, Bill Smith, Richard Fuggle, Habib Khoury, Sam Pillai, John Boyle

Briefing Paper

Briefing Paper	Team of Writers
Russia and Newly Independent States	Lilia K. Malik, Nikolai I. Koronkevich, Irina S. Zaitseva, Elena A. Barabanova

Thematic Reviews and Contributing Papers

I.1. Social impact of large dams: equity and distributional issues

Lead Writer	**William Adams**
Contributing Paper	**Writer**
Social Impacts of an African Dam: Equity and Distributional Issues in the Senegal River Valley.	Adrian Adams
Downstream Impact of Dams.	William Adams
Dams and Benefit Sharing. A Submission from Hydro-Quebec.	Dominique Egré, Joseph Milewski
Assessing the Project – Social Impacts and Large Dams.	Hugh Brody
Report on the Social Impact of Dams: Distributional and Equity Issues – Latin American Region.	Carmen Ferradas
Some Evidence on Overall Distributional and Equity Impacts.	Pablo Gutman
Balancing Pain and Gains – A Perspective Paper on Gender and Large Dams.	Lyla Mehta, Bina Srinivasan
Social Impacts of Large Dams: The China Case.	Lubiao Zhang

I.2. Dams, indigenous people and vulnerable ethnic minorities

Lead Writer	**Marcus Colchester**
Contributing Paper	**Writer**
Dams, Indigenous People and Vulnerable Ethnic Minorities: A Case Study on the Ibaloy People and the Agno River Basin, Province of Benguet, Philippines.	Jaqueline Carino
The Chixoy Dam in Guatemala: The Maya Achi Genocide. The Story of Forced Resettlement.	Jaroslave Colajacomo
A Case Study on the Proposed Epupa Hydro Power Dam in Namibia.	Andrew Corbett
The Resettlement of Indigenous People Affected by the Bakun Hydro-Electric Project, Sarawak, Malaysia.	Gabungan
Lake Winnipeg Regulation Churchill-Nelson River Diversion Project and the Crees of Northern Manitoba, Canada.	Luke Hertlein
Land Acquisition Act and Impact on Tribal Development in India.	Manisha Marwaha
Operationalisation of Free Prior Informed Consent.	Lyla Mehta, Maria Stankovitch
Dams and Tribal People in India.	Amrita Patwardhan
The Alta-Case in Norway. A Story about How Another Hydroelectric Dam-Project was Forced through in Norway.	Ande Somby

I.3. Displacement, resettlement, rehabilitation, reparation and development

Lead Writers — **Leopoldo Bartolome, Chris de Wet, Harsh Mander, Vijay Nagaraj**

Contributing Paper	Writer
The Experience with Dams and Resettlement in Argentina.	Leopoldo Bartolome, Christine Danklmaier
Displacement, Resettlement, Rehabilitation, Repatriation and Development. African Experience.	Chris de Wet
Displacement, Resettlement, Rehabilitation, Repatriation and Development – China Report.	Jun Jing
Displacement, Policy and Law in India.	Ravi Hemadre, Harsh Mander, Vijay Nagraj
Displacement, Resettlement, Rehabilitation, Repatriation and Development. The Mexican Case.	Scott Robinson

II.1. Dams, ecosystem functions and environmental restoration

Lead Writers — **Mike Acreman, Ger Bergkamp, Patrick Dugan, Jeff McNeely**

Contributing Paper	Writer
Dams and Biological Diversity – Establishing Strategic Linkages under the Conventions.	Asheline Appleton
Ecosystem Impacts of Large Dams.	Mike Acreman, Matthew McCartney, Caroline Sullivan
Managed Flood Releases from Reservoirs – Issues and Guidance.	Mike Acreman, Edward Barbier, Martin Birley, Kenneth Campbell, Frank Farquharson, Nicholas Hodgson, Jeremy Lazenby, Matthew McCartney, John Morton, David Smith, Caroline Sullivan
Capacity and Information Base Requirements for Effective Management of Fish Biodiversity, Fish Stocks and Fisheries Threatened or Affected by Dams during the Project Cycle.	Garry Bernacsek
International Mechanisms for Avoiding, Mitigating and Compensating the Impacts of Large Dams on Aquatic and Related Ecosystems and Species.	John Bizer
Definition and Implementation of Instream Flows.	Cate Brown, Jackie King, Rebecca Tharme
Information Needs for Appraisal and Monitoring of Ecosystem Impacts.	Cate Brown, Jackie King
Large Dams and Freshwater Fish Biodiversity.	John Craig
Biodiversity Impacts of Large Dams.	John Craig, Nick Davidson, Don McAllister, Dianne Murray, Mary Seddon
Biodiversity Impacts of Large Dams: Waterbirds.	Nick Davidson, Simon Delany
Fundamental Legal and Ethical Principles in Adjudging the Merits of Development Projects.	Charles DiLeva
The Influence of Dams on River Fisheries.	Donald Jackson, Gerd Marmulla

Dams and Fish Migration.	Michel Larinier
A Review of Guidance and Criteria for Managing Reservoirs and Associated Riverine Environments to Benefit Fish and Fisheries.	Steve Miranda
Molluscan Biodiversity and the Impact on Large Dams.	Mary Seddon
Report on the Conference on Hydrological and Geochemical Processes in Large Scale River Basins, 15–19 November, 1999, Manaus, Brazil.	Leonard Sklar

II.2. Dams and global change

Lead Writers — **Nigel Arnell, Mike Hulme, Luiz Pinguelli Rosa, Marco Aurelio dos Santos**

Contributing Paper	Writer
An Analysis of the Linkages between the UNFCCC Legal Regime and Dams.	Albert Mumma

III.1. Economic, financial and distributional analysis

Lead Writers — **Alec Penman, Robert Unsworth**

Contributing Paper	Writer
Methods for Valuation of Flood Control Benefits.	Colin Green
Resettlement Costs.	Pablo Gutman
Distributional Analysis.	Pablo Gutman
Survey of Multilateral Bank Practice on Financial and Economic Analysis of Large Dams.	Anneli Lagman
Methods for Valuation of Impact of Hydropower Projects.	Anil Markandya
Review Paper I - Review Paper on Financial, Economic, and Distributional Analysis .	Michelle Manion, Bruce McKenney, Robert Unsworth
Example of SAM Analysis in the Republic of South Africa.	David Mullins
Methodological Approach for the Distributional Effectiveness of Large Dams.	Kyra Naudascher-Jankowski
Review Paper II - Financial, Economic and Distributional Analysis.	Alec Penman
Methods for Valuation of Irrigation Benefits.	Douglas Southgate

III.2. International trends in project financing

Lead Writer — **Per Ljung**

Contributing Paper	Writer
Database Support.	Lily Donge
Hydropower Dams.	Chris Head
Multipurpose Dams.	Chris Head
Note on Financial Instruments and Incentives.	Michael Kelly
Financing Statistics, Trends and Policies of International Financial Institutions.	Hilary Sunman

IV.1. Electricity supply and demand management options

Lead Writers — **Maritess Cabrera, Anton Eberhard, Michael Lazarus, Thierry Lefevre, Donal O'Leary, Chella Rajan**

Contributing Paper	Writer
Renewable Options.	Glynn Morris
Major Trends in Energy Development.	Donal O'Leary
Demand Management.	Roger Peters
Life Cycle Analysis.	Bjorn Svenson
Rural and Appropriate Energy.	Rona Wilkinson
Submissions Contributing to the Thematic.	International Cogenereration Alliance, International Hydropower Association, International Atomic Energy Association, American Solar Energy Society, European Wind Energy Association, National Hydrogen Association, Gas Research Institute

IV.2. Irrigation options

Lead Writer — **K. Sanmuganathan**

Contributing Paper	Writer
Privatisation of Infrastructure Hydraulic Work – Chilean Experience.	Pablo Anguita Salas
Future Approaches towards Taking up Dam Projects.	S. Char
Developing Irrigation Options for Small Farmers.	Keith Frausto
Some Drainage Options.	Alfred Heuperman
Options Assessment and the Planning System in the IBIS, Pakistan.	Khalid Hussain
Biotechnology in Semi Arid Tropics.	ICRISAT
Contributions from the Latin American Experience.	Hector Maletta
Potential for Improvements of Water Harvesting.	Dieter Prinz, Anupam Singh
Irrigation and Agriculture Experience and Options in Israel.	Yehuda Shevah
Options in Agricultural Policy.	Laurence Smith
Assessment of Irrigation Options in India.	Himanshu Thakkar

IV.3. Water supply options

Lead Writers — **Colin Fenn, David Sutherland**

Contributing Paper	Writer
Water Efficiency Case Studies from California.	Mary Ann Dickinson
Contributions Relating to Rainwater Harvesting.	John Gould
Supporting Note.	Allan Lambert
Contributions on Community Based Systems.	Jon Lane
Example of Demand Management from South Africa.	Guy Preston
Supporting Note.	Philip Turner

IV.4. Flood control and management options

Lead Writer **Colin Green**

Contributing Paper	Writer
Draft ICOLD Bulletin on Dams and Floods.	Luis Berga
A Review of the Role of Dams in Flood Management (draft).	Patrick Hawker
Why Multipurpose Dams Function in Japan.	Minoru Kuriki
Support to Lead Writer.	Dennis Parker
Support to Lead Writer.	Sylvia Tunstall
Assessment of Flood Management Options.	Johannes van Duivendijk
Flood Action Plan in Bangladesh.	Herb Wiebe

IV.5. Operation, monitoring and decommissioning of dams

Lead Writers **Peder Hjorth, Charles Howard, Kuniyoshi Takeuchi**

Contributing Paper Writer	
Managing for Unforeseen Consequences of Large Dam Operations.	K. Betts, Michael Falter, Peter Goodwin
Dams in Spain.	Enrique Garcia
Hume and Dartmouth Dams, Murray Darling Basin, Australia.	Brian Haisman
Macquarie Marshes, Murray Darling Basin, Australia.	Brian Haisman
Operation, Monitoring and Rehabilitation of Dams/ Reservoirs in Japan.	Joji Harada, Kuniyoshi Takeuchi
Operation, Monitoring and Decommissioning of Dams.	Peder Hjorth
A Report on Large Dams in India.	V. Jauhari
US Federal Energy Regulatory Commission.	Thomas Russo
Operation, Monitoring and Decommissioning of Dams.	Geoffrey Simms
Ex Post Evaluation of Dams and Related Water Projects.	James Wescoat
Flushing of Sediments from Reservoirs.	Rodney White

V.1. Planning approaches

Lead Writers **David Nichols, Theo Stewart, David von Hippel**

Contributing Paper	Writer
Planning Survey.	Daud Beg
Regional Integrated Resource Planning.	Catherine Fedorsky
Planning Survey.	Matthias Finger
Planning Survey.	J. Karmacharya
Planning Survey.	Don Moore
First Draft of Planning Approaches Thematic.	David Nichols, Theo Stewart, David von Hippel
Planning Survey.	Miguel Nucete
Planning Survey.	Girish Sant
Multiple Criteria Decision Analysis.	Theo Stewart

V.2. Environmental and social assessment for large dams

Lead Writer	Barry Sadler

Contributing Paper	Writer
Social Impact Assessment.	Frank Vanclay
Environmental and Social Impact Assessment for Large Dams – Thematic Review from the Point of View of Developing Countries.	Iara Verocai

V.3. River basins – institutional frameworks and management options

Lead Writer	Peter Millington

Contributing Paper	Writer
Dams in the Context of Transboundary/International Waters.	Len Abrams
Los Consejos de Cuenca en el Desarrollo de las Presas en México.	Enrique Castelan Crespo
Transboundary Impacts of Dams: Conflict Prevention Strategies.	Fiona Curtin
Review of the Role of River Basin Organizations in Latin America.	Luis Garcia
Water Resources National Policy in Brazil.	Raymundo Garrido
Large Dams, Transboundary Waters, Conflicts.	Ramaswamy Iyer
Dams on Transboundary Rivers.	Erik Mostert
River Basins: Institutional Framework and Management Options for Latin America.	Cecilia Tortajada
Possible Approach to the Management of Dams on International and Inter-Provincial Rivers.	Anthony Turton
Development and Transboundary Waters: Obstacles and Opportunities.	Aaron Wolf

V.4. Regulation, compliance and implementation

Lead Writers	Angela Cropper, Mark Halle, Danny Bradlow, John Scanlon

Contributing Paper	Writer
Report on International and Comparative Water Law Applicable to Large Dam Construction.	Daniel Bradlow, Gabriel Eckstein
Human Rights and Development.	Balakrishnan Rajagopal
Implementing World Commission on Dams Guidelines within an International Certification System.	Tom Rotherham
Export Credit Agencies.	Lori Udall
World Bank Inspection Panel.	Lori Udall
Transparency and Corruption Prevention when Building Large Dams.	Michael Wiehen

V.5. Participation, negotiation and conflict management

Lead Writer	**Bruce Stedman**
Contributing Paper	**Writer**
Support to Lead Writer.	Tisha Greyling
Support to Lead Writer.	Anne Randmer
Support to Lead Writer.	Vanchai Vatanasapt
Support to Lead Writer.	Arch Isabel Viana

Working Papers

Dams and human health

Writers	Robert Bos, William Jobin, Martin H. Birley, P.V. Unnikrishnan, M'barack Diop

Dams and cultural heritage management

Writers	Steven Brandt, Fekri Hassan

Cross-Check Survey

The Cross-Check Survey – Methodology, Findings and Lessons Learned.	WCD Secretariat

Submissions

The WCD received 947 submissions from 79 countries that were reflected in the Case Study, Thematic Reviews and Regional Consultations. The full list of submissions is available on the WCD website at www.dams.org.

Annex V

Dams, Water and Energy – A Statistical Profile

This Annex has two main sections: data on the world population of dams and regional profiles of large dams. The regional profiles present largely the statistical data that are available; social and environmental data are limited. Readers interested in further details on regional and national trends and on the issues under debate should refer to the reports on the WCD Regional Consultations, which are available at www.dams.org.

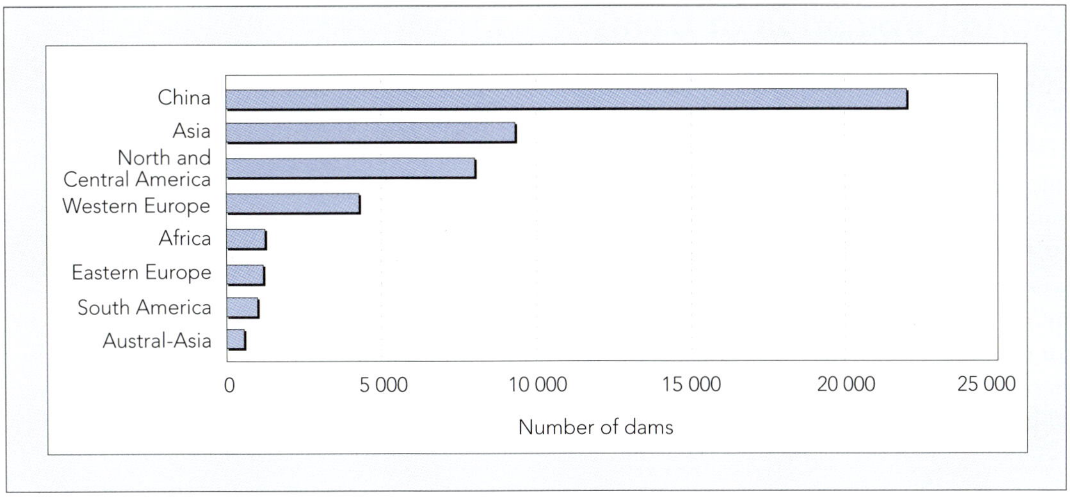

Figure V.1 Regional distribution of large dams in 2000

Source: WCD compilation of various sources and ICOLD, 1998.

Figure V.2 Cumulative commissioning of large dams in the 20th century

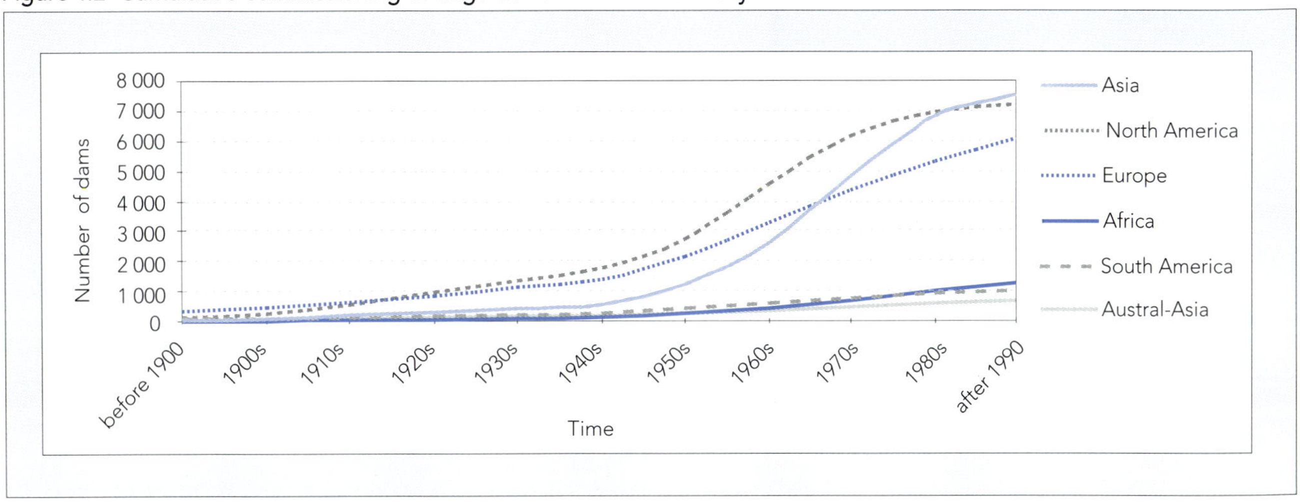

Source: ICOLD 1998, excluding over 90% of large dams in China.

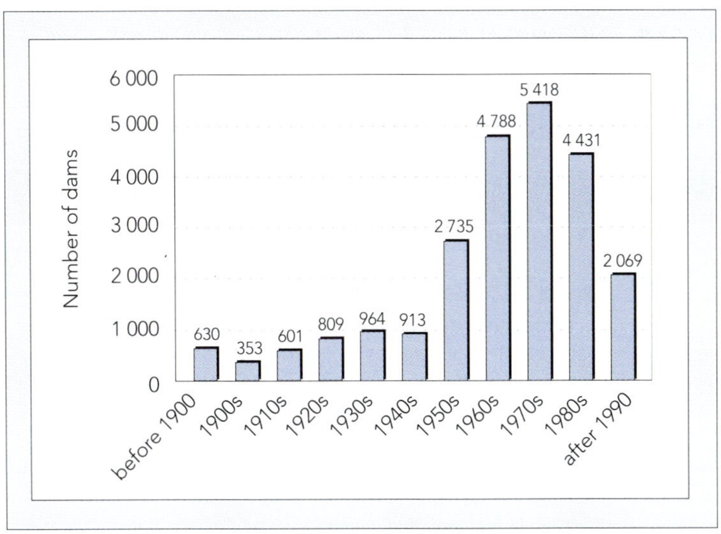

Figure V.3 Commissioning of large dams globally, by decade in the 20th century

Source: ICOLD, 1998, excluding over 90% of large dams in China.

World Population of Dams

This section is based primarily on the ICOLD *World Register of Dams* (ICOLD 1998). This voluntary register contains information on 25 420 large dams, which is a partial list, since member countries reported 41 413 dams in 1996. The register is constrained by the following: (1) it provides information on a limited set of parameters for each large dam such as location, commissioning date, purpose, height, reservoir size, spillway capacity, etc.; (2) it is incomplete for certain countries, the most significant being China with only 1 855 of the estimated 22 000 large dams registered; (3) gaps for other countries such as the Russian Federation are likely to bias the sample in similar ways; (4) data for the 1990s are under-reported to an unknown extent and also contain dams that have not been completed; and (5) entries are for a dam and not a reservoir – therefore, care must be taken in finding average reservoir volume capacities and surface areas where more than one dam is associated with a particular reservoir.

Table V.1 Top 20 countries by number of large dams

	Country	ICOLD World Register of Dams 1998	Other sources	Percent of total dams	Cumulative percent
1	China	1 855	22 000	46.2	46.2
2	United States	6 375	6 575	13.8	60.0
3	India	4 011	4 291	9.0	69.0
4	Japan	1 077	2 675	5.6	74.6
5	Spain	1 187	1 196	2.5	77.1
6	Canada	793	793	1.7	78.8
7	South Korea	765	765	1.6	80.4
8	Turkey	625	625	1.3	81.7
9	Brazil	594	594	1.2	82.9
10	France	569	569	1.2	84.1
11	South Africa	539	539	1.1	85.2
12	Mexico	537	537	1.1	86.3
13	Italy	524	524	1.1	87.4
14	United Kingdom	517	517	1.1	88.5
15	Australia	486	486	1.0	89.5
16	Norway	335	335	0.7	90.2
17	Germany	311	311	0.7	90.9
18	Albania	306	306	0.6	91.5
19	Romania	246	246	0.5	92.0
20	Zimbabwe	213	213	0.4	92.4
	Others	3 558	3 558	7.0	100.0
	Total	25 423	47 655	100.0	

Notes: There are different estimates for the number of large dams in each country. The World Register of Dams maintained by ICOLD is voluntary. By updating information on the top five countries, the WCD estimates there may be up to 48 000 large dams worldwide. As the higher estimate cannot be confirmed by more than one source, however, the WCD uses the figure of ' over 45 000' . The sources used are noted below. China is the major factor in the estimate.

China:	■ ICOLD (1998) has 1 855 dams. It notes the actual number of large dams in China may be over 20 000. ICOLD (2000) has 4 434 dams. ■ WCD China country study indicates there are over 84 000 human-made lakes, of which an estimated 22 000 have large dams by ICOLD' s definition, as reported in the WCD China Country Review. According to the Institute for Agricultural Economics (Zhang, 2000), by the end of 1999 there were 22 104 dams higher than 15m; of these, 17 526 dams were 15–30m high and 4 578 dams were over 30m (including 32 higher than 100m). There were also 320 dams under construction, 23 of which were more than 100m high. Approximately 45% of all dams were for irrigation. ■ IJHD (2000) indicates 26 094.
United States	■ ICOLD (1998) has 6 375 dams. ■ IJHD (2000) quotes an updated number of 6 575 large dams above 15 m and 75 000 dams in operation. ■ The National Inventory of Dams (USACE, 2000) indicates there were 6 390 large dams in 1996.
India	■ ICOLD (1998) has 4 011 large dams. ■ WCD India Country Review quotes 4 291 according to the National Register for Large Dams (CWC, 1994).
Japan	■ ICOLD (1998) has 1 077 large dams registered, noting that only those above 30 m in height were reported. ■ Japan Dam Almanac (Ministry of Construction, Japan, 1994), updated annually, estimates 2 675 large dams. ■ IJHD (2000) estimates 2 560 dams.
Spain	■ ICOLD (1998) has 1 187. ICOLD (Berga et al, 2000) quotes updated number of 1 196 dams. ■ IJHD (2000) estimates there are 906 large dams (above 30 m).
Russian Federation	■ ICOLD (1998) has 91; mostly hydropower dams were reported. ■ Russia has 400 human-made reservoirs larger than 3 million m^3 (a large dam, using ICOLD' s definition), according to the Hydroproject Institute, cited in the WCD Briefing Paper on the Russian Federation and NIS.

Functions of large dams, by region

Figure V.4 Hydropower dams, by region

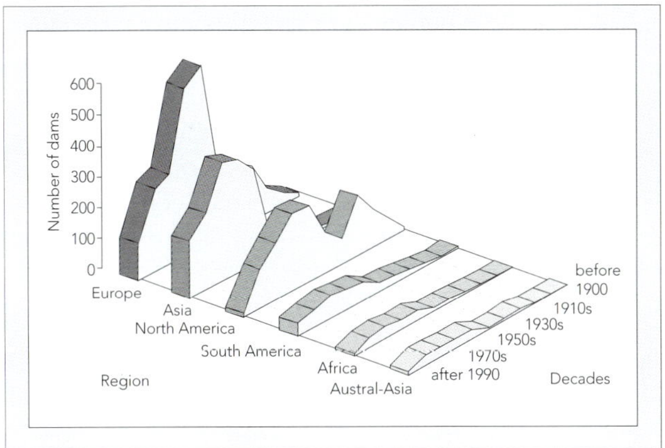

Source: ICOLD, 1998.

Figure V.5 Irrigation dams, by region

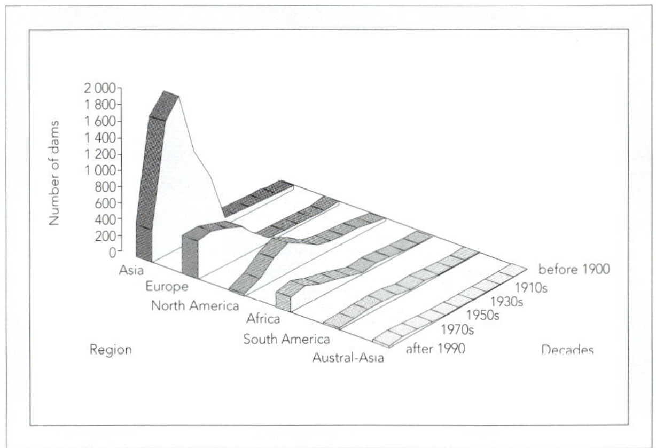

Source: ICOLD, 1998.

Figure V.6 Water supply dams, by region

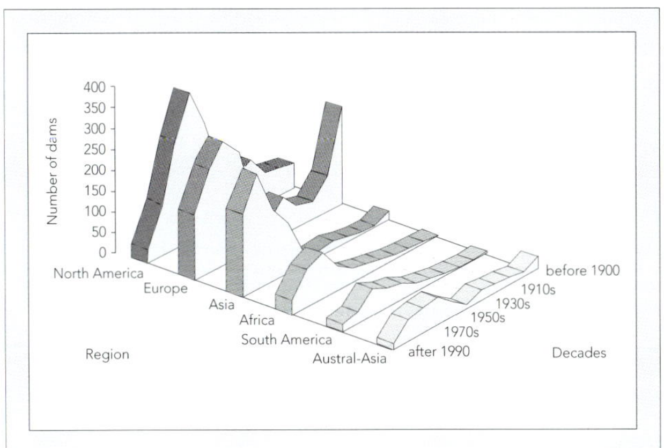

Source: ICOLD, 1998.

Figure V.7 Flood control dams, by region

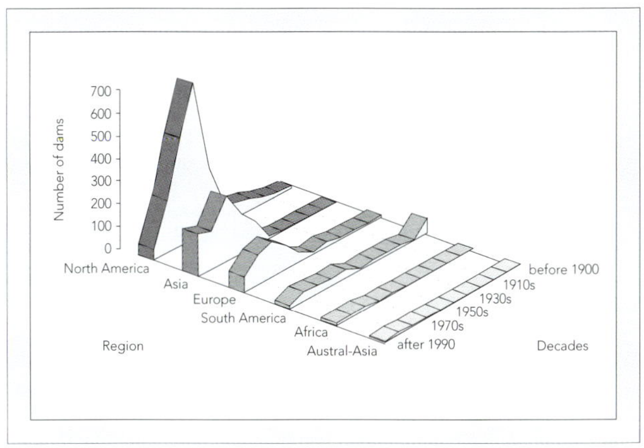

Source: ICOLD, 1998.

Figure V.8 Multipurpose dams, by region

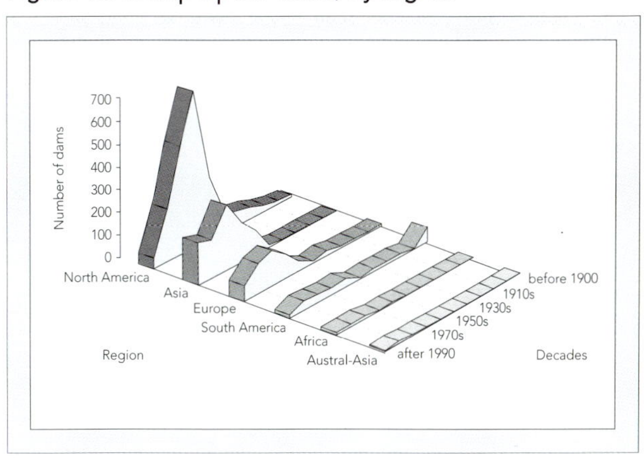

Source: ICOLD, 1998.

Note: Over 90% of dams in China are excluded as time series data are not available. Categorisation of 140 countries of the six regions as used in ICOLD (1998) is as follows:

***Africa**: Algeria, Angola, Benin, Botswana, Burkina Faso, Cameroon, Congo, Côte d'Ivoire, Dem. Rep. of Congo, Egypt, Ethiopia, Gabon, Ghana, Guinea, Kenya, Lesotho, Liberia, Libya, Madagascar, Malawi, Mali, Mauritius, Morocco, Mozambique, Namibia, Nigeria, Senegal, Seychelles, Sierra Leone, South Africa, Sudan, Swaziland, Tanzania, Togo, Tunisia, Uganda, Zambia, and Zimbabwe;*

***Asia**: Afghanistan, Bangladesh, Brunei, Cambodia, China, India, Iran, Iraq, Japan, Jordan, Kazakhstan, Kyrgyzstan, Laos, Latvia, Lebanon, Malaysia, Myanmar, Nepal, North Korea, Pakistan, Philippines, Saudi Arabia, Singapore, South Korea, Sri Lanka, Syria, Taiwan/China, Tajikistan, Thailand, Uzbekistan, and Viet Nam;*

***Austral-Asia**: Australia, Fiji, Indonesia, New Zealand, and Papua-New Guinea;*

***Europe**: Albania, Armenia, Austria, Azerbaijan, Belgium, Bosnia-Herzegovina, Bulgaria, Croatia, Cyprus, Czech Republic, Denmark,*

Finland, France, Georgia, Germany, Greece, Hungary, Iceland, Ireland, Italy, Lithuania, Luxembourg, Macedonia, Moldova, Netherlands, Norway, Poland, Portugal, Romania, Russian Federation, Slovakia, Slovenia, Spain, Sweden, Switzerland, Turkey, Ukraine, United Kingdom, and Yugoslavia;

North America: *Antigua, Canada, Cuba, El Salvador, Haiti, Honduras, Jamaica, Mexico, Nicaragua, Trinidad & Tobago, and United States; and*

South America: *Argentina, Bolivia, Brazil, Chile, Colombia, Costa Rica, Dominican Republic, Ecuador, Guatemala, Guyana, Panama, Paraguay, Peru, Suriname, Uruguay, and Venezuela.*

Physical attributes of large dams

These figures show the distribution of dams by height, reservoir surface area (km^2) and reservoir volume (million m^3) globally and across the regions. These parameters have an influence on the use and operation of dams and on the nature and scope of the impacts.

Figure V.9 Global distribution of dam heights (m)

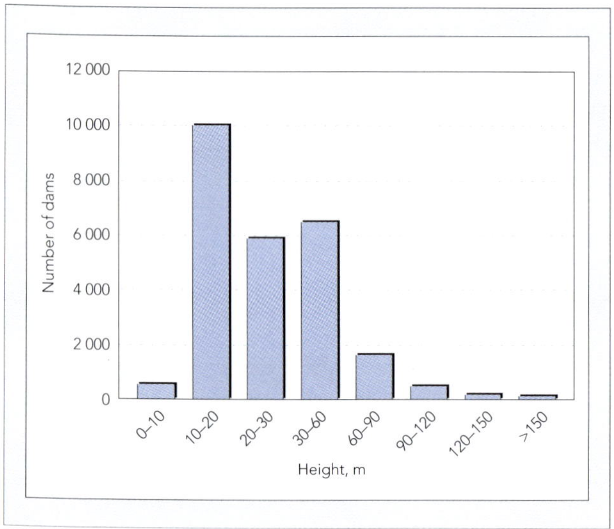

Source: ICOLD, 1998.

Figure V.10 Regional distribution of dam heights (m)

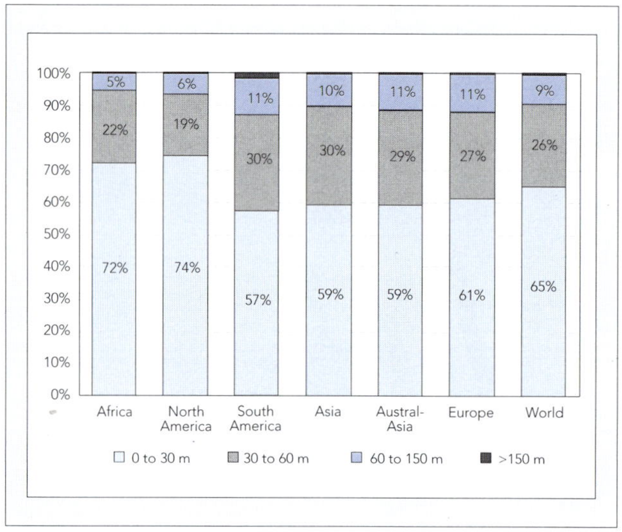

Source: ICOLD, 1998.

Figure V.11 Global distribution of reservoir surface area (km^2)

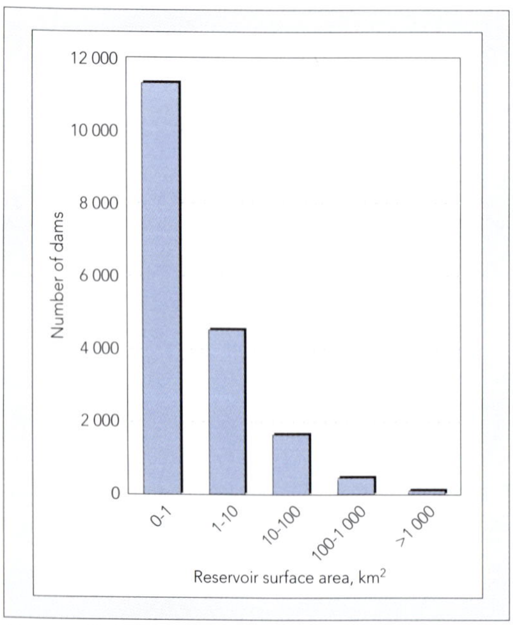

Source: ICOLD, 1998.

Figure V.12 Regional distribution of reservoir surface area (km^2)

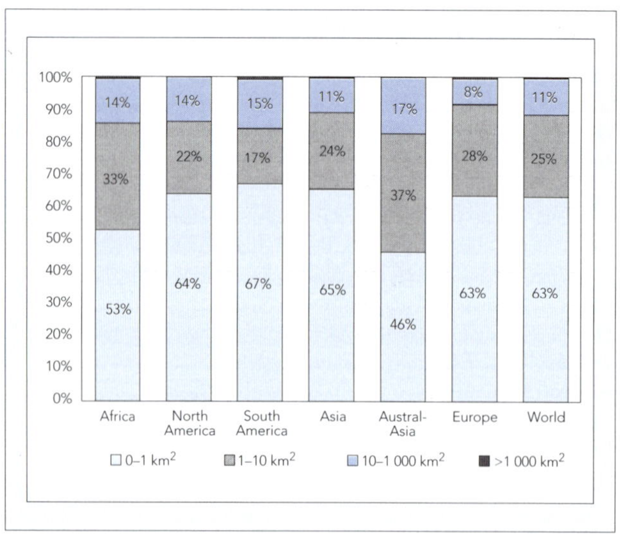

Source: ICOLD, 1998.

Figure V.13 Global distribution of reservoir volume (million m³)

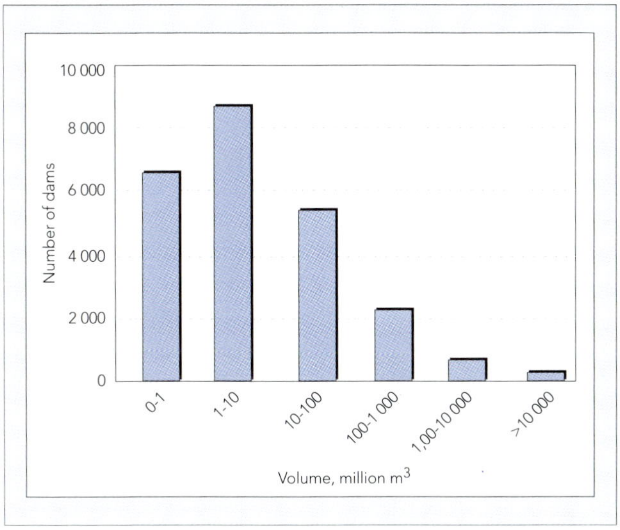

Source: ICOLD, 1998.

Figure V.14 Regional distribution of reservoir volume (million m³)

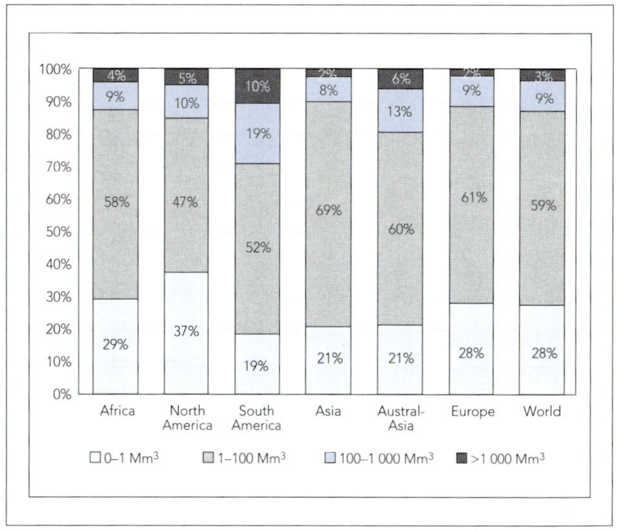

Source: ICOLD, 1998.

Figure V.15 Share of world population of dams, by country

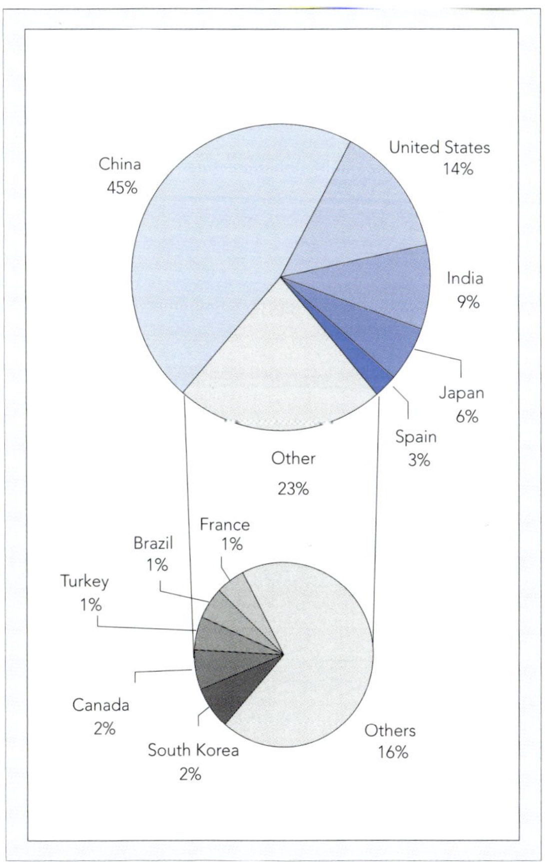

Source: See Table V.1.

Table V.2 The 10 countries with the most large dams

	By number of large dams	By function			
		Irrigation	Water supply	Flood control	Hydropower
1	China	China	United States	China	China
2	United States	India	United Kingdom	United States	United States
3	India	United States	Spain	Japan	Canada
4	Spain	Korea	Japan	Brazil	Japan
5	Japan	Spain	Australia	Germany	Spain
6	Canada	Turkey	Thailand	Romania	Italy
7	Korea	Japan	South Africa	Mexico	France
8	Turkey	Mexico	Brazil	Korea	Norway
9	Brazil	South Africa	France	Canada	Brazil
10	France	Albania	Germany	Turkey	Sweden

Note: This table shows that China, India and the United States have outpaced the world in building large dams, based on ICOLD 1998 and WCD correction for China.

Table V.3 Dams in 140 countries

Regions and countries	Number of dams
Africa	
South Africa	539
Zimbabwe	213
Algeria	107
Morocco	92
Tunisia	72
Nigeria	45
Côte d'Ivoire	22
Angola	15
Dem. Rep. of Congo	14
Kenya	14
Namibia	13
Libya	12
Madagascar	10
Cameroon	9
Mauritius	9
Burkina Faso	8
Ethiopia	8
Mozambique	8
Lesotho	7
Egypt	6
Swaziland	6
Ghana	5
Sudan	4
Zambia	4
Botswana	3
Malawi	3
Benin	2
Congo	2
Guinea	2
Mali	2
Senegal	2
Seychelles	2
Sierra Leone	2
Tanzania	2
Togo	2
Gabon	1
Liberia	1
Uganda	1
Total	1 269
Western Europe	
Spain	1 196
France	569
Italy	524
United Kingdom	517
Norway	335
Germany	311
Sweden	190
Switzerland	156
Austria	149
Portugal	103
Finland	55
Cyprus	52
Greece	46
Iceland	20
Ireland	16
Belgium	15
Denmark	10
Netherlands	10
Luxembourg	3
Total	4 277
South America	
Brazil	594
Argentina	101
Chile	88
Venezuela	74
Colombia	49
Peru	43
Ecuador	11
Bolivia	6
Uruguay	6
Paraguay	4
Guyana	2
Suriname	1
Total	979
Eastern Europe	
Albania	306
Romania	246
Bulgaria	180
Czech Republic	118
Poland	69
Yugoslavia	69
Slovakia	50
Slovenia	30
Croatia	29
Bosnia-Herzegovina	25
Ukraine	21
Lithuania	20
Macedonia	18
Hungary	15
Latvia	5
Moldova	2
Total	1 203
North and Central America	
United States	6 575
Canada	793
Mexico	537
Cuba	49
Dominican Republic	11
Costa Rica	9
Honduras	9
Panama	6
El Salvador	5
Guatemala	4
Nicaragua	4
Trinidad & Tobago	4
Jamaica	2
Antigua	1
Haiti	1
Total	8 010
Asia	
China	22 000
India	4 291
Japan	2 675
South Korea	765
Turkey	625
Thailand	204
Indonesia	96
Russia	91
Pakistan	71
North Korea	70
Iran	66
Malaysia	59
Taipei, China	51
Sri Lanka	46
Syria	41
Saudi Arabia	38
Azerbaijan	17
Armenia	16
Philippines	15
Georgia	14
Uzbekistan	14
Iraq	13
Kazakhstan	12
Kyrgyzstan	11
Tajikistan	7
Jordan	5
Lebanon	5
Myanmar	5
Nepal	3
Viet Nam	3
Singapore	3
Afghanistan	2
Brunei	2
Cambodia	2
Bangladesh	1
Laos	1
Total	31 340
Austral-Asia	
Australia	486
New Zealand	86
Papua New Guinea	3
Fiji	2
Total	577

Source: Based on ICOLD, 1998, IJHD, 2000 and other sources. Regional categories do not match the ICOLD classification given for Figures V.4 – V.8.

Table V.4 Number of dams under construction, selected countries

By number of large dams		By function
India	960[a] 695[b] 16[c] – above 60 meters	Irrigation
China	280[a] 90[c] – above 60 meters	Flood control Hydropower, including pumped storage
Turkey	193[a] 209[c] – more the 60 meters	Irrigation, hydropower and water supply
South Korea	132[a] 4[c] – more the 60 meters	Multi-purpose: hydropower, flood management Water supply
Japan	463[d] 90[a] 51[c] – more than 60 meters 13[c] – more the 100 meters	Multi-purpose flood control with hydropower and water supply
Iran	48[c] – above 60 meters	Irrigation, single and multi-purpose

Note: Other countries currently building dams over 60 meters include Spain (10), Algeria (7), Italy (9), Romania (8), Brazil (6), Venezuela (5) and the Russian Federation (5).[c]

[a] ICOLD, 1997
[b] WCD India Country Study
[c] IJHD, 2000
[d] Ministry of Construction, Japan, 1999

Regional Profiles of Large Dams

Table V.5 Summary of regional statistics on large dams

	World[a]	Europe	Asia	North and Central America	South America	Africa	Austral-Asia
Total number of large dams with China	25 420[a] ~48 000[a]	5 480	31 340	8 010	979	1 269	577
Average height[b](m)	31	33	33	28	37	28	33
Average reservoir[b] area (km^2)	23	7	44	13	30	43	17
Avg. reservoir capacity[b] (million m^3)	269	70	268	998	1 011	883	205
Technically feasible hydroelectric potential[c] (TWh/year)	14 370	1 225	6 800	1 660	2 665	1 750	270
Annual Hydroelectric Production[c] (TWh/year)	2 643	552	753	700	534	62	42
Exploited technically feasible hydroelectric potential[c] (%)	>18%	>45%	>11%	>42%	>20%	>3.5%	

[a] *The primary source of data is ICOLD 1998, but the regional divisions in this Table and in Figures V.16 through V.27 follow those described in Table V.3. The 1998 ICOLD Register has 25 420 dams registered. Reporting depends on the member countries. Table V.1 indicates how the global estimate of nearly 48 000 large dams is arrived at, with the main issue being the number of dams in China.*

[b] *The ICOLD 1998 database was used to calculate the average dam height, reservoir capacity and surface area by region.*

[c] *IJHD 2000. Technical Feasibility is based on the conversion of all river head and flow in the major rivers in region into energy.*

Western and Eastern Europe: hydropower in the North and irrigation in the South led dam development in Europe

At the turn of the twentieth century, most large dams in Europe were found in the United Kingdom (220 dams).[1] By 2000, more than 4 277 large dams were registered in Western Europe; Spain built the most (about 1 200), followed by France, Italy and the United Kingdom, each with more than 500 large dams. Until 1960, France and Italy were building dams at the same rate as

Spain. East European countries have about 1 200 large dams; Albania (306 dams, almost all for irrigation) and Romania (246 dams, for all purposes) have built the most in the region.

As shown in Figure V.17, hydropower and then irrigation and water supply have been the main purposes of dams in Europe. There is marked contrast in reservoir use (and importance) across Europe, which reflects topography, rainfall and national polices, particularly on hydropower. Numerous hydropower reservoirs – often located in mountainous regions and in Nordic countries – are distinguished from the generally smaller irrigation and water supply reservoirs located in lowland and southern regions of Europe.[2] Approximately one-quarter of dams in Europe are multi-purpose. Hydropower provides more than half of the electricity supply in several European countries (86% in Albania and 90% in Iceland, for example) and more than 99% in Norway.[3]

Dam construction and hydropower development passed its peak in the 1960s in many parts of Europe. At present there is a focus on refurbishment and upgrading dams and adapting to new regulations. In 2000, some 2 460 MW of additional hydroelectric generating capacity was under construction in 22 countries, with the most significant new dam development under way in Bosnia and Bulgaria. According to industry sources, other active hydropower developers are Germany (in the eastern part), Greece, Iceland, Italy, Macedonia, Portugal, Slovenia and the Ukraine.[4] Refurbishment is the main focus in many East European countries, and plans for additional dams for power and flood control have also been made. Spain is most active in dam building overall for other purposes and is now implementing a number of multi-purpose schemes. Spain has plans for further dam construction for drought management, though there is debate on these plans.

Figure V.16 Large dams commissioned per decade in Europe

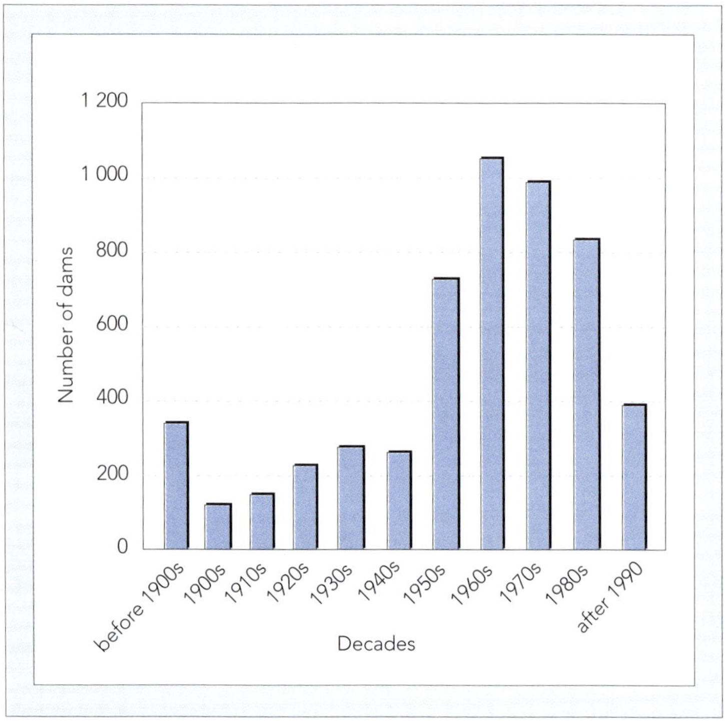

Source: ICOLD, 1998.

Figure V.17 Breakdown by purpose of dams in Europe

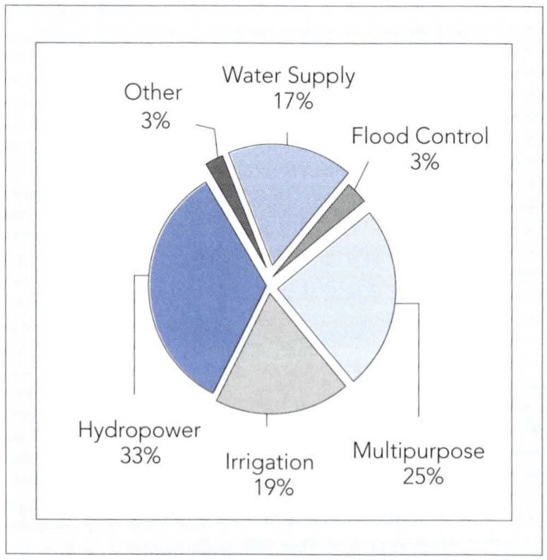

Source: ICOLD, 1998.
Note: Rates of dam commissioning in the 1990s are underreported.

Asia: the most active region today for dam development – and historically for irrigation

Asia as a region, and particularly China and India, is one of the most active in the world in terms of the number of dams in operation, under construction and planned. China, with half the world's large dams and an active construction programme, dominates the picture. ICOLD (1998) has 8 500 large dams listed in the Asian region, but WCD estimates there are more than 30 000.

Most large dams in Asia were built for irrigation, followed by hydropower, flood control and water supply functions. A quarter are multi-purpose. These tend to be the larger projects. There are nevertheless large differences across Asia in the purpose and type of dams. The primary purposes for dams built today include irrigation in India and Turkey; flood control and power, including pumped storage, in China; flood management and hydro-pumped storage in Japan; and irrigation and power supply in Iran. Hydropower provides more than 50% of the national electricity supply in nine Asian countries. It represents 19% of the total power generation in China, 25% in India, and 19% in the Russian Federation. The balance is largely based on coal-fired generation.

India, China, Turkey, Japan and Iran are among the most active dam-building countries overall. The overall rate of large dam building peaked in Asia in the 1970s–1980s at over 200 dams per year. Statistics (excluding China) show the pace of construction slowed in the 1990s, reflecting multiple trends, including a focus on improving existing surface irrigation infrastructure. Still, in 2000 more than 83 000 MW of additional hydroelectric generating capacity was under construction in 23 countries. The

Figure V.18 Large dams commissioned per decade in Asia

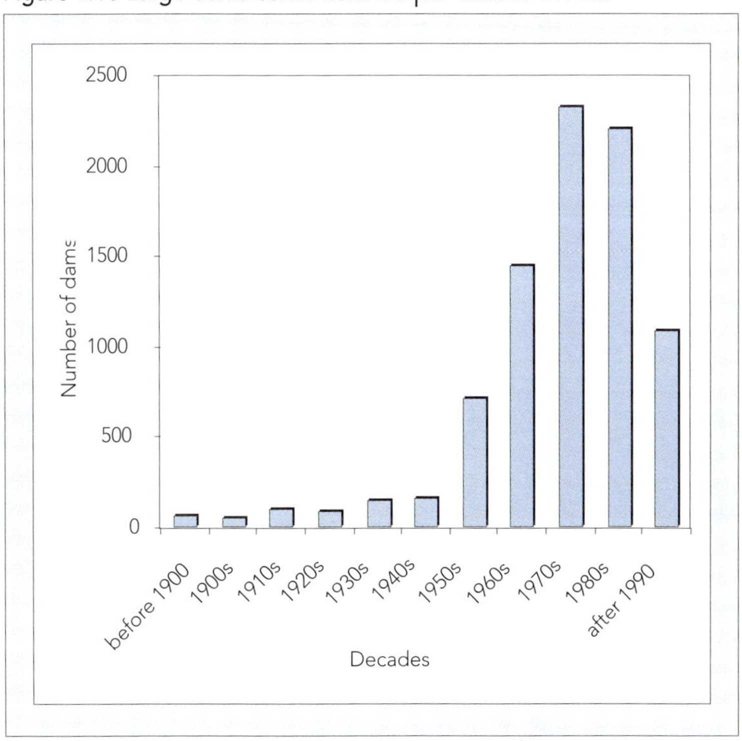

Source: ICOLD, 1998.

Figure V.19 Breakdown by purpose of dams in Asia

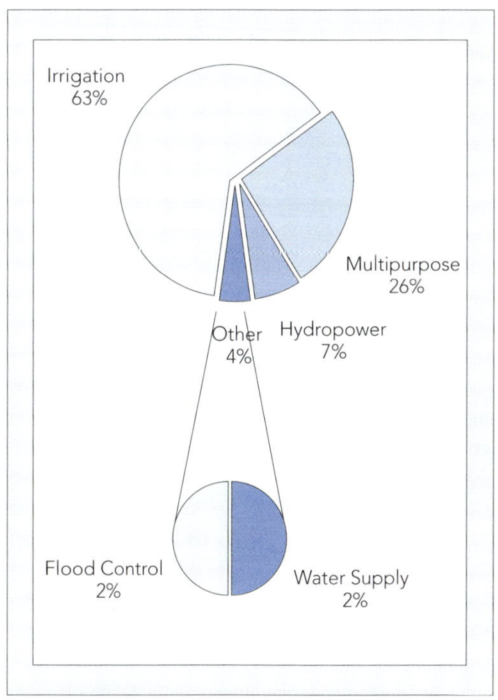

Source: ICOLD 1998. Note: Rates of dam commissioning in the 1990s are underreported. Figures above exclude China

majority of the development is in China, followed by India, Indonesia and Iran.

The Russian Federation, where economic and institutional restructuring are under way, has focused on completing large projects started under the former political systems but abandoned in the 1990s. Other priorities include the rehabilitation of large dams in operation. Russia is planning to construct five new large dams. Democratisation and the emergence of NGOs have led to greater involvement and public debate on water and energy, though active participation of non-governmental interests in decision-making is limited as yet.

A discussion of the issues around the development of dams in the Asian region may be found on the WCD website in the report on the South Asia Consultation held in Colombo, Sri Lanka, in December 1998 and the East and Southeast Asia Consultation in Hanoi, Vietnam, in February 2000.

North and Central America: 80% of the region's dams are in the United States

Of more than 8 000 large dams in North and Central America, about four-fifths (6 575) are in the United States. About 40% of these are multi-purpose. Hydropower provides over 50% of the electricity supply in 7 of the 15 countries in the region, including Canada, Guatemala and Honduras. Canada has the highest hydropower generation in the world; combined with the U.S., North America produces over a quarter of the world's hydropower.[5] Flood management and recreation have become principal uses of large dams in the United States, and operations are increasingly determined by those seasonal requirements. There are large variations in climate in the region, which determine the predominant use of dams locally. In southern and arid areas of the region, irrigation and water supply are more important.

Figure V.20 Large dams commissioned per decade in North and Central America

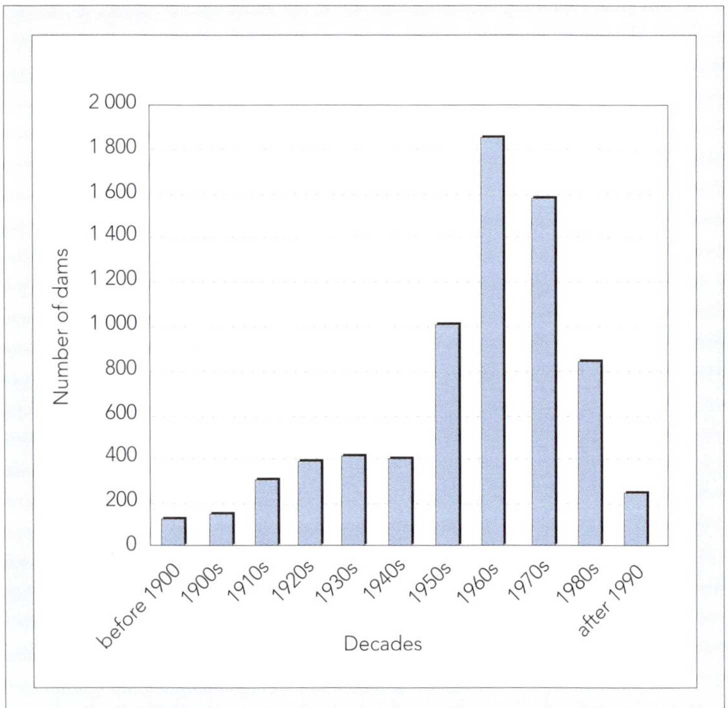

Source: ICOLD, 1998. Note: Rates of dam commissioning in the 1990s are underreported.

Figure V.21 Breakdown by purpose of dams in North and Central America

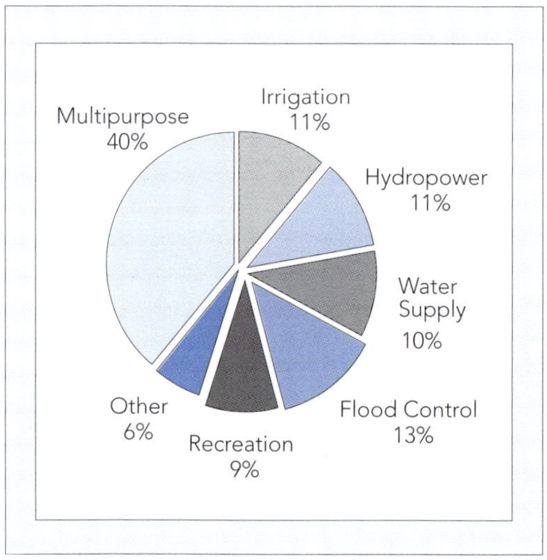

Source: ICOLD, 1998.

Dam commissioning in the region increased dramatically after the Second World War and peaked in the late 1960s at about 180 dams per year. Commissioning rates have recently decreased to about 40 dams per year (in 1990–95). In the United States, the rate of decommissioning has now exceeded the rate of construction of newly licensed large dams.

No new hydropower development is currently planned in the United States, and some 30 000 MW of existing hydropower is due to be relicensed in the next 20 years. The major activities in the United States and Canada occur around refurbishment, upgrading, optimisation of the operation of dams and, in the United States, decommissioning. Most notably, over 400 dams of all sizes, though mostly small, have been decommissioned in the United States. The only major dam reported under construction in the United States is in Puerto Rico (multi-purpose); a dam for a new water supply reservoir in California (Diamond Valley Lake) was completed in 1999. Canada has a large hydropower potential – and while more dams are proposed, especially in Quebec, Labrador and Newfoundland, there is uncertainty over such developments.

Central America has more active dam-building programmes or plans. In 2000, some 2 124 MW of hydroelectric generating capacity was under construction in five countries. High-head small hydropower is actively pursued. Among those active in the region are Guatemala, El Salvador, Costa Rica and Honduras. Mexico has 540 dams in operation and no major dams currently under construction. While more dams are planned, industry considers the outlook uncertain, given the restructuring of the power sector.[6]

South America: hydropower has driven dam building in the region, with Brazil leading

Almost two-thirds of the 979 large dams in South America are in Brazil (594 in 1998). The primary purposes for building large dams

Figure V.22 Large dams commissioned per decade in South America

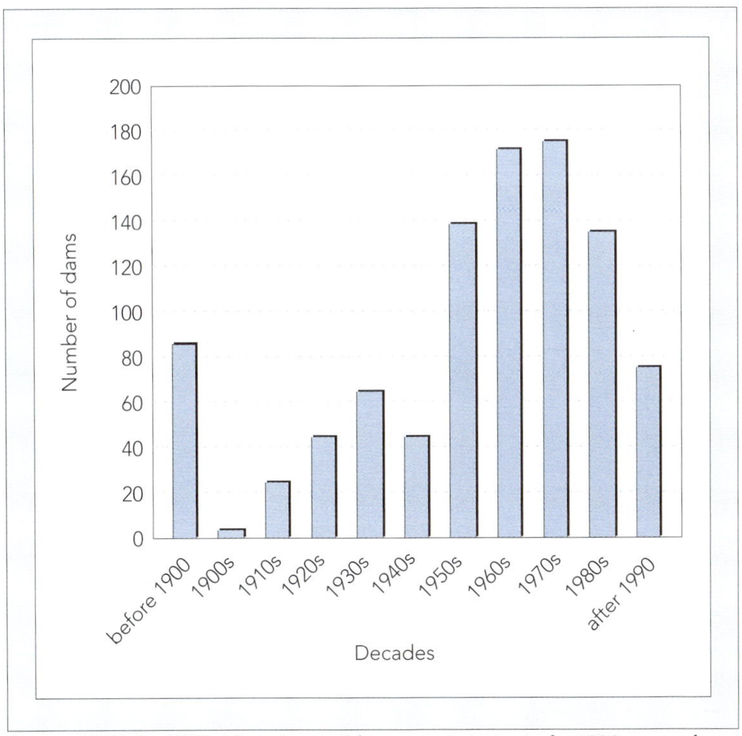

Source: ICOLD, 1998. Note: Rates of dam commissioning in the 1990s are underreported.

Figure V.23 Breakdown by purpose of dams in South America

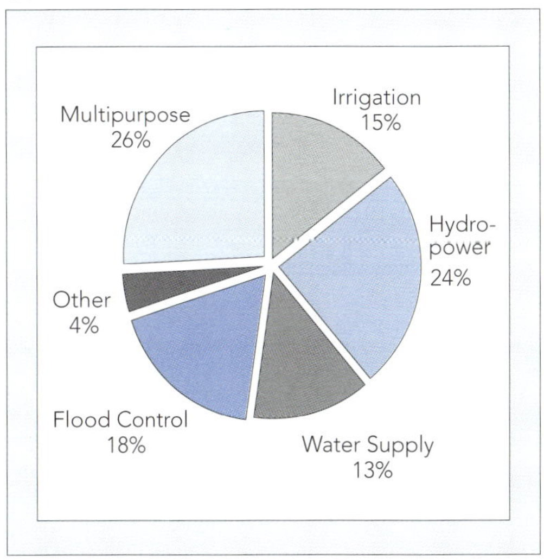

Source: ICOLD, 1998.

have been hydropower generation and flood control. Among multi-purpose dams, irrigation, flood control and water supply functions are important. The peak of the dam building in South America took place in 1960-79, when an average of 17 dams per year were commissioned.

The region's most active hydropower developers have been Brazil – which generates over 93% of its electricity from hydropower – Venezuela (73%), Ecuador (68%), Chile (57%) and Colombia (68%).[7] Hydropower supplies more than half of electricity generation in 10 of the 12 countries in the region that have dams, including Paraguay (nearly 100%) and Peru (74%).

As of 2000, some 18 000 MW of additional hydropower capacity were under construction in 10 countries in Latin America.[8] Despite the large hydropower potential of the region, the expansion of natural gas networks, regional interconnection of power grids and the restructuring and privatisation of the power industry have made the prospects for large hydropower dams more uncertain. Industry projections indicate that further development in the short to medium term is likely to occur around medium and small hydropower dams.[9]

A discussion of the issues around the development of dams in the region may be found on the WCD website in the report on the Latin America Consultation held in São Paulo, Brazil, in August 1999.

Africa: irrigation and hydropower have been the main drivers for dam building

Of the over 1 269 large dams in Africa, South Africa (with between 539 and 791) and Zimbabwe (with between 213 and 233) together account for over 60% of the dams on the continent.[10] Irrigation is the single largest purpose for building large dams in Africa, especially in the northern and southern regions that have large arid or semi-arid zones. In central and other parts of

Figure V.24 Large dams commissioned per decade in Africa

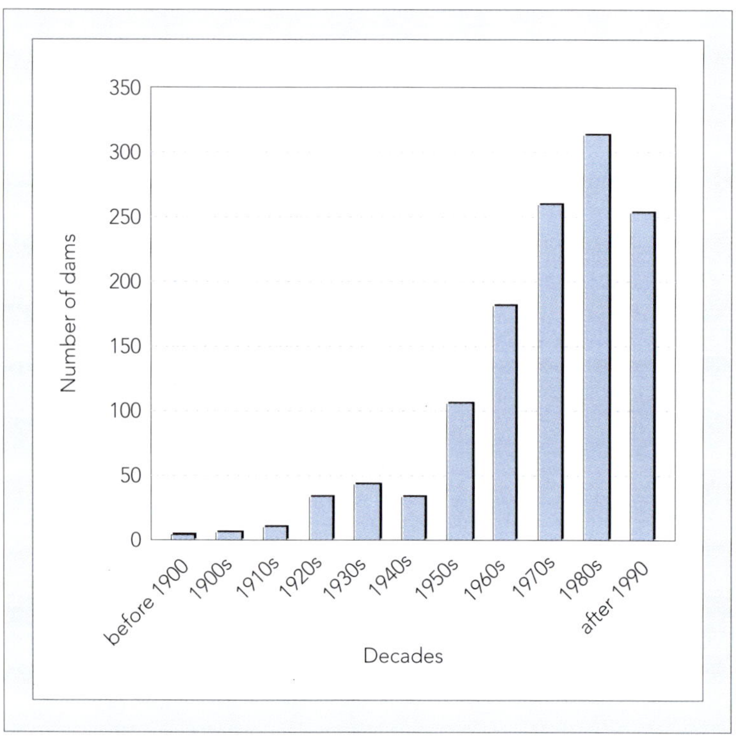

Source: ICOLD, 1998. Note: Rates of dam commissioning in the 1990s are underreported.

Figure V.25 Breakdown by purpose of dams in Africa

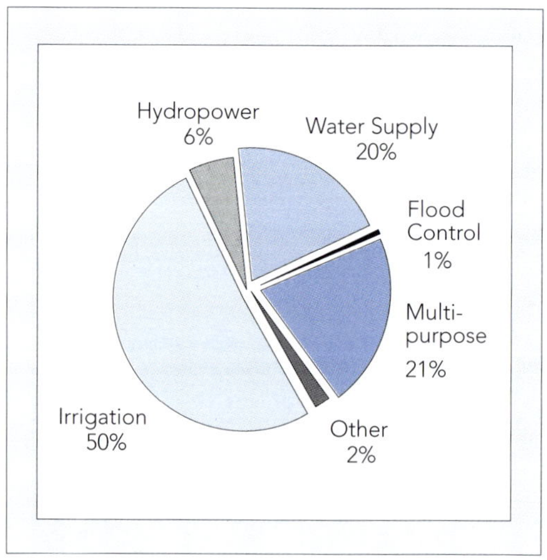

Source: ICOLD, 1998.

Africa, which are less arid, hydropower is the primary reason for dam building.

In South Africa, large dams have a capacity equivalent to 50% of mean annual river flow. Most dams have irrigation and water supply as their main uses (just 1.9% of electricity generation is hydropower). In Zimbabwe, 87% of the large dams are embankment dams and hydropower accounts for 17% of electrical supply, mainly from the Kariba dam jointly operated with Zambia. The Southern African Power Pool Arrangement will have a major impact on the region and its investment in power supply infrastructure in the future. In Africa hydropower contributes more than 80% of electricity production in 18 countries, and over 50% in 25 countries. Droughts in East Africa in the 1990s affected power generation significantly in those regions where reservoirs were drawn down.

Irrigation and water resources development projects are under way in northern Africa, especially in Tunisia, Morocco (where 7 large dams are under construction) and Algeria (with 13 dams being built). As of 2000, more than 2 000 MW of additional hydroelectric generating capacity was under construction in 17 countries. Zimbabwe and South Africa have a number of irrigation and water supply large dams under construction.

A discussion of the issues around the development of dams on the African continent may be found on the WCD website in the report on the Africa and Middle East Consultations held in Cairo, Egypt, in December 1999.

Austral-Asia: close to half the dams were built for water supply

Of the nearly 577 dams in the region, the majority (486) are located in Australia. New Zealand has 86. Almost 50% of the large dams in the region were built as single-purpose water supply dams. Hydropower and irrigation were the next most important reasons for dam

Figure V.26 Large dams commissioned per decade in Austral-Asia

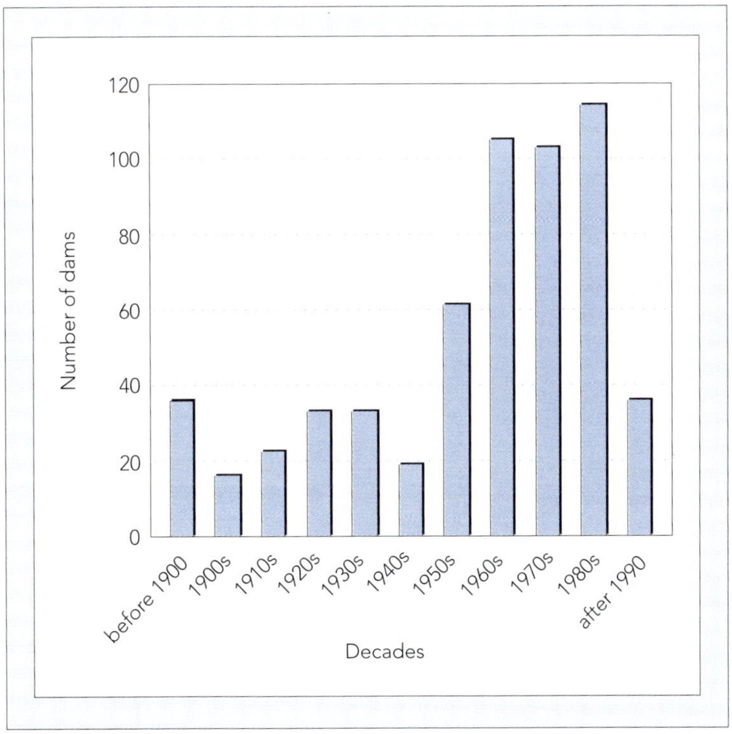

Source: ICOLD, 1998. Note: Rates of dam commissioning in the 1990s are underreported.

Figure V.27 Breakdown by purpose of dams in Austral-Asia

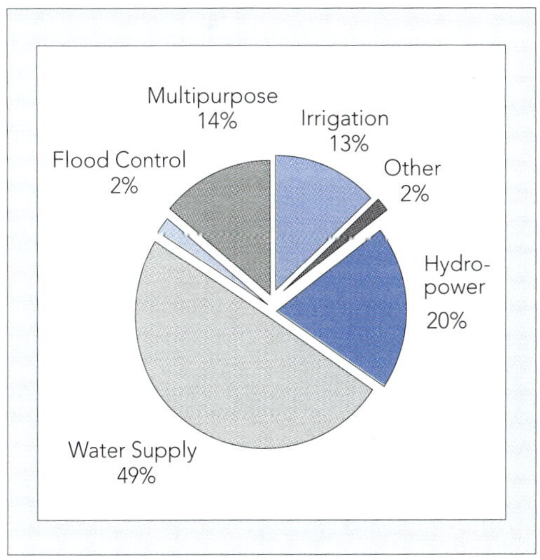

Source: ICOLD, 1998.

building. Hydropower accounts for more than 50% of electricity supply in four countries in the region (Fiji, New Zealand, Papua New Guinea and New Caledonia).

Dam building peaked in Australia and New Zealand in the 1980s (at about 10 large dams per year) and slowed dramatically in the 1990s. At present no new major dams or large hydropower projects are under way in Australia and New Zealand, though New Zealand recently completed a water supply dam. Both Australia and New Zealand have reformed the water and power sector regulatory framework, and restructured or privatised water management and power sector institutions. While there are several new dams being carried as options in longer-term planning, it appears unlikely they will go ahead in the foreseeable future.[11]

Australia and New Zealand do have several projects under way to heighten and refurbish existing dams and are generally focusing on improvement, safety and optimising the operation of existing dams. Current trends in Austral-Asia are towards small-scale hydropower development within the region's island states and towards water supply projects as demand grows.

Endnotes

1 This classification of Europe includes Western and Eastern Europe but excludes the Russian Federation and Turkey, which are reported under the Asia Region.

2 EEA, 1999.

3 Albania and Iceland from IJHD 2000; Norway from UNDP et al, 2000.

4 IJHD, 2000.

5 UNDP et al, 2000.

6 IJHD, 2000.

7 Brazil from IEA, 2000.

8 IJHD, 2000.

9 Ibid.

10 Lower figures from ICOLD, 1998; higher figures from IJHD, 2000.

11 IJHD, 2000.

Annex VI
United Nations Declarations

The Right to Development

The General Assembly,

Having considered the question of the right to development decides to adopt the Declaration on the Right to Development on December 4, 1986, the text of which is annexed to the present resolution.

ANNEX

Declaration on the Right to Development

The General Assembly,

Bearing in mind the purposes and principles of the Charter of the United Nations relating to the achievement of international co-operation in solving international problems of an economic, social, cultural or humanitarian nature, and in promoting and encouraging respect for human rights and fundamental freedoms for all without distinction as to race, sex, language or religion,

Recognising that development is a comprehensive economic, social, cultural and political process, which aims at the constant improvement of the well-being of the entire population and of all individuals on the basis of their active, free and meaningful participation in development and in the fair distribution of benefits resulting therefrom,

Considering that under the provisions of the Universal Declaration of Human Rights everyone is entitled to a social and international order in which the rights and freedoms set forth in that Declaration can be fully realised,

Recalling the provisions of the International Covenant on Economic, Social and Cultural Rights and of the International Covenant on Civil and Political Rights,

Recalling further the relevant agreements, conventions, resolutions, recommendations and other instruments of the United Nations and its specialised agencies concerning the integral development of the human being, economic and social progress and development of all peoples, including those instruments concerning decolonisation, the prevention of discrimination, respect for and observance of, human rights and fundamental freedoms, the maintenance of international peace and security and the further promotion of friendly relations and co-operation among States in accordance with the Charter,

Recalling the right of peoples to self-determination, by virtue of which they have the right freely to determine their political status and to pursue their economic, social and cultural development,

Recalling also the right of peoples to exercise, subject to the relevant provisions of both International Covenants on Human Rights, full and complete sovereignty over all their natural wealth and resources,

Mindful of the obligation of States under the Charter to promote universal respect for and observance of human rights and fundamental freedoms for all without distinction of any kind such as race, colour, sex, language, religion, political or other opinion, national or social origin, property, birth or other status,

Considering that the elimination of the massive and flagrant violations of the human rights of the peoples and individuals affected by situations such as those resulting from colonialism, neo-colonialism, apartheid, all forms of racism and racial discrimination, foreign domination and occupation, aggression and threats against national sovereignty, national unity and territorial integrity and threats of war would contribute to the establishment of circumstances propitious to the development of a great part of mankind,

Concerned at the existence of serious obstacles to development, as well as to the complete fulfilment of human beings and of peoples, constituted, inter alia, by the denial of civil, political, economic, social and cultural rights, and considering that all human rights and fundamental freedoms are indivisible and interdependent and that, in order to promote development, equal attention and urgent consideration should be given to the implementation, promotion and protection of civil, political, economic, social and cultural rights and that, accordingly, the promotion of, respect for and enjoyment of certain human rights and fundamental freedoms cannot justify the denial of other human rights and fundamental freedoms,

Considering that international peace and security are essential elements for the realisation of the right to development,

Reaffirming that there is a close relationship between disarmament and development and that progress in the field of disarmament would considerably promote progress in the field of development and that resources released through disarmament measures should be devoted to the economic and social development and well-being of all peoples and, in particular, those of the developing countries,

Recognising that the human person is the central subject of the development process and that development policy should therefore make the human being the main participant and beneficiary of development,

Recognising that the creation of conditions favourable to the development of peoples and individuals is the primary responsibility of their States,

Aware that efforts at the international level to promote and protect human rights should be accompanied by efforts to establish a new international economic order,

Confirming that the right to development is an inalienable human right and that equality of opportunity for development is a prerogative both of nations and of individuals who make up nations,

Proclaims the following Declaration on the Right to Development:

Article 1

1. The right to development is an inalienable human right by virtue of which every human person and all peoples are entitled to participate in, contribute to, and enjoy economic, social, cultural and political development, in which all human rights and fundamental freedoms can be fully realised.
2. The human right to development also implies the full realisation of the right of peoples to self-determination, which includes, subject to the relevant provisions of both International Covenants on Human Rights, the exercise of their inalienable right to full sovereignty over all their natural wealth and resources.

Article 2

1. The human person is the central subject of development and should be the active participant and beneficiary of the right to development.
2. All human beings have a responsibility for development, individually and collectively, taking into account the need for full respect for their human rights and fundamental freedoms as well as their duties to the community, which alone can ensure the free and complete fulfilment of the human being, and they should therefore promote and protect an appropriate political, social and economic order for development.
3. States have the right and the duty to formulate appropriate national development policies that aim at the constant improvement of the well-being of the entire population and of all individuals, on the basis of their active, free and meaningful participation in development and in the fair distribution of the benefits resulting therefrom.

Article 3

1. States have the primary responsibility for the creation of national and international conditions favourable to the realisation of the right to development.
2. The realisation of the right to development requires full respect for the principles of international law concerning friendly relations and co-operation among States in accordance with the Charter of the United Nations.
3. States have the duty to co-operate with each other in ensuring development and eliminating obstacles to development. States should realise their rights and fulfil their duties in such a manner as to promote a new international economic order based on sovereign equality, interdependence, mutual interest and co-operation among all States, as well as to encourage the observance and realisation of human rights.

Article 4

1. States have the duty to take steps, individually and collectively, to formulate international development policies with a view to facilitating the full realisation of the right to development.
2. Sustained action is required to promote more rapid development of developing countries. As a complement to the efforts of developing countries, effective international co-operation is essential in providing these countries with appropriate means and facilities to foster their comprehensive development.

Article 5

States shall take resolute steps to eliminate the massive and flagrant violations of the human rights of peoples and human beings affected by situations such as those resulting from apartheid, all forms of racism and racial discrimination, colonialism, foreign domination and occupation, aggression, foreign interference and threats against national sovereignty, national unity and territorial integrity, threats of war and refusal to recognise the fundamental right of peoples to self-determination.

Article 6

1. All States should co-operate with a view to promoting, encouraging and strengthening universal respect for and observance of all human rights and fundamental freedoms for all without any distinction as to race, sex, language or religion.
2. All human rights and fundamental freedoms are indivisible and interdependent; equal attention and urgent consideration should be given to the implementation, promotion and protection of civil, political, economic, social and cultural rights.
3. States should take steps to eliminate obstacles to development resulting from failure to observe civil and political rights, as well as economic, social and cultural rights.

Article 7

All States should promote the establishment, maintenance and strengthening of international peace and security and, to that end, should do their utmost to achieve general and complete disarmament under effective international control, as well as to ensure that the resources released by effective disarmament measures are used for comprehensive development, in particular that of the developing countries.

Article 8

1. States should undertake, at the national level, all necessary measures for the realisation of the right to development and shall ensure, inter alia, equality of opportunity for all in their access to basic resources, education, health services, food, housing, employment and the fair distribution of income. Effective measures should be undertaken to ensure that women have an active role in the development process. Appropriate economic and social reforms should be carried out with a view to eradicating all social injustices.
2. States should encourage popular participation in all spheres as an important factor in development and in the full realisation of all human rights.

Article 9

1. All the aspects of the right to development set forth in the present Declaration are indivisible and interdependent and each of them should be considered in the context of the whole.
2. Nothing in the present Declaration shall be construed as being contrary to the purposes and principles of the United Nations, or as implying that any State, group or person has a right to engage in any activity or to perform any act aimed at the violation of the rights set forth in the Universal Declaration of Human Rights and in the International Covenants on Human Rights.

Article 10

Steps should be taken to ensure the full exercise and progressive enhancement of the right to development, including the formulation, adoption and implementation of policy, legislative and other measures at the national and international levels.

Universal Declaration of Human Rights

Adopted and proclaimed by General Assembly resolution 217 A (III) of 10 December 1948

On December 10, 1948 the General Assembly of the United Nations adopted and proclaimed the Universal Declaration of Human Rights the full text of which appears in the following pages. Following this historic act the Assembly called upon all Member countries to publicise the text of the Declaration and "to cause it to be disseminated, displayed, read and expounded principally in schools and other educational institutions, without distinction based on the political status of countries or territories."

PREAMBLE

Whereas recognition of the inherent dignity and of the equal and inalienable rights of all members of the human family is the foundation of freedom, justice and peace in the world,

Whereas disregard and contempt for human rights have resulted in barbarous acts which have outraged the conscience of mankind, and the advent of a world in which human beings shall enjoy freedom of speech and belief and freedom from fear and want has been proclaimed as the highest aspiration of the common people,

Whereas it is essential, if man is not to be compelled to have recourse, as a last resort, to rebellion against tyranny and oppression, that human rights should be protected by the rule of law,

Whereas it is essential to promote the development of friendly relations between nations,

Whereas the peoples of the United Nations have in the Charter reaffirmed their faith in fundamental human rights, in the dignity and worth of the human person and in the equal rights of men and women and have determined to promote social progress and better standards of life in larger freedom,

Whereas Member States have pledged themselves to achieve, in co-operation with the United Nations, the promotion of universal respect for and observance of human rights and fundamental freedoms,

Whereas a common understanding of these rights and freedoms is of the greatest importance for the full realisation of this pledge,

Now, Therefore THE GENERAL ASSEMBLY proclaims THIS UNIVERSAL DECLARATION OF HUMAN RIGHTS as a common standard of achievement for all peoples and all nations, to the end that every individual and every organ of society, keeping this Declaration constantly in mind, shall strive by teaching and education to promote respect for these rights and freedoms and by progressive measures, national and international, to secure their universal and effective recognition and observance, both among the peoples of Member States themselves and among the peoples of territories under their jurisdiction.

Article 1

All human beings are born free and equal in dignity and rights. They are endowed with reason and conscience and should act towards one another in a spirit of brotherhood.

Article 2

Everyone is entitled to all the rights and freedoms set forth in this Declaration, without distinction of any kind, such as race, colour, sex, language, religion, political or other opinion, national or social origin, property, birth or other status. Furthermore, no distinction shall be made on the basis of the political, jurisdictional or international status of the country or territory to which a person belongs, whether it be independent, trust, non-self-governing or under any other limitation of sovereignty.

Article 3

Everyone has the right to life, liberty and security of person.

Article 4

No one shall be held in slavery or servitude; slavery and the slave trade shall be prohibited in all their forms.

Article 5

No one shall be subjected to torture or to cruel, inhuman or degrading treatment or punishment.

Article 6

Everyone has the right to recognition everywhere as a person before the law.

Article 7

All are equal before the law and are entitled without any discrimination to equal protection of the law. All are entitled to equal protection against any discrimination in violation of this Declaration and against any incitement to such discrimination.

Article 8

Everyone has the right to an effective remedy by the competent national tribunals for acts violating the fundamental rights granted him by the constitution or by law.

Article 9

No one shall be subjected to arbitrary arrest, detention or exile.

Article 10

Everyone is entitled in full equality to a fair and public hearing by an independent and impartial tribunal, in the determination of his rights and obligations and of any criminal charge against him.

Article 11

(1) Everyone charged with a penal offence has the right to be presumed innocent until proved guilty according to law in a public trial at which he has had all the guarantees necessary for his defence.

(2) No one shall be held guilty of any penal offence on account of any act or omission, which did not constitute a penal offence, under national or international law, at the time when it was committed. Nor shall a heavier penalty be imposed than the one that was applicable at the time the penal offence was committed.

Article 12

No one shall be subjected to arbitrary interference with his privacy, family, home or correspondence, nor to attacks upon his honour and reputation. Everyone has the right to the protection of the law against such interference or attacks.

Article 13

(1) Everyone has the right to freedom of movement and residence within the borders of each state.

(2) Everyone has the right to leave any country, including his own, and to return to his country.

Article 14

(1) Everyone has the right to seek and to enjoy in other countries asylum from persecution.

(2) This right may not be invoked in the case of prosecutions genuinely arising from non-political crimes or from acts contrary to the purposes and principles of the United Nations.

Article 15

(1) Everyone has the right to a nationality.

(2) No one shall be arbitrarily deprived of his nationality nor denied the right to change his nationality.

Article 16

(1) Men and women of full age, without any limitation due to race, nationality or religion, have the right to marry and to found a family. They are entitled to equal rights as to marriage, during marriage and at its dissolution.

(2) Marriage shall be entered into only with the free and full consent of the intending spouses.

(3) The family is the natural and fundamental group unit of society and is entitled to protection by society and the State.

Article 17

(1) Everyone has the right to own property alone as well as in association with others.

(2) No one shall be arbitrarily deprived of his property.

Article 18

Everyone has the right to freedom of thought, conscience and religion; this right includes freedom to change his religion or belief, and freedom, either alone or in community with others and in public or private, to manifest his religion or belief in teaching, practice, worship and observance.

Article 19

Everyone has the right to freedom of opinion and expression; this right includes freedom to hold opinions without interference and to seek, receive and impart information and ideas through any media and regardless of frontiers.

Article 20

(1) Everyone has the right to freedom of peaceful assembly and association.

(2) No one may be compelled to belong to an association.

Article 21

(1) Everyone has the right to take part in the government of his country, directly or through freely chosen representatives.

(2) Everyone has the right of equal access to public service in his country.

(3) The will of the people shall be the basis of the authority of government; this will shall be expressed in periodic and genuine elections which shall be by universal and equal suffrage and shall be held by secret vote or by equivalent free voting procedures.

Article 22

Everyone, as a member of society, has the right to social security and is entitled to realisation, through national effort and international co-operation and in accordance with the organisation and resources of each State, of the economic, social and cultural rights indispensable for his dignity and the free development of his personality.

Article 23

(1) Everyone has the right to work, to free choice of employment, to just and favourable conditions of work and to protection against unemployment.

(2) Everyone, without any discrimination, has the right to equal pay for equal work.

(3) Everyone who works has the right to just and favourable remuneration ensuring for himself and his family an existence worthy of human dignity, and supplemented, if necessary, by other means of social protection.

(4) Everyone has the right to form and to join trade unions for the protection of his interests.

Article 24:

Everyone has the right to rest and leisure, including reasonable limitation of working hours and periodic holidays with pay.

Article 25

(1) Everyone has the right to a standard of living adequate for the health and well-being of himself and of his family, including food, clothing, housing and medical care and necessary social services, and the right to security in the event of unemployment, sickness, disability, widowhood, old age or other lack of livelihood in circumstances beyond his control.

(2) Motherhood and childhood are entitled to special care and assistance. All children, whether born in or out of wedlock, shall enjoy the same social protection.

Article 26

(1) Everyone has the right to education. Education shall be free, at least in the elementary and fundamental stages. Elementary education shall be compulsory. Technical and professional education shall be made generally available and higher education shall be equally accessible to all on the basis of merit.

(2) Education shall be directed to the full development of the human personality and to the strengthening of respect for human rights and fundamental freedoms. It shall promote understanding, tolerance and friendship among all nations, racial or religious groups, and shall further the activities of the United Nations for the maintenance of peace.

(3) Parents have a prior right to choose the kind of education that shall be given to their children.

Article 27

(1) Everyone has the right freely to participate in the cultural life of the community, to enjoy the arts and to share in scientific advancement and its benefits.

(2) Everyone has the right to the protection of the moral and material interests resulting from any scientific, literary or artistic production of which he is the author.

Article 28

Everyone is entitled to a social and international order in which the rights and freedoms set forth in this Declaration can be fully realised.

Article 29

(1) Everyone has duties to the community in which alone the free and full development of his personality is possible.

(2) In the exercise of his rights and freedoms, everyone shall be subject only to such limitations as are determined by law solely for the purpose of securing due recognition and respect for the rights and freedoms of others and of meeting the just requirements of morality, public order and the general welfare in a democratic society.

(3) These rights and freedoms may in no case be exercised contrary to the purposes and principles of the United Nations.

Article 30

Nothing in this Declaration may be interpreted as implying for any State, group or person any right to engage in any activity or to perform any act aimed at the destruction of any of the rights and freedoms set forth herein.

The Rio Declaration

Rio Declaration on Environment and Development

The United Nations Conference on Environment and Development,

Having met at Rio de Janeiro from 3 to 14 June 1992,

Reaffirming the Declaration of the United Nations Conference on the Human Environment, adopted at Stockholm on 16 June 1972, a/ and seeking to build upon it,

With the goal of establishing a new and equitable global partnership through the creation of new levels of co-operation among States, key sectors of societies and people,

Working towards international agreements which respect the interests of all and protect the integrity of the global environmental and developmental system,

Recognising the integral and interdependent nature of the Earth, our home,

Proclaims that:

Principle 1

Human beings are at the centre of concerns for sustainable development. They are entitled to a healthy and productive life in harmony with nature.

Principle 2

States have, in accordance with the Charter of the United Nations and the principles of international law, the sovereign right to exploit their own resources pursuant to their own environmental and developmental policies, and the responsibility to ensure that activities within their jurisdiction or control do not cause damage to the environment of other States or of areas beyond the limits of national jurisdiction.

Principle 3

The right to development must be fulfilled so as to equitably meet developmental and environmental needs of present and future generations.

Principle 4

In order to achieve sustainable development, environmental protection shall constitute an integral part of the development process and cannot be considered in isolation from it.

Principle 5

All States and all people shall co-operate in the essential task of eradicating poverty as an indispensable requirement for sustainable development, in order to decrease the disparities in standards of living and better meet the needs of the majority of the people of the world.

Principle 6

The special situation and needs of developing countries, particularly the least developed and those most environmentally vulnerable, shall be given special priority. International actions in the field of environment and development should also address the interests and needs of all countries.

Principle 7

States shall cooperate in a spirit of global partnership to conserve, protect and restore the health and integrity of the Earth's ecosystem. In view of the different contributions to global environmental degradation, States have common but differentiated responsibilities. The developed countries acknowledge the responsibility that they bear in the international pursuit to sustainable development in view of the pressures their societies place on the global environment and of the technologies and financial resources they command.

Principle 8

To achieve sustainable development and a higher quality of life for all people, States should reduce and eliminate unsustainable patterns of production and consumption and promote appropriate demographic policies.

Principle 9

States should cooperate to strengthen endogenous capacity-building for sustainable development by improving scientific understanding through exchanges of scientific and technological knowledge, and by enhancing the development, adaptation, diffusion and transfer of technologies, including new and innovative technologies.

Principle 10

Environmental issues are best handled with participation of all concerned citizens, at the relevant level. At the national level, each individual shall have appropriate access to information concerning the environment that is held by public authorities, including information on hazardous materials and activities in their communities, and the opportunity to participate in decision-making processes. States shall facilitate and encourage public awareness and participation by making information widely available. Effective access to judicial and administrative proceedings, including redress and remedy, shall be provided.

Principle 11

States shall enact effective environmental legislation. Environmental standards, management objectives and priorities should reflect the environmental and development context to which they apply. Standards applied by some countries may be inappropriate and of unwarranted economic and social cost to other countries, in particular developing countries.

Principle 12

States should cooperate to promote a supportive and open international economic system that would lead to economic growth and sustainable development in all countries, to better address the problems of environmental degradation. Trade policy measures for environmental purposes should not constitute a means of arbitrary or unjustifiable discrimination or a disguised restriction on international trade. Unilateral actions to deal with environmental challenges outside the jurisdiction of the importing country should be avoided. Environmental measures addressing transboundary or global environmental problems should, as far as possible, be based on an international consensus.

Principle 13

States shall develop national law regarding liability and compensation for the victims of pollution and other environmental damage. States shall also cooperate in an expeditious and more determined manner to develop further international law regarding liability and compensation for adverse effects of environmental damage caused by activities within their jurisdiction or control to areas beyond their jurisdiction.

Principle 14

States should effectively cooperate to discourage or prevent the relocation and transfer to other States of any activities and substances that cause severe environmental degradation or are found to be harmful to human health.

Principle 15

In order to protect the environment, the precautionary approach shall be widely applied by States according to their capabilities. Where there are threats of serious or irreversible damage, lack of full scientific certainty shall not be used as a reason for postponing cost-effective measures to prevent environmental degradation.

Principle 16

National authorities should endeavour to promote the internalisation of environmental costs and the use of economic instruments, taking into account the approach that the polluter should, in principle, bear the cost of pollution, with due regard to the public interest and without distorting international trade and investment.

Principle 17

Environmental impact assessment, as a national instrument, shall be undertaken for proposed activities that are likely to have a significant adverse impact on the environment and are subject to a decision of a competent national authority.

Principle 18

States shall immediately notify other States of any natural disasters or other emergencies that are likely to produce sudden harmful effects on the environment of those States. Every effort shall be made by the international community to help States so afflicted.

Principle 19

States shall provide prior and timely notification and relevant information to potentially affected States on activities that may have a significant adverse transboundary environmental effect and shall consult with those States at an early stage and in good faith.

Principle 20

Women have a vital role in environmental management and development. Their full participation is therefore essential to achieve sustainable development.

Principle 21

The creativity, ideals and courage of the youth of the world should be mobilised to forge a global partnership in order to achieve sustainable development and ensure a better future for all.

Principle 22

Indigenous people and their communities and other local communities have a vital role in environmental management and development because of their knowledge and traditional practices. States should recognise and duly support their identity, culture and interests and enable their effective participation in the achievement of sustainable development.

Principle 23

The environment and natural resources of people under oppression, domination and occupation shall be protected.

Principle 24

Warfare is inherently destructive of sustainable development. States shall therefore respect international law providing protection for the environment in times of armed conflict and cooperate in its further development, as necessary.

Principle 25

Peace, development and environmental protection are interdependent and indivisible.

Principle 26

States shall resolve all their environmental disputes peacefully and by appropriate means in accordance with the Charter of the United Nations.

Principle 27

States and people shall cooperate in good faith and in a spirit of partnership in the fulfilment of the principles embodied in this Declaration and in the further development of international law in the field of sustainable development.

Annex VII

Members of the World Commission on Dams

Professor Kader Asmal, Chair, is a prominent member of President Thabo Mbeki's Cabinet as Minister of Education of the Republic of South Africa. Under Nelson Mandela, he was the Minister of Water Affairs and Forestry and led the fundamental review and reform of South Africa's water resource management policy. Prior to his return from exile in 1990, Professor Asmal was a law professor at Trinity College Dublin for 27 years, specialising in human rights, labour and international law. From 1980 to 1986 he was Dean of the Faculty of Arts (Humanities). He was also founder of the British Anti-Apartheid Movement as well as the Irish Anti-Apartheid Movement in 1963, and its chairperson until 1991.

In 1983, Professor Asmal received the Prix UNESCO for the advancement of human rights. In 1993, he became a member of the negotiating team of the African National Congress at the Multi-Party Negotiating Forum, and in May 1994 was elected to the National Assembly. In 1996, he was awarded the Gold Medal Award for conservation from the World Wide Fund for Nature–South Africa. He is also a patron of the Global Water Partnership. On 22 March 2000 – *World Water Day* – Professor Asmal was awarded the 10th Stockholm Water Prize by the Stockholm Water Foundation. His selection to chair the World Commission on Dams was the result of a comprehensive global search process and consultation with participants at the Gland workshop.

Lakshmi Chand Jain, Vice Chair, has served on India's National Planning Commission and Planning Boards of several states and was appointed High Commissioner for India to South Africa for 1997–99. He was a member of the Government of India's Independent Committee to report on selected aspects of the Sardar Sarovar Project. He has also served as the Chairperson of the Industrial Development Services, a techno-economic consultancy organisation in India, for 30 years. L.C. Jain received the prestigious Ramon Magsaysay Award for Public Service in 1989.

Donald J. Blackmore is the Chief Executive of the Murray-Darling Basin Commission, in Canberra, Australia. He has brought principles of environmentally sustainable water management to a major river basin initially focused on irrigation and hydroelectric power generation. Since 1990 he has served as Director and Deputy Chairperson of Australia's Land & Water Resources Research & Development Corporation. He also was a member of the International Advisory Panel for the Aral Sea.

Joji Carino's work began as an activist and analyst of indigenous peoples' issues in her native Philippines, particularly in relation to dam projects in the Cordillera region. Over 25 years she has worked as an active campaigner and advocate of indigenous peoples' human rights. She now works for the Tebtebba Foundation (Indigenous Centre for International Policy, Research, and Education). Well known for defending the interests of tribal and other indigenous peoples and minorities, she has effectively carried her agenda into global fora.

José Goldemberg is a professor at the University of São Paulo, Brazil, and has been recognized for his work on the future of energy globally. He was the chairman and CEO of the Energy Company of the State of São Paulo. He has served as Rector of his University, and as Secretary of Science and Technology for the Federal Government of Brazil and as Minister of Education. He currently serves in senior capacities with the International Energy Initiative and the Intergovernmental Panel on Climate Change. His most recent work has been as Chairperson of the World Energy Assessment.

Judy Henderson trained professionally as a medical doctor. She served as Chair of Oxfam International, was a board member of the Environmental Protection Agency of New South Wales, Australia, and a former board member of Greenpeace International. She has a distinguished record of involvement in social and environmental issues internationally. She is currently a board member of the Ethical Investment Agency.

Göran Lindahl is the President and CEO of ABB Ltd., a global technology group with headquarters in Zurich. He is a member of the Advisory Board for the Alliance for Global Sustainability, deputy chairman of the Prince of Wales Business Leaders Forum and on the advisory board of the World Childhood Foundation. He is also a frequent speaker at the World Economic Forum. An electrical engineer by profession, he has been involved in many major electricity infrastructure projects, including large hydroelectric schemes. As the head of ABB, Göran Lindahl leads one of the world's largest industrial companies. The ABB Group employs about 160,000 people in more than 100 countries.

Deborah Moore was until recently Senior Scientist at Environmental Defense, a US-based NGO, where she continues as Consulting Scientist working to protect living rivers worldwide. In the western United States, Moore has worked with Native American communities and the U.S. Congress to design and promote innovative water rights and river restoration arrangements. Internationally, she has contributed to many global water policy fora, including the Dublin Conference on Water and the Environment, and analysed the performance of large-scale river development projects and alternatives in Asia and Latin America.

Medha Patkar graduated in physical sciences and did postgraduate research in social sciences. She was a member of the faculty at the Tata Institute of Social Sciences before founding the Narmada Bachao Andolan (Struggle to Save the Narmada River) in India, a people's movement against the construction of large dams on the Narmada River and for alternatives in water, power and development. She is a founding member and National Convenor of the National Alliance of People's Movements. She is internationally recognised as a campaigner for human and political rights.

Thayer Scudder is an emeritus professor of anthropology at the California Institute of Technology. His work over 40 years on socioeconomic issues associated with river basin development has been definitive in the field. His work in Africa is best known, but he has undertaken studies of sustainable resource use in all parts of the world with a focus on resettlement and socioeconomic issues related to infrastructure development. He has also served on a number of independent review panels for dam projects in Africa and Asia.

Jan Veltrop worked with the Harza Engineering Company from 1954 to 1994 except for a three-year period when he served as Dean of the Faculty of Engineering at the University of Nigeria. He was Chairman of the U.S. Committee on Large Dams (1981-82) and President of the International Commission on Large Dams (1988-91). At Harza he was Chief Engineer, member of the Board of Directors, and retired as Senior Vice President. He worked on many world-class hydroprojects such as Mangla, Tarbela, Guri, Yacyreta, Karun-I, Ertan and Bath County pumped-storage; received ASCE's Rickey Medal in 1997 for his contributions in the field of hydroelectric engineering; and was elected a member of the National Academy of Engineering in 1998.

Achim Steiner *(WCD Secretary-General, ex-officio Commissioner)* has served as an advisor on international development policy as well as economic planning and natural resources management. During his career he has worked for both governmental and non-governmental organisations, with extended assignments in India, Pakistan, Germany, Zimbabwe/Southern Africa, the United States, and Vietnam. Most recently, he served as Senior Policy Advisor for Global Policy with IUCN in Washington and Chief Technical Advisor with the Mekong River Commission/GTZ, based in Hanoi, Viet Nam.

Annex VIII

A Profile of the WCD Secretariat

Programme Staff

(on September 1, 2000)
Secretary-General
Achim Steiner – Germany

Bruce Aylward – United States
Jeremy Bird* – United Kingdom
Christopher Clarke – South Africa
Lawrence Haas* – Canada
Saneeya Hussain* – Pakistan
Madiodio Niasse – Senegal
S. Parasuraman* – India
Corli Pretorius – South Africa
Jamie Skinner – United Kingdom
James Workman – United States

* Team Leader

Finance & Administration Staff

(on September 1, 2000)

Solly Fazel – South Africa
Jacques Coetzee – South Africa
Lucia Hickman – South Africa
Noluthando Magadla – South Africa
Yumna McCann – South Africa
Pamela Morris – South Africa
Nadia Richards – South Africa
Phumla Yeki – South Africa

Research Fellows

The WCD Secretariat was fortunate to have a large number of research fellows contributing to the work programme during the life of the Commission. The time they spent at the Commission ranged from one month to nine months.

Mark Cassidy – Australia
Sophia Chan – Canada
Luis Paulo Ferraz – Brazil
Nicolas Gutman – Argentina
Laurence Haller – Switzerland
Huynh Thuba – Viet Nam
Anneli Lagman – Philippines
Khutso Madubanya – South Africa
Sandi Nielsen – Zimbabwe
Pumeza Nodada – South Africa
Alice Ojwang – Kenya
Saule Ospanova – Kazakhstan
Sarah Porter – United States
Chaminda Rajapakse – Sri Lanka
Manrique Rojas – Costa Rica
Sohini Sengupta – India
Jason Switzer – Canada
Pamela Wallace – Malawi
Webster Whande – Zimbabwe

Temporary Staff

The Commission would like to recognise the contribution of the following individuals who joined the Secretariat for various periods of time:

Programme Staff

Kate Dunn – Canada
Pat Govender – South Africa
Sanjeev Khagram – United States
Elisabeth Monosowski* – Brazil
Eric Noel – Australia
Engelbertus Oud – The Netherlands

Finance & Administration Staff

Nikki Boltman – South Africa
Kathy Bova – South Africa
Michel Cozien – South Africa
Salih Hendricks – South Africa
Ian Fennell – United Kingdom
Marshay Mackay – South Africa
Ashleigh Meier – Zimbabwe

Consultants and Advisors

The WCD also wishes to acknowledge the contribution of the following individuals for their special inputs as consultants and advisors to the Secretariat and the Commissioners:

Jeremy Berkoff – United Kingdom
Maggie Black – United Kingdom
Dick Cloete – South Africa
Tim Cullen – United Kingdom
Robert Goodland – United States
Pablo Gutman – Argentina
Mark Halle – Switzerland
Minoru Kuriki – Japan
Donal O'Leary – Ireland
Katarzyna Ronda – South Africa
John Scanlon – Australia
Mark Svendson – United States
Linda Starke – United States
Allan Taylor – South Africa
Martin ter Woort – Canada

Index

The index does not cover the Executive Summary (pages xxvi-xxxvii) or the glossary (Annex II). Page numbers given in italics indicate illustrations, figures or tables.

A

B

C

U

V

W

X

Y

Z